Genetic Resources for Viticulture

Genetic Resources for Viticulture

Guest Editors

Antonio Amores-Arrocha
Ana Jiménez-Cantizano

MDPI

Basel • Beijing • Wuhan • Barcelona • Belgrade • Novi Sad • Cluj • Manchester

Guest Editors
Antonio Amores-Arrocha
Department of Chemical Engineering and Food Technology
Vegetal Production Area and Wine and Agri-food Research Institute (IVAGRO)
University of Cadiz
Cadiz
Spain

Ana Jiménez-Cantizano
Department of Chemical Engineering and Food Technology
Vegetal Production Area and Wine and Agri-food Research Institute (IVAGRO)
University of Cadiz
Cadiz
Spain

Editorial Office
MDPI AG
Grosspeteranlage 5
4052 Basel, Switzerland

This is a reprint of the Special Issue, published open access by the journal *Horticulturae* (ISSN 2311-7524), freely accessible at: https://www.mdpi.com/journal/horticulturae/special_issues/20U0B55L45.

For citation purposes, cite each article independently as indicated on the article page online and as indicated below:

Lastname, A.A.; Lastname, B.B. Article Title. *Journal Name* **Year**, *Volume Number*, Page Range.

ISBN 978-3-7258-4019-9 (Hbk)
ISBN 978-3-7258-4020-5 (PDF)
https://doi.org/10.3390/books978-3-7258-4020-5

Contents

About the Editors

Antonio Amores-Arrocha

Antonio Amores-Arrocha currently serves as a Professor in the Department of Chemical Engineering and Food Technology, Plant Production Area, and Researcher in the Wine and Agri-food Research Institute (IVAGRO), at the University of Cadiz. He received his PhD in Agrifood Resources at the University of Cadiz in 2018. His research interests mainly include bee pollen, alcoholic fermentation, white winemaking, red winemaking, volatile compounds in wines, sensory analysis of wines, physical-chemical analysis of wines, morphological, agronomical, and physiological characterization of grapevine varieties, and the study of winemaking climate change adaptation strategies.

Ana Jiménez-Cantizano

Ana Jiménez-Cantizano currently serves as a Professor in the Department of Chemical Engineering and Food Technology, Plant Production Area, and Researcher in the Wine and Agri-food Research Institute (IVAGRO) at the University of Cadiz. She received her PhD from the University of Cadiz in 2014. Her research interests mainly include the genetic, morphological, agronomical, physiological, and oenological characterization of *Vitis vinifera* grapevines, in addition to the study of viticultural strategies for adaptation to climate change.

horticulturae MDPI

Editorial

Genetic Resources for Viticulture

Antonio Amores-Arrocha * and Ana Jiménez-Cantizano *

Department of Chemical Engineering and Food Technology, Vegetal Production Area, University of Cadiz, Agrifood Campus of International Excellence (ceiA3), IVAGRO, Puerto Real, P.O. Box 40, 11510 Cadiz, Spain
* Correspondence: antonio.amores@uca.es (A.A.-A.); ana.jimenezcantizano@uca.es (A.J.-C.)

Grapevine (*Vitis vinifera* L.) is one of the most widely cultivated and economically valuable horticultural crops globally [1]. According to the International Organisation of Vine and Wine (OIV), the world's vineyard surface stands at 7.2 million hectares [2], distributed mainly between latitudes 4° and 51° in the Northern Hemisphere (NH) and between 6° and 45° in the Southern Hemisphere (SH) [3]. In the last 3 years, there has been a 2.3% decrease in the global vineyard surface area, driven by the removal of vineyards in major vine-growing regions (including all grape types: wine, table, and dried grapes) across both hemispheres.

This reduction was mainly motivated by extreme climatic conditions and widespread fungal diseases that severely impacted many vineyards worldwide, culminating in a historically low global wine production of 237 million hectoliters. This marked a 10% drop from 2022 and represented the lowest output since 1961, and a decrease of 2.6% in the consumption of wine [2]. These numbers show that viticulture is facing new challenges such as climate change, diseases, pests, and the need to produce sustainably, adapting to consumer demand and taste [4,5].

In this context, the genetic resources available for viticulture could be of particular interest. Internationally, there are approximately 253 institutions holding plant material, and 23,000 cultivars are registered according to the *Vitis* International Varieties Catalogue (VIVC) [6]. This is evidence of high levels of recorded biodiversity, which provide a crucial reservoir of genetic heterogeneity, providing a reservoir of valuable allelic combinations that can offer genetic resistance or tolerance to both biotic and abiotic stresses [7].

In this Special Issue entitled "Genetic Resources for Viticulture", we would like to include studies related to the identification and agronomic, physiological, and oenological characterization of new and local grapevine varieties, rootstocks, clones, and interspecific hybrids as sources of genetic resources to meet the challenges of viticulture.

Received: 10 March 2025
Accepted: 21 March 2025
Published: 27 March 2025

Citation: Amores-Arrocha, A.; Jiménez-Cantizano, A. Genetic Resources for Viticulture. *Horticulturae* **2025**, *11*, 362. https://doi.org/10.3390/horticulturae11040362

1. An Overview of Published Articles

The articles by Fort et al. (2023, 2024) (contributions 1, 2, and 3, respectively) explore the genetic diversity of grapevines (*Vitis vinifera* L.) cultivated in the Canary volcanic archipelago, located in European overseas territory (near Western Sahara). The origin of the vine in the Canary Islands comes from the introduction of European varieties in the 15th century. These islands are home to a great diversity of vines due to their evolution and adaptation to these habitats. In these studies, the volcanic islands of La Gomera, Fuerteventura, and El Hierro have been prospected, collecting a total of 247 samples of grapevine plants, which were genotyped using 20 SSRs. In addition, a parentage study was carried out with the different genotypes obtained. The results of these research articles revealed notable genetic diversity among Canary Islands strains, suggesting a long history of cultivation and adaptation to island conditions. An outstanding discovery was the identification of local varieties exclusive to each island, representing a valuable genetic heritage. These native varieties could harbor genes for disease resistance, drought tolerance, or unique organoleptic characteristics, making them a genetic resource of great interest for varietal improvement. Likewise, the articles made it possible to trace kinship relationships between the Canary varieties and other varieties from different wine regions, providing information on the origin and dispersion of the vines in the archipelago. The authors concluded these articles by pointing out the importance of the Canary Islands as reservoirs of genetic diversity of the vine. These results can not only contribute to improving knowledge of the history and evolution of viticulture in the archipelago, but also be of great interest for the conservation of this genetic heritage. The identification and characterization of local varieties from the Canary Islands could open up new perspectives of interest in the recovery and valorization of wine heritage, as well as for the creation of new high-quality wines adapted to local conditions.

González et al. (2023) (contribution 4) focused on revealing the rich genetic diversity of vines (*Vitis vinifera* L.) grown on the islands of Ibiza and Formentera, belonging to the Balearic archipelago (Spain). By combining genetic data with ethnobotanical information, the authors managed to decipher the complex history of these varieties and establish phylogenetic relationships between them. The set of results obtained after the analysis of microsatellite markers and descriptions of morphological and ampelometric characteristics, together with the ethnobotanical study through interviews with local winegrowers to collect information on the cultivated varieties, local names, and their traditional use, revealed a great diversity of genetics between the strains grown on both islands. This suggests a long history of cultivation and selection, as well as an adaptation to the agroclimatic conditions of each island. Through this research, the researchers identified new synonyms and homonyms for some varieties, which could help clarify the nomenclature and avoid confusion in their identification. Likewise, new kinship relationships were established between some varieties, which makes it possible to reconstruct the evolutionary history of viticulture in Ibiza and Formentera. In conclusion, the authors indicated that these findings are not only relevant to the scientific community, but could also be of interest to local viticulture, opening avenues to the recovery of native varieties, in addition to the diversification of the region's wines.

The research by Rahimi et al. (2023) (contribution 5) systematically described the habitats, growth habits, morphology, and anatomy of widely spread wild grapevine populations growing in two distinct habitats in north Israel (Beit Tsaida and Banias River sites). The results of this work show the identification of plants belonging to *Vitis vinifera* L. subsp. *Sylvestris*. These new genetic resources should be conserved for the preservation of the species.

Using morphometric techniques, domestication indices, multivariate analysis, and Bayesian hypothesis testing, Valera et al. (2023) (contribution 6) identify grapevine seeds found in archaeological sites in the ancient Upper Euphrates. Archaeological seeds were compared with a pull of 782 grape seed samples provided by various European and U.S. institutions and entities. Digital imaging techniques and statistical analyses were used to measure various morphological characteristics of the seeds, such as length, width, and thickness. Using Bayesian analysis, statistical models were constructed to compare the archaeological seeds with reference samples of modern and ancient varieties.

In the seventh article, Bhattarai et al. (2023) (contribution 7) focused on characterizing the volatile compound profile of different muscadine grape varieties, a species native to North America with great viticultural potential. To this end, different volatile compounds responsible for the characteristic aromas and flavors were identified and quantified in different varieties of muscadine grapes, as well as in hybrids between muscadine and other vine species. The results showed a diversity of volatile compounds (terpenes, esters, alcohols, ketones, and sulfur compounds) associated with odorant series corresponding to fruity, floral, citrus, green, and spicy odors. In conclusion, this work contributes to a better understanding of the aromatic potential of grapes of this species. The results revealed a significant variation in the composition of aromatic volatile compounds between the different muscadine genotypes studied, as well as distinctive composition patterns depending on the variety and the ripening stage. The results of this study are valuable for muscadine breeding programs, consumer preference studies, and the development of metabolic markers to evaluate the ripening quality of grapes.

The article by Ruiz-García et al. (2023) (contribution 8) focuses on the characterization of new vine varieties obtained from crosses of the Monastrell variety, a red grape native to the region of Murcia, Spain. These new genotypes ("Calblanque", a white genotype, and "Calnegre", "Gebas", and "Myrtia", red genotypes) have been specifically developed to adapt to the hot and dry climatic conditions of this region and to improve the quality of the wines produced. The red genotypes were selected for their phenolic quality, which was much higher than that of the parents, and for their different harvest dates that allow staggered harvesting and cultivation in different areas. "Calblanque" was selected for its good balance of acidity and aromatic profile. The attributes of these new varieties could allow them to better adapt to the effects of climate change on the quality of grapes and wine in warm areas. Wines made with these new varieties showed greater aromatic complexity and better tannic structure, suggesting great oenological potential. In conclusion, these results indicate that the new varieties obtained from Monastrell crosses represent a valuable tool for wine growers, since they would allow them to adapt to the new environmental conditions generated by climate change in the region of Murcia and produce high-quality wines.

The research works of Rodríguez-Torres et al. (2023) (contribution 9) and Jiménez-Cantizano et al. (2023) (contribution 10) deal with the genetic characterization of local vine varieties, contributing to the preservation and valorization of the wine heritage of Andalusia (Southern Spain). To this end, both articles used microsatellite markers (SSR) for the genetic characterization of the vine and to determine kinship relationships between the different samples.

Rodríguez-Torres et al. (2023) (contribution 9) examined 98 samples from six Andalusian wine-growing areas. To do this, they used microsatellite marker analysis, identifying 33 different genotypes, of which 20 corresponded to varieties already described (11 of them are from six minority cultivars in Andalusia: "Rojal Tinto", "Beba", "Zurieles", "Rome", "Hebén", "Mollar Cano", "Listán Prieto", "Listán del Condado", "Jarrosuelto", "Negra Dorada", and "Mantúo de Pilas"), while the other 12 profiles did not match previously identified varieties. In addition to the genetic characterization, they carried out a health analysis using an ELISA test to detect the presence of viruses (vine fan virus, vine spot virus, and viruses associated with vine leaf curl) due to the requirement of healthy clones of the new varieties for their authorization for cultivation in Spain.

Meanwhile, Jiménez-Cantizano et al. (2023) (contribution 10) analyzed 49 vine accessions collected in the districts of four provinces of Andalusia (Spain). In total, 30 different genotypes were identified, 22 of which corresponded to known varieties and 8 to new genotypes. These results confirm the high genetic diversity present in local Andalusian varieties. As conclusions of this work, the eight new genotypes identified in this work have not yet been identified and could represent ancient local varieties in danger of extinction. These new cultivars could be used to make original wines and, therefore, these genotypes have been preserved in the "Rancho de la Merced" germplasm bank (Andalusia, Spain). The results of this work show that there is significant biodiversity in old vineyards located in the region of Andalusia of cultivars not yet exploited, which may be of great interest to the wine industry.

The article by Arslan et al. (2023) (contribution 11) focused on analyzing the genetic diversity and population structure of the Kara grape variety, a native variety of Anatolia. To this end, an analysis of 49 Kara grape cultivars from six regional subpopulations in Turkey was carried out, using microsatellite markers (SSR) together with ampelographic characterization. The results of the molecular analyses showed high genetic diversity, possibly explained by the adaptation of the vine to different environmental conditions throughout its history and by selection processes. Four synonyms and five homonymies were also identified. According to the results of the ampelographic analysis, it was determined that the shape of the berry and the density of upright hairs on the young shoot/density of lying hairs at the end of the young shoot were the determining characteristics between the different cultivars. The conclusions of this article indicate that the genetic characterization of the Turkish Kara grape germplasm has been obtained using SSR for the first time. These results could be very useful for the development of other genetic characterization studies of grapes and will contribute to wine research in other areas such as the improvement, protection, and registration of varieties.

In addition to this, this Special Issue includes articles of great interest in the viticulture sector, such as the resistance of the vine against diseases such as *Xylella fastidiosa*. The article by Martínez et al. (2023) (contribution 12) focuses on evaluating the resistance of different European grapevine cultivars and rootstocks to the bacterium *Xylella fastidiosa* subsp. fastidiosa, a pathogen that causes serious damage to crops. This research acquires special relevance due to the growing threat that this bacterium poses to viticulture worldwide. Using molecular and phenotypic techniques, the researchers evaluated the plant's response to infection, analyzing bacterial colonization in different tissues, the expression of genes related to the immune response, and the appearance of symptoms. The results showed a significant difference in resistance to this disease between the different cultivars and rootstocks evaluated. This could allow the identification of varieties and rootstocks that are more tolerant to bacteria, which could be used for the development of more sustainable and effective management strategies.

In parallel, Yin et al. (2023) (contribution 13) have deepened our understanding of the agronomic characteristics of rootstocks, a key aspect for the selection of resistant plant material adapted to different environmental conditions. Rootstocks are vine varieties used as a base for grafting other varieties, and their choice can significantly influence the yield and quality of the vines. For this, characteristics such as the multiplication rate, the time of sprouting and maturation, and vegetative growth were studied, in addition to analyzing the relationship between the vigor of the rootstocks and the inheritance of this trait through the parents, in 31 vine rootstocks. The results revealed wide variability between their ease of propagation and their phenology, suggesting a wide range of adaptations to different environmental conditions. This diversity is crucial for the selection of suitable rootstocks and for optimizing wine production in various regions, allowing the vine to adapt to specific climatic conditions and improving the efficiency of cultivation systems.

2. Conclusions

The vine (*Vitis vinifera* L.) has been cultivated by humans since ancient times, forming a fundamental pillar of the culture and economy of numerous civilizations. Its wide geographical distribution and its adaptation to diverse climates and soils have given rise to extraordinary genetic diversity. This Special Issue focuses on exploring the genetic diversity of the vine in different wine-growing regions, with a particular emphasis on those territories where viticulture has been an ancient activity in the Mediterranean.

The articles compiled in this volume address various topics, from the genetic characterization of local varieties to the evaluation of disease resistance and adaptation to extreme climatic conditions. Through a combination of molecular, morphological, and statistical analysis, the authors offer a comprehensive view of the genetic complexity of the grapevine and its relationship with environmental and historical factors.

A highlight of this volume is the attention paid to local and indigenous varieties. These varieties, often adapted to specific conditions and carrying a unique genetic heritage, represent an invaluable resource for modern viticulture. The articles presented here reveal the richness and diversity of these varieties, as well as their potential for obtaining high-quality wines and adaptation to the challenges of climate change.

Likewise, this volume highlights the importance of international collaboration in wine research. The articles presented here are the result of the collaboration of researchers from different countries, which has allowed a broader and deeper vision of the genetic diversity of the vine worldwide to be obtained.

In conclusion, this Special Issue offers a valuable contribution to the knowledge of grapevine genetic diversity and its implications for viticulture. The results presented here are relevant from both a scientific and practical perspective and can serve as a basis for future research and applications in the wine sector.

As a final note, we would like to emphasize that this Special Issue is distinguished by its focus on wine regions with a long history of grape cultivation, but which have been less studied compared to other, better-known regions. By exploring the genetic diversity of these regions, this volume can contribute to filling a gap in the scientific literature and offer a new perspective on grapevine evolution and adaptation. Furthermore, the combination of different methodological approaches and the integration of data from various sources make this Special Issue an essential reference for researchers, wine growers, and all those interested in the genetic diversity of the vine and its conservation.

Conflicts of Interest: The author declares no conflicts of interest.

List of Contributions

1. Fort, F.; Lin-Yang, Q.; Valls, C.; Sancho-Galán, P.; Canals, J.M.; Zamora, F. Analysis of the Diversity Presented by *Vitis vinifera* L. in the Volcanic Island of La Gomera (Canary Archipelago, Spain) Using Simple Sequence Repeats (SSRs) as Molecular Markers. *Horticulturae* **2024**, *10*, 14. https://doi.org/10.3390/horticulturae10010014.
2. Fort, F.; Lin-Yang, Q.; Valls, C.; Sancho-Galán, P.; Canals, J.M.; Zamora, F. Characterisation and identification of Vines from Fuerteventura (Canary Volcanic Archipelago (Spain)) using simple sequence repeat markers. *Horticulturae* **2023**, *9*, 1301. https://doi.org/10.3390/horticulturae9121301.
3. Fort, F.; Lin-Yang, Q.; Suárez-Abreu, L.R.; Sancho-Galán, P.; Canals, J.M.; Zamora, F. Study of Molecular Biodiversity and Population Structure of *Vitis vinifera* L. ssp. *vinifera* on the Volcanic Island of El Hierro (Canary Islands, Spain) by Using Microsatellite Markers. *Horticulturae* **2023**, *9*, 1297. https://doi.org/10.3390/horticulturae9121297.
4. González, R.; Vargas, A.M.; Garnatje, T.; Vallès, J.; de Andrés, M.T. Exploring Diversity among Grapevines Varieties (*Vitis vinifera* L.) in Ibiza and Formentera (Balearic Islands, Spain) Using Microsatellite Markers, Ampelographic Methods and an Ethnobotanical Approach. *Horticulturae* **2023**, *9*, 1307. https://doi.org/10.3390/horticulturae9121307.
5. Rahimi, O.; Berger, J.Z.; Shtein, I.; Kher, M.M.; Frumin, S.; Hübner, S.; Weiss, E.; Drori, E. Wild Grapevine (*Vitis vinifera* L. subsp. *sylvestris* (CC Gmelin) Hegi)—Novel Species to the Israeli Flora. *Horticulturae* **2023**, *9*, 998. https://doi.org/10.3390/horticulturae9090998.
6. Valera, J.; Matilla-Seiquer, G.; Obón, C.; Alcaraz, F.; Rivera, D. Grapevine in the ancient upper euphrates: Horticultural implications of a bayesian morphometric study of archaeological seeds. *Horticulturae* **2023**, *9*, 803. https://doi.org/10.3390/horticulturae9070803.

7. Bhattarai, G.; Giannopoulos, O.; Corn, R.N.; McAvoy, C.E.; Deltsidis, A.; Worthington, M.L.; Conner, P.J. Analysis of the Aroma Volatile Profile of Muscadine Grape Germplasm by Headspace Solid-Phase Microextraction Coupled with Gas Chromatography-Mass Spectrometry. *Horticulturae* **2023**, *9*, 1054. https://doi.org/10.3390/horticulturae9091054.
8. Ruiz-García, L.; Fernández-Fernández, J.I.; Martínez-Mora, C.; Moreno-Olivares, J.D.; Giménez-Bañón, M.J.; Fernández-López, D.J.; Bleda-Sánchez, J.A.; Gil-Muñoz, R. Characterization of New Grapevine Varieties Cross-Bred from Monastrell, Authorized for Winemaking in the Warm Region of Murcia (South-Eastern Spain). *Horticulturae* **2023**, *9*, 760. https://doi.org/10.3390/horticulturae9070760.
9. Rodríguez-Torres, I.; Martín Carrillo, A.; Ramírez MD, P.; Gómez Gálvez, F.J.; Velasco Arjona, L.; Padilla, C.; Cretazzo, E. Genotypic and Sanitary Characterization of Minority Grapevine Varieties Prospected in Andalusia, Spain. *Horticulturae* **2023**, *9*, 759. https://doi.org/10.3390/horticulturae9070759.
10. Jiménez-Cantizano, A.; Puig-Pujol, A.; Arroyo-García, R. Identification of *Vitis vinifera* L. Local Cultivars Recovered in Andalusia (Spain) by Using Microsatellite Markers. *Horticulturae* **2023**, *9*, 316. https://doi.org/10.3390/horticulturae9030316.
11. Arslan, N.; Yılmaz Baydu, F.; Hazrati, N.; Yüksel Özmen, C.; Ergönül, O.; Uysal, T.; Yaşasın, A.S.; Özer, C.; Boz, Y.; Kuleyin, Y.S.; et al. Genetic Diversity and Population Structure Analysis of Anatolian Kara Grapevine (*Vitis vinifera* L.) Germplasm Using Simple Sequence Repeats. *Horticulturae* **2023**, *9*, 743. https://doi.org/10.3390/horticulturae9070743.
12. Martínez, S.; Lacuesta, M.; Relloso, J.B.; Aragonés, A.; Herrán, A.; Ortiz-Barredo, A. European Grapevine Cultivars and Rootstocks Show Differential Resistance to *Xylella fastidiosa* Subsp. *fastidiosa*. *Horticulturae* **2023**, *9*, 1224. https://doi.org/10.3390/horticulturae9111224.
13. Yin, Y.; Han, B.; Li, M.; Jia, N.; Liu, C.; Sun, Y.; Wang, Y.; Gao, Q.; Guo, Z.. Multiplication, Phenological Period and Growth Vigor of Thirty-One Grapevine Rootstocks and the Role of Parentage in Vigor Heredity. *Horticulturae* **2023**, *9*, 241. https://doi.org/10.3390/horticulturae9020241.

References

1. Alston, J.M.; Sambucci, O. Grapes in the World Economy. In *The Grape Genome*; Cantu, D., Walker, M.A., Eds.; Springer: Cham, Switzerland, 2019; pp. 1–24.
2. International Organisation of Vine and Wine (OIV). *State of the World Vine and Wine Sector-OIV, 2023*; OIV: Dijon, France, 2024.
3. Tonietto, J. Les Macroclimats Viticoles Mondiaux et l'Influence du Mésoclimat sur la Typicité de la Syrah et du Muscat de Hambourg dans le sud de la France: Méthodologie de Caractérisation. Ph.D. Thesis, Ecole Nationale Supérieure Agronomique, Montpellier, France, 1999; 233p.
4. Gutiérrez-Gordillo, S.; Pérez-González, J.M.; Sancho-Galán, P.; Amores-Arrocha, A.; Palacios, V.; Jiménez-Cantizano, A. Physiological, Enological and Agronomic Characterization of Pedro Ximénez Grapevine Cultivar under Organic Farming in a Warm Climate Zone. *Agronomy* **2023**, *13*, 1732. [CrossRef]
5. Sancho-Galán, P.; Amores-Arrocha, A.; Pérez-González, J.M.; Gutiérrez-Gordillo, S.; Palacios, V.; Jimenez-Cantizano, A. New types of organic wines with Pedro Ximenez variety grown in the Marco de Jerez Sherry region. *BIO Web Conf.* **2023**, *68*, 02037. [CrossRef]
6. Vitis International Variety Catalogue. Available online: www.vivc.de (accessed on 1 March 2025).
7. Magon, G.; De Rosa, V.; Martina, M.; Falchi, R.; Acquadro, A.; Barcaccia, G.; Portis, E.; Vannozzi, A.; De Paoli, E. Boosting grapevine breeding for climate-smart viticulture: From genetic resources to predictive genomics. *Front. Plant Sci.* **2023**, *14*, 1293186. [CrossRef] [PubMed]

horticulturae

Article

Analysis of the Diversity Presented by *Vitis vinifera* L. in the Volcanic Island of La Gomera (Canary Archipelago, Spain) Using Simple Sequence Repeats (SSRs) as Molecular Markers

Francesca Fort [1,*,†], Qiying Lin-Yang [1,†], Carla Valls [1], Pau Sancho-Galán [2], Joan Miquel Canals [1] and Fernando Zamora [1]

1 Grupo de Tecnología Enológica (TECNENOL), Department of Biochemistry and Biotechnology, Faculty of Oenology, Rovira i Virgili University, Sescelades Campus, C/Marcel lí Domingo, 1, E-43007-Tarragona, Spain; qiying.lin@estudiants.urv.cat (Q.L.-Y.); carlavlls813@gmail.com (C.V.); jmcanals@urv.cat (J.M.C.); fernando.zamora@urv.cat (F.Z.)

2 Department of Chemical Engineering and Food Technology, Science Faculty, University of Cadiz, Agrifood Campus of International Excellence (CeiA3), Wine and Food Research Institute (IVAGRO), P.O. Box 40, 11510 Puerto Real, Spain; pau.sancho@uca.es

* Correspondence: mariafrancesca.fort@urv.cat

† These authors contributed equally to this work.

Abstract: La Gomera Island is one of the areas of our planet where the phylloxera plague never arrived. To measure the genetic diversity of the vine after more than 500 years (inter- and intravarietal variability) of adaptation to this new environment, a prospection was carried out. For this purpose, 120 samples were collected and genotyped with 20 SSRs. A total of 52 unique profiles were found corresponding to 4 new varieties (Coello blanca, Barrerita negra, Malvasia periquin gomerae, Verdello gomerae), 9 individuals identical to the most widespread profile, and 39 individuals that presented variations (1 corresponding to a mutation of a new variety (Verdello gomerae de Monacal) and 38 corresponding to variations of known varieties, some of which included cases of triallelism or quadriallelism). The population of local vines in La Gomera Island is considered to be the most unique in the Canary Islands to date. It is hypothesised that the grapevine varieties Malvasia periquin gomerae and Verdello gomerae are possibly the most unique and that the Barrerita negra variety may have resulted from an interspecific crossbreeding. The Coello blanco variety (admixed) seems to have a strong Central European influence. Finally, we propose that the prime name for the Albillo forastero variety, which was arbitrarily imposed by the scientific community, be changed to the more widespread and better-known name in La Gomera Island and the Canary Archipelago, which is Forastera gomerae.

Keywords: *Vitis vinifera* L.; SSR; microsatellite; diversity; volcanic; Canary Islands; La Gomera

Citation: Fort, F.; Lin-Yang, Q.; Valls, C.; Sancho-Galán, P.; Canals, J.M.; Zamora, F. Analysis of the Diversity Presented by *Vitis vinifera* L. in the Volcanic Island of La Gomera (Canary Archipelago, Spain) Using Simple Sequence Repeats (SSRs) as Molecular Markers. *Horticulturae* **2024**, *10*, 14. https://doi.org/10.3390/horticulturae10010014

Academic Editor: Stefano Poni

Received: 2 November 2023
Revised: 12 December 2023
Accepted: 16 December 2023
Published: 22 December 2023

1. Introduction

Nowadays, the cultivation of *Vitis vinifera* L. varieties is one of the oldest and most relevant worldwide. The FAO (Food and Agriculture Organization of the United Nations, Washington, DC, USA) affirms that the most economically valued crop is the vine for the vinification of medium- and high-quality wines [1]. Furthermore, the NASS (National Agricultural Statistics Service) reaffirmed its importance as the sixth most economically valuable crop in 2021 in the United States [2]. The global production volume of fresh grapes in 2021, according to the OIV (International Organisation of Vine and Wine), was 74.8 million tons, of which more than 20% came from Europe. Specifically, and in this order, Italy, Spain, and France led European production by a wide margin, representing 79% of the total production [3].

However, if one examines the cultivated varieties' biodiversity on a global scale, it can be seen that almost the same varieties are grown in most countries, occupying between

70 and 90% of the cultivated area. In this sense, these varieties are already known as international varieties [1]. This fact highlights the need for further exploration and study of the rest of the varieties. It is, therefore, about searching for unique and interesting molecular profiles (MP-SSRs) that not only help to expand the range of wines offered in the current market or mitigate the effects of climate change but also give advantages to other relevant factors, whether biotic (pests and/or diseases) and/or abiotic (salinity, etc.).

A starting point for this purpose is undoubtedly La Gomera Island. This island is located in European overseas territory (near Western Sahara); it belongs to the volcanic archipelago of the Canary Islands and Macaronesia (a group of Atlantic volcanic islands) (Figure 1) [4].

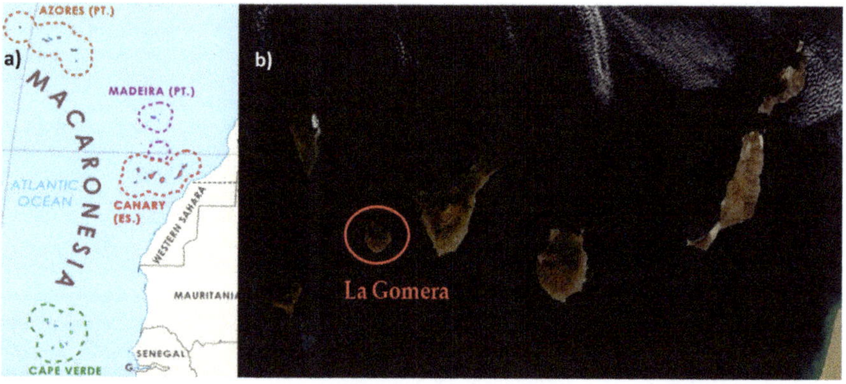

Figure 1. (**a**) Macaronesia map [5]. (**b**) Canary Islands archipelago and La Gomera Island [6].

It is now widely accepted that the vine was not a part of the existing crops in the Canary Islands until its introduction in the 15th century. This belief was challenged by the discovery of fossilised *Vitis* seeds in archaeological excavations [7]. However, *Vitis vinifera* ssp. *sylvestris*, like much of the endemic and fossilised flora of the islands, does not appear today, as it disappeared afterward for unknown reasons. It is for this reason that grapevines in the Canary Islands come from those cultivated (domesticated) varieties introduced by Europeans. Thus, a low diversity of vines compared to other regions of the planet could be expected, but the opposite is true. Due to the phylloxera plague (*Daktulosphaira vitifoliae*) that hit Europe at the end of the 19th century, the loss of *Vitis vinifera* L. varieties was much more accentuated on the mainland than in the Canary Islands, which remained unaffected by this plague. It is for this reason that the Canary Islands, including La Gomera Island, are one of the last strongholds of the European vine varieties that existed before this plague [8]. As a result, vines have evolved and adapted to this new habitat for more than 500 years (through natural crosses, mutations, and selection (natural and anthropogenic)).

La Gomera Island has a similar climate to the islands of La Palma and El Hierro (hot, semiarid climates, BSh Köppen) with incident precipitation in high areas and scarce precipitation near the coast [9]. Furthermore, if its orography is observed, it can be concluded that apart from making possible the humid microclimate of Garajonay Park (laurel forest dating from the Tertiary period, which is also a World Heritage Site), the only areas suitable for agricultural cultivation are the mountain's slopes and ravines [8]. These areas were initially unsuitable for vine cultivation. In these circumstances, the traditional farmers overcame this problem with heroic viticulture based on the construction of dry-stone terraces (Figure 2). These were staggered stone constructions without cement (mortar) on the slopes, which made planting possible. Under these conditions, when the crops took root, they strengthened the construction and prevented the soil from sliding away [10].

Figure 2. Staggered dry-stone terrace in La Gomera [10].

Grape-berry production in La Gomera Island is currently one of the most important crops in the island's agricultural sector. Within the area for permanent crops (citrus and noncitrus fruit trees, vineyards, olives, etc.), 139.7 hectares are reserved for dry crops. Specifically, 117.9 hectares are reserved for vine cultivation, almost all of which are used for winemaking [11]. In 2020, the total grape production was 53,576 kg, of which 38,834 kg corresponded to white grapes. Examined by variety, the Albillo forastero is the most important variety in this category. Comprising more than 82% of the total amount of white grapes of the island [12] and 85% of the wine-growing area of the island of La Gomera, Albillo forastero is the most cultivated grapevine variety [13]. It is currently known that this variety comes from a cross between the Andalusian variety Palomino fino (known in the Canary Islands as Listan blanco) and the Portuguese variety Verdelho branco (known in the Canary Islands as Verdello). This cross also produced another Canary Islands variety known as Albillo criollo (a local variety from La Palma Island) [14].

The island of La Gomera has had its own Appellation d'Origine Contrôlée (AOC) since 2003, which accepts 31 varieties for winemaking [12]. The most curious thing is that among the accepted varieties, there is no Albillo forastero, the prime name (PN) found in the *Vitis* International Variety Catalogue (VIVC) [15], a worldwide database for grapevine varieties. On the other hand, the synonymous names of this variety, Forastera blanca and Forastera gomera, are found. Moreover, on La Gomera Island and in the Canary Archipelago (the only place in the world where this variety can be found), the name Albillo forastero has never been used to refer to this local variety. It is also observed that the term "local" is always used and not the term "autochthonous". Accepting the proposal of Dr. Manna Crespan (2nd OENOVITI International Symposium, 2014) [16], the term autochthonous is not used because the main variety of the island of La Gomera did not originate on the island itself, but it is the island's inhabitants who have preserved it until the present day, making it unique in the world.

Historically, grapevine varieties' characterisation and identification were based on the description and comparison of morphological characters. The science that allowed these comparative studies of *Vitis vinifera* L.'s phenotypes to be carried out is ampelography [17]. Two of the most widespread errors generated by this phenotypic science (which is markedly subjective, influenced by the abiotic factors of the environment and by the state of health of the individual to be analysed) are homonyms and synonyms. A term is said to be a homonym when it is used to name two or more vine varieties. This is the case for the term Ugni blanc, which is used to name three different varieties (the Italian variety Trebiano toscano, the Spanish variety Viura, and the Portuguese variety Douradinha). On the other hand, a term is qualified as a synonym when it is one of the different names of the same variety. In this case, a variety has a PN and one or more synonymous names. As an example to illustrate this concept, for the most cultivated variety on La Gomera Island, the PN is Albillo forastero and the four synonymous names registered in the VIVC are Forastera, Forastera blanca, Forastera gomera, and Gomera [15]. Of all the names by which the same variety is known, the most widespread or the one most commonly used in the place where the variety possibly originated or where it has been preserved is chosen as the PN [18,19]. This is not the case in this example. The names most commonly used on La Gomera Island and in the Canary Islands are Forastera blanca and Forastera gomera, and it is the international community that has arbitrarily imposed the term Albillo forastero, unknown to all the inhabitants of La Gomera and the rest of the Canary Islands, as the PN. In order to avoid the appearance of errors of this and other kinds, from the 1990s onwards, different molecular marker techniques were implemented for the characterisation and identification of grapevine varieties based on the study of the genotype (which is practically invariable). Currently, the most widely implemented and therefore used are SSRs (simple sequence repeats) or microsatellites and SNPs (single-nucleotide polymorphisms) [19].

This work aimed to study the intervarietal and intravarietal diversity of the island of La Gomera existing at the time of sampling using the SSR technique. In addition, another objective was to detect possible errors in terminology and to find synonyms and homonyms, if any. The last objective was to verify the uniqueness of the population of the La Gomera grapevine population. For this purpose, a population structure study was carried out, comparing the population of local varieties of La Gomera with those of other populations in the Canary Islands and the rest of the world.

2. Materials and Methods

2.1. Plant Material

In order to explore the *Vitis vinifera* L. biodiversity extent, 120 vine shoot samples were collected in different municipalities of La Gomera Island (Agulo, Alajeró, Hermigua, San Sebastián de La Gomera, Valle Gran Rey, and Vallehermoso) through a mass selection process carried out by the local winegrowers and supervised by the Regulatory Council of the Vinos de La Gomera AOC technicians. Once harvested, shoots were stored at −20 °C until processing. Table S1 shows detailed information on the analysed grape varieties.

2.2. DNA Extraction and Purification

The purified DNA was obtained using a specific method for foliar and recalcitrant tissues such as wood and seeds [20,21]. This is an adaptation based on the protocol of Fort et al. [22], which was improved by adding PVP (polyvinylpyrrolidone) to the extraction buffer and by adding two chloroform washes. With the help of a Thermo Fisher® Scientific NanoDrop TM 1000 spectrophotometer and using the electrophoresis technique, the levels of purity, integrity, and concentration of the nucleic acid were precisely evaluated.

2.3. Microsatellites

The twenty SSRs that the Investigación en Tecnología Enológica (TECNENOL) research group uses for genotyping samples from the Canary Islands were also tested on this occasion (VVS2, VVS3, VVS29 [23]; VVMD5, VVMD6, VVMD7 [24]; VVMD27, VVMD28, VVMD36 [25]; VrZAG21, VrZAG47, VrZAG62, VrZAG64, VrZAG79, VrZAG83 [26]; VvUCH11, VvUCH12, VvUCH19 [27]; scu06vv [28]; VChr19a [29]). Of these, there are two SSRs that are not independent, as they analyse the same area of the genome with different primers. These are VrZAG47 and VVMD27 [30]. Furthermore, in this SSR "kit" used by TECNENOL, there are 7 SSRs that coincide with some of the 9 SSRs accepted by the international scientific community [31]. In Table S2, the values of the allelic lengths of the SSRs considered as international standards corresponding to the unique MP-SSRs of this grapevine population can be seen.

2.4. DNA Amplification

For the satellite region amplification, an AmpliTaq DNA Polymerase kit (Applied Biosystems, Foster City, California, USA) was used with a final reaction volume of 12.5 μL. The amounts of each reagent broken down per well were as follows: 1.25 μL of buffer, 2 μL of dNTPs, 0.125 μL of deionised formamide, 0.0625 μL of Taq polymerase, and 4 ng of DNA. In addition to 1 μM of each primer (1.25 μL), with the particularity that the forward primer (Fw) was labelled with a specific fluorochrome (6-FAM: VVS3, VVMD7, VVMD28, VVMD36, VrZAG47, VrZAG62, VrZAG83, VvUCH11, and VvUCH19; HEX: VVS2, VVS29, VVMD6, VVMD27, VrZAG21, VrZAG79, and VChr19a; NED: VVMD5, VrZAG64, scu06vv, VvUCH12). Seven thermocycling blocks were performed according to the different annealing temperatures (*Ta*) (Table S3), and for all of them, the thermocycling conditions were as follows: (a) a first stage of 5 min at 95 °C; (b) a second stage of 40 cycles: 45 s at 95 °C during, 30 s at the corresponding Ta, and 1 min 30 s at 72 °C; and (c) a third stage of 7 min at 72 °C. An Applied Biosystems 2720 Thermal Cycler was used for this process (Foster City, CA, USA).

2.5. Amplified Fragments Length Measurement

Measurements of the polymerase chain reaction (PCR) amplified fragments was performed by capillary electrophoresis with an ABI PRISM 3730® genetic analyser (Applied Biosystems, Foster City, CA, USA). The preparation of the amplified plates was carried out by adding to each well the corresponding amplification product, 20.5 μL of deionised formamide, and 0.25 μL of the internal marker GeneScan ROXTM 500 (Applied Biosystems, Foster City, CA, USA). Each plate was then denatured by a thermocycling regime at 95 °C for 3 min. Peak Scanner Software 2.0 (Applied Biosystems, NJ, USA) was used to size the amplified fragments.

2.6. Data Analysis

GenAlEx 6.5 software was used for different purposes [32,33]. Data input files were prepared according to this program using the Excel program 2016 (Microsoft 365, USA). Firstly, the efficiency and effectiveness of the SSR kit used by TECNENOL was evaluated. For this purpose, the following statistical parameters were measured: Na (number of different alleles), Ne (number of effective alleles: alleles that are transmitted to the next generation), Ho (observed heterozygosity: the computed heterozygotes), diversity index or He (expected heterozygosity: estimation of the heterozygotes that the population under study could have), F (fixation index: parameter that measures the goodness of homozygotes), and PI (probability of identity: the probability that two MP-SSRs with the same SSR value are the same variant). Secondly, it allowed us to find all the MP-SSRs that matched each other in order to eliminate redundant information. Assignment tests to check the sample distribution goodness of fit in different populations were also performed using this program. GenAlEx 6.5 bases this strategy on the allele frequency of each accession. It also allowed us to calculate a logarithmic probability value of this accession for each subpopulation using the allele frequencies of the respective subpopulations and assign an individual to the subpopulation with the highest logarithmic probability value [34]. Thirdly, based on the standardised covariance of the genetic distances calculated for the codominant markers, GenAlEx 6.5 enabled two-dimensional graphical representations to be made for a set of populations and also for a set of individuals belonging to different populations. Finally, it made it possible to calculate the coefficient of genetic differentiation between populations assuming the infinite allele model (Fst).

Python Data [35], applying Matplotlib strategy, and MEGA version 7 [36], applying the neighbour-joining approach [37], were used to make the three-dimensional graphical representations of PCoA, the phylogenetic trees, and the circular dendrograms.

The structure of the different populations that emerged from this study was explored using the cluster analysis method based on models implemented in the program Structure 2.3 [38,39]. This program calculates a probability value for a predetermined number of K populations (or clusters) and assigns the part of each individual's genome derived from each cluster. Population structure was tested from K = 1 to K = 7 for the local population of La Gomera Island and for the Canary Islands population and from K = 1 to K = 9 for the world population. Ten independent replications were performed, composed of 1.000.000 Markov chain Monte Carlo (MCMC) steps after discarding the first 100.000. It was assumed that the current populations' allele frequencies were correlated and that they could have originated from more than one ancestral population. The most probable value of K was determined according to the method of Evanno et al. [40]. The parameter q defines what proportion of an individual's genome belongs to the different predefined clusters (K). The membership of a population to a cluster was accepted for mean values of $q \geq 0.85$.

3. Results

3.1. SSR Polymorphism

Once the 120 samples were genotyped, the first data normalisation was carried out by eliminating 68 identical profiles (including the 2 "sports": Bermejuela negra and Malvasia rosada) (Tables S1 and S2). From the remaining 52 MP-SSRs, which corresponded to 19 varieties, the relevant statistics were calculated to determine the goodness of fit of the 20 SSRs used. Table S4 presents the results of the main statistics. The total number of alleles in the unique MP-SSR population was 181, with a mean value of 9.1. The SSRs with the highest number of alleles were VVMD28 with 16 alleles and VVMD36 with 13 alleles. The SSRs with the lowest number of alleles were VVS3 with four alleles, UCH19 with five alleles, and VVS29 with six alleles. The mean number of effective alleles was 4.4, with the SSRs VVMD28, VVMD27, and VVMD36 showing the best values (8.32, 6.76, and 6.72, respectively), and the worst results were shown by the SSRs VVS29 with 1.41 alleles, VVS3 with 2.17 alleles, and VVS3 with 2.17 alleles passing to the next generation, respectively. The mean Ho (78.8% heterozygotes) was higher than the He (72.4%). The highest percentages of heterozygotes were observed in the SSRs VVMD36 with 98.1% heterozygotes; VVS2 with 96.2%; and ZAG79, VVMD28, and SCU06 with 94.2%. In contrast, the lowest values corresponded to VVS29 with 30.8%, UCH19 with 51.9%, and VVS3 with 55.8%. The highest values of the diversity index, also known as He, were for SSRs VVMD28, ZAG47, VVMD27, and VVMD36 (88%, 86.9%, 85.2%, and 85.1%, respectively), and the lowest values were again for SSRs VVS29 with 29.1%, UCH19 with 51.9%, and VVS3 with 53.9%. Only four SSRs had an F index with zero or positive values; these were UCH19 (F = 0.000), ZAG47 (F = 0.005), ZAG83 (F = 0.112), and UCH12 (F = 0.144). The lowest PI was shown by SSRs VVMD28, ZAG47, VVMD27, and VVMD36 with values of 2.55×10^{-2}, 3.04×10^{-2}, 3.86×10^{-2}, and 3.95×10^{-2}, respectively, and the highest PI was shown by SSRs VVS29, VVS3, and UCH19 with values of 5.16×10^{-1}, 3.07×10^{-1}, and 2.68×10^{-1}. The accumulative PI for the 20-SSR kit reached a value of 3.1×10^{-21}.

3.2. Grapevine Variety Analysis

All of the original and conclusive information concerning the 120 accessions can be found in Table S1. In addition, it presents in detail the similarity of the MP-SSR of a given accession concerning the most widespread MP-SSR according to the TECNENOL database, even specifying which allele presents the variation. With all of this information and the possibility of comparing it with the VIVC database, a study was carried out at both the molecular and terminological scales. In Table S2 and for the 52 unique MP-SSRs, the numerical values of the allelic lengths measured for the 7 international SSRs available to the TECNENOL research group are also presented. Furthermore, the accessions in these tables (Tables S1 and S2) correspond to 19 varieties, of which 5 are local varieties from the island of La Gomera (Albillo forastero, Barrerita negra, Coello blanca, Malvasia periquin gomerae, Verdello gomerae); also, 5 are local varieties from the rest of the Canary Archipelago (Bermejuela, Listan negro, Torrontes volcanico, Uva de Año, Verijadiego), and the remaining 9 correspond to foreign varieties from the Canary Islands: 3 from Spain (Mollar cano, Palomino Fino, Tempranillo tinto), 2 from Portugal (Caracol, Verdelho branco), 1 from France (Alicante Henri Bouschet), 1 from Greece (Muscat of Alexandria), 1 from the Balkan Peninsula (Malvasia Dubrovacka), and 1 from the United States of America (Ruby cabernet). Mutations should also be highlighted, whether colour (sport) or numerical (see Table S1). There is a colour mutation that corresponds to a sport widely spread throughout the Canary Islands (Malvasia Dubrovacka rosada), another that corresponds to a sport from the island of La Gomera (Bermejuela negra), and an individual that presents a numerical variation with respect to a variety from the island of La Gomera (Verdello gomerae de Monacal).

In both Tables S1 and S2, it can be seen that the grouping corresponding to the Albillo forastero variety is made up of 46 individuals and is, therefore, the most numerous. Among them, 26 accessions were identical to the most widespread MP-SSRs in the TECNENOL database (identities), 16 individuals presented variations in their allelic length in one allele, 3 accessions showed variations in two alleles, and 1 sample presented variation in three alleles. These results defined 10 different MP-SSRs for this variety. One MP-SSR corresponded to an identity with the most widespread profile in our database. This profile was named with its PN according to the VIVC, which in this case corresponded to the term Albillo forastero. Six MP-SSRs showed allelic length variation in one allele: (a) The variation in VVS3-1 (numerical change in the first SSR allele VVS3), which is known as Forastera de la Isla Redonda and was previously published by Fort et al. [41]. (b) The VVS3-2 variation that was called Forastera blanca de Agulo. (c) The mutation in VVS29-2, which is known as Forastera blanca de Vallehermoso. In addition, there are two mutated profiles in this same allele but with different allelic lengths that are marked with *, but all are known with the same term. (d) The variation in UCH12-2 known in La Gomera as Forastera blanca roquillos. Two individuals showed variation in their allelic length in two alleles of the same SSR (VVS3-1, VVS29-2), but as in the previous case, a numerical difference in allelic length was detected for VVS29-2. Both specimens are known as Forastera blanca Simancas. The last profile detected for the variety Albillo forastero corresponded to an accession with variation in allelic length in three alleles (VVS3-1, VVS29-1, VVS29-2) known on the island of La Gomera as Forastera blanca tamargada.

The only representative of the French variety Alicante Henri Bouschet, known in La Gomera as Alicante tintilla, had a mutation in one allele (VVS29-2).

The local Canarian variety Bermejuela cluster consisted of seven members, two identities, two mutations in one allele, one mutation in two alleles, one mutation in four alleles, and a colour mutation with the same MP-SSR as the most widespread genetic profile in the TECNENOL database (these cases are known as "sport"). This colour mutation is known in La Gomera Island as Marmajuelo negro [14]. This grouping provided four different MP-SSRs: (a) the identity itself known by the name Bermejuela; (b) the variation in one allele (VVS3-2), a profile already described by TECNENOL in El Hierro Island and named Bermajuelo del Echedo [41], which is also known on the island of La Gomera by the synonymous name Marmajuelo corto (Table S1); (c) the variation in two alleles (VVS3-2, SCU06-1) known as Marmajuelo de Vallehermoso; and (d) the variation in four alleles (VVS3-1, VVS3-2, VChr19a-1, VChr19a-2) known in La Gomera as Marmajuelo de Valle bajo.

The Portuguese variety Caracol was recorded under the erratic name Forastera negra and did not show any variation in its MP-SSR.

There are eighteen entries of the local Canarian variety Listan negro, of which half are identities, eight have one variation in one allele, and the last one is mutated in two alleles. This group of individuals gives rise to six different MP-SSRs, from the identity named Listan negro to the profiles mutated on one allele which give rise to four different MP-SSRs: (a) variation in VVS3-1 which is known as Listan negro de Hermigua; (b) two variations, one in VVS2-2, for which the profile is already published under the name Listan negro santanero [42], and one in VVS29-2, for which the profile is also published under the name Listan negro de la corona [42]; (c) variation in VVMD6-1 known in La Gomera as Listan negro de Vallehermoso; and finally (d) the accession registered as Listan negro de lo Machado, which corresponds to an MP-SSR with variations in VVS3-1 and ZAG83, the last variation being a case of triallelism.

The five members of the Malvasia Dubrovacka variety are distributed in one identity, two individuals mutated in one allele, and another mutated in three alleles of which one is triallelic; the last component, a rosé sport very widespread in the Canary Islands and unique in the world, is the Malvasia rosada. For this variety, there are four MP-SSRs corresponding to an identity known as Malvasia Dubrovacka, a variation in VVS2-2 known as Malvasia blanca de Agulo, a variation in VVS29-2 known as Malvasia blanca de Vallehermoso, and finally the Malvasia blanca piedra gorda with three variations in VVS3-2, VVS29-2, and SCU06, the latter variation being a case of triallelism.

In La Gomera Island, three specimens of the Mollar cano variety were collected, one identity with the same name and two individuals with one variation in one allele. These were the Negramoll de Vallehermoso (VVS3-1) and the Mulata del macayo (SCU06-2) mutations.

In contrast, for the Muscat of Alexandria cluster, only individuals with variations were present. Two accessions with one variation in one allele and two with variations in two alleles displayed a total of three MP-SSRs, with one variation being the accession Moscatel de Hermigua (VVS29-2) and two variations being the mutations Moscatel de la caleta (VVS3-1, VVS29-2) and Moscatel de la caleta fino (VVS29-2, SCU06-2).

The grouping of the Palomino fino grapevine variety, known in the Canary Islands by the synonymous name Listan blanco, contained 21 individuals. Of these, 13 were identities. Six had one variation in an allele, and of these, two were triallelic. In addition, one individual had two variations, one of which was triallelic, and the last one had four variations, one of which was also triallelic. In total, six different MP-SSRs were computed: (a) The identity. (b) An accession, Listan blanca chicharrera, with a variation in VVS3-1, which was already published [42]. In addition, Listan blanco de Hermigua, with a variation in VVS3-2 and a variation in the form of triallelism in SSR ZAG83, was already published in the Lanzarote Island prospection as Listan blanco de la bodega [42]. (c) The profile known as Listan blanco de Vallehermoso, presenting variations in VVS29-2 and ZAG83 (triallelic). (d) The entry registered as Listan blanco de espina with four variations (VVS3-1, ZAG79-1, ZAG79-2, ZAG83 (triallelic)).

A single entry corresponded to a Ruby cabernet mutated in three alleles (VVS29-2, VVMD27-1, VVMD36-1) and known as Ruby cabernet ingenio.

The Spanish variety Tempranillo tinto was represented by two mutated individuals, one on an allele (VVS29-2), called Negra mora, and the other, known as Tempranillo de Vallehermoso, with a very rare case of quadriallelism in the SSR (SCU06).

The local variety of Lanzarote Island, Torrontes volcanico, was represented by three accessions, one of which was an identity while the other two had variation in one allele. The mutation known as Torrontes volcanico montoro had a variation in UCH12-2, and Torrontes volcanico machado had a variation in SCU06-1.

The variety also originating from Lanzarote, Uva de año, presented only one component (Uva de año montoro), which presented two variations in two different alleles (VVS3-1, VVS29-2).

With only one component, there were also the following: (a) The Portuguese Verdelho branco with a case of triallelism in the SSR UCH19 (Verdello blanco del Corte). (b) The Verijadiego grapevine variety from El Hierro Island with an identity. (c) Three of the four new varieties: the variety Barrerita negra, also described by Rodríguez-Torres [43] as unknown no. 2; the variety Coello blanca; and Malvasia periquin gomerae, which presented a case of quadriallelism in the SSR (SCU06). (d) The last new variety, which presented two individuals, of which one was a mutation of the other in two alleles. These were the variety Verdello gomerae and its mutation Verdello gomerae de Monacal (VVS3-1, ZAG83-1).

3.3. Genetic Structure of the Grapevine Population of the Island of La Gomera

In order to carry out the study of the grapevine population uniqueness in La Gomera Island, a second data normalisation was carried out. Table S2 shows the 52 unique MP-SSRs corresponding to 19 varieties, which means that 33 individuals were left out because they were genetically very close to each other and could alter the final result of the corresponding study (63.5%). In this sense, the representatives of the 19 varieties mentioned above remained, which did not always correspond to identities, but for the cases of the Alicante Herni Bouschet (with a similarity to the most widespread MP-SSR of 97.5%, i.e., with a variation of only one allele), Muscat of Alexandria (97.5%), Ruby cabernet (92.5% (with a variation of three alleles)), Tempranillo tinto (97.5%), Uva de año (95% (with a variation of two alleles)), and Verdelho branco (97.5%), these were represented by individuals with variations. In order to obtain the best population distribution in different ancestral populations (K), distributions by population from one to seven were tested, with the best distribution being in two populations (K = 2) (Figure S1). Each individual from the population of La Gomera Island was distributed according to a statistical parameter q (which indicates the percentage of its inferred genome that belongs to one of these populations) in these two proposed populations. In this study, the arbitrary percentage of 85% [44] was chosen, so that values for $q \geq 85\%$ correspond to pure individuals belonging to a given population and those with a value of $q < 85\%$ are admixed individuals for the same population. Figure 3 shows the best distribution for the La Gomera Island population in two populations (POP1 and POP2). Figure S2 shows the values for the parameter q as well as the origin of the members of each population considered. Thus, POP1 consisted of four individuals and represented 21% of the total population on the island. Of these, three individuals were pure, two were new varieties (Malvasia periquin gomerae and Verdello gomerae), and one was the Balkan variety Malvasia Dubrovacka. The population's fourth member, and admixed, was the Spanish variety Tempranillo tinto (97.5%). In contrast, for POP2 (79% of the island population), all individuals were found to be pure. This grouping included (a) three varieties from La Gomera, of which two were new (Barrerita negra and Coello blanca) and one was the well-known Albillo forastero; (b) five Canary Islands varieties (Bermejuela, Listan negro, Torrontes volcanico, Uva de año (95%), and Verijadiego); (c) two Spanish varieties (Mollar cano and Palomino fino); (d) two Portuguese varieties (Caracol and Verdelho branco (97.5%)); (e) one Greek variety (Muscat of Alexandria (97.5%)); (f) one French variety (Alicante Henri Bouschet (97.5%)); and (g) one American variety (Ruby cabernet (92.5%)).

Apart from locating correctly in each population (POP1 and POP2) the different members of the La Gomera Island population, the Structure 2.3 program allowed us to detect the admixed individuals that, in another standardisation, were eliminated, with the aim of constructing the main coordinate plots (PCoA) without interference. Thus, the only admixed individual in both populations, Tempranillo tinto (97.5%), was eliminated, leaving La Gomera Island with 18 components distributed in two populations. An assignment test was carried out using the GenAlEx 6.5 program, resulting in a 100% goodness of fit.

The two- and three-dimensional PCoA representations are presented in Figure 4.

Coordinate 1 (with a goodness of fit of 15.11%) and coordinate 2 (with a goodness of fit of 12.71%) in Figure 4a give a total reliability of 27.82%, separating the 18 grapevine varieties into four quadrants. The upper-right quadrant contains all of the POP1 varieties, with a clear influence from the Eastern Mediterranean Basin (specifically from the Balkan Peninsula (BP)). Of the three POP1 varieties, it is the local ones from the island of La Gomera that are significantly distant from the Malvasia Dubrovacka, located in the lower part of the quadrant. POP2, on the other hand, is distributed throughout the rest of the quadrants. In the upper-left quadrant, the Coello blanca and Verdelho branco varieties (97.5%) are also positioned far away from the rest. In the lower-right quadrant are the Canary Island varieties Torrontes volcanico and Uva de año (95%), the Spanish Mollar cano, and the American Ruby cabernet. The rest of the POP2 representatives are located in the lower-left quadrant. In the three-dimensional representation of the population of La Gomera, a higher goodness is obtained (36.96%), but the same distribution is maintained with some significant changes (Figure 4b). Two important shifts were detected: the approximation of Muscat of Alexandria closer to the BP-influenced varieties and the approach of the local variety from La Gomera, Albillo forastero, closer to the pairing of Coello blanca and Verdelho branco (97.5%). It should also be noted that in both presentations, the local variety Barrerita negra continued to stay away from the rest of the varieties in its quadrant.

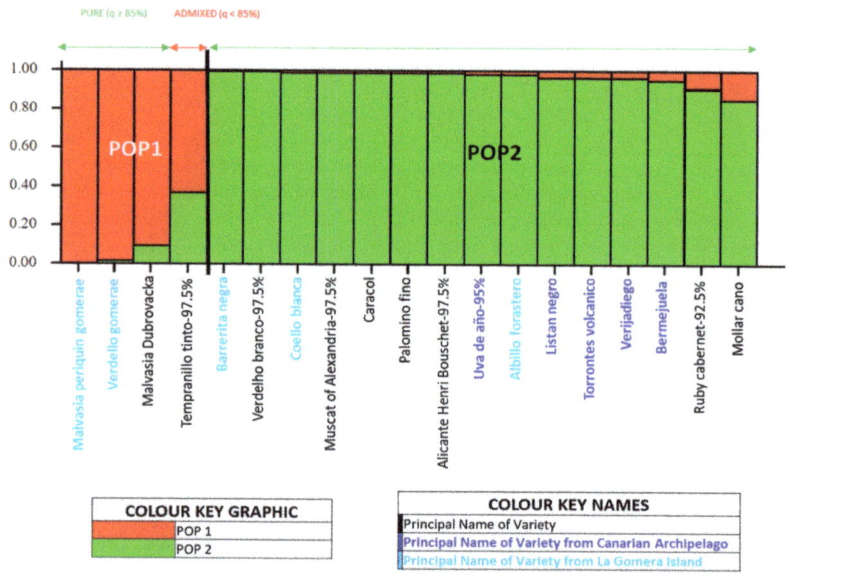

Figure 3. La Gomera grapevine varieties' populations (unique MP-SSRs). Structure 3.2 diagram: K = 2 distribution for pure and admixed individuals.

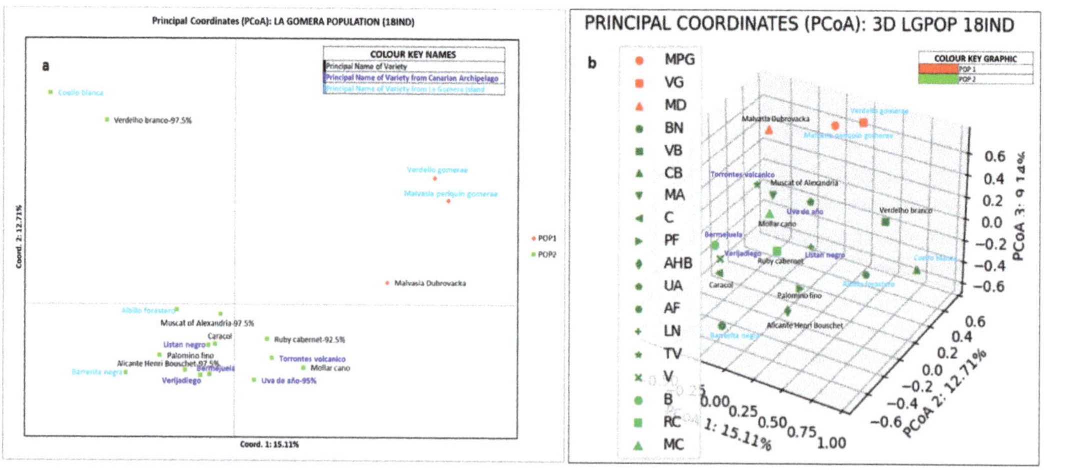

Figure 4. PCoA representations of the grapevine varieties' populations from La Gomera Island normalised for K = 2. (**a**) Two-dimensional representation of the 2 populations by individual, and (**b**) three-dimensional representation of the 2 populations by individual.

3.4. Relation of La Gomera Grapevine Population with respect to the Canary Archipelago Population

Once the above results were obtained, it was considered appropriate to begin a comparative study of the local varieties of La Gomera Island with respect to the local varieties of the Canary Islands in order to gauge the uniqueness of the population of this first island and, furthermore, to see if any other noteworthy characteristics emerged that should be taken into account. The starting point was a total population of 36 individuals, 5 of which were local varieties from La Gomera (the Albillo forastero variety was added to the 4 new varieties).

The same strategy was used as in the previous section. Thus, the Structure 2.3 program was used to obtain the best distribution for the population of Canary Islands varieties. After testing up to seven different distributions, it was the one corresponding to K = 5, i.e., distributing the population of 36 individuals in five ancestral populations, which gave the smallest error and therefore was the best of all the distributions (Figure S3). Figure 5 shows the graphical representation of the Structure 2.3 program, displaying the distribution of all individuals by ancestral population (as a function of the q statistic) and by pure and admixed varieties. The q numerical values and the statistical data associated with the whole study are shown in Figure S4. Figure 5 and Figure S4 show that POP1 consists of a total of seven varieties (19%) almost equally distributed between the islands of Lanzarote and El Hierro, of which five are pure (71%) and two are admixed (29%). POP2 is made up of nine components that represent 25% of the total Canary Islands population, of which five are pure varieties (56%) and four are admixed (44%). This cluster includes the interspecific cross known as Bienmesabe tinto (Canary Islands (IC)), in addition to the variety from Fuerteventura Island, which has been postulated so far as another possible interspecific cross [45], alongside other varieties from all over the archipelago that usually appear as very singular MP-SSRs in other TECNENOL research group studies. The local Gomeran variety, Barrerita negra (pure), is also included in this grouping. POP3 is constituted by eight members (22%) mainly from Lanzarote Island, which are related very closely or completely to the original variety of BP Malvasia Dubrovacka. It consists of four pure and four admixed individuals. In this group are present as pure individuals the new local varieties of the island of La Gomera, Malvasia periquin gomerae and Verdello gomerae. The POP4 group with six components represents 17% of the total and is formed almost exclusively by varieties from El Hierro Island, of which three are pure and three are admixed. Finally, POP5, also with six components distributed in three pure and three admixed varieties, is characterised by the fact that all the pure individuals come from IC and the two local varieties present are admixed (Albillo forastero and Coello blanca).

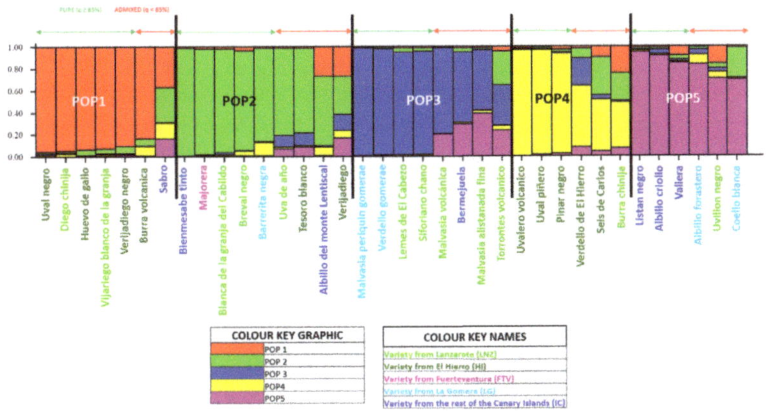

Figure 5. Canary grapevine varieties' populations (unique molecular profiles). Structure 2.3 diagram: K = 5 distribution for pure and admixed individuals.

Once the pure varieties were known, the data were standardised by eliminating the admixed varieties, with the aim of optimising the graphical representations by means of PCoA. In this way, from a population of 36 Canary Islands varieties, a population of 20 was obtained. It should be pointed out once again that under these new circumstances, the population of La Gomera local grapevines was reduced to three individuals (Barrerita negra, Malvasia periquin gomerae, and Verdello gomerae). An assignment test was carried out to check the reliability of the location of each pure individual to its corresponding population. This was 90%. Figure 6 shows the two- and three-dimensional PCoA representations. Figure 6(a.1) shows the two-dimensional PCoA representation by population with a goodness of 69.11%, corroborated by Figure 6(a.2), where the matrix of the Fst statistic for each pair of populations is shown. From this information, it can be seen that POP1 and POP5 are the closest populations occupying a single quadrant, the lower-left one. Also close to them is POP2, positioned in the upper-left quadrant, very close to coordinate 2. In the right quadrants and very distant from this group of populations are POP4 in the upper quadrant and POP3 in the lower quadrant, the latter being the most distant of them all. Figure 6(a.3) shows the two-dimensional representation by PCoA, now by individual. In this case, the distribution is practically the same as that resulting from the population plot but with a rotation on both axes. Thus, POP1 and POP5 appear again very close to each other in the upper-right quadrant (POP5 more dispersed in the centre of the intersection between the axes), while POP2 is close to them and distributed in the lower-right quadrant. Again in the opposite quadrants are POP3 in the upper-left quadrant and POP4 in the lower-left quadrant. In a specific grapevine variety analysis (Figure 6(a.3,b)), the position of the variety Bienmesabe tinto, which is significantly distant from the rest of the group, should be highlighted. The varieties Majorera, Barrerita negra, Malvasia periquin gomerae, and Verdello gomerae are also far from the rest (Figure 6b) although to a lesser extent. While the Majorera variety in Figure 6(a.3) (two-dimensional representation by individual (24.37% goodness of fit)) appears indistinguishable from the rest of the varieties, in Figure 6b (three-dimensional representation by individual (33.87% goodness)), it appears distant from the main group. In both figures, the three local varieties of La Gomera Island also appear in separate locations from the group of Canary Islands *vinifera*, with the variety Barrerita negra occupying a position very close to the variety Bienmesabe tinto. Also, the varieties Malvasia periquin gomerae and Verdello gomerae appear in Figure 6b, standing out from the rest and appearing close to the Majorera grapevine variety.

3.5. Relation of La Gomera Grapevine Population with Respect to the World Population

In order to complete and give consistency to the hypotheses set out in the Discussion section, the possibility of comparing the population of 5 local vines from La Gomera Island with the collection of 309 individuals of *Vitis vinifera* ssp. *vinifera* from 22 countries of the world from the TECNENOL database [41,42,45–48], always genotyped with the same 20 SSRs, was considered. This comparative study was carried out using two different strategies. In the first, a purely genetic criterion was taken into account, and in the second, a geographical criterion (country of origin of each variety) was also used.

To implement the first criterion, the Structure 2.3 program was used again so that the best distribution for the 314 varieties under study was sought, in this case between one and nine different distributions. Figure S5 shows that the best distribution for the world population analysed was K = 2, so the components were divided into two populations (Figure 7a and Figure S6). POP1 consists of 183 individuals (58%), mostly from Italy, France and Central Europe, Greece, the Balkan Peninsula, and the Eastern Mediterranean. A total of 88% were pure (161 individuals) and 12% were admixed (22 individuals). POP2, on the other hand, is a population mainly of Spanish origin made up of 131 individuals (42%), of which 103 are pure (79%) and 28 are admixed (21%). The detection of admixed individuals is essential for the necessary data standardisation and thus for the continuation of this study. Therefore, the 22 admixed varieties of POP1 and the 28 admixed varieties of POP2 were removed. In total, 50 individuals were eliminated, reducing the study population to 264 varieties, with a goodness of assignment of 100%. The Canary Islands varietal population was entirely located in POP2 (Figure S6), and of the 36 varieties from the Canary Islands, only 3 were found to be admixed: Malvasia alistanada fina, Albillo criollo, and Coello blanco (new variety from La Gomera Island).

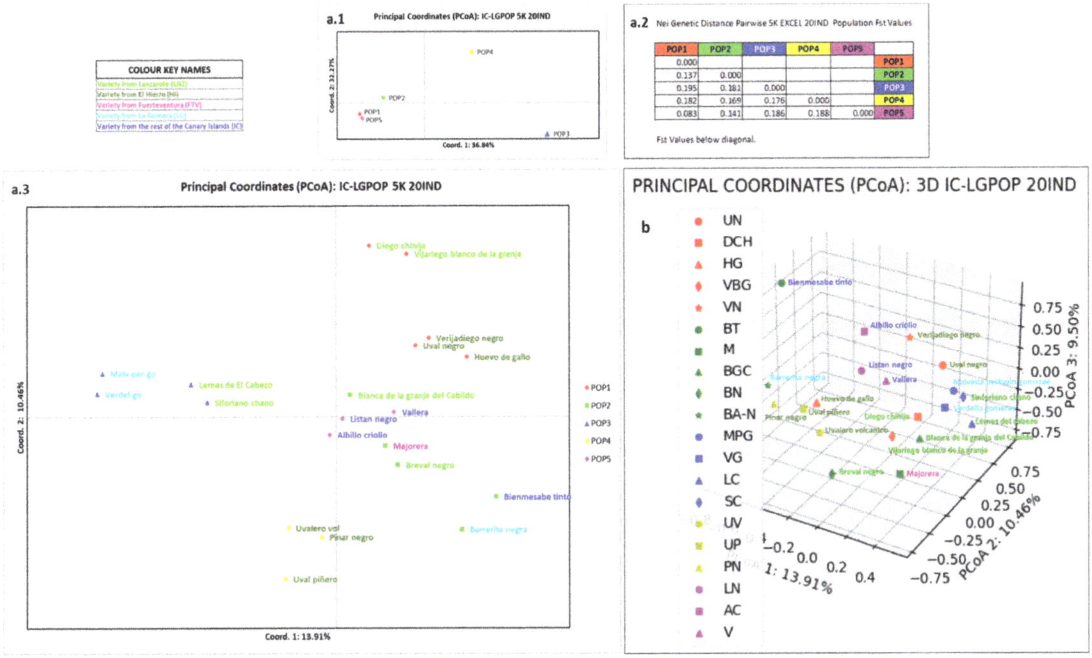

Figure 6. Population of varieties of the Canary Islands (20 pure varieties). (**a.1**) Two-dimensional representation of the 5 populations of the Canary Islands by population. (**a.2**) Values of the Fst statistic for each population. (**a.3**) Two-dimensional representation of the 5 populations by individual. (**b**) Three-dimensional representation of the 5 populations by individual.

Figure 7b,c show the circular dendrogram and the phylogenetic tree of the 264 pure individuals distributed in two populations. It can be clearly seen in both cases how the population of varieties from La Gomera Island occupies a branch by itself (either the branch or the variety name coloured in pink). These are the new varieties Malvasia periquin gomerae and Verdello gomerae as well as the already-known Albillo forastero. The Barrerita negra variety is found together with the rest of the Canary Islands varieties in the continuous branch. It should be remembered that the Coello blanca variety was eliminated as it was an admixed variety. Therefore, (a) the first main branch contains the La Gomera Island population, (b) the second main branch is made up of varieties from the Canary Archipelago, where the varieties from El Hierro Island predominate as well as the variety from Fuerteventura Island known as Majorera, and (c) in one of the sub-branches of the third main branch (origin of the Spanish varieties and also of the rest of the world), there is the other half of the Canary Islands varieties, dominated by the varieties from the island of Lanzarote, together with the interspecific cross Bienmesabe tinto.

The two- and three-dimensional PCoA representations are shown in Figure 8. In order to see the extent of the uniqueness of La Gomera and the Canary Islands, both populations were extracted from POP2 and studied as independent populations.

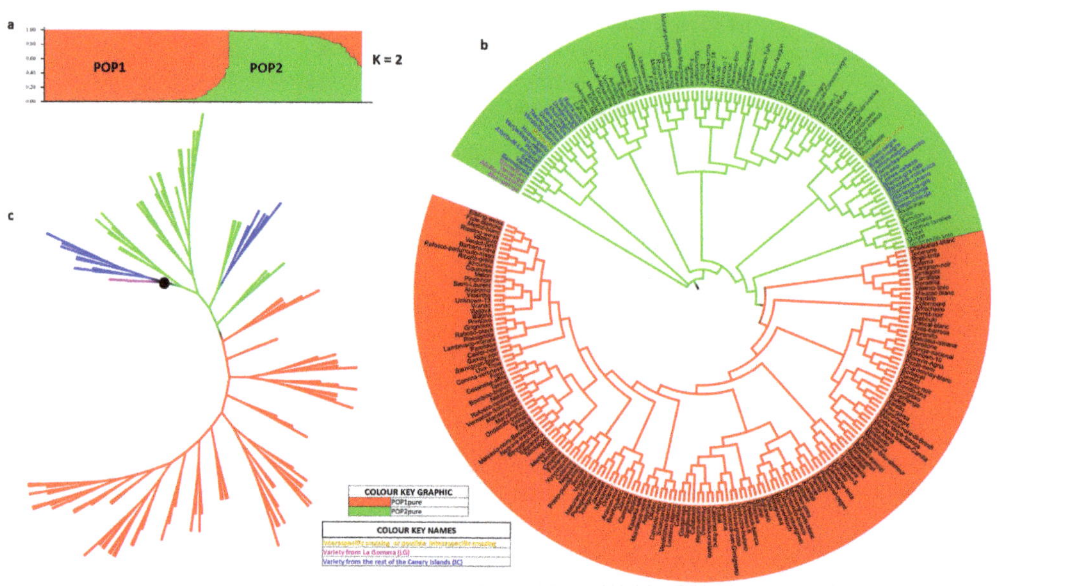

Figure 7. World population (314 individuals) distributed in 2 populations. (**a**) Graphical representation of K = 2 according to Structure 2.3 (with pure and admixed individuals). (**b**) Circular neighbour-joining dendrogram of the world population's 264 pure individuals. (**c**) Pure individuals' world population phylogenetic tree.

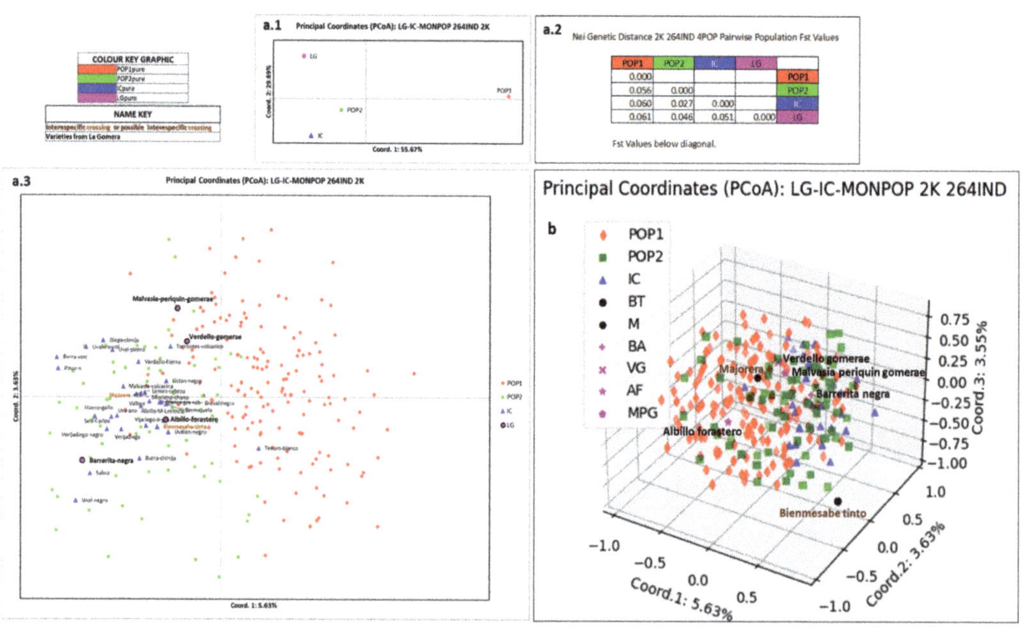

Figure 8. PCoA representation of La Gomera, Canary Islands, and worldwide populations normalised for K = 2. (**a.1**) Two-dimensional representation of the 4 populations per population. (**a.2**) Values of the Fst statistic for each population. (**a.3**) Two-dimensional representation of the 4 populations per individual. (**b**) Three-dimensional representation of the 4 populations per individual.

In Figure 8(a.1,a.2) (with a goodness of fit of 85.56%), it can be clearly seen how the La Gomera Island population, occupying the upper-right quadrant, is very distant from both POP2 and IC, which occupy the lower-left quadrant. In Figure 8(a.3,b), corresponding to the representations by individual, it is possible to see the total overlap of the three populations. Perhaps the most relevant fact is that in Figure 8b, the Bienmesabe tinto variety once again moves away from the vinifera pool, and the Majorera variety is apparently integrated with the rest of vinifera.

Once the geographical component was introduced into this study to consolidate the results obtained so far, the distribution of the 314 varieties was observed in seven geographical areas arbitrarily defined according to the country of origin of each variety published in the VIVC [15,49]. It was decided to distribute by area and not by country, as there were countries with only one representative. In this way, the seven areas that were the object of our study were defined as follows (Figure S7): EASTMED-CAU (Algeria, Cyprus, Georgia, Israel, Lebanon, Tunisia, and Turkey), BP (Bosnia and Herzegovina, Bulgaria, Croatia, Greece, Serbia, Slovenia, and Montenegro), ITA (Italy), FRA-CEU (Austria, France, Germany, Hungary, and Switzerland), IP (Spain and Portugal), IC (Canary Islands), and LG (La Gomera Island). Using this distribution, an assignment test was performed with the GenAlEx 6.5 program, which gave a goodness of fit of 61%. This test helped to locate all those varieties that were misplaced. The EASTMED-CAU population, which accounted for 4% of the total, was composed of 33% pure individuals and 67% admixed individuals. In BP, 9% of the world population was found to be half pure and half admixed. ITA had 23% of the representatives, of which 51% were pure and 49% were admixed. FRA-CE had 19% of the population, with 72% pure and 28% admixed varieties. IP was the largest population with 33% of varieties, of which 65% were pure and 35% were admixed. IC accounted for 10% of the world total, and this population was distributed in 75% pure individuals and 20% admixed individuals. Finally, LG, with the smallest population (2% of the total), presented 40% pure individuals (Malvasia periquin gomerae and Verdello gomerae) and 60% admixed individuals (Albillo forastero, Barrerita negra, and Coello blanca). Once the admixed or misplaced varieties were located, they were eliminated (123 individuals), leaving a starting population of 191 pure or well-placed varieties. Thus, with the populations of well-located individuals, a second allocation test was carried out, which showed a goodness of fit of 91%.

Figure 9 shows the circular diagram and the corresponding phylogenetic tree for the world population under the geographical criterion. In both images, it can be seen, in a significant way, how LG has a singular and outstanding character with respect to both the IC varieties and the varieties of the rest of the world.

It is also worth noting that the IC population is distributed in three sub-branches (Figure 9b). In the first, the LG and the Majorera varieties are separated and are almost entirely made up of individuals from the island of El Hierro (Figure 9a). The second sub-branch, linked to the previous ones, is made up of individuals from Lanzarote Island, and the third sub-branch, now linked to peninsular individuals and where the Bienmesabe tinto variety is located, is made up of the rest of the varieties from Lanzarote and varieties from the rest of the Canary Islands. Another remarkable fact is the dichotomy between PI varieties. This population is spread over the three main branches of both figures, the main one being the one most linked to IC. Figure 10 shows the results of the graphical representations using PCoA.

Figure 10(a.1,a.2) with a reliability of 76.74% show the final distribution of the seven world populations. LG is shown to be significantly distant from the rest of the populations, occupying the lower-right quadrant. The rest of the populations occupy the left quadrants. On the other hand, IC and IP, which occupy the upper-left quadrant, are close to each other but differ from the rest of the world populations, which are distributed in the lower-left quadrant. Figure 10(a.3) does not give relevant information, as both LG and IC, and the varieties highlighted as being either possible interspecific crosses or as being a true interspecific cross, overlap with IP. Finally, Figure 10b shows that the Bienmesabe tinto variety (a true interspecific cross) is significantly displaced from the viniferas and that the pure varieties of LG together with Majorera (a possible interspecific cross) are slightly different from the rest of the viniferas.

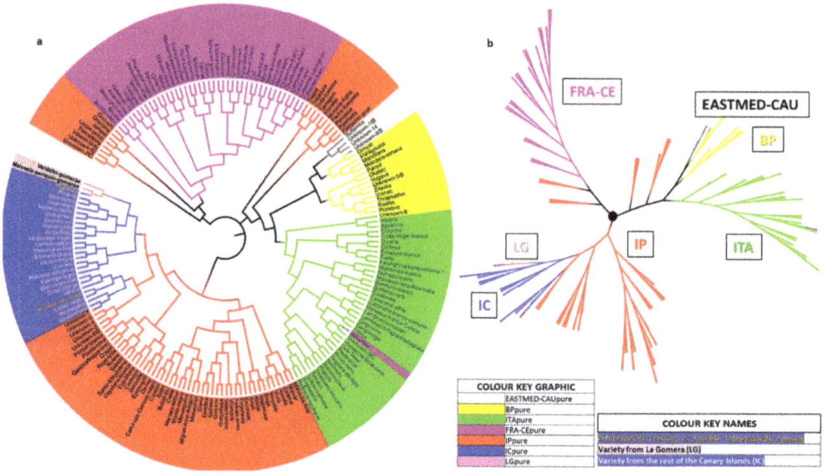

Figure 9. World population (191 individuals) distributed in populations corresponding to 7 geographical areas. (**a**) Circular neighbour-joining dendrogram of the 191 pure individuals of the world population, highlighting the Bienmesabe tinto, Majorera, and LG locations. (**b**) Phylogenetic tree of 7 populations' distribution with all their individuals.

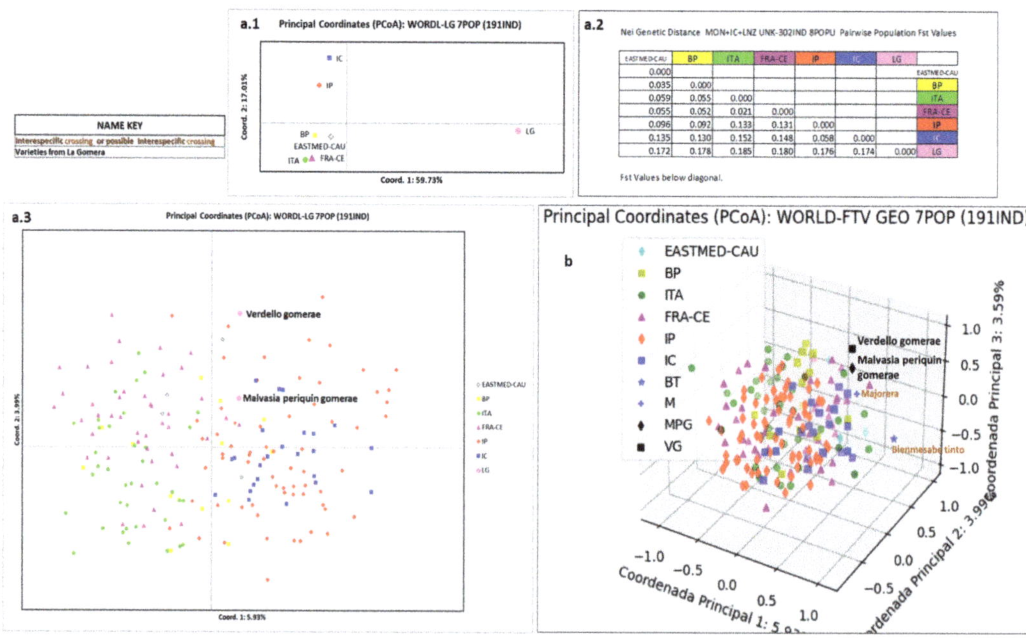

Figure 10. PCoA representation of La Gomera, Canary Islands, and world populations for the geographical criterion. (**a.1**) Two-dimensional representation of the 7 populations per population. (**a.2**) Values of the Fst statistic for each population. (**a.3**) Two-dimensional representation of the 7 populations per individual. (**b**) Three-dimensional representation of the 7 populations per individual.

4. Discussion

One of the main objectives of this study was to analyse the MP-SSRs of the 120 samples obtained during the La Gomera Island prospection with the 20 SSRs usually analysed by the TECNENOL research group. Once the MP-SSRs were obtained, they were compared with the TECNENOL database [41,42,45–48] and with specialized books on Canary Islands grapevine varieties [14,43,50]. Finally, the MP-SSRs that were unknown were compared with a worldwide database (*Vitis* International Variety Catalogue (VIVC)) [15,49].

The discovery of four new MP-SSRs and an individual with variations concerning one of these new profiles (a mutation), as well as 29 individuals with variations in their MP-SSRs of the most widespread MP-SSR, gives an idea of the degree of evolution of *Vitis vinifera* ssp. *vinifera* on La Gomera Island, also known as Colombina Island. More than 500 years of natural selection, natural crosses, and mutations, as well as anthropogenic selection, have been the driving force behind the adaptation of this subspecies.

As in the rest of the Canary Archipelago, La Gomera was never attacked by *phylloxera*, and for this reason, its adaptive evolution was never interrupted [51,52]. Therefore, one of the objectives of the Discussion section is to justify the uniqueness of the local wine-growing population of La Gomera.

4.1. SSR Polymorphism

The first step is to assess the good functionality of the tools used for the analysis, in this case the chosen SSRs. For this purpose, six statistical parameters were used, among which the allelic diversity (Na and He), the goodness of homozygote or F index (i.e., that a homozygote is real and has not lost alleles), and the probability of identity (PI) were estimated (Table S4). A values comparison of the main statistics with respect to the different published works can be a very complex task as there are often variables that are difficult to handle. These are, for example, the number of starting samples, the number and type of SSRs used for the corresponding genotyping, and the uniformity of the population studied, which always depends on the final objective of each study [53]. Because of this, the reference values from island studies are taken as reference values in order to minimise differences. Moita et al. [54] studied a 34-grapevine population from the Azores archipelago with 14 SSRs; Avramidou et al. [55] worked with 56 accessions from the island of Crete, genotyped with 13 SSRs; and TECNENOL prospected three islands with the same 20 SSRs: Lanzarote with 223 samples analysed (statistical study with 99 unique MP-SSRs) [42], El Hierro with 87 accessions genotyped (statistical study with 46 unique MP-SSRs) [41], and Fuerteventura with 40 samples analysed (statistical study with 15 unique MP-SSRs) [45]. In the La Gomera Island grapevine populations study, 120 samples were genotyped, and the statistical study was carried out to determine the goodness of the chosen kit, with 52 unique MP-SSRs. When the Na and/or He (parameters that indicate allelic diversity) are compared, it can be deduced that as the number of individuals in the population which are not of the same variety increases and the number of SSRs tested also increases, the values found will be higher. On the other hand and in general terms, the higher the number of SSRs, the lower the accumulative PI. Therefore, with lower values in the accumulative PI, the ability of a given SSR kit to distinguish two apparently identical samples will show the high discrimination capacity of the chosen SSR kit. The populations of Crete Island [55] and the population of the Azores archipelago [54] showed higher accumulative PI values than those found in La Gomera Island and the rest of the islands surveyed by TECNENOL. This is probably not due to the malfunctioning of the SSR kit used but to the low number of SSRs tested (only 13 or 14 SSRs). Regarding the F index (probability that a kit presents null alleles), the La Gomera population only presents two SSRs with an index higher than 0.01, a threshold above which homozygotes are considered doubtful [56]. In this sense, this work presents the best values of all the studies consulted. Therefore, and comparing the results found on La Gomera Island with the rest of the islands prospected by the TECNENOL research group, it can be concluded that the SSR kit used continues to give good results and is therefore effective and efficient for this research's purposes, as the main parameters do not show major differences (except for the island of Fuerteventura, where the low number of members of the population means that many of the parameters show different indices from the usual ones). For La Gomera Island, the best-performing SSRs were VVMD28, VVMD36, VVMD27, and ZAG47, and the worst-performing SSRs were, once again, VVS29, VVS3, and UCH19 (Table S4).

4.2. Grapevine Variety Analysis

The population of accessions collected in the La Gomera Island prospection turned out to be fairly uniform, since in a first standardisation it went from 120 individuals (Table S1) to 52 unique MP-SSRs (Table S2). This meant the removal of 68 individuals (56.67%) from the population because they possessed the same genetic profile either to an identity 708 or to a mutation. This reduction was slightly higher than that found on the island of Lanzarote [42], with 55% of individuals discarded in the first data normalisation, which was so far the highest (El Hierro had 52.9% [41] and Fuerteventura had 37.5% [45]). This population standardisation of the island of La Gomera also included the elimination of two sports unique in the world, Malvasia rosada [14,43,48–50] and Bermejuela tinta [14]. It is also interesting to note that these 52 unique MP-SSRs correspond to representatives of 19 varieties, which demonstrates the existence of a very significant intravarietal variability (63%).

Of these 19 varieties, 3 correspond to individuals described for the first time, the varieties Coello blanca, Malvasia periquin gomerae, and Verdello gomerae, and another individual that was described by Rodríguez-Torres in 2013 [43] as unknown no. 2, which in this work is recorded as Barrerita negra. It should be noted that for the variety Verdello gomerae, another sample was also collected which turned out to be a mutation in two alleles of this new variety, registered as Verdello gomerae de Monacal. The consideration of an MP-SSR as corresponding to a new variety is somewhat arbitrary for the scientific community. Different authors established different thresholds, for example, Ibañez et al. [57] (SSR (2 alleles/26), 92%), Vélez [58] (SSR (2 alleles/18), 89%), and Cabezas et al. [59] (SNP, 90%). For this work, the threshold was set at a difference of 5 alleles out of the 40 studied for 20 SSRs, which corresponded to an MP-SSR similarity of 87.5% with respect to the most widespread MP-SSR in the TECNENOL database.

Continuing with Tables S1 and S2, it can be stated that in this study, six errors were detected in the registration of accession names with respect to their MP-SSRs, probably due to a lack of knowledge by the vine grower (in red), and 14 entries registered were identified as unknown individuals (pale green). Also, 29 new MP-SSRs for known varieties (intense green) and six more profiles were recorded that were already published in other TECNENOL studies (Forastera de la Isla Redonda and Bermajuelo del Echedo [41]; Listan negro santanero, Listan negro de la corona, Listan blanca chicharrera, and Listan blanco de la bodega [42]). In addition, cases of tri- and quadruple allelism were detected, probably due to hybridisation or periclinal chimeras. This phenomenon was previously described by other authors [60–62], and individuals showing triallelic SSRs were also found in the Lanzarote Island prospection [42], one of them (Listan blanco de la bodega) coinciding with sample codes 1111 and 3111 from the La Gomera prospection (ZAG83). The samples with triallelic SSRs were Listan negro de lo Machado and Listan blanco de Vallehermoso, also in the SSR ZAG83; Malvasia blanca piedra gorda (SCU06); Listan blanco de espina (ZAG83); and Verdello blanco del Corte (UCH19). Finally, the samples Tempranillo de Vallehermoso and the new variety Malvasia periquin gomerae showed a quadriallelic SSR (SCU06) [60].

In the lexicography section, the inclusion in the VIVC of the PN of the four new varieties (Barrerita negra, Coello blanca, Malvasia periquin gomerae, and Verdello gomerae) is proposed. It is also proposed that the new MP-SSR names corresponding to mutations of known varieties be included, as the VIVC database has been doing. In this way, the Pinot noir variety mutant is known by the proper name Pinot meunier, and in the case of the sports, these would also receive their own name, for example, Garnacha blanca (berry colour mutation) and Garnacha peluda (hairy Garnacha mutation), names that differ from the PN of the variety, which is Garnacha tinta (red Garnacha). Therefore,

29 more names should also be included (Forastera blanca de Agulo, Forastera blanca de Vallehermoso, Forastera blanca roquillos, Forastera blanca simancas, Forastera blanca tamargada, Alicante tintilla, Marmajuelo de Vallehermoso, Marmajuelo de Valle bajo, Listan negro de Hermigua, Listan negro de Vallehermoso, Listan negro de lo Machado, Malvasia blanca de Agulo, Malvasia blanca de Vallehermoso, Malvasia blanca piedra gorda, Negramoll de Vallehermoso, Mulata del macayo, Moscatel de Hermigua, Moscatel de la caleta, Moscatel de la caleta fino, Listan blanco de Hermigua, Listan blanco de Vallehermoso, Listan blanco de espina, Ruby cabernet ingenio, Negra Mora, Tempranillo de Vallehermoso, Torrontes volcanico montoro, Torrontes volcanico machado, Uva de año montoro, and Verdello blanco del Corte), as should the new variety mutation (Verdello gomerae de Monacal) and the new sport PN (Bermejuela negra). The inclusion of three new synonyms is also proposed: Marmajuelo Corto as a synonym of Bermajuelo del Echedo [41], Marmajuelo negro as a synonym of Bermejuela negra, and Moscatel Blanca as a synonym of Moscatel de Hermigua. Finally, there are two synonyms registered for another variety that are proposed to be included in a given variety. This is the case of the term Listan negra, a synonym registered for the variety Listan prieto, now proposed as a synonym of Listan negro, and the case of the name Malvasía blanca, a synonym registered for the variety Alarije, now proposed as a synonym of Malvasía Dubrovacka [49].

4.3. Genetic Structure of the Grapevine Population of the Island of La Gomera

The population of varieties found in the La Gomera Island prospection, as mentioned above, was 19 varieties. Four of these were new local varieties from La Gomera: Coello blanca, Barrerita negra, Malvasia periquin gomerae, and Verdello gomerae. The remaining 15 varieties turned out to be well-known varieties (Albillo forastero, Alicante Henri Bouschet, Bermejuela, Caracol, Listan negro, Malvasia Dubrovacka, Mollar cano, Muscat of Alexandria, Palomino fino, Ruby cabernet, Tempranillo tinto, Torrontes volcanico, Uva de año, Verdelho branco, and Verijadiego).

The most relevant conclusions after observing the figures related to this section (Figures 3, 4a,b, S1 and S2) are as follows: (a) The new local varieties Malvasia periquin gomerae and Verdello gomerae have a possible influence from the BP (Eastern Mediterranean). These varieties present a very characteristic and singular MP-SSR since in all the graphical representations by PCoA they appear close to the variety Malvasia Dubrovacka (chlorotype A, existing but not widespread in the BP [63]) but significantly distant from the rest (Figure 4), thus forming a single grouping (POP1) (Figure 3). (b) The MP-SSR of the Coello blanca variety probably has a Central European influence. It appears (Figure 4) also close to the Portuguese variety Verdelho branco (chlorotype D), which is a cross between the Central European variety of unknown origin and chlorotype D, Savagnin blanc, and an unknown variety. It is not uncommon that many varieties from the north of the Iberian Peninsula (Spain and Portugal) have as a progenitor the Savagnin blanc variety, which emigrated from the centre of Europe to establish and proliferate in this area through the Camino de Santiago [64]. Both varieties, Coello blanca and Verdelho branco, also have an MP-SSR far from the rest of the varieties that make up the population of La Gomera, probably due to their Central European origin (Figure 4). (c) The best-known local variety of La Gomera Island is the Albillo forastero. This, alongside the local variety from La Palma Island, Albillo criollo, is the result of a cross between the Spanish variety Palomino fino (chlorotype D) and the Portuguese variety Verdelho branco (chlorotype D) [14,63]. In Figure 4b, it is located between its progenitors but close to the other local Gomeran variety, Coello blanca. (d) The last local variety of La Gomera, known as Barrerita negra, is also singular as it is located equidistant from the previous ones but significantly distant from the group of viniferas. This time, in Figure 4a,b, it is positioned close to the Alicante Henri Bouschet variety (chlorotype A), a cross made by Louis Bouschet in 1829 [49] with the French varieties Teinturier (chlorotype A and a cross between Savagnin blanc (chlorotype D) and an unknown variety) and Aramon noir (chlorotype D and a cross between the French variety Oliven (chlorotype D) and the Central European variety of unknown origin Heunisch weiss (chlorotype C)). Therefore, it is possible that the MP-SSRs of the local varieties from La Gomera Island are not at all conventional, at least not the new local varieties.

Finally, the reliability rates of 27.82% for the two-dimensional PCoA representation and 36.96% for the three-dimensional PCoA representation (Figure 4) should be noted. The overall performance of the PCoA representations decreases as we increase the number of samples to be represented, sometimes reaching no more than 10% in studies with global populations. This is due to the fact that when working with 20 SSRs, 40 allelic values will define the position of a point (variety) in the PCoA plot. This means that the ideal representation, without error and therefore with 100% goodness of fit, would be a 40-dimensional representation. Given the impossibility of executing graphical representations with so many dimensions, the scientific community has to assume as good the reduction to two or three dimensions and their intrinsic errors, a fact that makes us accept this loss of reliability of the varieties represented by PCoA [65,66]. The solution to this problem is to assume trends, not accuracies.

4.4. Relation of La Gomera Grapevine Population with Respect to the Canary Archipelago Population

A comparative study was carried out between the local varieties of La Gomera Island and the local varieties of the rest of the Canary Islands in order to see if the affinities and trends between them provided relevant information. To this end, a population structure study was carried out using Structure 2.3. This study proposed the distribution of the 36 Canary Islands local varieties in five ancestral populations (Figures 5 and S4). Although very little information on crosses or on the chlorotypes of these varieties is available, trends can be observed at first sight in this distribution. Furthermore, once the standardisation was performed by eliminating the admixed individuals, Figure 6 as a whole also provides relevant information.

In this sense, the POP3 population (Figures 5 and S4) is the most remote population (Figure 6(a.1,a.2)); as it is made up of representatives closely related to Malvasia Dubrovacka, it is the most Eastern Mediterranean Basin (BP)-influenced population. This population includes Malvasia volcanica, the best-known local variety on the island of Lanzarote and an offspring of Malvasia Dubrovacka [14,42]. The other parent, the Canary Islands variety Bermejuela, also belongs to this group. The rest of the components of this group, in all the studies carried out by the TECNENOL research group [41,42,45–48], have always been related either to Malvasia Dubrovacka or to oriental varieties. Malvasia periquin gomerae and Verdello gomerae form part of this group as pure and local varieties of La Gomera Island.

The POP4 population, also positioned significantly away from the rest and occupying a single quadrant (Figure 6(a.1,a.2)), is a pure population constituted by El Hierro Island individuals. Nearly all of its components (83.33%) are individuals from El Hierro Island. As it is one of the archipelago's westernmost islands, its relationship with the Azores and Madeira was strong [51,52], so much so that the Verdello de El Hierro variety (admixed) is the result of a cross between the local variety of El Hierro Island, Verijadiego (admixed with POP2), and the Portuguese Alfrocheiro (the result of a cross between the Central European variety Savagnin blanc and an unknown variety) [14]. There is no local variety from the island of La Gomera in this grouping. The POP2 population, positioned in the upper-left quadrant very close to the axis and also very close to POP1 and POP5 (Figure 6(a.1,a.2)), is constituted by a group of very diverse grapevine varieties (Figures 5 and S4). It can be observed in this cluster the variety Bienmesabe tinto, described as an interspecific cross, and also the variety Majorera (a possible interspecific cross) [45] as well as the Gomeran variety Barrerita negra. Other varieties that appear in previous studies close to Bienmesabe tinto are those from Lanzarote Island, Blanca de la granja del Cabildo and Breval negro [45]. Also noteworthy in this group are the variety Tesoro blanco from El Hierro, always admixed, and vinifera but with a very distant origin from the population of the Canary Islands [41]. It is interesting to note how Barrerita negra is always positioned very close to the variety Bienmesabe tinto (Figure 6(a.3,b)), thus allowing us to hypothesise about a possible interspecific origin of this variety from the island of La Gomera. The POP5 population is a cluster characterised by the varieties that are part of the Spanish variety Palomino fino progeny [14,43,50]. Individuals such as Listan negro (Palomino fino × Mollar cano), Albillo criollo de la Isla de La Palma, and its sister of both parents, Albillo forastero from La Gomera Island (Palomino fino × Verdelho branco (Savagnin blanc × unknown)) [14,43,50]. In this grouping and as an admixed variety, the Gomeran variety Coello blanco also appears. In Figure 4a,b, this variety appears very close to the Portuguese variety Verdelho branco, and therefore it is hypothesised that it might have a Central European origin. Finally, the POP1 population is presented as a population whose components come almost equally from the islands of El Hierro and Lanzarote (Figures 5 and S4). Furthermore, this population occupies the lower-left quadrant, very close to POP5 (Figure 6).

In conclusion, the uniqueness of the new local varieties of La Gomera Island is confirmed for the second time, with the Albillo forastero variety having a more integrated MP-SSR in the vinifera group, i.e., not so unique. It is hypothesised that the MP-SSRs of the Coello blanco grapevine variety have a strong Central European influence, and those of the Malvasia periquin gomerae and Verdello gomerae varieties have a marked influence from BP (Eastern Mediterranean Basin). Finally, it is also hypothesised about the possibility that the Barrerita negra variety comes from an interspecific cross.

4.5. Relation of La Gomera Grapevine Population with Respect to the World Population

The next step was to broaden the comparative universe and to see how the new local varieties from La Gomera Island would behave, and in turn, how the La Gomera population as a whole would behave. Also, the opportunity to observe the Bienmesabe tinto and Majorera varieties' behaviour related or possibly related to other species of the genus *Vitis*.

A purely genetic strategy and a genetic–geographical strategy were used. In both strategies, admixed varieties from La Gomera were removed. The one that was always eliminated was the white Coello variety, because it was either admixed or badly located. This variety could have an MP-SSR with Central European influence due to its proximity in Figure 4a,b to the Portuguese variety Verdelho branco (Savagnin blanc × unknown). In addition, in the genetic–geographical strategy, the varieties Barrerita negra and Albillo forastero were also eliminated.

Another fact common to both strategies was the remarkable singularity of the La Gomera Island population in the circular dendrograms, phylogenetic trees, and PCoA graphs by population (Figures 7b,c, 8(a.1,a.2), 9a,b and 10(a.1,a.2)). In all these cases, it is clear how the La Gomera population (a) forms a main branch (Figure 7b,c) when the population is made up of the local varieties Malvasia periquin gomerae, Verdello gomerae, and Albillo forastero, forming part of the genetic strategy; (b) forms a sub-branch (Figure 9a,b) when the population consists of the grapevine varieties Malvasia periquin gomerae and Verdello gomerae; or (c) occupies a quadrant on its own away from the other populations in both the genetic and genetic–geographical strategies when the population consists only of varieties from La Gomera Island, Malvasia periquin gomerae and Verdello gomerae (Figures 8(a.1,a.2) and 10(a.1,a.2)). This may highlight the fact that the uniqueness of LG is given by the MP-SSR of either the two varieties or one of them. With regard to the Bienmesabe tinto and Majorera varieties, it should be noted that in all cases, they appear quite separate but with the common feature that Bienmesabe tinto is more closely related to the varieties from Lanzarote Island, while the Majorera variety appears closely linked to the varieties from El Hierro Island.

In the case of the three-dimensional PCoA by individuals' representations, in Figure 8b, it is apparently only the variety Bienmesabe tinto that moves away from the whole, while in Figure 10b, both the varieties Bienmesabe tinto and Majorera as well as the two local varieties of LG that make up this population (Malvasia periquin gomerae and Verdello gomerae) are significantly separated from the rest.

Finally, another relevant fact related to IP should be analysed. In Figure 9a,b, IP presents a fragmented population in such a way that a couple of IP varieties' sub-branches are in the main branch that gives rise to the FRA-CEU population, and another couple of IP sub-branches are located in another of the three main branches, the one that gives rise to the EASTMED-CAU, BP, and ITA varieties. It is clear that the Iberian Peninsula over the years had major stages in its history that were linked to a strong bidirectional transfer, in this case of wood or vine seeds. From the arrival of the Phoenicians and Greeks to the conquest by the Crown of Aragon of the Balkan territories via central-southern Italy [67], the Iberian Peninsula had close bonds with the eastern part of the Mediterranean Basin. This fact could well justify the influence on the MP-SSRs of certain varieties by the EASTMED-CAU, BP, and ITA populations. The previously mentioned case of the Camino de Santiago should also be remembered, which perfectly justifies the Central European influence on the MP-SSRs of the northern varieties of IP [64].

5. Conclusions

The conclusions reached after genotyping 120 samples from the prospection of the volcanic island of La Gomera are diverse and interesting. As usual, the kit of 20 SSRs performed efficiently and effectively, although not all of them had the same qualitative level. Four new varieties from La Gomera Island (Coello blanca, Barrerita negra, Malvasia periquin gomerae, and Verdello gomerae) and a mutation of the variety Verdello gomerae, known as Verdello gomerae de Monacal, are presented for the first time and proposed for inclusion in the VIVC. For the first time, 29 unique MP-SSRs corresponding to variations of known varieties were also computed and are also proposed for inclusion in the global database. Additionally, six errors were detected, 14 unknown varieties were identified, and individuals with cases of triallelism and quadriallelism were described.

About the lexicography, four PNs are presented (Coello blanca, Barrerita negra, Malvasia periquin gomerae, and Verdello gomerae), as is one name of a mutation of a new local variety (Verdello gomerae de Monacal), the inclusion of the sport Bermejuela negra, and the registration of twenty-nine names of new mutations corresponding to known varieties. Additionally, three new synonymous names are presented: Marmajuelo Corto as a synonym of Bermajuelo del Echedo; Marmajuelo negro as a synonym of Bermejuela negra (PN of the sport); and Moscatel Blanca as a synonym of Moscatel de Hermigua. Finally, there are two synonyms registered for another variety that we propose be included in a given variety. This is the case of Listan negra, a synonym registered for the variety Listan prieto, now proposed as a synonym of Listan negro, and Malvasía blanca, a synonym registered for the variety Alarije, now proposed as a synonym of Malvasía Dubrovacka.

The population of local grapevines from La Gomera is thought to be the most unique in the Canary Islands to date. It is hypothesised that varieties with a strong influence from the Eastern Mediterranean Basin, Malvasia periquin gomerae and Verdello gomerae, are possibly the most unique. The white Coello variety, which is admixed, seems to have a strong Central European influence. It is hypothesised that the black Barrerita variety as well as the Fuerteventura Majorera variety may be the result of interspecific crossbreeding, as the Bienmesabe tinto variety turned out to be.

Supplementary Materials: The following supporting information can be downloaded at: https://www.mdpi.com/article/10.3390/horticulturae10010014/s1, Table S1: Information on 120 accessions from La Gomera (original and conclusive) [15,41–43,45–49]; Table S2: Unique molecular profile of the 52 accessions collected in La Gomera Island prospection. International SSRs coinciding with the TECNENOL SSRs [15,41–43,45–49]; Table S3: Annealing temperature (Ta), range (in base pairs) where the peaks have to appear in the electropherogram and type of fluorescence label of the 20 SSRs used by TECNENOL; Table S4: Characteristics of the twenty microsatellite markers used in this study; Figure S1: The four steps of the graphical method of Evanno et al. [40], allowing the estimation of the true number of ancestral K groups for a population with 19 individuals from La Gomera Island; Figure S2: Genetic structure of the La Gomera Island population. Distribution K = 2 (individuals belonging to each group or population). Details of the ratio of pure and admixed individuals according to the value of q (pure ($q \geq 85\%$) and admixed ($q < 85\%$)); Figure S3: The four steps of the graphical method of Evanno et al. [40], allowing the estimation of the true number of ancestral K groups for a population with 36 individuals from Canary Islands collection (IC including La Gomera Island); Figure S4: Genetic structure of the Canary Islands population (36 varieties). Distribution K = 5 (individuals belonging to each group or population). Details of the ratio of pure and admixed individuals according to the value of q (pure ($q \geq 85\%$) and admixed ($q < 85\%$)); Figure S5: The four steps of the graphical method of Evanno et al. [40], allowing the estimation of the true number of ancestral K groups for a population with 314 individuals from the TECNENOL database (including La Gomera Island); Figure S6: Genetic structure of the world population. Distribution K = 2 (individuals belonging to each group or population). Details of the proportion of pure and admixed individuals as a function of q value. Nationalities that make up pure and admixed groups; Figure S7. Genetic structure of the world population (191 individuals). Distribution in 7 geographical areas. Detail of the ratio of well-assigned (pure) and misassigned (admixed) individuals. Nationalities that make up each of the groups: EASTMED-CAU (Algeria, Cyprus, Georgia, Israel, Lebanon, Tunisia, and Turkey), BP (Bosnia and Herzegovina, Bulgaria, Croatia, Greece, Serbia, Slovenia, and Montenegro), ITA (Italy), FRA-CEU (Austria, France, Germany, Hungary, and Switzerland), IP (Spain and Portugal), IC (Canary Archipelago), and LG (La Gomera Island).

Author Contributions: Conceptualization, F.F. and Q.L.-Y.; methodology, F.F., Q.L.-Y. and C.V.; software, P.S.-G., J.M.C. and F.Z.; validation, F.F., Q.L.-Y. and C.V.; formal analysis, Q.L.-Y. and C.V.; investigation, F.F. and Q.L.-Y.; resources, F.F. and Q.L.-Y.; data curation, F.F., Q.L.-Y. and C.V.; writing—original draft preparation, F.F., Q.L.-Y. and P.S.-G.; writing—review and editing, F.F. and P.S.-G.; visualization, P.S.-G., J.M.C. and F.Z.; supervision, F.F., Q.L.-Y., C.V., P.S.-G., J.M.C. and F.Z. All authors have read and agreed to the published version of the manuscript.

Funding: This project has been financed by the Cabildo Insular de isla de La Gomera through the Consejo Regulador de la Denominación de Origen Vinos de La Gomera.

Data Availability Statement: Data are contained within the article and Supplementary Materials.

Acknowledgments: The authors are grateful to Javier Ibañez and the viticulture specialists Armenia Mendoza and Nancy Melo for their valuable support. We would also like to thank Núria Boronat, Rosa Pastor, Isabel Araque, Braulio Esteve-Zarzoso, Laia Fañanás, and Santiago Moreno for their support in the laboratory.

Conflicts of Interest: The authors declare no conflict of interest.

References

1. Wolkovich, E.M.; García de Cortázar-Atauri, I.; Morales-Castilla, I.; Nicholas, K.A.; Lacombe, T. From Pinot to Xinomavro in the world's future wine-growing regions. *Nat. Clim. Chang.* **2018**, *8*, 29–37. [CrossRef]
2. NASS. 2021. Available online: https://quickstats.nass.usda.gov/results/9E7FD873-AD79-3190-8C62-0B7B559A0F68 (accessed on 23 April 2023).
3. Análisis Annual del Sector Vitivinícola Mundial en 2021. Organización Internacional de la Viña y el Vino (OIV). Available online: https://www.oiv.int/sites/default/files/documents/OIV_Analisis_anual_del_sector_vitivinicola_mundial_en_2021.pdf (accessed on 17 October 2023).
4. INE. 2021. Available online: https://www.ine.es/jaxiT3/Datos.htm?t=2910 (accessed on 7 July 2023).
5. Wikimedia Commons. 2014. Available online: https://commons.wikimedia.org/w/index.php?curid=33566121#file (accessed on 7 July 2023).
6. NASA. 2011. Available online: https://www.flickr.com/photos/gsfc/6630087415/in/photostream/ (accessed on 7 July 2023).
7. Arco, M.J.; Atiénzar, E.; Rosario, M.C.; Arco, M.M.; González, C.; Arco, M.C. El Menceyato de Icod en el poblamiento de Tenerife: D.; Gaspar, Las Palomas y Los Guanches. Sobre el poblamiento y las estrategias de alimentación vegetal entre los Guanches. *Eres* **2000**, *9*, 67–129.
8. Hidalgo, J.; Hidalgo, L. La Filoxera. In *Tratado de Viticultura*, 2nd ed.; Mundi-Prensa: Madrid, Spain, 2019; Volume 1.
9. Climate Data. 2021. Available online: https://es.climate-data.org/europe/espana/la-gomera-10273/ (accessed on 7 July 2023).
10. Elías, L.V. *El Paisaje del Viñedo en las Islas Canarias*; PASOS: Revista de turismo y Patrimonio Cultural; Asociación Canaria de Antropología: Tenerife, Spain, 2013; Volume 11, p. 152. Available online: http://www.pasosonline.org/Publicados/pasosoedita/PSEdita11.pdf (accessed on 7 July 2023).
11. Instituto Canario de Estadística. 2021. Available online: https://www3.gobiernodecanarias.org/istac/statistical-visualizer/visualizer/data.html?resourceType=dataset&agencyId=ISTAC&resourceId=E01135A_000004&version=1.0# (accessed on 8 July 2023).
12. Consejo Regulador de la Gomera. 2020. Available online: https://vinoslagomera.com/ (accessed on 8 July 2023).

13. Pliego de Condiciones de la Denominación de Origen Protegida de Vinos "La Gomera". 2013. Available online: https://vinoslagomera.com/wp-content/uploads/2013/03/Pliego_de_condiciones_DOP_de_vinos_La_Gomera.pdf (accessed on 8 July 2023).
14. Rodriguez-Torres, I. Albillo Forastero. In *Variedades de vid Cultivadas en Canarias. Descriptores Morfológicos. Caracterización Morfológica, Molecular, Agronómica y Enológica*, 2nd ed.; Instituto Canario de Investigaciones Agrarias, Ed.; Gobierno de Canarias: Santa Cruz de Tenerife, Spain, 2018. Available online: https://www.icia.es/icia/download/Publicaciones/Variedades_Vid_Canarias.pdf (accessed on 8 July 2023).
15. Maul, E.; Röckel, R. Vitis International Variety Catalogue (VIVC): A cultivar database referenced by genetic profiles and morphology. In *BIO Web of Conferences*; EDP Sciences: Hulis, France, 2015; Volume 5, p. 01009. [CrossRef]
16. Crespan, M. Autochthonous varieties and more used vines, link to genetic. In Proceedings of the 2nd OENOVITI International Symposium, Session I: Genetic Characterization, Phenotyping and Pedigrees, Geisenheim, Germany, 3–5 November 2014; pp. 7–11. Available online: https://www.calameo.com/books/003503752f575f124dc11 (accessed on 11 July 2023).
17. This, P.; Lacombe, T.; Thomas, M.R. Historical origins and genetic diversity of wine grapes. *Trends Genet.* **2006**, *22*, 511–519. [CrossRef]
18. Lopes, M.S.; Sefc, K.M.; Eiras-Dias, E.; Steinkellner, H.; Laimer Câmara Machado, M.; Câmara Machado, A. The use of microsatellites for germplasm management in a Portuguese grapevine collection. *Theor. Appl. Genet.* **1999**, *99*, 733–739. [CrossRef]
19. Sefc, K.M.; Lefort, F.; Grando, M.S.; Scott, K.D.; Steinkellner, H.; Thomas, M.R. Microsatellite Markers for Grapevine: Tools for Cultivar Identification and Pedigree Reconstruction. In *Microsatellite Markers for Grapevine: A State of the Art*; Roubelakis-Angelakis, K.A., Ed.; Molecular Biology & Biotechnology of the Grapevine; Springer: Dordrecht, The Netherlands, 2001. [CrossRef]
20. Marsal, G.; Baiges, I.; Canals, J.M.; Zamora, F.; Fort, F. A fast, efficient method for extracting DNA from leaves, stems, and seeds of *Vitis vinifera* L. *Am. J. Enol. Vitic.* **2011**, *62*, 376–381. [CrossRef]
21. Marsal, G.; Boronat, N.; Canals, J.M.; Zamora, F.; Fort, F. Comparison of the efficiency of some of the most usual DNA extraction methods for woody plants in different tissues of *Vitis vinifera* L. *J. Int. Sci. Vigne Vin* **2013**, *47*, 227–237. [CrossRef]
22. Fort, F.; Hayoun, L.; Valls, J.; Canals, J.M.; Arola, L.; Zamora, F. A new and simple method for rapid extraction and isolation of high-quality RNA from grape (*Vitis vinifera*) berries. *J. Sci. Food Agric.* **2008**, *88*, 179–184. [CrossRef]
23. Thomas, M.R.; Scott, N.S. Microsatellite repeats in grapevine reveal DNA polymorphisms when analyzed as sequence-tagged sites (STSs). *Theor. Appl. Genet.* **1993**, *86*, 985–990. [CrossRef]
24. Bowers, J.E.; Dangl, G.S.; Vignani, R.; Meredith, C.P. Isolation and characterization of new polymorphic simple sequence repeat loci in grape (*Vitis vinifera* L.). *Genome* **1996**, *39*, 628–633. [CrossRef]
25. Bowers, J.E.; Dangl, G.S.; Meredith, C.P. Development and characterization of additional microsatellite DNA markers for grape. *Am. J. Enol. Vitic.* **1999**, *50*, 243–246. [CrossRef]
26. Sefc, K.M.; Regner, F.; Turetschek, E.; Glössl, J.; Steinkellner, H. Identification of microsatellite sequences in *Vitis riparia* and their applicability for genotyping of different *Vitis* species. *Genome* **1999**, *42*, 367–373. [CrossRef]
27. Scott, K.D.; Eggler, P.; Seaton, G.; Rosseto, M.; Abblet, E.M.; Lee, L.S.; Henry, R.J. Analysis of SSRs derived from grape ESTs. *Theor. Appl. Genet.* **2000**, *100*, 723–726. [CrossRef]
28. Lefort, F.; Kyvelos, C.; Zervou, M.; Edwards, K.; Roubelakis-Angelakis, K. Characterization of new microsatellite loci from *Vitis vinifera* and their conservation in some *Vitis* species and hybrids. *Mol. Ecol. Resour.* **2002**, *2*, 20–21. [CrossRef]
29. Cipriani, G.; Spadotto, A.; Jurman, I.; Di Gaspero, G.; Crespan, M.; Meneghetti, S.; Frare, E.; Vignani, R.; Cresti, M.; Morgante, M.; et al. The SSR-based molecular profile of 1005 grapevine (*Vitis vinifera* L.) accessions uncovers new synonymy and parentages and reveals a large admixture amongst varieties of different geographic origins. *Theor. Appl. Genet.* **2010**, *121*, 1569–1585. [CrossRef]
30. Dalbó, M.A.; Ye, G.N.; Weeden, N.F.; Steinkellner, H.; Sefc, K.M.; Reisch, B.I. A gene-controlling sex in grapevines is placed on a molecular marker-based genetic map. *Genome* **2000**, *43*, 333–340. [CrossRef]
31. This, P.; Jung, A.; Boccacci, P.; Borrego, J.; Botta, R.; Costantini, L.; Crespan, M.; Dangl, G.S.; Eisenheld, C.; Ferreira-Monteiro, F.; et al. Development of a standard set of microsatellite reference alleles for the identification of grape cultivars. *Theor. Appl. Genet.* **2004**, *109*, 1448–1458. [CrossRef]
32. Paetkau, D.; Calvert, W.; Stirling, I.; Strobeck, C. Microsatellite analysis of population structure in Canadian polar bears. *Mol. Ecol.* **1995**, *4*, 347–354. [CrossRef]
33. Peakall, R.; Smouse, P.E. GenAlEx 6.5: Genetic analysis in Excel. Population genetic software for teaching and research-an update. *Bioinformatics* **2012**, *28*, 2537–2539. [CrossRef]
34. Paetkau, D.; Slade, R.; Burden, M.; Estoup, A. Genetic assignment methods for the direct, real-time estimation of migration rate: A simulation-based exploration of accuracy and power. *Mol. Ecol.* **2004**, *13*, 55–65. [CrossRef]
35. VanderPlas, J. Three-Dimensional Plotting in Matplotlib. In *Python Data Science Handbook*; O'Reilly Media, Inc.: Sevastopol, CA, USA, 2016. Available online: https://jakevdp.github.io/PythonDataScienceHandbook/04.12-three-dimensional-plotting.html (accessed on 21 August 2023).
36. Kumar, S.; Stecher, G.; Tamura, K. MEGA7: Molecular Evolutionary Genetics Analysis version 7.0 for bigger datasets. *Mol. Biol. Evol.* **2016**, *33*, 1870–1874. [CrossRef]
37. Saitou, N.; Nei, M. The neighbor-joining method: A new method for reconstructing phylogenetic threes. *Mol. Biol. Evol.* **1987**, *4*, 406–425. [CrossRef] [PubMed]

38. Pritchard, J.K.; Stephens, M.; Donnelly, P. Inference of population structure using multilocus genotype data. *Genetics* **2000**, *155*, 945–959. [CrossRef] [PubMed]
39. Falush, D.; Stephens, M.; Pritchard, J.K. Inference of population structure using multilocus genotype data: Linked loci and correlated allele frequencies. *Genetics* **2003**, *164*, 1567–1587. [CrossRef] [PubMed]
40. Evanno, G.; Regnaut, S.; Goudet, J. Detecting the number of clusters of individuals using the software structure: A simulation study. *Mol. Ecol.* **2005**, *14*, 2611–2620. [CrossRef] [PubMed]
41. Fort, F.; Lin-Yang, Q.; Suárez-Abreu, L.R.; Sancho-Galán, P.; Canals, J.M.; Zamora, F. Study of Molecular Biodiversity and Population Structure of *Vitis vinifera* L. ssp. *vinifera* on the Volcanic Island of El Hierro (Canary Islands, Spain) by Using Microsatellite Markers. *Horticulturae* **2023**, *9*, 1297. [CrossRef]
42. Fort, F.; Marsal, G.; Mateo-Sanz, J.M.; Pena, V.; Canals, J.M.; Zamora, F. Molecular characterisation of the current cultivars of *Vitis vinifera* L. in Lanzarote (Canary Islands, Spain) reveals nine individuals which correspond to eight new varieties and two new sports. *Oeno One* **2022**, *56*, 281–295. [CrossRef]
43. Rodriguez-Torres, I. *Variedades de vid Cultivadas en Canarias. Descriptores Morfológicos. Caracterización Morfológica, Molecular, Agronómica y Enológica*, 1st ed.; Instituto Canario de Investigaciones Agrarias, Gobierno de Canarias: Santa Cruz de Tenerife, Spain, 2013.
44. Bacilieri, R.; Lacombe, T.; Cunff, L.L.; Di Vecchi-Staraz, M.; Laucou, V.; Genna, B.; Perós, J.P.; This, P.; Boursiquot, J.M. Genetic structure in cultivated grapevines is linked to geography and human selection. *BMC Plant Biol.* **2013**, *13*, 25. [CrossRef] [PubMed]
45. Fort, F.; Lin-Yang, Q.; Valls, C.; Sancho-Galán, P.; Canals, J.M.; Zamora, F. Characterisation and Identification of Vines from Fuerteventura (Canary Volcanic Archipelago (Spain)) Using Simple Sequence Repeat Markers. *Horticulturae* **2023**, *9*, 1301. [CrossRef]
46. Marsal, G.; Mateo, J.M.; Canals, J.M.; Zamora, F.; Fort, F. SSR analysis of 338 accessions planted in Penedes (Spain) reveals 28 unreported molecular profiles of *Vitis vinifera* L. *Am. J. Enol. Vitic.* **2016**, *67*, 466–470. [CrossRef]
47. Marsal, G.; Bota, J.; Martorell, A.; Canals, J.M.; Zamora, F.; Fort, F. Local cultivars of *Vitis vinifera* L. in Spanish Islands: Balearic Archipelago. *Sci. Hortic.* **2017**, *226*, 122–132. [CrossRef]
48. Marsal, G.; Mendez, J.J.; Mateo-Sanz, J.M.; Ferrer, S.; Canals, J.M.; Zamora, F.; Fort, F. Molecular characterization of *Vitis vinifera* L. local cultivars from volcanic areas (the Canary Islands and Madeira) using SSR markers. *Oeno One* **2019**, *4*, 667–680. [CrossRef]
49. Maul, E.; Röckel, F. Vitis International Variety Catalogue. 2015. Available online: http://www.vivc.de (accessed on 7 July 2023).
50. Zerolo, J.; Cabello, F.; Espino, A.; Borrego, J.; Ibañez, J.; Rodriguez-Torres, I.; Muñoz-Organero, G.; Rubio, C.; Hernández, M. *Variedades de Vid de Cultivo Tradicional en Canarias*, 1st ed.; Instituto Canario de Calidad Agroalimentaria, Gobierno de Canarias: Santa Cruz de Tenerife, Spain, 2006.
51. Macías, A.M. Colonización y viticultura. El caso de las Canarias, 1350–1550. In *Douro: Estudos y Documentos*; Revistas da Facultade de Letras da Universidade do Porto: Porto, Portugal, 2002; Volume 13. Available online: https://ojs.letras.up.pt/index.php/dou/article/view/12586 (accessed on 7 July 2023).
52. Macías, A. El paisaje vitícola de Canarias. Cinco siglos de historia. *Ería* **2005**, *68*, 351–364. [CrossRef]
53. Aliquo, G.; Torres, R.; Lacombe, T.; Boursiquot, J.M.; Laucou, V.; Gualpa, J.; Fanzone, M.; Sari, S.; Pérez-Peña, J.; Prieto, J.A. Identity and parentage of some South American grapevine cultivars present in Argentina. *Aust. J. Grape Wine Res.* **2017**, *23*, 452–460. [CrossRef]
54. Moita, A.; Santos, R.; Catarina, A. Unraveling the origin of *Vitis vinifera* L. Verdelho. *Aust. J. Grape Wine Res.* **2018**, *24*, 450–460. [CrossRef]
55. Avramidou, E.V.; Masaoutis, I.; Pitsoli, T.D.; Kapazoglou, A.; Pikraki, M.; Trantas, E.A.; Nikolantonakis, M.; Doulis, A.G. Analysis of Wine-Producing *Vitis vinifera* L. Biotypes, Autochthonous to Crete (Greece), Employing Ampelographic and Microsatellite Markers. *Life* **2023**, *13*, 220. [CrossRef]
56. Costantini, L.; Monaco, A.; Vouillamoz, J.F.; Forlani, M.; Grando, M.S. Genetic relationships among local Vitis vinifera cultivars from Campania (Italy). *Vitis* **2005**, *44*, 25–34. [CrossRef]
57. Ibañez, J.; De Andrés, M.T.; Molino, A.; Borrego, J. Genetic study of key Spanish grapevine varieties using microsatellite. analysis. *Am. J. Enol. Vitic.* **2003**, *54*, 22–30. [CrossRef]
58. Vélez, M.D.; Ibáñez, J. Evaluation of the uniformity and stability of Microsatellite markers in grapevine. *Acta Hortic.* **2009**, *827*, 163–168. [CrossRef]
59. Cabezas, A.; Ibañez, J.; Lijavetzky, D.; Vélez, D.; Bravo, G.; Rodríguez, V.; Carreño, I.; Jermakow, A.M.; Carreño, J.; Ruiz-García, L.; et al. A 48 SNP set for grapevine cultivar identification. *BMC Plant Biol.* **2011**, *11*, 153. [CrossRef]
60. Riaz, S.; Garisson, K.E.; Dangl, G.S.; Boursiquot, J.M.; Meredith, C.P. Genetic divergence and chimerism within ancient asexually propagated winegrape cultivars. *J. Am. Soc. Hortic. Sci.* **2002**, *127*, 508–514. [CrossRef]
61. Pelsy, F. Molecular and cellular mechanisms of diversity within grapevine varieties. *Heredity* **2010**, *104*, 331–340. [CrossRef] [PubMed]
62. Grigoriou, A.; Tsaniklidis, G.; Hagidimitriou, M.; Nikoloudakis, N. The Cypriot indigenous grapevine germplasm is a multi-clonal varietal mixture. *Plants* **2020**, *9*, 1034. [CrossRef] [PubMed]
63. Arroyo-Garcia, R.; Ruiz-García, L.; Bolling, L.; Ocete, R.; López, M.A.; Arnold, C.; Ergul, A.; Söylemezoğlu, G.; Uzun, H.I.; Cabello, F.; et al. Multiple origins of cultivated grapevine (*Vitis vinifera* L. ssp. *sativa*) based on chloroplast DNA polymorphisms. *Mol. Ecol.* **2006**, *15*, 3707–3714. [CrossRef] [PubMed]

64. Casanova, J.; Mozas, P.; Ortiz, J.M. Ampelography and microsatellite DNA analysis of autochthonous and endangered grapevine cultivars in the province of Huesca (Spain). *Span. J. Agric. Res.* **2011**, *9*, 790–800. [CrossRef]
65. Emanuelli, F.; Lorenzi, S.; Grzeskowiak, L.; Catalano, V.; Stefanini, M.; Troggio, M.; Myles, S.; Martinez-Zapater, J.M.; Zyprian, E.; Moreira, F.M.; et al. Genetic diversity and population structure assessed by SSR and SNP markers in a large germplasm collection of grape. *BMC Plant Biol.* **2013**, *13*, 39. [CrossRef]
66. Dong, Y.; Duan, S.; Xia, Q.; Liang, Z.; Dong, X.; Margaryan, K.; Musayev, M.; Goryslavets, S.; Zdunić, G.; Bert, P.-F.; et al. Dual domestications and origin of traits in grapevine evolution. *Science* **2023**, *379*, 892–901. [CrossRef]
67. García-Muñoz, S.; Lacombe, T.; De Andrés, M.T.; Gaforio, L.; Muñoz-Organero, G.; Laucou, V.; This, P.; Cabello, F. Grape varieties (*Vitis vinifera* L.) from the Balearic Islands: Genetic characterization and relationship with Iberian Peninsula and Mediterranean Basin. *Genet. Resour. Crop Evol.* **2012**, *59*, 589–605. [CrossRef]

horticulturae

MDPI

Article

Characterisation and Identification of Vines from Fuerteventura (Canary Volcanic Archipelago (Spain)) Using Simple Sequence Repeat Markers

Francesca Fort [1,*], Qiying Lin-Yang [1], Carla Valls [1], Pau Sancho-Galán [2], Joan Miquel Canals [1] and Fernando Zamora [1]

1 Enological Technology Group (TECNENOL), Department of Biochemistry and Biotechnology, Faculty of Oenology, Rovira i Virgili University, Sescelades Campus, C/Marcel·lí Domingo, 1, 43007 Tarragona, Spain; qiying.lin@estudiants.urv.cat (Q.L.-Y.); carlavlls813@gmail.com (C.V.); jmcanals@urv.cat (J.M.C.); fer-nando.zamora@urv.cat (F.Z.)

2 Department of Chemical Engineering and Food Technology, Science Faculty, University of Cadiz, Agrifood Campus of International Excellence (CeiA3), Wine and Food Research Institute (IVAGRO), P.O. Box 40, 11510 Puerto Real, Spain; pau.sancho@uca.es

* Correspondence: mariafrancesca.fort@urv.cat; Tel.: +34-97755-8799

Abstract: A prospection was carried out on Fuerteventura Island in order to explore the existing biodiversity of *Vitis vinifera* ssp. *vinifera* in almost desert conditions (<120 mm per year). For this purpose, 40 individuals were collected and genotyped with 20 SSRs. Nine known varieties and one unknown variety, named Majorera on the island, were identified. In addition, four new mutations were found in the varieties Listan negro and Listan prieto, respectively. Thirteen unknown individuals and five erratic accessions were identified. Seven new names are proposed for inclusion in the world databank (one main name (Majorera), one new synonym for Listan negro (Hoja moral), four new mutation names (Listan prieto de Antigua, Listan prieto de Vega, Hoja moral de El Rosario and Hoja moral de Betancuria) and a new synonymy for Muscat Hamburg (Moscatel), which is very widespread on the island). Finally, the possibility is raised that the new Majorera variety is not a pure vinifera but the product of an interspecific cross, as has happened with the variety found on the island of La Palma, Bienmesabe tinto. Once again, the Canary Archipelago shows itself to be a possible centre for the creation of biodiversity for the cultivated vine.

Keywords: *Vitis vinifera* L.; SSR; microsatellite; climate change; volcanic; Canary Islands; Fuerteventura Island

Citation: Fort, F.; Lin-Yang, Q.; Valls, C.; Sancho-Galán, P.; Canals, J.M.; Zamora, F. Characterisation and Identification of Vines from Fuerteventura (Canary Volcanic Archipelago (Spain)) Using Simple Sequence Repeat Markers. *Horticulturae* **2023**, *9*, 1301. https://doi.org/10.3390/horticulturae9121301

Academic Editor: Caixi Zhang

Received: 16 October 2023
Revised: 22 November 2023
Accepted: 24 November 2023
Published: 4 December 2023

1. Introduction

One of the main problems affecting vine cultivation (one of the oldest and most economically important crops in the world) in the 21st century is climate change [1]. *Vitis vinifera* L. species, like all other living organisms, will suffer significant genetic erosion. However, this will not be the only erosion that vine has suffered throughout its history. Already in the 19th century, this plant species suffered the consequences of the *Phylloxera* plague, causing a high mortality of vines and the disappearance of most of its varieties worldwide (except for certain areas of the planet such as Crete, the Canary Islands, Chile...) [2]. During the 20th century, the problem lay in the excessive use of a small number of varieties, which came to be considered international (such as Cabernet sauvignon, Chardonnay, Pinot noir...) [1]. In the 21st century, the problem will be climate change in its main aspects: the increase in average temperatures during vegetative growth and in some regions, the frequency and severity of frosts during spring and autumn, very hot days during the summer, extreme rain or hail events and widespread spatial and temporal drought [3], so much so that, at the inauguration of the 43rd World Congress of Vine and Wine (Mexico, November 2022), its Director General stated: "...this year's harvest was characterised by extreme heat and a record drought that accelerated ripening in vineyards all over the world... this was the worst drought in the last 500 years" [4]. What is most striking is that these seasonal (punctual) and devastating phenomena due to large-scale climatology, which has been defined as climatic variability, are centred in narrow areas of our planet, which happen to correspond mostly to wine-growing areas. Hannah et al.'s [5] predictions are not encouraging at all. They predict a gradual disappearance of the current wine-growing areas towards more northern areas of the planet, substantially changing the global landscape by 2050 and beyond. The effects of climate change on vines have already been observed for 50 years. Above all, what is being observed is an earlier phenological stage (budbreak, flowering, veraison and harvest) and shorter intervals between these stages, with the latter being shortened by 6 to 17 days, depending on the variety and location [1]. Thus, for the most widely planted varieties on our planet (known as international varieties), the production of a quality product from their harvest will be endangered and jeopardised.

Focusing on this problem analysis, it can be stated that adaptation responses to climate change in the world of wine can vary from those implemented at a winery level, which can be relatively easy, cheap and immediate, but which provide a low adaptation potential, or those implemented at a vineyard level, which can be more expensive and can lead to more difficulties, but which have a higher adaptation potential. Strategies that could be implemented at a vineyard level, in turn, are divided into two. There are those with medium-term effects and medium adaptation potential, based on vineyard management and aimed at mitigating radiation, leaf and bunch temperature and water deficit. There are also those with a high potential for adaptation to climate change effects; therefore, long-term effects will be based on changing the location of vineyards and/or replacing old varieties (often international varieties) with others that are much better adapted to the new changing conditions [6]. Obviously, the implementation of the latter will require a high level of investment and a change in legislation (especially at the level of appellation d'origine contrôlée (AOC)). For many vine growers, a change in vineyard location will be an impossible task, but it will be possible to opt for a change to different grapevine varieties in the same location. This would be a matter of using a new varietal range that has a wide temperature niche for vegetative growth and ripening, adapted to the new circumstances, and is also efficient in water management [1,7]. To this end, and as advised by Wolkovich et al. [1], current efforts are being made to study both the intervarietal and intra-varietal variability of the current marketed crops and, among them, *Vitis vinifera* L.

This study's main objective is to prospect areas that are much drier than normal for a wine-growing area, with very low rainfall, to locate high-temperature and drought-resistant vines. In Spain, there are territories where these conditions are found, as is the case of the Canary Islands (Figure 1). The Canary Islands are one of the five archipelagos of Macaronesia, a group of volcanic islands located in the Atlantic Ocean. The easternmost islands of this archipelago have desert conditions, as they are only a few kilometres from the Sahara Desert, as is the case of Fuerteventura, which is only 97 km away [8].

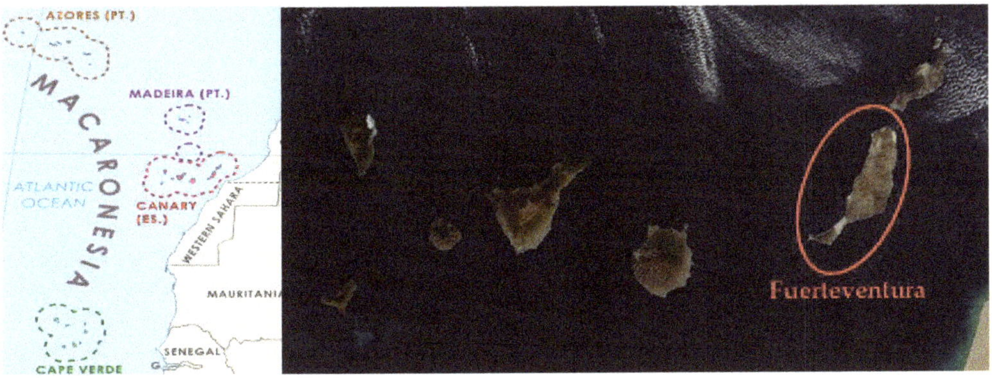

Figure 1. (**Left**) Macaronesia map [9]. (**Right**) Canary Archipelago and Fuerteventura Island [10].

Fuerteventura is an island situated at the coordinates 28°24′ N 14°00′ W and is the archipelago's second-largest island in terms of area [11]. According to the Koppen climate classification, the climate is hot desert (BWh) in most of Fuerteventura Island, with very low rainfall that is always below 120 mm per year. Due to the island's low altitude (between 0 and 813 m), this island does not retain humid air masses, as is the case on other archipelago islands [12]. Fuerteventura is the Canary Island with the smallest surface area dedicated to vine cultivation (about 10 ha) and is the only island without an AOC [13]. Paradoxically, the first written document in which the first wine was produced in the Canary Islands was in Fuerteventura [14].

This scarcity of vine cultivation has been maintained throughout history by the presence of *gavias* (Figure 2). *Gavias* are structures built by Fuerteventura inhabitants to retain and make the most of the scarce precipitation water that falls on the island. Thus, the walls and surroundings of these structures are where most of the vines are found, planted with the aim of providing consistency to the structure's containment dam [15].

Figure 2. Detail of a *gavia* on the Fuerteventura Island [15].

Thus, and given all the above, the aim of this research was to carry out a prospection of this arid island, which has never suffered from the *Phylloxera* plague (like the rest of the Canary Islands), in order to identify the different individuals currently forgotten in the *gavias* or in old abandoned plantations. At the same time, a study of the island's vine varieties names was also carried out, as well as, whenever possible, a study of the population structure.

2. Materials and Methods

2.1. Plant Material

Forty vine samples from different locations on Fuerteventura Island were analysed (woody tissue). Prospection was carried out by vine growers and hunters by means of a mass selection process. Samples were collected either in the bush (barren land) or on the slopes of *gavias*. Once collected, they were stored at −20 °C until analysis. Detailed information about the accessions analysed is given in Table S1.

2.2. DNA Extraction and Purification

DNA extraction was carried out using a protocol developed by the Tecnología Enológica (TECNENOL) research group [16,17]. It is an adaptation of Fort et al.'s [18] protocol, specific for RNA procurement. It was also improved by introducing polyvinylpyrrolidone (PVP) into the extraction buffer and adding another chloroform wash. The NanoDrop TM 1000 Spectrophotometer (Thermo Fisher® Scientific, Waltham, MA, USA) was used to determine the extraction yield and purity of the extracted DNA.

2.3. Microsatellites

The 20 SSRs used for the genotyping of the samples from Fuerteventura Island were previously chosen for their discrimination capacity and polymorphism. Of all of them, 7 match (Table S2) with the 9 that were proposed by the international community as reference genetic markers (VVS2, VVMD5, VVMD7, VVMD25, VVMD27, VVMD28, VVMD32, VrZAG62, VrZAG79) [19]. SSR "kit", with which the TECNENOL group has been working for several years, comprises: VVS2, VVS3, VVS29 [20]; VVMD5, VVMD6, VVMD7 [21]; VVMD27, VVMD28, VVMD36 [22]; VrZAG21, VrZAG47, VrZAG62, VrZAG64, VrZAG79, VrZAG83 [23]; scu06vv [24]; VvUCH11, VvUCH12, VvUCH19 [25]; VChr19a [26]. Microsatellites VrZAG47 and VVMD27 are not independent loci, i.e., they amplify the same area of the genome. The difference between the two is the design of their primers [27].

2.4. Polymerase Chain Reaction (PCR)

In order to carry out the PCR, an Applied Biosystems 2720 Thermal Cycler (Foster City, CA, USA) was used. Amplification was carried out from 4 ng of DNA and 1 μM of each primer with a fluorescent dye attached to the *primer Forward* (Fw) (6-FAM: VVS3, VVMD7, VVMD28, VVMD36, VrZAG47, VrZAG62, VrZAG83, VvUCH11 y VvUCH19; HEX: VVS2, VVS29, VVMD6, VVMD27, VrZAG21, VrZAG79 y VChr19a; NED: VVMD5, VrZAG64, scu06vv, VvUCH12) by using AmpliTaq DNA Polymerase kit (Applied Biosystems, Foster City, CA, USA). Thermocycling conditions were: a first stage at 95 °C for 5 min, a second phase with 40 cycles at: 95 °C for 45 s; the annealing temperature for each SSR (48.1 °C, 50 °C, 52 °C, 54.4 °C, 56.1 °C, 56.6 °C and 58 °C (Table S3)) for 30 s; 72 °C for 1 min and a half; and a last stage at 72 °C for 7 min.

2.5. Measurement of Amplified Fragment Lengths

In addition to the PCR product, formamide and the internal marker GeneScan ROXTM 500 (Applied Biosystems, Foster City, CA, USA) was used to prepare the plates for measuring the length of the amplified fragments. Furthermore, these plates were subjected to thermal denaturation (3 min at 95 °C). Fragment separation was performed by capillary electrophoresis with an ABI PRISM 3730® genetic analyser (Applied Biosystems, Foster

City, CA, USA). Finally, Peak Scanner Software (Applied Biosystems, NJ, USA) was used to interpret the electrophenograms.

2.6. Data Analysis

GenAlEx 6.5 software [28,29] was used to assess the reliability of the chosen SSR kit. For this purpose, the statistics Na (number of different alleles), Ne (number of effective alleles. These are the alleles that are transmitted to the next generation), Ho (observed heterozygosity. These are the heterozygotes computed), He (expected heterozygosity. Estimation of the heterozygotes that the population under study might have. Also known as diversity index), F (fixation index. Parameter measuring the goodness of homozygosity) and PI (probability of identity. Probability that two molecular profiles (MP-SSR) with the same SSR values are the same variety) were used. Furthermore, the same software was used to identify MP-SSRs identical to each other (data normalisation) and, in addition, the assignment test was performed. This last strategy is based on the allele frequency of each entry and calculates a log probability value of this entry for each subpopulation using the allele frequencies of the respective subpopulations. Thus, an individual is assigned to the subpopulation with the highest log likelihood value [30]. To examine the genetic relationships between the populations found, two-dimensional principal coordinate analysis (PCoA) was used, both for populations and for populations of individuals based on the standardised covariance of the genetic distances calculated for the co-dominant markers. Finally, the value of Fst, the coefficient of genetic differentiation between populations, was calculated assuming the infinite allele model.

For the three-dimensional representation of PCoA, the Python Data programme [31] was used, applying the Matplotlib strategy. The programme MEGA version 7 [32] was used for the phylogenetic trees and circular dendrograms (neighbour-joining method [33]).

The assessment of the population structure, which entailed the best combination of populations and the identification of admixed individuals, was carried out using the Structure 2.3. [34,35] programme. This model-based software uses a Bayesian clustering method in which several ancestral populations (K) are assumed to be present, each characterised by a set of allele frequencies at each *locus*. Individuals in the sample are assigned to populations (clusters), or conjointly to more populations if their genotypes indicate that they are admixed. All loci are assumed to be independent and each population K is assumed to follow Hardy–Weinberg equilibrium. Posterior probabilities were estimated using the Markov chain Monte Carlo (MCMC) method. MCMC chains were run with a 100,000 burn-in period followed by 1,000,000 iterations using a model allowing for admixture and correlated allele frequencies. Structure was run at least ten times by setting K from 1 to 7 (1 to 9 for global varieties), and an average likelihood value, L (K), was calculated across all runs for each K. The mean log probability of the data for each K was calculated to determine the most appropriate number of clusters, and the value of K for which this probability was highest was selected. The ΔK was then calculated using the method proposed by Evanno et al. [36]. ΔK is a quantity based on the rate of change in the log probability of the data between successive K values.

As has been usual in previous studies carried out by the same research group, the 40 MP-SSR obtained were compared with their extensive database [37–41] in order to incorporate new individuals and new mutations, if any. In addition, the existing literature on Canary Island varieties was reviewed in depth [39–43], as well as the main database worldwide, *Vitis* International Variety Catalog (VIVC) [44]). This exhaustive search for information allowed us not only to cross-reference the MP-SSRs but also to verify the accuracy of the names registered as names of origin. Thus, the original information provided by the winegrowers, the conclusive information of each accession, the percentage of similarity with a certain sample from the TECNENOL database and the SSRs in which variability has been detected, among other characteristics, are shown in Table S1. Additionally, information on the unique MP-SSRs for this population and the values found for 7 reference SSRs can be found in Table S2.

3. Results

Only 40 individuals were collected from the prospection on the island of Fuerteventura, which has no viticultural tradition, and these were analysed with TECNENOL's own 20 SSRs.

3.1. SSR Polymorphism

The first objective was to find the accessions whose MP-SSRs were identical for all SSRs. We found 25 MP-SSRs identical to other profiles in the population or to each other. Thus, the population of unique profiles in Fuerteventura Island was 15 individuals (Table S2).

Table S4 presents the values for the main statistics that allowed us to affirm whether the chosen SSR kit was effective and efficient for characterising and identifying the Fuerteventura Island grapevine population. The total number of alleles found for this population was 134 (Na). The mean was 6.7 alleles per SSR, with SSRs VVS3, UCH19 and VVS29 showing the fewest alleles (2, 2 and 3, respectively) and SSRs VVMD28 and ZAG79 showing the most alleles (10 each). The mean number of alleles passing to the next generation (Ne) was 4.1, with SSRs UCH19 and VVS29 showing the lowest values (1.14 and 1.23, respectively), and SSRs VVMD36 and ZAG47 showing the highest values (6.62 and 6.82, respectively). The mean value of observed heterozygosity (Ho = 0.807) was significantly higher than the mean value of expected heterozygosity, also known as the genetic diversity index (He = 0.689). The lower values corresponded to SSRs UCH19 and VVS29 (Ho = 0.13 and 0.2, respectively; He = 0.12 and 0.18, respectively) and the higher values corresponded, for Ho, to SSRs VVS2, ZAG62, ZAG79 and VChr19a (Ho = 1) and, for He, to SSRs ZAG47 and VVMD36 (with a value of 0.85). Only two SSRs showed positive values for the F index. These were SSR ZAG83, with a value very close to 0 (0.015), and VVMD6, with a larger value (0.11). Finally, it should be noted that the accumulative identity probability for the population of 15 vines on the island of Fuerteventura was 7.5×10^{-20}, with SSR VVMD36 having the lowest value (3.9×10^{-2}) and SSR UCH19 having the highest value (7.7×10^{-1}).

3.2. Grapevine Variety Analysis

Table S1 shows that the 40 accessions analysed corresponded to nine identified varieties and 1 corresponded to a new variety. All of them had a Spanish origin except for two: the Greek variety Muscat from Alexandria and the English variety Muscat Hamburg. Furthermore, within the group of Spanish varieties, it was found that Listan negro and Torrontes volcanico varieties are local varieties of the Canary Archipelago.

The grapevine variety with the most representatives found in the prospection of Fuerteventura Island was Listan prieto, with 18 accessions. In this group, three different MP-SSRs were described: the MP-SSR coinciding with the most extended in the TECNENOL database, which corresponded to four individuals; a mutated MP-SSR in the first allele of the SSR VVMD28 (VVMD28-1), which included 11 components (with the representative that entered with the name Listan prieto de Antigua (the new name for this mutation)); and four samples with two mutations (VVS3-2 and VVMD28-1) whose representative was entered under the name Listan prieto de Vega (the new name for this mutation).

The local Canary Islands variety Listan negro group contained seven entries, and four different MP-SSRs were identified in this grouping. In Fuerteventura Island, this variety is known as Hoja moral (new synonymy). Of these four different MP-SSRs, four accessions corresponded to the most widespread MP-SSR and were registered under the name of the new synonymy of Hoja moral: one MP-SSR was registered under the name of Hoja moral de El Rosario, which mutated in VVUCH12-2; one sample was registered under the name of Hoja moral de Betancuria, which presented variation in VVZAG21-1; and, finally, another sample was entered under the name of Hoja moral, which presented a mutation in VVS3-2.

The MP-SSR of the Greek variety Muscat of Alexandria was also identified in five accessions from Fuerteventura Island. All the components were found to be identical to the most widespread MP-SSR. Three of them were registered under the name Moscatel, the synonymy registered in the VIVC for this variety, and the other two individuals were entered as unknown.

The Spanish white grapevine variety, Airén, was represented by three accessions registered with names such as unknown and erroneous names Moscatel and Listan prieto, and presented a single MP-SSR identical to the most widespread. The Andalusian variety Palomino fino was represented in this collection by two accessions mutated in VVS3-1 and entries with the erroneous name Listan prieto. The MP-SSRs with only one representative were the Andalusian-Extremeña (SW Spain) grapevine variety Beba, which entered under the term Burra blanca, the Andalusian Mollar cano, which was registered as unknown, the English Muscat Hamburg, which entered under the name Moscatel, the Canarian Torrontes volcanico, entered as unknown and the new variety, registered under the name Majorera, which corresponded to a new MP-SSR.

3.3. Fuerteventura Grapevine Population Genetic Structure

From the 15 unique profiles found in Fuerteventura Island (Table S2), another data normalisation was performed in order to initiate a population structure study. Mutant individuals for a given variety were eliminated from this collection. Thus, the population under study had 10 members, corresponding to the 10 varieties described above (Tables S1 and S2). All of them belonged to the most extended profiles, except for the Palomino fino variety, whose two representatives had the same mutation (97.5% similarity), so a mutated individual had to be taken as a representative.

The use of the Structure 2.3. programme to find the best distribution of this 10-member collection in different populations was impossible due to the low number of individuals that made up the collection. For this reason, a principal coordinate analysis (PCoA) representation was performed directly to observe the position taken by these individuals according to their genetic distances (Figure 3).

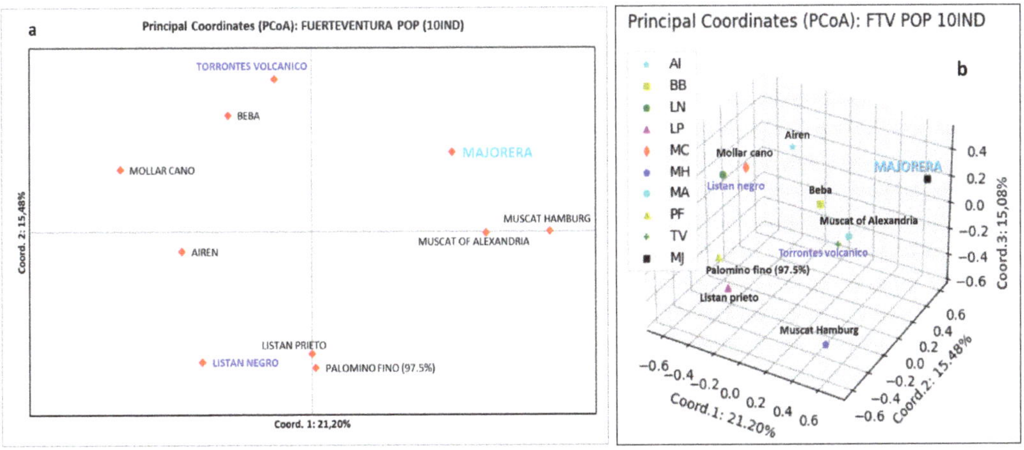

Figure 3. Fuerteventura Island population (unique and non-redundant molecular profiles). (**a**) Two-dimensional representation using PCoA. (**b**) Three-dimensional representation using PCoA. The names in navy blue correspond to local varieties of the Canary Islands. The name in light blue corresponds to the new local variety. The names in black correspond to foreign varieties of the archipelago.

Figure 3a shows how coordinate 1 with a 21.2% goodness of fit separates the Majorera (Canary Islands), Muscat of Alexandria (Greek) and Muscat Hamburg (English) varieties from the rest, with Palomino fino and Listan prieto being very close to the axis or above it, respectively. Coordinate 2 (15.4% goodness of fit), on the other hand, places in the upper quadrants the Spanish varieties Mollar cano, Beba, Torrontes volcánico (from Lanzarote Island) and the new variety Majorera (from Fuerteventura Island), while the two Muscat varieties are above the axis. The new Majorera grapevine variety is practically located alone in the upper right quadrant. Figure 3b shows a three-dimensional representation using PCoA, with a distribution very similar to the previous one. Once again, the Majorera variety stands out from the rest of the individuals from the Fuerteventura Island collection. In this image, it can also be seen how the English grapevine variety Muscat Hamburg is significantly different from both the Majorera variety and the rest.

3.4. Majorera Variety Relation with Respect to Canary Archipelago Grapevine Population

As a strategy to confirm the uniqueness of the Majorera variety MP-SSR, a study in which this variety was related to the rest of the varieties of the Canary Islands was carried out.

The Structure 2.3. programme was used to find the best ancestral population distribution from 1 to 7. Figure S1 shows the results of this study, with K = 6 being the best distribution, followed at a very short distance by K = 4. The graphs generated by the Structure 2.3. programme for these distributions are presented in Figure S2. It can be seen in this figure, as for K = 6, that there is no clear distinction between the ancestral populations POP1 and POP2. On the other hand, the definition of the four ancestral populations for K = 4 is clear and sharp. The result observed for K = 6 in Figure S2 is confirmed numerically in Table S5. This table shows in detail the membership of each of the 32 individuals of the Canary Archipelago population to one of the six ancestral populations defined in Structure 2.3. The results show that POP1 and POP2 are not well defined, their individuals all being admixed ($q < 85\%$), but, in addition, POP3 also has all its individuals admixed. This fact makes it impossible to continue the study using the six-population distribution. For this reason, the K = 4 distribution was used, as it had given a result very slightly different from the previous one (Figure S1), and, in addition, Figure S2 shows the graph of this distribution with its four well-defined ancestral populations. The numerical results supporting Figure S2 are presented in Figure S3 and graphically represented in Figure 4. For K = 4, the 32 Canarian varieties are grouped into four populations, each with pure individuals ($q \geq 85\%$) and admixed ($q < 85\%$). In POP1, 34% of individuals (11 varieties), 4 pure individuals (36%) and 7 admixed (64%), are assigned. It is a very diverse population. The first pure member of POP1 is the red Bienmesabe tinto variety (rest of the Canary Islands (IC)), followed by the Majorera variety (Fuerteventura Island (FTV)) and the Lanzarote Island (LNZ) varieties Blanca de la Granja del Cabildo and Breval negro. The seven crossbred varieties are from all the islands. POP2, which has 16% of the total population (five varieties), with three pure individuals (60%) and two admixed individuals (40%), is a pure population from El Hierro island (HI). With one more individual than POP2 (19%), the third population (POP3) is presented with five pure individuals (83%) and one admixed (17%). This population is made up of individuals from HI and individuals from LNZ. POP4 is another large population with 10 members (31%), of which 3 will be pure (30%) and the remaining 7 admixed (70%). This population will also be characterised by being very diverse but with a large presence of pure IC individuals.

Normalising the data, i.e., leaving out the admixed varieties (17 varieties), the two- and three-dimensional PCoA graphical representations of the 15 pure individuals were carried out, as well as a phylogenetic tree representation (Figure 5).

Figure 5a shows the two-dimensional distribution of the four ancestral populations that have generated the Canary Islands population, with a 93% goodness of fit (assignment test). Coordinate 1 (46.98%) separates POP4 from the rest, while coordinate 2 (41.7%) places POP4 and POP2 in the upper quadrants. Overall, POP2, POP3 and POP4 are distributed far apart and equidistant while POP1 is positioned close to POP3. The Fst parameter confirms this arrangement numerically (Figure 5b). Figure 5c presents the two-dimensional representation of PCoA, but now by individuals. It can be seen how the location of the populations broken down by individuals has redistributed the four populations, causing

POP2 to move to the right quadrants. It can be seen that the Majorera variety is positioned next to the red Bienmesabe tinto variety. These two varieties, together with the Breval negro variety, are located alone in the upper left quadrant. In fact, in this representation, each population occupies a different quadrant. Figure 5d shows the three-dimensional plot of the 15 Canarian individuals using PCoA. In this representation, with one more dimension, the Majorera variety behaviour is again significantly different from the rest of the collection. Another variety that is also very distant from the rest is Bienmesabe tinto. Both belong to POP1 together with Breval negro and the variety Blanca de la Granja del Cabildo. Finally, the global Canary Islands population is shown in Figure 5e, represented by a phylogenetic tree. It shows the phylogenetic relationship between the four populations, with three main branches from which POP1, POP2 and, from the third branch, POP3 and POP4 derive.

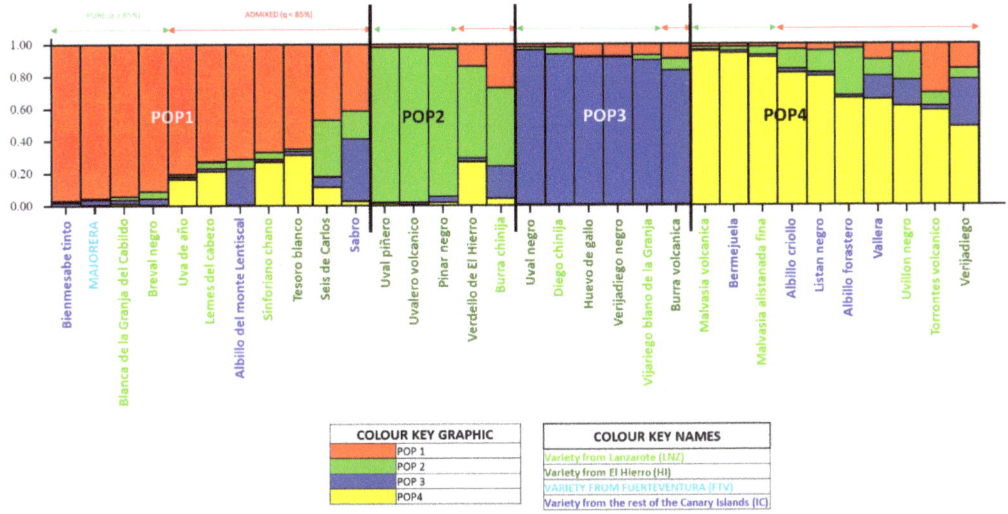

Figure 4. Population of Canary Islands varieties (unique and non-redundant molecular profiles). Diagram Structure 3.2. 4K distribution for pure and admixed individuals.

3.5. Majorera Variety Relation with Respect to World Population

In this section, the Majorera variety was compared with 308 vinifera varieties from 22 countries around the world (TECNENOL database [37–41]). To perform this, two strategies were carried out: one purely genetic, and a second in which a geographical parameter was introduced.

As far as the first strategy is concerned, the genetic structure was studied using Structure 2.3. The population now consists of 309 individuals to be distributed according to the best value of K (from 1 to 9). In this case, the best distribution was for K = 2 (Figure S4) followed by K = 7. The vinifera world population was divided into two ancestral populations, POP1 and POP2, with 182 (59%) and 127 (41%) members, respectively (Figure 6a and Figure S5). Specifically, in Figure S5, the q values for each variety for a given population are presented. Thus, POP1 has 88% pure individuals (161 individuals out of a total population of 309) and 12% admixed (21/309). This is a population with an abundance of representatives from Italy (51 pure individuals and 4 admixed out of a total of 72 Italians), France (39 pure and 5 admixed out of 49), Spain (24 pure and 4 admixed out of 106), Portugal (12 pure and 4 admixed out of 22) and Greece (11 pure out of 14), among others. On the other hand, POP2, with 98 pure individuals (77%) and 29 admixed (23%), hosts individuals from Spain (67 pure and 11 admixed out of 106 individuals), Italy (10 pure and 7 admixed out of 72), France (4 pure and 1 admixed out of 49), Portugal (2 pure and 4 admixed out of 22) and Greece (3 pure out of 14), among others. The IC population (with 32 varieties) is entirely positioned in POP2, with 30 pure and 2 admixed individuals (Albillo criollo and Malvasia alistanada fina) (Figure S5). From these results, we eliminated 50 varieties with a q value lower than 85%.

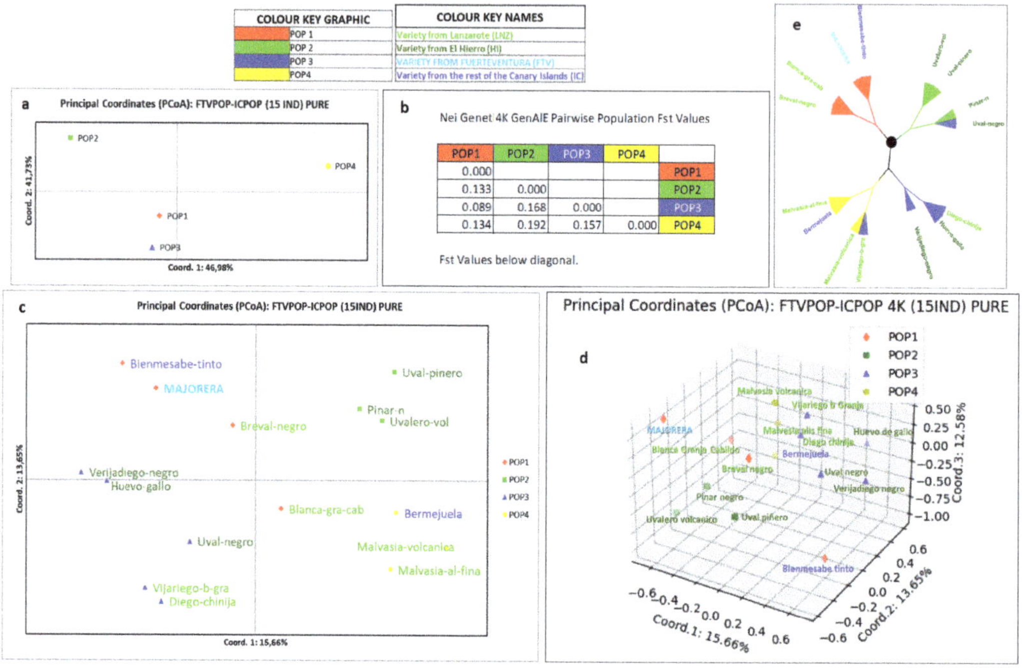

POP1	POP2	POP3	POP4	
0.000				POP1
0.133	0.000			POP2
0.089	0.168	0.000		POP3
0.134	0.192	0.157	0.000	POP4

Figure 5. Population of Canary Islands grapevine varieties (15 pure varieties). (**a**) Two-dimensional representation of the 4 Canarian populations by population. (**b**) Values of the Fst statistic for each population. (**c**) Two-dimensional representation of the 4 populations by individuals. (**d**) Three-dimensional representation of the 4 populations by individuals. (**e**) Phylogenetic tree of the 4 populations by individuals. The names in navy blue correspond to Canary Islands varieties, the name in light blue corresponds to the Fuerteventura Island variety, the names in acid green correspond to Lanzarote Island varieties and the names in dark green correspond to El Hierro island varieties.

Figure 6b shows the circular dendrogram of this K = 2 distribution for the 259 pure varieties after the application of an assignment test that gave a goodness of fit equal to 100%. Interestingly, four varieties assigned to POP1 were incorporated into POP2 (Chasselas blanc (FRA), Tenerone (ITA), Parellada (ESP) and Rojal tinta (ESP)), and the POP2 variety Valent blanc (ESP) was incorporated into POP1. Otherwise, the distribution is as expected according to the Structure 2.3. programme and the GenAlEx 6.5 assignment test. The varieties Bienmesabe tinto and Majorera are located in the second major branch from the origin (Figure 6b,c) and in different clusters. The IC grapevine varieties are distributed over almost all branches and arms of POP2.

The 259 individuals distribution by PCoA representation is shown in Figure 7. To highlight the role of the IC population and new Majorera variety, a new population (POP3) corresponding to IC was constructed and extracted from POP2. Applying the assignment test, it gave a goodness of fit of 93% (data not shown). Figure 7(a.2,a.3) show the distribution of these three populations by population. It can be seen how coordinate 1 (76.65%) separates the populations of Spanish origin (POP2 and POP3 (IC)) from POP1, which corresponds to the clustering of most of the varieties from the rest of the world. In turn, coordinate 2

(23.35%) isolates POP1 and POP3 (IC) in the upper quadrants. The statistical parameter (Fst) supporting this positioning is shown in Figure 7(a.3). The two-dimensional distribution by individuals (Figure 7(a.1)) shows how coordinate 1 (5.68%) again separates POP2 and POP3 (IC) from POP1, but now coordinate 2 (3.68%) leaving the individuals of all the populations distributed in the upper and lower quadrants indistinctly. Bienmesabe tinto and Majorera variety locations overlap with the populations of POP2 and POP3 (IC). The three-dimensional representation (Figure 7b) shows broadly the distribution described for Figure 7(a.1) except that the varieties Bienmesabe tinto and Majorera are separated from the rest of the varieties of both POP2 and POP3 (IC).

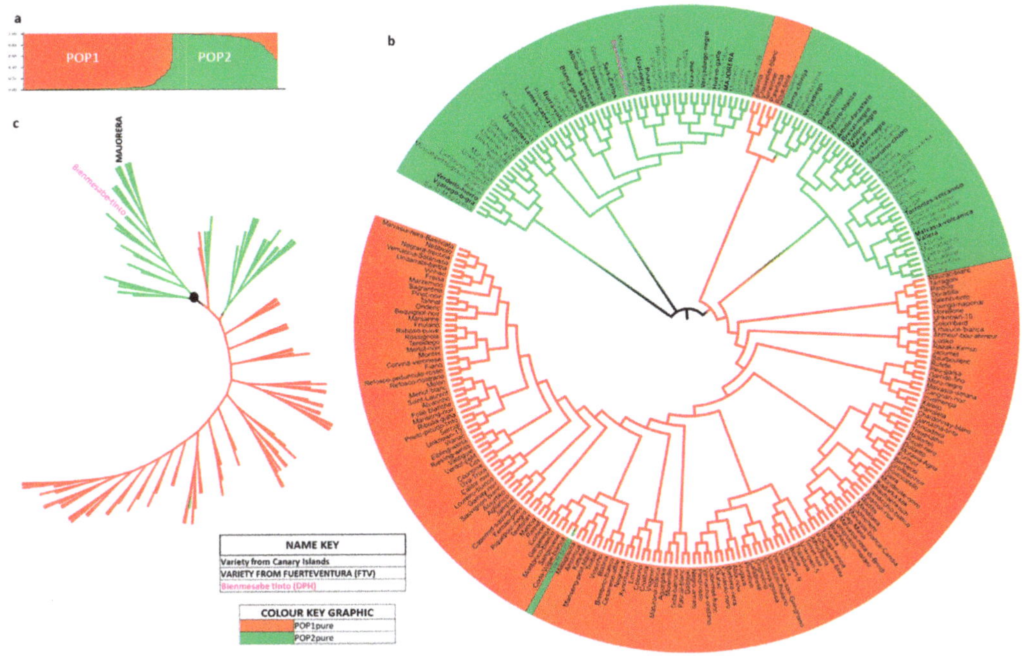

Figure 6. The world population (259 individuals) distributed into 2 populations. (**a**) Graphical representation of K = 2 according to Structure 2.3. (with pure and admixed individuals); (**b**) neighbour-joining circular dendrogram of the 259 world population pure individuals, highlighting the location of Bienmesabe tinto (pink and bold), Majorera (capital and bold) and IC varieties (bold). (**c**) Phylogenetic tree corresponding to the population of 259 pure individuals, highlighting the Bienmesabe tinto and Majorera varieties.

As for the distribution using a geographical component, in this case, the 309 varieties were grouped geographically, 307 belonging to *Vitis vinifera* ssp. *vinifera*, the Bienmesabe tinto variety and the Majorera variety. Five populations were formed: EASTMED-CAU (Algeria, Cyprus, Georgia, Israel, Lebanon, Tunisia and Turkey), BP (Bosnia and Herzegovina, Bulgaria, Croatia, Greece, Serbia, Slovenia and Montenegro), ITA (Italy), FRA-CEU (Austria, France, Germany, Hungary and Switzerland) and IP (Spain and Portugal). The Canary Islands population was extracted from the latter (IC).

Using the GenAlEx 6.5 programme, an assignment test was carried out in order to find out which individuals were poorly placed in each group (admixed), and to be able to eliminate them (data normalisation). The 309 individuals' goodness of distribution was 61% (data not shown). The final population of well-placed individuals with a goodness of fit of 91% (data not shown) and spread over six geographic areas dropped to 188 varieties.

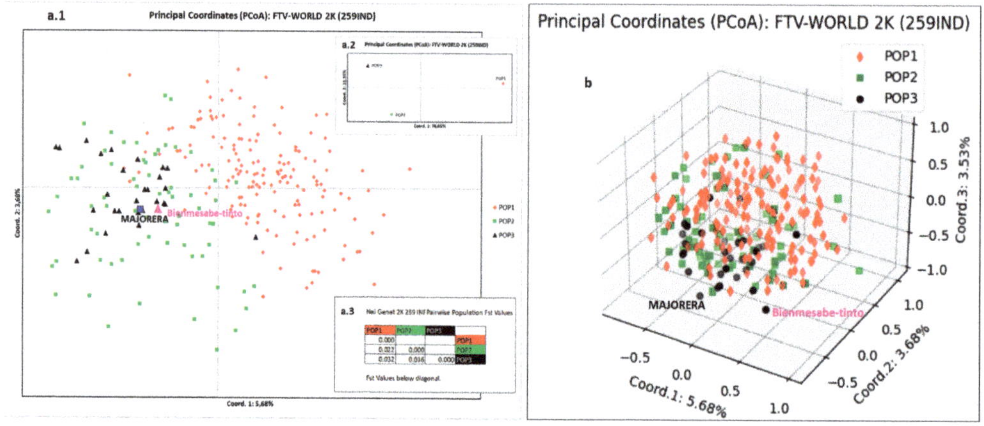

Figure 7. PCoA representation of Canary Islands population and the wold varieties normalised for K = 2. (**a.1**) Two-dimensional representation of the 3 populations by individuals, (**a.2**) two-dimensional representation of the 3 populations by population, (**a.3**) Fst statistic values for each population. (**b**) Three-dimensional representation of the 3 populations by individuals. POP3 corresponds to IC.

Figure S6 presents the characteristics of each geographical population. Thus, EASTMED-CAU, with 4% of the population (12 individuals), presented four pure individuals (33%), with a Turkish variety and the rest of the unknown ones that were previously assigned to this grouping, as well as eight admixed (67%) corresponding to the rest of this cluster nationalities. BP, with 9% of the population (28 varieties), showed 50% of pure individuals, with Greek and Croatian varieties dominating, and 50% of admixed varieties coming mostly from Greece. The Italian cluster (ITA), with 24% of the world population (73 samples), was also almost equally distributed, with 51% of pure varieties (37 individuals) and 49% of admixed (36 individuals). The French-European cluster, with 60 varieties, represented 24% of the world population, with 72% pure (43 varieties) and 28% mixed (17 individuals), with French varieties dominating in both cases. The most numerous grouping corresponded to IP, with 34% of the components of the world population (104 individuals), divided between 63% of pure individuals (66 varieties) and 37% of admixed (38 samples). In this case, it should be noted that, with Spanish varieties dominating in both cases (pure and mixed), a proportionally greater crossbreeding was observed in Portuguese varieties. And, lastly, IC, corresponded to 10% of the total, with 24 pure varieties (75%) and 8 mixed varieties (25%).

Figure 8 shows the circular dendrogram and the phylogenetic tree of this population of pure varieties. Both representations show the 188 varieties distributed in three main branches. The first one corresponds to the EASTMED-CAU, BP and ITA populations, with two small independent IP arms. The second corresponds to the IP population itself, with the Canary Islands population and, diluted in this population, Bienmesabe tinto and Majorera varieties. Finally, in the third large branch, the FRA-CEU population is present, and, in the same way as in the first large branch, there are two small independent branches with IP varieties. Bienmesabe tinto and Majorera varieties are found in the same branch as the IC varieties but in different clusters.

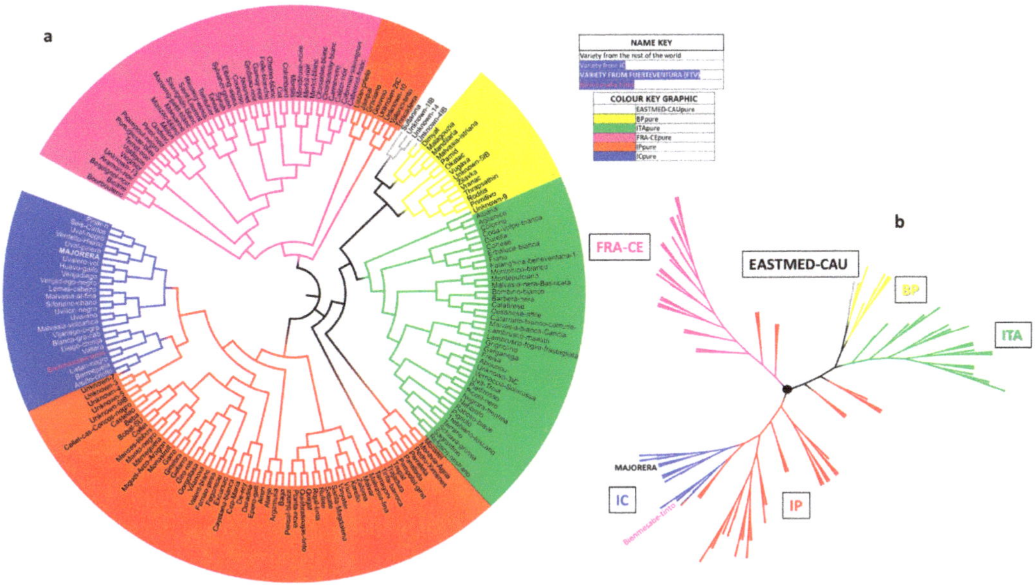

Figure 8. World population (188 individuals) distributed in populations corresponding to six geographical areas. (**a**) Circular neighbour-joining dendrogram of the 188 pure individuals of the world population, highlighting the location of the Bienmesabe tinto and Majorera varieties; (**b**) phylogenetic tree of the distribution of these 6 populations with all their individuals.

Figure 9 corresponds to the graphical representations using two-dimensional PCoA with populations and with individuals (Figure 9a), and three-dimensional PCoA with individuals (Figure 9b). In the two-dimensional representation by populations (Figure 9(a.2)), it can be seen how coordinate 1 (42.02%) separates the IP and IC populations from the rest, with these being located in the right quadrants. The IC population is significantly differentiated from the rest, occupying a solitary quadrant, in the same way as the ITA and IP populations. Coordinate 2 (18.09%) separates the most singular populations in the upper quadrants, these being the most distant, ITA and IC. The Fst statistic confirms the arrangement of each population in the corresponding quadrants (Figure 9(a.3)). The two-dimensional individual distribution looks somewhat similar to that of populations (Figure 9(a.1)). Coordinate 1 (6.03%) clearly separates the Spanish populations (IP and IC) from those of ITA and FRA-CE, while the populations of EASTMED-CAU and BP are distributed along the central part of the two coordinate axes. Coordinate 2 (3.98%) has no effect on these populations. Bienmesabe tinto and Majorera varieties are located overlapping with POP2 and IC. Again, the three-dimensional representation shows a significant differentiation of the Bienmesabe tinto and Majorera varieties with respect to the rest of the varieties worldwide (Figure 9b).

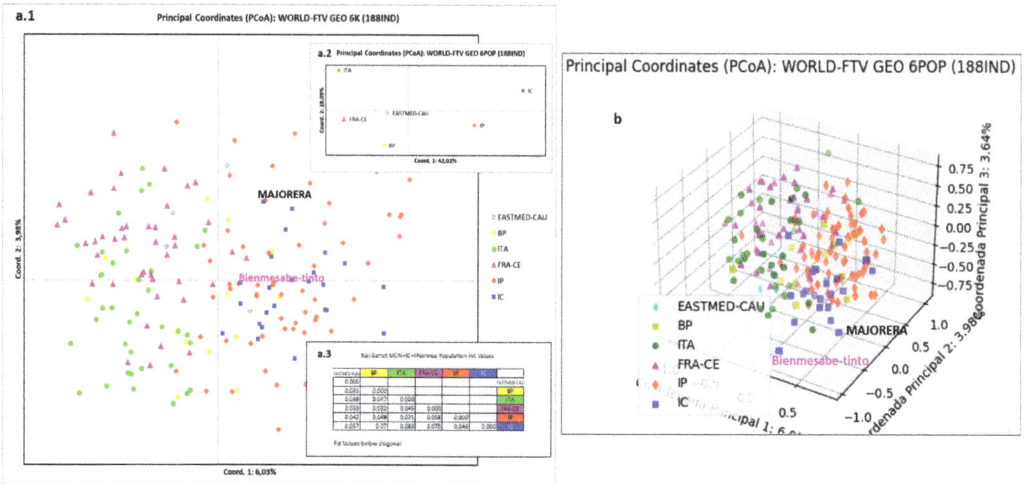

Figure 9. PCoA representation of the Canary Islands population and of the world varieties by geographical criterion. (**a.1**) Two-dimensional representation of the 6 populations by individuals, (**a.2**) two-dimensional representation of the 6 populations by population, (**a.3**) values of the Fst statistic for each population. (**b**) Three-dimensional representation of the 6 populations by individuals.

4. Discussion

After showing the results obtained in the Fuerteventura Island prospection, in this section, the singularity and possible origin of the new MP-SSR found is going to be justified, among other aspects.

4.1. SSR Polymorphism

As usual in TECNENOL studies, the 20 SSRs used to genotype the samples in question proved to be efficient and effective. In this sample population from Fuerteventura Island, the SSRs with the best performance were the ZAG79 (Na: 10; Ho: 1; He: 0.833; F: −0.2; PI: 4.8×10^{-2}), VVMD27 (Na: 9; Ho: 0.933; He: 0.836; F: −0.117; PI: 4.7×10^{-2}), ZAG47 (Na: 9; Ho: 0.933; He: 0.849; F: −0.099; PI: 4.0×10^{-2}) and VVMD36 (Na: 8; Ho: 0.933; He: 0.853; F: −0.094; PI: 3.9×10^{-2}). But, there were also SSRs that did not meet the desired expectations, such as UCH19 (Na: 2; Ho: 0.133; He: 0.124; F: −0.071; PI: 7.7×10^{-1}), VVS29 (Na: 3; Ho: 0.2; He: 0.184; F: −0.084; PI: 6.7×10^{-1}) and VVS3 (Na: 2; Ho: 0.733; He: 0.491; F: −0.493; PI: 3.8×10^{-1}). However, all SSR joint actions did provide very satisfactory results.

Before pointing out the statistical parameters that give confidence in the chosen SSRs, it should be borne in mind that the values found for each population depend on three conditioning factors: (a) the number of SSRs used, (b) the number of samples analysed and (c) the proximity of the population samples to be analysed, which is why it is difficult to compare with other authors [45–52]. It should be noted that the sample under study is a population belonging to an almost desert island where viticulture has been lost over the years, which is why only 40 samples have been collected. This number is similar to the number of individuals analysed in the study on Andalusian grape varieties carried out by Jiménez-Cantizano et al. [52] (49 samples), which also used 20 SSRs, so it is a good basis for comparison. In this case, 15 MP-SSRs were found out of the 40 individuals prospected, and 30 different genotypes were obtained in the Andalusian varieties study. From these data, when contrasting the coincident statistical parameters to compare the goodness of the SSRs used (Table S4), it is observed that all the parameter values of this research show significantly lower results, with the exception of Ho. While the mean Ho value of the present work was 0.803 (80.3% of heterozygotes), in the Andalusian varieties study, it was 0.763 (76.3% of heterozygotes); thus, the Fuerteventura Island population had fewer homozygous individuals (almost 5% fewer). On the other hand, this study's accumulative IP reached the value of $7{,}5 \times 10^{-20}$, 44.4% lower than the value obtained in the work on Andalusian varieties (1.7×10^{-19}) for 20 SSRs in both cases. With respect to the other works consulted [45–51], the values were always significantly lower (except for PI) and uneven, mainly due to the low number of samples analysed and not due to a loss of efficiency and efficacy of the SSRs used.

4.2. Grapevine Variety Analysis

From the 40 accessions registered for the Fuerteventura Island population, once the first data normalisation was carried out, the Fuerteventura population was reduced to 15 individuals with a unique MP-SSR. These corresponded to 10 varieties, of which 9 were known and registered in the databases (Airén, Beba, Listan negro, Listan prieto, Mollar cano, Muscat Hamburg, Muscat of Alexandria, Palomino fino and Torrontes volcánico) and 1 was unknown (Majorera). The difference in five individuals was due to the existence of intra-varietal variability, i.e., four new mutations and one mutation already described in the Lanzarote Island study (whose name is Listan negro santanero) [40] had been described for known varieties in this collection. Apart from this, it must be said that the Palomino fino variety (in this work) was also represented by a mutant described in two previous studies since the population of Fuerteventura Island did not have the most widespread profile for this variety. This mutant genetic profile was found on the island of Lanzarote, where it received the name Listan blanca chicharrera (published name) [40], which has also been found in el Hierro island [41]. Another case of mutation described previously occurred in the Andalusian variety Mollar cano. This variety was represented by a mutation found on the Island of Lanzarote [40] known as Mollar bonilla.

Overall, in the Fuerteventura population characterisation and identification study, 13 samples were identified that were entered with the name unknown and 5 accessions with erratic names were detected, which have been perfectly identified. With regard to lexicography, it is suggested that the term Majorera be included in the VIVC database as the main name of the new variety in this collection. It is also proposed to include the new synonymy Hoja moral, with which the Listan negro variety is known throughout the island, as well as the names of the mutations found for this variety and also for the Listan prieto variety (Listán prieto de Antigua, Listan prieto de Vega, Hoja moral de El Rosario and Hoja moral de Betancuria), since, as has been pointed out in previous research articles [40,41], the VIVC database gives proper names to mutations of the Pinot varieties noir, Chasselas blanc and Trousseau noir, among others. The term Moscatel (synonymy registered to designate the Muscat of Alexandria variety and well used by the Fuertevetura island winegrowers) is detected to name the red variety Muscat Hamburg and is proposed as a new synonymy for this English red Muscat designed by Mr. Seward Snow [44]. A very unusual case occurs with the name of the accession Beba. It entered under the name of Burra blanca, a synonymy registered for the Airen variety and often used in the Canary archipelago to designate the Beba variety (a case also found in the study of El Hierro island [41]). For this reason, it is considered an erratic name and not a synonymy, since, if it were considered a synonymy, we would be creating a case of homonymy.

4.3. Fuerteventura Grapevine Population Genetic Structure

The aim of this section is basically to find out about the Fuerteventura varieties' collection behaviour, paying special attention to the new local variety.

Figure 3a shows how coordinate 1 separates the varieties originating from or with more influence from the east of the Mediterranean Basin from the rest (of Spanish origin). On the other hand, coordinate 2 places in the upper quadrants those varieties known to have chlorotype A (characteristic of the Iberian Peninsula) [53]. It is also known that the female progenitor of these same varieties is the variety that originated in North Africa (according to Pierre Galet [54,55]), known by the names of Heben (main name) or Gibi (very widespread synonym). In addition, the Canary Islands variety Torrontes volcánico and the Majorera variety, of which neither their chlorotype nor their progenitors are known, are also found at this location. The exception is the variety Airen, which is located in the lower left quadrant, very close to the axis. Airen, which also has chlorotype A, has the Heben variety as its female progenitor. Its location so close to the previous varieties would justify the fact that it is considered to belong to the same group and, therefore, within the *modus operandi* error of all the graphical representations of this study and similar works. The location of each individual on a graph would be due to the confluence of 40 numerical data corresponding to the two alleles of each of the 20 SSRs. Under these conditions, only a 40-dimensional graphical representation would be error-free. It is for this reason that the reduction from 40 to 2 or at best 3 dimensions leads to these deviations. In the lower quadrants, and thus on the same axis or below the axis of coordinate 2, there are mostly varieties with chlorotype D. These are Listan prieto, Palomino fino, Muscat Hamburg and, with chlorotype B, Muscat of Alexandria. But, the latter variety is a cross between the Greek varieties Heptaliko (chlorotype B) and Muscat á petits grains blancs, also with chlorotype D, which means that the Muscat of Alexandria has Balkan Peninsula ancestry, where this chlorotype is mostly from. Muscat Hamburg is a cross between the Greek variety Muscat of Alexandria and the Italian variety Schiva grossa (chlorotype D), carried out by the English hybridiser Seward Snow [44]. Finally, we can also see in the lower quadrants the Canary Island variety Listan negro, of which the chlorotype is unknown but its cross is known. This variety's progenitors are the Spanish varieties Palomino fino (chlorotype D) and Mollar cano (chlorotype A). As far as the new variety Majorera is concerned, as it is found practically only in the upper right quadrant, it can be concluded that it has a very characteristic MP-SSR and is influenced by the east of the Mediterranean Basin. Figure 3b broadly reproduces the same distribution as Figure 3a for the sample collection under study. Again, the Majorera variety is one of those moving away from the group, together with the surprising position of Muscat Hamburg, which, in the two-dimensional graph, appeared next to its parental, Muscat of Alexandria. Listan negro also suffers a shift between its parentals: whereas, in Figure 3a, it was located close to the Palomino fino, now it is closer to the Mollar cano. Evidently, the reliability of the varieties' position in this study improves with a dimension increase. In this particular case, the accumulative reliability percentage in two dimensions is 36.68% (the sum of the percentages of coordinates 1 and 2 (Figure 3a)) but, for three dimensions, the goodness of representation increases to 51.76% (sum of the three coordinates in Figure 3b).

Finally, it can be concluded that the new local variety from Fuerteventura has a significantly different MP-SSR from the rest of the viniferas prospected on the island. The differences found in the "signature" variety Muscat Hamburg may be due to the Italian influence of its MP-SSR, thus differentiating it from the rest of the members of this population with Spanish influence.

4.4. Majorera Variety Relation with Respect to Canary Archipelago Grapevine Population

In order to obtain a better understanding of the Majorera variety behaviour and to check whether it continued to be significantly different, in this case from the varieties of the archipelago itself, a population genetics study was carried out. To this end, the 32 Canary Islands varieties were divided into six populations. This proposal was rejected because of the dysfunction of POP1 with respect to POP2 and because POP3, although well defined as such, did not have pure individuals. The proposal to divide the Canary Islands grapevine population into four ancestral populations was then taken. The membership of each individual to a given population was based on the q value, which is defined as the measure of the link of an individual with an ancestral population based on its genetic similarity (percentage of its genome inferred that belongs to the group [56]) and is ordered from the largest to the smallest value (Table S5). Seventeen admixed individuals were detected and eliminated, and the Canary Island population consisted of 15 varieties, which were divided into these four populations. POP1 held the varieties Bienmesabe tinto (Island of La Palma), Majorera (Island of Fuerteventura), Breval negro (Island of Lanzarote) and Blanca de la Granja del Cabildo (Island of Lanzarote) (Figure 4 and Figure S3). Bienmesabe tinto is a variety that was published in 2019 by TECNENOL [39] as a new variety of *Vitis vinifera* ssp. *vinifera*, and, after checking its MP-SSR in the VIVC database, did not give any match. Currently, consulting this database, it can be seen that it has been registered as the result of a cross between species of the genus *Vitis*, i.e., as a direct hybrid producer (DPH), having been previously described by other research groups as such [57,58]. In this sense, the fact that this MP-SSR, which shows the Bienmesabe tinto variety, is positioned so far away from the other grape varieties is of great value. Based on these data and given Majorera's proximity to this variety and how far it is located from the rest of *vinifera*, it is possible that it is also a DPH. POP2 is presented as a cluster closely related to HI varieties and, in POP3, the HI and LNZ varieties share prominence. On the other hand, in POP4, we find the progeny of Malvasia Dubrovacka and Bermejuela, Malvasia volcanica and a new variety, also from LNZ, which is known as Malvasia alistanada fina. Malvasia Dubrovacka has a chlorotype A, which is very frequent in the Iberian Peninsula but rare in the Balkan Peninsula [49,53], although several studies group its MP-SSR with Malvasia profiles from the Balkans [59]. Palomino fino progeny, Listan negro (Palomino fino × Mollar cano), Albillo criollo and Albillo forastero (Palomino fino × Verdelho branco) are also located in this population. This is important because of the chlorotype D of the Palomino fino variety, which is closely related to the eastern Mediterranean basin [49,53]. Therefore, it can be concluded that this cluster groups varieties (pure or admixed) with Eastern Mediterranean influence.

Figure 5 shows contradictory information regarding the relationship that these populations may have. While, in Figure 5a–e, it can be seen that POP1 and POP2 are markedly differentiated in all images, in the case of POP3 and POP4, it depends on whether the graph is represented by individuals or populations. In the case of population representations (Figure 5a), POP3 and POP4 appear distant and, in the case of individual representations, they are related. Clearly, the reliability of this representation has decreased with respect to the representation by populations. Nevertheless, trends can be observed. It can also be seen how POP1 presents all its individuals as dispersed, while the rest of the populations maintain all their individuals as grouped together as a sample of the relationship between them, clearly differentiating one population from the other.

With regard to the variety under study, Majorera, it can be concluded in this section that it has a significantly different MP-SSR from the rest of the Canary Islands varieties, with the exception of the Bienmesabe tinto variety. This uniqueness may be due to the nature of its progenitors. There is a possibility that this variety is a DPH, i.e., an interspecific cross. However, the fact that, in the three-dimensional representation (Figure 5d), it is very distant from the Bienmesabe tinto variety may mean that its parents (interspecific or not) are not very closely related to the Bienmesabe tinto variety parentals.

4.5. Majorera Variety Relation with Respect to World Population

The availability of a complete database for *Vitis vinifera* ssp. *vinifera* in the TECNENOL research group allowed us to compare the MP-SSR of the Majorera variety with 308 other varieties from 22 countries around the world and, in this way, attempt to confirm the indications that point to this genetic profile as an interspecific cross.

The first strategy that was tested was purely genetic. Structure 2.3 prioritised the distribution K = 2, forming two populations that are markedly differentiated: POP1 (with mostly varieties from the rest of the world) and POP2 (with mostly Spanish individuals and from which the Canarian population was extracted (POP3)) (Figure 6b,c and Figure 7a,b). This behaviour is observed in all TECNENOL works; in the same way, the CI population singularity with respect to the Spanish population and also with the world population is also observed in all of them (Figure 7(a.2,a.3)) when working with total populations. However, in the representations with all individuals in the form of a circular dendrogram (Figure 6b), phylogenetic tree (Figure 6c) and two-dimensional PCoA (Figure 7(a.1)), it is no longer possible to demonstrate the uniqueness of IC with respect to POP2. If the behaviour of the varieties Bienmesabe tinto and Majorera is observed in these graphical representations, with the former being a DPH and the latter having the potential to be one, there are no significant differences with respect to the rest of the varieties either. Bienmesabe tinto and Majorera varieties' locations overlap with the populations of POP2 and POP3 (IC) without giving relevant information. Only in the three-dimensional representation (Figure 7b) do these two varieties appear as significantly distant from the rest, especially the Bienmesabe tinto variety. It should be noted that an additional dimension adds reliability to the final result. In this case, the reliability went from 9.36% in the two-dimensional representation to 12.88% in the three-dimensional graph, which represents an increase of 27.33%. Despite this, the final goodness of fit is still significantly low and similar to that described by most authors for PCoA in two dimensions [59,60].

In addition, another strategy was tested. This included a geographic component. Thus, this world population was grouped by geographical areas as proposed by other authors [53,56], according to the origin of the variety registered in the VIVC.

In this particular case, the 309 varieties, 307 belonging to *Vitis vinifera* ssp. *vinifera*, the Bienmesabe tinto variety as DPH and the Majorera variety that could have originated from another interspecific cross, were grouped into six areas. The objective was to see what influence the geographical component had in supporting the possibility that the only MP-SSR on the island of Fuerteventura was not a pure vinifera.

When distributing the world population into six geographical areas, the global behaviour observed was similar to the previous one. The representation by population, once again, shows the uniqueness not only of the IC varieties but also the uniqueness of the PI varieties with respect to the rest of the varieties in the world (Figure 8a,b and Figure 9a.2,a.3,b). In addition, the singularity between IC and PI can also be observed, although they are closely related. In contrast, the individual representation in Figure 9(a.1) shows no differentiation between IC and PI. Figure 8a,b show how some PI varieties are located in branches belonging to the EAST-MED-CAU, BP and ITA areas, and in the one belonging to the FRA-CE area. This transposition may well be due to the influence of the MP-SSRs in question on these reference areas. Possibly, this influence is due not only to the conquest of new territories by the Crown of Aragon, which in the Middle Ages reached as far as the Balkan Peninsula [61], but also to the Camino de Santiago (FRA-CE) [62]. It is obvious to think that, in these situations, there was a bidirectional transfer of varieties. Regarding the varieties Bienmesabe tinto and Majorera, again it will be the three-dimensional PCoA representation (Figure 9b) that distinguishes these varieties significantly from the rest of the world varieties.

It must also be noted that, for both strategies, (a) coordinate 1 of Figure 7(a.1) and Figure 9(a.1) is the one that separates the specific populations for each case, while Coordinate 2 does not have a notable role or does not have any paper; (b) the major Spanish populations (peninsular and insular), as well as the Bienmesabe tinto and Majorera varieties, overlap without distinction (Figure 7(a.1) and Figure 9(a.1)); (c) it is always the third dimension collected in Figures 7b and 9b that is able to differentiate between Bienmesabe tinto and Majorera varieties, and consequently show their uniqueness.

In conclusion, it can be hypothesised that there is a possibility that the Majorera variety from Fuerteventura Island is a DPH. This will have to be confirmed in further studies.

5. Conclusions

After carrying out this study, it can be concluded that, once again, the SSR kit used for this study worked correctly. Regarding the MP-SSR, this study has described: (a) 10 different variety profiles, where 9 correspond to known varieties and 1 corresponds to a new variety; (b) 4 new mutations of known varieties (Listan prieto from Antigua, Listan prieto from Vega, Hoja moral from El Rosario and Hoja moral from Betancuria); (c) identified mutations of known varieties that have been described in other works (Listan negro santanero, Mollar bonilla, Listan blanca chicharrera); and (d) 13 MP-SSR of accessions registered as unknown that have been identified. At a lexicographic level, five erratic names have been detected. Furthermore, this work proposes the inclusion in the global database (VIVC) of seven new names: one main name of a variety (Majorera); one new name synonymous with the Canarian variety Listan negro (Hoja moral); four names of mutations (Listan prieto from Antigua, Listan prieto from Vega, Hoja moral from El Rosario and Hoja moral from Betancuria); and the new name synonymous with the Moscatel variety Hamburg (Muscat). The uniqueness of the population of Canary Islands varieties as a potential centre for creating biodiversity for cultivated vines is also confirmed. Finally, the possibility is raised that the new variety from the island of Fuerteventura derives from an intraspecific crossing.

Supplementary Materials: The following supporting information can be downloaded at: https://www.mdpi.com/article/10.3390/horticulturae9121301/s1, Table S1: Information (original and conclusive) on 40 accessions from Fuerteventura Island. Table S2: Results of 7 international SSRs of 15 genetic profiles from Fuerteventura that correspond to 10 different varieties. The yellow box indicates a mutated allele. Table S3: Annealing temperature (*Ta*) of SSR. Table S4: Characterisation of the twenty microsatellite markers used in this study. Figure S1: The four steps of the graphical method of Evanno et al., (2005), allowing for the estimation of the true number of ancestral K groups for a population with 32 individuals from the Canary Islands collection (IC, including Fuerteventura Island). Figure S2: Plots of the best distributions of the 32 individuals of the Canary Islands (including Fuerteventura) from Structure 2.3. programme for 4K and 6K. Table S5: Genetic structure of the Canary Islands population. Distribution K = 6 (individuals belonging to each group or population). Breakdown of the ratio of pure and admixed individuals according to the q value (pure ($q \geq 85\%$) and admixed ($q < 85\%$)). Figure S3: Genetic structure of the Canary Islands population (32 varieties). Distribution K = 4 (individuals belonging to each group or population). Breakdown of the ratio of pure and admixed individuals according to the q value (pure ($q \geq 85\%$) and admixed ($q < 85\%$)). Status of each population. Figure S4: The four steps of the graphical method of Evanno et al., (2005), allowing for the estimation of the true number of ancestral K groups for a population of 309 varieties from the TECNENOL database. Figure S5: Genetic structure of the world population (309 individuals). Distribution K = 2 (individuals belonging to each group or population). Breakdown of the ratio of pure and admixed individuals according to the q value (pure ($q \geq 85\%$) and admixed ($q < 85\%$)). Status of each population. Figure S6: Genetic structure of the world population (188 individuals). Distribution in 6 geographical areas. Breakdown of the well-assigned (pure) and wrong-assigned (admixed) individuals. Nations that compose each of the groups: EASTMED-CAU (Algeria, Cyprus, Georgia, Israel, Lebanon, Tunisia and Turkey), BP (Bosnia and Herzegovina, Bulgaria, Croatia, Greece, Serbia, Slovenia and Montenegro), ITA (Italy), FRA-CEU (Austria, France, Germany, Hungary and Switzerland), IP (Spain and Portugal) and IC (Canary archipelago).

Author Contributions: Conceptualisation, F.F. and Q.L.-Y.; methodology, F.F., Q.L.-Y. and C.V.; software, P.S.-G., J.M.C. and F.Z.; validation, F.F. and Q.L.-Y.; formal analysis, F.F., Q.L.-Y. and C.V.; investigation, F.F., Q.L.-Y. and C.V.; resources, F.F.; data curation, F.F. and Q.L.-Y.; writing—original draft preparation, F.F. and P.S.-G.; writing—review and editing, F.F. and P.S.-G.; visualisation, J.M.C. and F.Z.; supervision, F.F., Q.L.-Y., C.V., P.S.-G., J.M.C. and F.Z. All authors have read and agreed to the published version of the manuscript.

Funding: This project has been funded by Cabildo Insular de la Isla de Fuerteventura.

Data Availability Statement: No new data were created or analyzed in this study. Data sharing is not applicable to this article.

Acknowledgments: The authors are grateful to Javier Ibañez and the viticulture specialists Alberto Gerardo González and Juan Manuel Brito for their valuable support. We would also like to thank Luis Ricardo Suárez de Abreu, Rosa Pastor, Isabel Araque, Braulio Esteve-Zarzoso, Laia Fañanás and Santiago Moreno for their support in the laboratory.

Conflicts of Interest: The authors declare no conflict of interest.

References

1. Wolkovich, E.M.; García de Cortázar-Atauri, I.; Morales-Castilla, I.; Nicholas, K.A.; Lacombe, T. From Pinot to Xinomavro in the world's future wine-growing regions. *Nat. Clim. Chang.* **2018**, *8*, 29–37. [CrossRef]
2. Togores, J.H.; Fernández-Cano, L.H. *Tratado de Viticultura*, 2nd ed.; Ediciones Mundi-Prensa: Madrid, Spain, 2019; Volume 1.
3. Jones, G.V. Climate Change: Observations, Projections, and General Implications for Viticulture and Wine Production. Economics Department, Oregon University. 2007, Paper No. 7, 15. Available online: https://www.researchgate.net/publication/253713937_Climate_Change_Observations_Projections_and_General_Implications_for_Viticulture_and_Wine_Production (accessed on 20 August 2023).
4. Roca, P. *Dos Bazas Para el Sector Vitivinícola en un Contexto Incierto, el Discurso de Pau Roca en el Congreso Mundial de la Viña y el Vino*; Tecnovino: Bilbao, Spain, 2023; Available online: https://www.tecnovino.com/dos-bazas-para-el-sector-vitivinicola-en-un-contexto-incierto-el-discurso-de-pau-roca-en-el-congreso-mundial-de-la-vina-y-el-vino/ (accessed on 20 August 2023).
5. Hannah, L.; Roehrdanz, P.R.; Ikegami, M.; Shepard, A.V.; Shaw, M.R.; Tabor, G.; Zhi, L.; Marquet, P.A.; Hijmans, R.J. Climate change, wine, and conservation. *Proc. Natl. Acad. Sci. USA* **2013**, *110*, 6907–6912. [CrossRef] [PubMed]
6. Coordinadora de Organizaciones de Agricultores y Ganaderos (COAG). Cambio Climático y Viñedo en España. Available online: https://adaptecca.es/sites/default/files/documentos/cambio_climatico_y_vinedo_en_espana_2016.pdf (accessed on 20 August 2023).
7. Bota, J.; Tomas, M.; Flexas, J.; Medrano, H.; Escalona, J.M. Differences among grapevine cultivars in their stomatal behavior and water use efficiency under progressive water stress. *Agric. Water Manag.* **2016**, *164*, 91–99. [CrossRef]
8. Wikipedia. Available online: https://es.wikipedia.org/wiki/Fuerteventura (accessed on 7 July 2023).
9. Wikimedia Commons. 2014. Available online: https://commons.wikimedia.org/w/index.php?curid=33566121#file (accessed on 7 July 2023).
10. NASA. 2011. Available online: https://www.flickr.com/photos/gsfc/6630087415/in/photostream/ (accessed on 7 July 2023).
11. Instituto Canario de Estadística (ISTAC). 2005. Available online: http://www.gobiernodecanarias.org:80/istac/publicaciones/pdfs/2006_005.pdf (accessed on 7 July 2023).
12. Climate Data. 2021. Available online: https://es.climate-data.org/europe/espana/fuerteventura-10238/ (accessed on 7 July 2023).
13. Lana, D. *Diagnóstico de la Viticultura en Canarias. Consejería de Agricultura*; Ganadería y Pesca: Sevilla, Spain, 2021; Available online: https://www.gobiernodecanarias.org/agricultura/detalle/Diagnostico-de-la-viticultura-en-Canarias/ (accessed on 7 July 2023).
14. Otamendi, J.J. *Historia del Vino de Lanzarote. Notas Para Su Estudio*; Bodegas El Grifo: Lanzarote, Spain, 2013; Available online: https://docplayer.es/9690813-Historia-del-vino-de-lanzarote-notas-para-su-estudio.html (accessed on 7 July 2023).
15. Las Gavias, un Tesoro en la Tierra de Fuerteventura y Lanzarote. In *Ciencia Canaria*; Agencia Canaria de Investigación, Innovación y Sociedad de la Información: Santa Cruz de Tenerife, Spain, 2017; Available online: https://www.cienciacanaria.es/secciones/a-fondo/865-las-gavias-un-tesoro-en-la-tierra-de-fuerteventura-y-lanzarote (accessed on 7 July 2023).
16. Marsal, G.; Baiges, I.; Canals, J.M.; Zamora, F.; Fort, F. A fast, efficient method for extracting DNA from leaves, stems, and seeds of *Vitis vinifera* L. *Am. J. Enol. Vitic.* **2011**, *62*, 376–381. [CrossRef]
17. Marsal, G.; Boronat, N.; Canals, J.M.; Zamora, F.; Fort, F. Comparison of the efficiency of some of the most usual DNA extraction methods for woody plants in different tissues of *Vitis vinifera* L. *J. Int. Sci. Vigne Vin* **2013**, *47*, 227–237. [CrossRef]
18. Fort, F.; Hayoun, L.; Valls, J.; Canals, J.M.; Arola, L.; Zamora, F. A new and simple method for rapid extraction and isolation of high-quality RNA from grape (*Vitis vinifera*) berries. *J. Sci. Food Agric.* **2008**, *88*, 179–184. [CrossRef]
19. This, P.; Jung, A.; Boccacci, P.; Borrego, J.; Botta, R.; Costantini, L.; Crespan, M.; Dangl, G.S.; Eisenheld, C.; Ferreira-Monteiro, F.; et al. Development of a standard set of microsatellite reference alleles for the identification of grape cultivars. *Theor. Appl. Genet.* **2004**, *109*, 1448–1458. [CrossRef]
20. Thomas, M.R.; Scott, N.S. Microsatellite repeats in grapevine reveal DNA polymorphisms when analyzed as sequence-tagged sites (STSs). *Theor. Appl. Genet.* **1993**, *86*, 985–990. [CrossRef]
21. Bowers, J.E.; Dangl, G.S.; Vignani, R.; Meredith, C.P. Isolation and characterization of new polymorphic simple sequence repeat loci in grape (*Vitis vinifera* L.). *Genome* **1996**, *39*, 628–633. [CrossRef]
22. Bowers, J.E.; Dangl, G.S.; Meredith, C.P. Development and characterization of additional microsatellite DNA markers for grape. *Am. J. Enol. Vitic.* **1999**, *50*, 243–246. [CrossRef]
23. Sefc, K.M.; Regner, F.; Turetschek, E.; Glössl, J.; Steinkellner, H. Identification of microsatellite sequences in *Vitis riparia* and their applicability for genotyping of different *Vitis* species. *Genome* **1999**, *42*, 367–373. [CrossRef] [PubMed]
24. Scott, K.D.; Eggler, P.; Seaton, G.; Rosseto, M.; Abblet, E.M.; Lee, L.S.; Henry, R.J. Analysis of SSRs derived from grape ESTs. *Theor. Appl. Genet.* **2000**, *100*, 723–726. [CrossRef]
25. Lefort, F.; Kyvelos, C.; Zervou, M.; Edwards, K.; Roubelakis-Angelakis, K. Characterization of new microsatellite loci from *Vitis vinifera* and their conservation in some *Vitis* species and hybrids. *Mol. Ecol. Resour.* **2002**, *2*, 20–21. [CrossRef]
26. Cipriani, G.; Spadotto, A.; Jurman, I.; Di Gaspero, G.; Crespan, M.; Meneghetti, S.; Frare, E.; Vignani, R.; Cresti, M.; Morgante, M.; et al. The SSR-based molecular profile of 1005 grapevine (*Vitis vinifera* L.) accessions uncovers new synonymy and parentages and reveals a large admixture amongst varieties of different geographic origins. *Theor. Appl. Genet.* **2010**, *121*, 1569–1585. [CrossRef] [PubMed]

27. Dalbó, M.A.; Ye, G.N.; Weeden, N.F.; Steinkellner, H.; Sefc, K.M.; Reisch, B.I. A gene-controlling sex in grapevines is placed on a molecular marker-based genetic map. *Genome* **2000**, *43*, 333–340. [CrossRef] [PubMed]
28. Paetkau, D.; Calvert, W.; Stirling, I.; Strobeck, C. Microsatellite analysis of population structure in Canadian polar bears. *Mol. Ecol.* **1995**, *4*, 347–354. [CrossRef] [PubMed]
29. Peakall, R.; Smouse, P.E. GenAlEx 6.5: Genetic analysis in Excel. Population genetic software for teaching and research-an update. *Bioinformatics* **2012**, *28*, 2537–2539. [CrossRef]
30. Paetkau, D.; Slade, R.; Burden, M.; Estoup, A. Genetic assignment methods for the direct, real-time estimation of migration rate: A simulation-based exploration of accuracy and power. *Mol. Ecol.* **2004**, *13*, 55–65. [CrossRef]
31. Three-Dimensional Plotting in Matplotlib (Python Data Science Handbook). Available online: https://jakevdp.github.io/PythonDataScienceHandbook/04.12-three-dimensional-plotting.html (accessed on 20 August 2023).
32. Kumar, S.; Stecher, G.; Tamura, K. MEGA7: Molecular Evolutionary Genetics Analysis version 7.0 for bigger datasets. *Mol. Biol. Evol.* **2016**, *33*, 1870–1874. [CrossRef]
33. Saitou, N.; Nei, M. The neighbor-joining method: A new method for reconstructing phylogenetic threes. *Mol. Biol. Evol.* **1987**, *4*, 406–425. [CrossRef]
34. Pritchard, J.K.; Stephens, M.; Donnelly, P. Inference of population structure using multilocus genotype data. *Genetics* **2000**, *155*, 945–959. [CrossRef] [PubMed]
35. Falush, D.; Stephens, M.; Pritchard, J.K. Inference of population structure using multilocus genotype data: Linked loci and correlated allele frequencies. *Genetics* **2003**, *164*, 1567–1587. [CrossRef] [PubMed]
36. Evanno, G.; Regnaut, S.; Goudet, J. Detecting the number of clusters of individuals using the software structure: A simulation study. *Mol. Ecol.* **2005**, *14*, 2611–2620. [CrossRef] [PubMed]
37. Marsal, G.; Mateo, J.M.; Canals, J.M.; Zamora, F.; Fort, F. SSR analysis of 338 accessions planted in Penedes (Spain) reveals 28 unreported molecular profiles of *Vitis vinifera* L. *Am. J. Enol. Vitic.* **2016**, *67*, 466–470. [CrossRef]
38. Marsal, G.; Bota, J.; Martorell, A.; Canals, J.M.; Zamora, F.; Fort, F. Local cultivars of *Vitis vinifera* L. in Spanish Islands: Balearic Archipelago. *Sci. Hortic.* **2017**, *226*, 122–132. [CrossRef]
39. Marsal, G.; Mendez, J.J.; Mateo-Sanz, J.M.; Ferrer, S.; Canals, J.M.; Zamora, F.; Fort, F. Molecular characterization of Vitis vinifera L. local cultivars from volcanic areas (the Canary Islands and Madeira) using SSR markers. *OENO One* **2019**, *4*, 667–680. [CrossRef]
40. Fort, F.; Marsal, G.; Mateo-Sanz, J.M.; Pena, V.; Canals, J.M.; Zamora, F. Molecular characterisation of the current cultivars of *Vitis vinifera* L. in Lanzarote (Canary Islands, Spain) reveals nine individuals which correspond to eight new varieties and two new sports. *OENO One* **2022**, *56*, 281–295. [CrossRef]
41. Fort, F.; Lin, Q.; Suárez-Abreu, L.R.; Sancho-Galán, P.; Canals, J.M.; Zamora, F. Study of Molecular Biodiversity and Population Structure of *Vitis vinifera* L. ssp. *vinifera* on the volcanic island of El Hierro (Canary Islands (Spain)) by using Microsatellite Markers. *Horticulturae*, 2024; submitted.
42. Zerolo, J.; Cabello, F.; Espino, A.; Borrego, J.; Ibañez, J.; Rodriguez-Torres, I.; Muñoz-Organero, G.; Rubio, C.; Hernández, M. *Variedades de Vid de Cultivo Tradicional en Canarias*, 1st ed.; Instituto Canario de Calidad Agroalimentaria, Gobierno de Canarias: Santa Cruz de Tenerife, Spain, 2006.
43. Rodriguez-Torres, I. Variedades de vid cultivadas en Canarias. Descriptores Morfológicos. In *Caracterización Morfológica, Molecular, Agronómica y Enológica*, 1st ed.; Instituto Canario de Investigaciones Agrarias, Gobierno de Canarias: Santa Cruz de Tenerife, Spain, 2018.
44. Maul, E.; Röckel, F. Vitis International Variety Catalogue. 2015. Available online: http://www.vivc.de (accessed on 7 July 2023).
45. Ibañez, J.; De Andrés, M.T.; Molino, A.; Borrego, J. Genetic study of key Spanish grapevine varieties using microsatellite analysis. *Am. J. Enol. Vitic.* **2003**, *54*, 22–30. [CrossRef]
46. Lopes, M.S.; Rodrigues dos Santos, M.; Eiras Dias, J.E.; Mendonça, D.; da Câmara Machadoa, A. Discrimination of Portuguese grapevines based on microsatellite markers. *J. Biotechnol.* **2006**, *127*, 34–44. [CrossRef]
47. Aliquo, G.; Torres, R.; Lacombe, T.; Boursiquot, J.M.; Laucou, V.; Gualpa, J.; Fanzone, M.; Sari, S.; Pérez-Peña, J.; Prieto, J.A. Identity and parentage of some South American grapevine cultivars present in Argentina. *Aust. J. Grape Wine Res.* **2017**, *23*, 452–460. [CrossRef]
48. Moita, A.; Santos, R.; Catarina, A. Unraveling the origin of *Vitis vinifera* L. Verdelho. *Aust. J. Grape Wine Res.* **2018**, *24*, 450–460. [CrossRef]
49. Žulj Mihaljević, M.; Maletić, E.; Preiner, D.; Zdunić, G.; Bubola, M.; Zyprian, E.; Pejić, I. Genetic Diversity, Population Structure, and Parentage Analysis of Croatian Grapevine Germplasm. *Genes* **2020**, *11*, 737. [CrossRef]
50. de Oliveira, G.L.; Niederauer, G.F.; de Oliveira, F.A.; Rodrigues, C.S.; Hernandes, J.L.; de Souza, A.P.; Moura, M.F. Genetic diversity, population structure and parentage analysis of Brazilian grapevine hybrids after half a century of genetic breeding. *Sci. Hortic.* **2023**, *311*, 111825. [CrossRef]
51. Arslan, N.; Yılmaz Baydu, F.; Hazrati, N.; Yüksel Özmen, C.; Ergönül, O.; Uysal, T.; Ya¸sasın, A.S.; Özer, C.; Boz, Y.; Kuleyin, Y.S.; et al. Genetic Diversity and Population Structure Analysis of Anatolian Kara Grapevine (*Vitis vinifera* L.) Germplasm Using Simple Sequence Repeats. *Horticulturae* **2023**, *9*, 743. [CrossRef]
52. Jiménez-Cantizano, A.; Puig-Pujol, A.; Arroyo-García, R. Identification of *Vitis vinifera* L. Local Cultivars Recovered in Andalusia (Spain) by Using Microsatellite Markers. *Horticulturae* **2023**, *9*, 316. [CrossRef]

53. Arroyo-García, R.; Ruiz-García, L.; Bolling, L.; Ocete, R.; López, M.A.; Arnold, C.; Ergul, A.; Söylemezoğlu, G.; Uzun, H.I.; Cabello, F.; et al. Multiple origins of cultivated grapevine (*Vitis vinifera* L. ssp. sativa) based on chloroplast DNA polymorphisms. *Mol. Ecol.* **2006**, *15*, 3707–3714. [CrossRef]
54. Galet, P. *Dictionnaire Encylcopédique des Cépages*, 1st ed.; Hachette: Paris, France, 2000.
55. Lacombe, T.; Boursiquot, J.M.; Laucou, V.; Dechesne, F.; Varès, D.; This, P. Relationships and genetic diversity within the accessions related to Malvasia held in the Domaine de Vassal grape germplasm repository. *Am. J. Enol. Vitic.* **2007**, *58*, 124–131. [CrossRef]
56. Bacilieri, R.; Lacombe, T.; Cunff, L.L.; Di Vecchi-Staraz, M.; Laucou, V.; Genna, B.; Perós, J.P.; This, P.; Boursiquot, J.M. Genetic structure in cultivated grapevines is linked to geography and human selection. *BMC Plant Biol.* **2013**, *13*, 25. [CrossRef]
57. Moreno-Sanz, P.; Loureiro, M.D.; Suárez, B. Microsatellite characterization of grapevine (*Vitis vinifera* L.) genetic diversity in Asturias (Northern spain). *Sci. Hortic.* **2011**, *129*, 433–440. [CrossRef]
58. Schneider, A.; Mainardi, G.; Raimondi, S. *Ampelografia Universale Storica Illustrata. I Vitigni del Mondo*, 1st ed.; L'Artistica: Savigliano, Italy, 2012.
59. Dong, Y.; Duan, S.; Xia, Q.; Liang, Z.; Dong, X.; Margaryan, K.; Musayev, M.; Goryslavets, S.; Zdunić, G.; Bert, P.-F.; et al. Dual domestications and origin of traits in grapevine evolution. *Science* **2023**, *379*, 892–901. [CrossRef] [PubMed]
60. Emanuelli, F.; Lorenzi, S.; Grzeskowiak, L.; Catalano, V.; Stefanini, M.; Troggio, M.; Myles, S.; Martínez-Zapater, J.M.; Zyprian, E.; Moreira, F.M.; et al. Genetic diversity and population structure assessed by SSR and SNP markers in large germplasm collection of grape. *BMC Plant Biol.* **2013**, *13*, 39. [CrossRef] [PubMed]
61. García-Muñoz, S.; Lacombe, T.; De Andrés, M.T.; Gaforio, L.; Muñoz-Organero, G.; Laucou, V.; This, P.; Cabello, F. Grape varieties (*Vitis vinifera* L.) from the Balearic Islands: Genetic characterization and relationship with Iberian Peninsula and Mediterranean Basin. *Genet. Resour. Crop Evol.* **2012**, *59*, 589–605. [CrossRef]
62. Casanova, J.; Mozas, P.; Ortiz, J.M. Ampelography and microsatellite DNA analysis of autochthonous and endangered grapevine cultivars in the province of Huesca (Spain). *Span. J. Agric. Res.* **2011**, *9*, 790–800. [CrossRef]

horticulturae

MDPI

Article

Study of Molecular Biodiversity and Population Structure of *Vitis vinifera* L. ssp. *vinifera* on the Volcanic Island of El Hierro (Canary Islands, Spain) by Using Microsatellite Markers

Francesca Fort [1,*], Qiying Lin-Yang [1], Luis Ricardo Suárez-Abreu [1], Pau Sancho-Galán [2], Joan Miquel Canals [1] and Fernando Zamora [1]

[1] Grupo de Tecnología Enológica (TECNENOL), Department of Biochemistry and Biotechnology, Faculty of Oenology, Rovira i Virgili University, Sescelades Campus, C/Marcel·lí Domingo, 1, 43007 Tarragona, Spain; qiying.lin@estudiants.urv.cat (Q.L.-Y.); luisricardo.suarez@estudiants.urv.cat (L.R.S.-A.); jmcanals@urv.cat (J.M.C.); fernando.zamora@urv.cat (F.Z.)

[2] Department of Chemical Engineering and Food Technology, Science Faculty, University of Cadiz, Agrifood Campus of International Excellence (CeiA3), Wine and Food Research Institute (IVAGRO), 11510 Puerto Real, Spain; pau.sancho@uca.es

* Correspondence: mariafrancesca.fort@urv.cat

Abstract: El Hierro island is postulated as the most biodiverse of the archipelago. To verify this hypothesis, the 87 individuals collected throughout the island were genotyped with 20 SSRs. As a result of this study, 28 varieties were described, 6 of which were new (Uval piñero, Uvalero volcánico, Pinar negro, Seis de Carlos, Tesoro blanco, Uval negro), and the first rose sport of the local Canary Islands variety Bermejuela was also found. Fifteen errors were detected in total. Eleven varieties were identified that were unknown to the vine growers and twenty individuals with variations (mutations) were found, of which two had already been described in a previous prospection in Lanzarote Island (intra-varietal variability). From this study, it is also proposed to incorporate 33 new names into the world database, corresponding mostly to the individuals and variations described for the first time, which represents a lexicographic enrichment. Finally, the singularity of the population of vines adapted to El Hierro island is demonstrated, not only with respect to the population of Canary Islands vines, but also with respect to the world population. The biodiversity and uniqueness of El Hierro and the Canary Archipelago reaffirm the proposal that the Canary Islands should be considered a world biodiversity centre.

Keywords: Vine (Vitis genre); SSR; characterisation; identification; volcanic; Canary Archipelago; El Hierro Island

Citation: Fort, F.; Lin-Yang, Q.; Suárez-Abreu, L.R.; Sancho-Galán, P.; Canals, J.M.; Zamora, F. Study of Molecular Biodiversity and Population Structure of *Vitis vinifera* L. ssp. *vinifera* on the Volcanic Island of El Hierro (Canary Islands, Spain) by Using Microsatellite Markers. *Horticulturae* **2023**, *9*, 1297. https://doi.org/10.3390/horticulturae9121297

Academic Editor: Paolo Sabbatini

Received: 2 October 2023
Revised: 15 November 2023
Accepted: 27 November 2023
Published: 30 November 2023

1. Introduction

Vitis vinifera L. is one of the oldest species in the Mediterranean basin, along with olive, wheat, and fig [1]. The European vine is a sarmentose vine belonging to the kingdom *Plantae*, division *Anthophyta* (*Magnoliophyta*), class *Magnoliopsida* (*Eudicotyledons*), subclass *Rosids*, order *Vitals*, family *Vitaceae*, subfamily *Viticoideae*, genus *Vitis*, and species *Vitis vinifera* L. according to the classical botanical classification [2]. Its genome is diploid with 19 chromosome pairs and an estimated size of 500 megabase pairs (Mbp) [1]. The species *Vitis vinifera* L. is divided into two subspecies, *Vitis vinifera* ssp. *sylvestris* (the wild form) and *Vitis vinifera* ssp. *vinifera* (the domesticated form). The wild vine is dioecious, with male and female species, whereas most modern cultivars (corresponding to the domesticated vine) are hermaphrodite plants. The vine is also highly heterozygous and requires vegetative propagation to maintain the differentiating characteristics of each variety [2].

Vitis vinifera spp. *vinifera* biodiversity originated at the end of the last glaciation of the Quaternary period (beginning of the Holocene) [3,4]. Wolkovich et al. [5] quantified intervarietal variability in this species from its origins to the present day in more than 6000 cultivars. Also, these authors report that 12 varieties (Cabernet Sauvignon, Chardonnay, Merlot, Pinot noir, Syrah, Sauvignon blanc, Riesling, Muscat à petits grains blanc, Gewürztraminer, Viognier, Pinot blanc, and Pinot gris) represent between 70–90% of the world's current vineyard area [5]. This fact is causing an important homogenisation of commercial wines, which together with wine sector legislation (especially the Appellation d'Origine Contrôlée (AOC)), which limits the number of varieties to be cultivated) is causing an important vine genetic erosion [6]. Additional factors are responsible for vine genetic erosion, such as *phylloxera* plague (which almost devastated the entire world vineyard at the end of the 19th and beginning of the 20th century) and, nowadays, alongside the homogenisation of commercial wines, climate change. Twenty-first century climate change effects will lead to the disappearance or displacement of many current wine-growing regions, along with a change in grape cultivar distribution. Additionally, consumers of AOC bottled wines are beginning to grow tired of the well-known international varieties, like Cabernet Sauvignon, Merlot, and Chardonnay. In this sense, those wine consumers seek to experience new sensory perceptions (experiential marketing) [7]. Consequently, emerging market trends emphasise the search for a typical character in wines, which directly contributes to the recovery of varietal biodiversity. Furthermore, the conservation and study of grape varieties diversity will undoubtedly be one of the potential solutions to mitigate the effects of climate change [5,8].

The Canary Islands are part (together with the Azores, Cape Verde, Madeira, and the Ilhas Selvagens) of Macaronesia, a biogeographical zone formed by this group of five archipelagos that stretches from southwest Europe to northwest Africa (Figure 1). The Canary Islands archipelago, formed by eight main islands (Tenerife, Fuerteventura, Gran Canaria, Lanzarote, La Palma, La Gomera, El Hierro y La Graciosa) and five minor islands (Alegranza, Islote de Lobos, Montaña Clara, Roque del Oeste y Roque del Este), are part of the Spanish national territory and is located off the north-western part of the African continent [9].

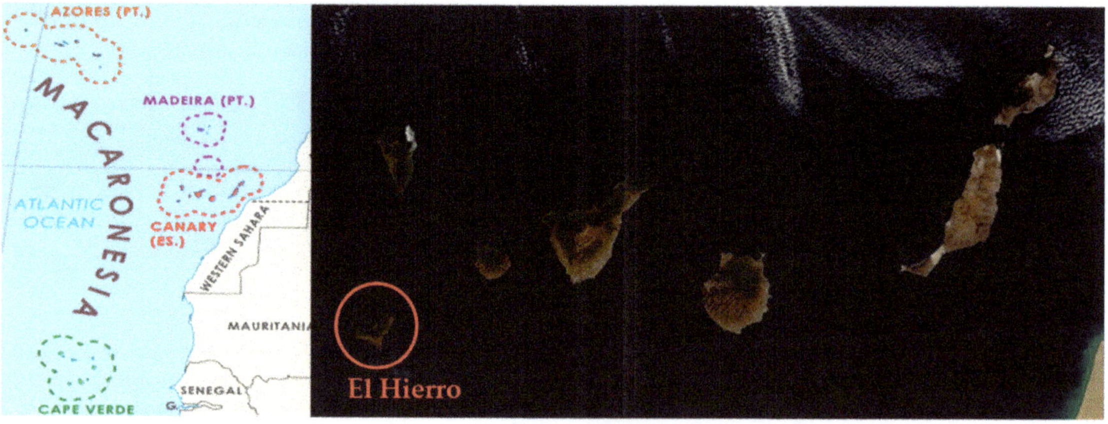

Figure 1. Macaronesia map (**left**) [10] and Canary Islands archipelago (**right**) [11].

The first vine varieties cultivated on the Macaronesian islands were introduced by the Spanish and Portuguese [12]. One of the most relevant historical events in the wine sector is that *phylloxera* (which devastated European vineyards and caused a drastic reduction in local varieties) never attacked the Canary Islands but did affect the Azores and Madeira archipelagos. This has allowed the appearance of new phenotypes due to the accumulation of genetic mutations over five centuries to enable the adaptation of new phenotypes. Therefore, many of the varieties of *Vitis vinifera* L. in the Canary Islands are the result not only of natural selection and mutations but also of natural crosses and anthropogenic selection [13].

The present work focuses on the study of grapevine varieties on El Hierro island of (Figure 2). This island is the Canary Archipelago's westernmost and southernmost point and is situated between parallels 27″38′ and 27″51′ north latitude. El Hierro is a volcanic origin island, with an estimated geological age of 1.2 million years, being the youngest of the Canary Islands [9].

Figure 2. El Hierro. Satellite view obtained from the NASA World Wind programme [14].

The climatic variation in each area of El Hierro island is conditioned by orography, but it is the clouds and their moisture that play a determining role. The trade winds, together with the Canary Current (a cold Gulf Stream branching that separates in the Azores), mean that the island does not have as arid a climate as the Sahara, which is at the same latitude [9].

El Hierro has all the soil and climatic characteristics required for quality viticulture. It is typical of the island to grow vines on terraces in order to make the greatest possible use of soil in very small areas which are subject to heavy erosion. The 203 hectares of vineyards in El Hierro are distributed as follows: 50% are in the municipality of Frontera, occupying the north-facing slopes of the Valle del Golfo, and the rest are distributed between the municipalities of Valverde (47.5 hectares) and El Pinar (51.5 hectares). A total of 86% of the vineyard area is located between 200 and 400 metres above sea level [15]. As mentioned above, *Vitis vinifera* L. has evolved for more than 500 years in the Canary Islands, resulting in unique varieties that allow the archipelago to be classified as one of the world's main centres of vine biodiversity [16]. In terms of biodiversity, El Hierro island is the most biodiverse in the Canary Islands (so far), and the following local varieties have been identified: Burra volcánica (White (W)) [16], Verijadiego (W) [16–18], Verdello de El Hierro (W) [17,18], Huevo de gallo (W) [16,18], Mollar cano rosado (Rose (Rs)) [16], and Verijadiego negro (Black (B) [16]).

The main objective of this study is to characterise and identify unknown autochthonous grapevine varieties in order to preserve the biodiversity, both inter- and intra-varietal, and to increase the range of wines that can be marketed in the future in this island. At the same time, if necessary, any errors in the identification of varieties and lexicographical errors that may appear will be corrected, and, finally, a study of the population structure will be carried out to see the extent of the uniqueness of the population of El Hierro individuals.

2. Materials and Methods

2.1. Plant Material

Eighty-seven samples (grapevine shoots) of different *Vitis vinifera* L. individuals were collected in different areas of El Hierro island by means of a mass selection strategy carried out by the local winegrowers. It was estimated that the best time to collect the samples would be during winter pruning (specifically, they were collected in January). On El Hierro, vineyards are planted with several varieties at the same time, following the Canary Islands traditional planting system. Once collected, grapevine shoots were stored at −20 °C until processing. Detailed information on the accessions analysed is shown in Table S1.

2.2. DNA Extraction and Purification

To extract the genetic material from the samples, the methodology proposed by Marsal et al. [19,20] was followed, with an adaptation based on the procedure of Fort et al. [21]. This method has been optimised by performing two chloroform washes, as this is more efficient in removing proteins. The quality of each extraction sample was evaluated with the help of the Thermo Fisher® Scientific NanoDrop TM 1000 Spectrophotometer (Waltham, MA, USA), which accurately measures the concentration and purity level of nucleic acids.

2.3. Simple Sequence Repeat (SSR) Markers

Grapevine samples were genotyped using 20 SSR markers, which were previously selected for their discrimination and polymorphism capacity based on previous studies: VVS2, VVS3, and VVS29 [22]; VVMD5, VVMD6, and VVMD7 [23]; VVMD27, VVMD28, and VVMD36 [24]; VrZAG21, VrZAG47, VrZAG62, VrZAG64, VrZAG79, and VrZAG83 [25]; SCU06vv [26]; VvUCH11, VvUCH12, and VvUCH19 [27]; VChr19a [28]. SSRs VrZAG47 and VVMD27 are not independent *loci*, meaning they amplify the same genome area. The difference between them is in the primer design [29]. Of all the SSRs, there are nine that the international scientific community [30,31] considers to be reference or international genetic markers: VVS2, VVMD5, VVMD7, VVMD25, VVMD27, VVMD28, VVMD32, VrZAG62, and VrZAG79. This research was carried out with 7 international SSRs plus 13 non-international SSRs, which were chosen based on their characteristics (Table S2).

2.4. DNA Amplification and Polymerase Chain Reaction (PCR)

An Applied Biosystems 2720 Thermal Cycler (Foster City, CA, USA) was used to perform PCR. This procedure was performed with 4 ng of DNA and 1 μM of each primer with a fluorescent dye attached to the forward primer (Fw) (6-FAM: VVS3, VVMD7, VVMD28, VVMD36, VrZAG47, VrZAG62, VrZAG83, VvUCH11, and VvUCH19; HEX: VVS2, VVS29, VVMD6, VVMD27, VrZAG21, VrZAG79, and VChr19a; NED: VVMD5, VrZAG64, scu06vv, and VvUCH12) using the Applied Biosystems AmpliTaq DNA Polymerase kit (Foster City, CA, USA). The thermocycling programme was as follows: 95 °C, 5 min; 40 cycles (95 °C, 45 s; Ta 30 s (Table S3); 72 °C, 1 min 30 s), and 72 °C, 7 min.

2.5. Amplified Fragments Length Measurement

For fragment measurement plates preparation, amplification products were mixed with 20.5 μL of deionised formamide and 0.25 μL of GeneScan ROXTM 500 internal marker (Applied Biosystems, Foster City, CA, USA). Each plate content was denatured with a thermocycling regime at 95 °C for 3 min. Fragments were separated by capillary electrophoresis with an ABI PRISM 3730® genetic analyser (Applied Biosystems, Foster City, CA, USA). Peak Scanner Software (Applied Biosystems, Sparta, NJ, USA) was used to measure the amplified fragments.

2.6. Data Analysis

To assess the reliability of the 20 SSRs used, the GenA-lEx 6.5 software [32,33] was employed. This software allows the study of six parameters: number of different alleles (Na), number of effective alleles (Ne), observed heterozygosity (Ho), expected heterozygosity (He), fixation index (F), and the probability of identity (PI). GenAlEx 6.5 has also allowed us to rule out identities, i.e., genetically identical individuals, as well as to detect mutations. Assignment tests based on allele frequency [34], also available in GenAlEx 6.5, were used for the first time to confirm the accessions belonging to each subpopulation generated by Structure 2.3. [35,36]. For each accession, a logarithmic probability value was calculated for each subpopulation using the allele frequencies of the respective subpopulations. An individual was assigned to the population with the highest logarithmic probability value. In addition, this software was also used to calculate populations' genetic differentiation using an analysis of molecular variance (AMOVA) with 999 dataset permutations for SSR genotypes, with the Fst (coefficient of genetic differentiation between populations) assuming the infinite allele model. Finally, two-dimensional principal coordinate analysis (PCoA) was used in GenAlEx 6.5 to further examine genetic relationships between populations based on the same SSR data, both for populations per se and for populations disaggregated by individuals. PCoA was based on the standardised covariance of genetic distances calculated for codominant markers.

Structure 2.3. software [35,36] has been used to assess population structure and identify crossbred individuals. This model-based software uses a Bayesian clustering method in which several ancestral populations (K) are assumed to be present, each characterised by a set of allele frequencies at each *locus*. Sample individuals are assigned to populations (clusters), or jointly to more populations if their genotypes indicate that they are admixed. All *loci* are assumed to be independent, and each population K is assumed to follow Hardy–Weinberg equilibrium. Posterior probabilities were estimated using the Markov chain Monte Carlo (MCMC) method. MCMC chains were run with a 100,000 burn-in period followed by 1,000,000 iterations using a model allowing for admixture and correlated allele frequencies. Structure was run at least ten times by setting K from 1 to 7 (1 to 9 for global varieties), and an average likelihood value, L (K), was calculated across all runs for each K. The mean log probability of the data for each K was calculated to determine the most appropriate number of clusters, and the value of K for which this probability was highest was selected. The ΔK was then calculated using the method proposed by Evanno et al. [37]. ΔK is a quantity based on the rate of change in the log probability of the data between successive K values.

Dendrograms and phylogenetic trees were constructed using the neighbour-joining method [38], using MEGA version 7 [39]. For the three-dimensional PCoA representations, the Matplotlib strategy was used using Python Data [40].

3. Results

The 87 grapevine shoots prospected were genotyped with the same 20 SSRs used in previous studies [6,16,41,42]. The aim was to compare the molecular profiles of SSRs (MP-SSRs) obtained with those that have been found and stored in a private research group database (TECNENOL: Grupo de Investigación en Tecnología Enológica).

3.1. SSR Polymorphism

Once the MP-SSRs of the whole population were obtained, identical profiles were searched for and, after the first data normalisation, a total of 41 individuals were eliminated (Tables S1 and S2). The remaining 46 unique MP-SSRs corresponded to 28 varieties of *Vitis vinifera* ssp vinifera. Thus, each variety would include individuals with variations with respect to the most widespread MP-SSR, but not the "sport" (colour or hairiness mutation so specific that molecular markers do not detect it), as they have the same MP-SSR.

To verify the goodness and efficiency of the 20 SSRs used, 6 parameters indicative of these traits were studied (Table S4). A total of 185 alleles (Na) were detected in this population, with a mean of 9.3 alleles. The SSR with the lowest number of alleles was VVS3 with 3 alleles, and the highest number was VVMD27 with 15 alleles. The mean number of effective alleles (Ne) was 5, ranging from 1.17 (VVS29) to 10.47 (VVMD27). The means of Ho and He were 0.796 and 0.737, respectively. The SSR VVS29 reached minimum values in both cases (Ho = 0.156 and He = 0.148). Likewise, the maximum value found for Ho was for VVS2 (0.978), while the highest value for He was shown by VVMD27 (0.904). Six SSRs had positive F values, although all of them were very close to 0 (VVMD6, VVMD27, VVMD28, VrZAG83, VvUCH11, and VvUCH19). Finally, the accumulative identity probability of the SSR set was 9.4 $\times$ 10^{-23}, defining a range between 7.3 $\times$ 10^{-1} for SSR VVS29, and 1.7 $\times$ 10^{-2} for SSR VVMD27.

3.2. *Grapevine Variety Analysis*

Grapevine variety analysis for their characterisation and identification was carried out at two levels: at the MP-SSR level and at the lexicographical level. To this end, an exhaustive bibliographic review (books, scientific articles, and databases) was carried out to find information on Canary Islands varieties [6,16–18,41,42]. The Vitis International Variety Catalogue (VIVC) database [43] was also used to verify the unknown individuals.

Tables S1 and S2 contain all relevant information on each genotyped individual. Table S1 contains both the original information provided by the winegrower and the conclusive information found at after matching MP-SSRs and names in the TECNENOL [6,16,41,42] and VIVC [43] databases. Information is also provided on the variations detected in each allele, the presence of tri-alleles, or the similarity percentage to the closest genome in the TECNENOL database. Table S2 also shows the seven SSRs values that coincide with the international SSRs [43].

Focusing on the 87 individuals analysis (Table S1), it can be seen that the Bermejuela variety was the one with the most entries. There were a total of 11, of which 7 corresponded to Bermejuela (W), with 5 identities with respect to the most widespread MP-SSR, and 2 mutations (Bermajuelo del Echedo (VVS3-2 (mutation of this SSR in the second allele)) and Bermajuelo del puerto (VVMD36-1 (mutation of this SSR in the first allele)). The remaining four entries corresponded to the new "sport", Bermejuela rosada (Rs), with only Bermajuelo rosado del tesoro showing variation (VVS2-2). The cluster of the local Herreña (term referring to autochthonous from El Hierro island) variety, Verijadiego, was much more uniform. Of its eight components, four individuals showed a MP-SSR identical to the most widespread one, and the remaining four showed a MP-SSR with a mutation in one allele (VVS3-2). Similarly, the Portuguese variety Alfrocheiro, which grouped six accessions, showed little variability (four with a MP-SSR identical to the most widespread, and the other two with a mutation in VVS3-2). The Listan negro and Vijariego blanco varieties, each consisting of five accessions, showed high variability. The Listan negro variety was very uniform, with only two MP-SSRs (four components identical to the most widespread MP-SSR, and one individual with a variation in VvUCH11-2), while the Andalusian variety Vijariego blanco was so variable that no component identical to the most widespread MP-SSR was recorded. Thus, the accessions Vijariego blanco from El Hierro and Burra blanca vary in VVMD36-1; the accession Eusebia presented two variations (VVS3-2 and VVMD36-1); the accession known as Diego de El Hierro showed three variations (VVMD28-1, VVMD36-1, and a case of tri-allelism in the SSR, SCU06vv), and finally, the accession known as Diego de Frontera also varied in three alleles (VVS3-2, VVMD36-1, and VvUCH12-2). With four accessions each, there are the following variations: (a) Samarrinho and Trousseau noir, without variations; (b) Mollar cano with three individuals with MP-SSR identical to the most widespread and one with a variation in VVS3-2; (c) Albillo forastero with one accession with a MP-SSR identical to the most widespread and three individuals that presented the same mutation (VVS3-1); (d) the Listan prieto variety showed one accession with a MP-SSR identical to the most widespread, two entries mutated on one allele (VVMD28-1) and one entry mutated on two alleles (VVS29-2 and VVMD28-1); (e) the Portuguese variety Molar showed one accession with a MP-SSR identical to the most widespread, two entries mutated on VvUCH12-1, and one entry mutated on VVS29-2. With three accessions, the varieties Muscat of Alexandria, Verdello de El Hierro and Malvasia fina have been obtained. The first two had all their components identical to the most widespread MP-SSR, and the last one had a mutated component in VVMD28-2. There was a group of five varieties with two components each: (a) the varieties Malvasia Dubrovacka and Sumoll, with all their components identical to the most widespread MP-SSR; (b) the varieties Palomino fino and Isabella, with a variation in one accession from the former in VVS3-1, and the latter with a tri-allelic in this same SSR; finally, (c) the variety Airen, which has all its accessions mutated in the same alleles (VVS3-2 and VVMD28-1). Finally, there is a group of eight varieties with only one representative and which do not present any variation with respect to the most widespread MP-SSR: the Spaniard south-west variety, Beba; the Lanzarote variety Uva de año, the Portuguese variety Verdelho branco, and the six new varieties described for the first time on the island of El Hierro (Uval piñero, Uvalero volcánico, Pinar negro, Seis de Carlos, Tesoro blanco, and Uval negro).

As noted in previous paragraphs, the 46 unique MP-SSRs (Table S2) found corresponded to 28 varieties, of which 6 were unknown. Of the remaining 22 varieties, 4 were Canary Islands varieties (Albillo forastero, Bermejuela, Listan negro and Uva de año) and, more concretely, 2 were varieties from the island of El Hierro already described by other authors (Verdello de El Hierro and Verijadiego) [17,18]; the remaining 16 were not from the Canary Islands archipelago. Regarding the latter, seven were Spanish (Airen (represented by a mutation), Beba, Listan prieto, Mollar cano, Palomino fino, Sumoll, and Vijariego blanco (represented by a mutation)), five were Portuguese (Alfrocheiro, Malvasia fina, Molar, Samarrinho, and Verdelho branco), and there were also the Greek Muscat of Alexandria, the French Trousseau noir, the Malvasia Dubrovacka of unknown origin (but located in the Balkan Peninsula), and the American variety corresponding to a direct producer hybrid (DPH), Isabella. A pink-coloured mutation ("sport") was also described for the first time for the Canary Islands variety Bermejuela, namely Bermejuela rosada. Rodríguez-Torres [18] also documented in his work another "sport" for the Bermejuela variety, namely Bermejuela tinta (black-violet), which was not found in this prospection. Eleven accessions were identified as "unknown", fifteen errors were detected, eighteen new mutations of known varieties were presented, and two mutations previously described on the island of Lanzarote were detected (Mollar bonilla corresponding to a mutation of the variety Mollar cano (VVS3-2) and Listan blanca chicharrera corresponding to a mutation of the variety Palomino fino (VVS3-1)) [42].

As far as variety names are concerned, 18 new names are proposed for the new mutations detected: Airen del pinar, Forastera de la isla Redonda, Baboso negro de Frontera, Bermajuelo del Echedo, Bermajuelo del puerto, Bermajuelo rosado del tesoro, Mierda de gallina, Listan negro del tesoro, Listan prieto chijo, Listan prieto herreño, Malvasia fina gabetera, Molar tintilla, Molar herreño, Verijadiego blanco de Frontrea, Vijariego blanco de El Hierro, Eusebia, and Diego de El Hierro y Diego de Frontera. In addition, one synonymous name registered for a given variety, used to name another variety, was detected. This is the case of the term Baboso blanco, used in El Hierro to refer to the Portuguese variety Samarrinho. Officially (VIVC), this term is a synonymy used to designate the white mutation of the French variety Trousseau noir, which is known under the name Bastardo blanco as the main name of this mutation ("sport" in this case). Five new synonymies are presented: Bermajuelo as a synonymy of El Hierro for the Canary Islands variety Bermejuela, Bermajuelo rosado de El Llano, Bermajuelo rosado, and Mulata rosada to name the new mutation described in this island for the first time (Bermejuela rosada), and Negra muelle as a new synonym of Listan negro. Finally, there is also a homonymous name in the case of the term Uval blanco, which is used in this island to name (as a synonym) both the Portuguese variety Malvasia fina and the local variety of this island, Verijadiego.

3.3. El Hierro Grapevine Population Genetic Structure

To carry out the genetic structure study of the accessions on El Hierro island, the Structure 2.3. programme was used carrying out an additional normalisation of the data. From the 46 unique MP-SSRs in Table S2, all individuals showing variations for the same variety were eliminated. Thus, only the most widespread MP-SSR was left, and in case there was none (Airen and Vijierego blanco), the individuals with the highest similarity to the most widespread MP-SSR were computed. In this sense, the el Hierro population consisted of 28 grapevine varieties. In order to know the best grapevine variety distribution in different populations (K), it was proposed that they be grouped into up to seven ancestral populations. Figure S1 shows the best distribution found after applying the correction of

Evanno et al. [37]. It can be seen that the best distribution corresponded to K = 3, i.e., it is proposed to group the 28 varieties into 3 ancestral populations. Figure 3a shows the 28 varieties from El Hierro distributed in 3 groupings. Applying the GenAlEx 6.5 assignment test for K = 3, it was seen that this distribution presented a goodness of assignment of 89%. In addition, it can be seen how each population groups their components into pure and admixed according to the q value. q is a measure of an individual's membership of a population based on its genetic similarity (percentage of its inferred genome belonging to the group [44]) and is ordered from highest to lowest (Table S5). This strategy allows a further normalisation of the data by allowing the detection and elimination of varieties with q-values < 85% (admixed), which would distort the final result of the study.

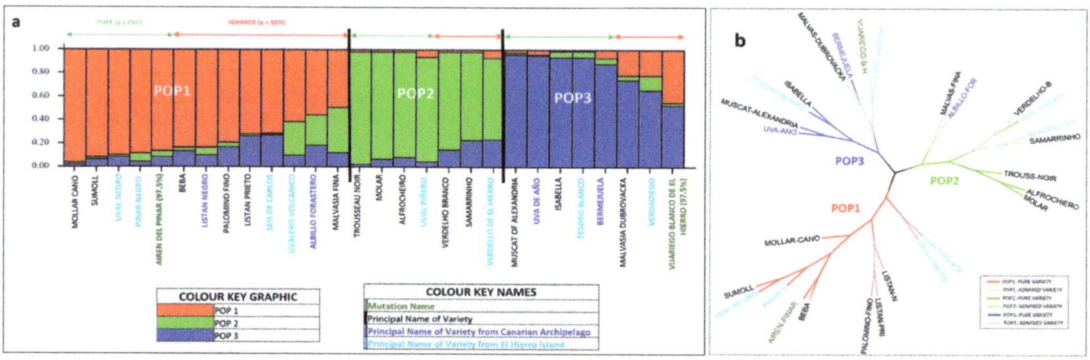

Figure 3. El Hierro grapevine varieties population (unique molecular profiles). (**a**) Structure 3.2 diagram: K = 3 distribution for pure and admixed individuals. (**b**) Phylogenetic tree of this distribution.

Thirteen varieties were grouped in POP1. Five were pure (q $\geq$ 85%; two of them from El Hierro) and the remaining were admixed. It can also be observed that 46% were Spanish varieties (seven varieties) and 8% were Portuguese (one variety). Additionally, 15% of the varieties were Canarian (two varieties) and 31% were exclusively from El Hierro island (four varieties). POP2 is a Portuguese group formed from seven varieties, four pure (57%) and three admixed (43%), of which two are Herreñas (one pure and one admixed). Finally, POP3 is shown with eight representatives with different origins, five pure (63%) and three admixed (37%). Figure 3b shows the phylogenetic tree for this population of 28 varieties. It clearly shows the three populations described and equidistant on three main branches. In each branch, the dichotomy between varieties considered pure and admixed can also be observed, with the exceptions of Uval pinero (pure variety of POP2) and Bermejuela (pure variety of POP3) which are grouped with the admixed varieties of their population.

The optimal representation of grapevine variety distribution from El Hierro by means of principal coordinates analysis (PCoA) required the elimination of the 14 admixed individuals from the population recommended by the Structure 2.3. program. Thus, the population of El Hierro was reduced to 14 pure representatives: (a) POP1 included the Spanish Mollar cano, Sumoll and a mutation of Airen, and Uval negro and Pinar negro from El Hierro; (b) POP2 included the French Trousseau noir, the Portuguese Molar and Alfrocheiro, and the Herreña Uval piñero; (c) POP3 featured the Greek variety Muscat of Alexandria, the Canary Islands varieties Uva de año and Bermejuela, the American DPH Isabella, the Malvasia Dubrovacka (Malvasia aromatica), and the variety Tesoro blanco from El Hierro. Figure 4 shows the two-dimensional (Figure 4a) and three-dimensional (Figure 4b) PCoA representations of this population.

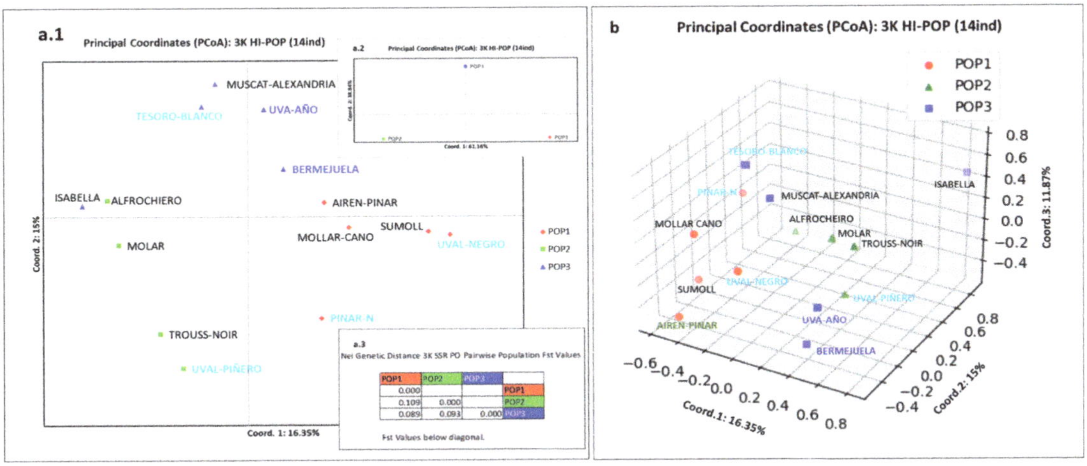

Figure 4. PCoA representations of the grapevine varieties population from El Hierro island normalised for K = 3. The names in navy blue correspond to Canarian varieties, the names in light blue correspond to varieties from el Hierro, and the names in black correspond to varieties from outside the archipelago. (**a.1**) Two-dimensional representation of the three populations by individuals, (**a.2**) two-dimensional representation of the three populations from el Hierro by population, and (**a.3**) values of the Fst statistic for each population. (**b**) Three-dimensional representation of the three populations by individuals.

Figure 4a.1 shows the population of pure varieties from El Hierro. It can be seen how coordinate 1 (with a goodness of fit of 16.35%) separates the Spanish varieties from the non-Spanish varieties, while coordinate 2 (with a goodness of fit of 15%) divides the varieties with an Eastern Mediterranean influence from those with a more Central European or Hispanic influence. In this way, POP1 is practically located in the lower right quadrant. Similarly, POP2 is located in the lower left quadrant, and POP3 occupies the central area of the upper quadrants. Other aspects to highlight are the position of the variety Tesoro blanco, which is very separate from the rest of the varieties on the island of El Hierro. The position of the DPH Isabella is also notable and very distant from the rest of the POP3 members. This arrangement of populations is also reproduced in Figure 4a.2, (representation of the populations with all their individuals grouped together). In addition, Figure 4a.3 shows the result of the analysis of molecular variance (AMOVA). Thus, the most distant populations are POP1 with respect to POP2, followed by POP2 with respect to POP3, and the closest are POP1 with respect to POP3. Finally, it should be noted that with one more dimension, DPH Isabella is distant from the rest, and the large dispersion of POP3 and the marked position of the varieties from El Hierro, which are widely scattered among them, are evident (Figure 4b).

3.4. El Hierro Grapevine Population Genetic Structure Respect to the World Population

The aim of this section is to see the extent of the uniqueness of the population of varieties on the island of El Hierro as a whole. To this end, in addition to the six new varieties described in this work (Uval piñero, Uvalero volcánico, Pinar negro, Seis de Carlos, Tesoro blanco, Uval negro), another five varieties described in previous works by different researchers will be added to the population of El Hierro. These are the varieties Burra volcanica [16], Huevo de gallo [16,18], Verdello de El Hierro [17,18], Verijadiego [16–18] and Verijadiego negro [16]. Therefore, from now on, the Herreña population will be formed by 11 varieties, which will be compared with a world population from the TECNENOL database, formed by 297 varieties (unique MP-SSRs of *Vitis vinifera* ssp vinifera) from 22 countries and analysed with the same 20 SSRs [6,16,41,42].

The procedure followed has been the same as in the previous section but involves trying to distribute a population of 308 unique MP-SSRs in up to 9 ancestral populations using Structure 2.3. Figure S2 shows the best distribution for the population under study which, in this case, corresponds to the value of K = 2. Figures S3 and S4 show the 308 varieties distributed in the 2 ancestral populations as a function of q. In this case, 181 MP-SSR (59%) were grouped in POP1, with a predominance of Italian (54 individuals out of a total population of 72 Italian (54/72)), French (44/49), and, to a lesser extent, Spanish (28/105), Portuguese (16/22), and Greek (11/14) varieties. This grouping had 161 pure components (89%), of which 51 were Italian, 39 were French, 24 were Spanish, 12 were Portuguese, and 11 were Greek, as well as 20 mestizos (11%), of which 3 were Italian, 5 were French, 4 were Spanish, and 4 were Portuguese. The remaining nationalities were represented mainly by Balkan, Eastern Mediterranean, and Central European countries. The remaining 127 components (41%) were placed in POP2, with a predominance of Spanish (77/105), Italian (18/72), Portuguese (6/22), and Greek (3/14) individuals. Of all the components of this group, 97 varieties were pure (76%), with 66 Spanish, 10 Italian, 2 Portuguese, and 3 Greek individuals, as well as 30 admixed varieties (24%) with 9 Spanish, 8 Italian, and 4 Portuguese varieties. In POP2, all the Canary Islands varieties were grouped together and, therefore, so too were the grapevine varieties from El Hierro, so that 11 varieties from Lanzarote, 11 varieties from El Hierro, and 7 from the rest of the Canary Islands were pure; the remaining two, the Albillo criollo variety from the island of La Palma and the Malvasia alistanada fina from Lanzarote, were found to be located next to the admixed individuals. It should be noted that, while most of the El Hierro varieties are very close together in the circular dendrogram (Figure S4b), there are two, the Uval piñero variety and the Tesoro blanco variety, which are far away from the main group. The Uval piñero variety is found in the first of the three main branches originating this circular dendrogram, while the Tesoro blanco variety is located in the first of the two branches originating POP1.

The identification of the 50 admixed individuals allowed us to perform the final data standardisation, and these were eliminated from the study. Thus, the population was left with 258 varieties. Applying the consequent assignment test, the goodness of fit of each variety in the two proposed populations (K = 2) was 100%. Figure 5a shows the Structure 2.3. diagram with pure and admixed individuals, and Figure 5b shows the circular dendrogram of the pure individuals. If the circular dendrograms are compared, with (Figure S4b) or without (Figure 5b), the overall result is practically the same. However, at a grapevine variety location level, small changes can be perceived due to the relocation of the pure individuals once the admixed individuals have been eliminated.

The two-dimensional and three-dimensional PCoA representation is presented in Figure 6. It should be noted that in order to analyse the uniqueness of the El Hierro and Canary Islands grapevine populations, these were extracted from POP2, preserving all their pure components (17 varieties for the Canary Islands (IC) and 11 varieties for the island of El Hierro (HI)).

In Figure 6(a.1,a.2) the clear separation between POP1 and POP2 with IC and HI is shown. In fact, it is coordinate 1 (with a goodness of fit of 5.69% for individuals, and 64.82% for populations), which practically separates them. The effect of coordinate 2 is not visible in Figure 6a.1 (with a goodness of fit of 3.70%), but it is visible in the population plot with a goodness of fit of 19.91% (Figure 6a.2), leaving the El Hierro population alone in the lower left quadrant. Figure 6a.3 confirms that the HI population is the most distant from the rest, followed by the IC population. The slight overlap of POP2, IC (located in the innermost zone) and HI (located in the outermost zone) shown in Figure 6a.1 is not reflected in Figure 6a.2, where these three populations occupying the left quadrants are perfectly separated (see also Figure 6a.3). Another fact to take into account is the distant position, not only with respect to HI but also to IC and POP2, of the variety Tesoro blanco (unknown number 6 [18]), which is practically located in POP1. Figure 6b shows the three-dimensional representation of PCoA for the population under study. This image also shows a separation between POP1 (practically located in the lower and innermost part) and POP2 (practically located in the upper front part). On the other hand, for the populations IC and HI, which are in the front part of the graphical representation, IC is in the lower part and HI in the upper and more external part. The HI variety Tesoro blanco is hidden by individuals of POP1 and POP2 (lower centre-right). However, Figure 7 (corresponding to another angle of the same three-dimensional representation above) shows the distant positions of both the Tesoro blanco and the Uval piñero varieties.

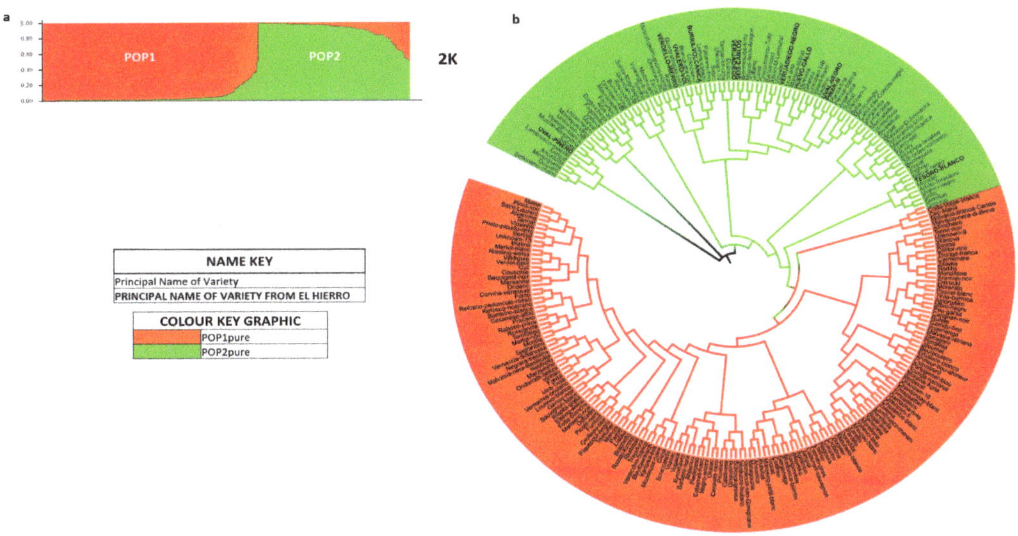

Figure 5. World population (258 individuals) distributed in two populations. (**a**) Graphical representation of K = 2 according to Structure 2.3. (with pure and admixed individuals). (**b**) Circular neighbour-joining dendrogram of the world population 258 pure individuals, highlighting the location from El Hierro (in capital letters and bold).

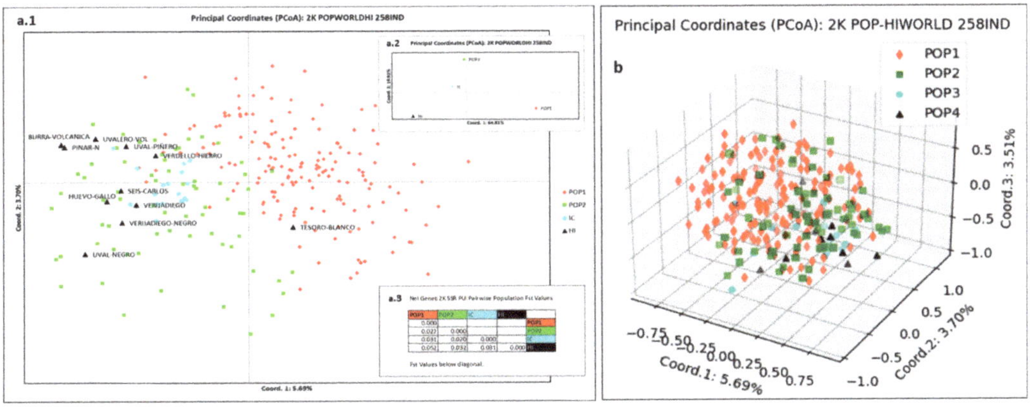

Figure 6. PCoA representation of the grapevine variety population from El Hierro, Canary Islands, and the world normalised to K = 2. (**a.1**) Two-dimensional representation of the four populations by individuals, (**a.2**) two-dimensional representation of the four populations by population, and (**a.3**) values of the Fst statistic for each population. (**b**) Three-dimensional representation of the four populations by individuals (in this representation, POP3 corresponds to IC, and POP4 corresponds to HI).

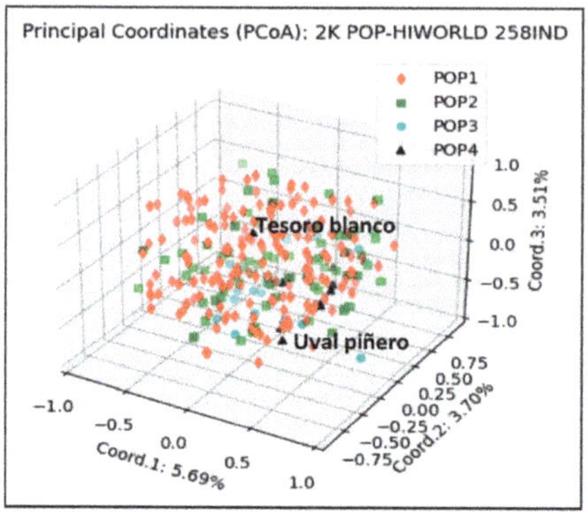

Figure 7. Three-dimensional representation of the four populations by individuals (in this representation, POP3 corresponds to IC and POP4 corresponds to HI). Visualisation under a different angle from the one presented in Figure 6b. The two El Hierro grapevine varieties with positions further away from the rest are highlighted.

In order to confirm the uniqueness observed for the population of varieties on El Hierro island, a study was carried out in which a geographical strategy was introduced as the main criterion. The aim was to group the varieties to create populations according to the country of origin registered in VIVC [43]. It was found that there were countries with a very low number of components, so the strategy of creating populations was extended to geographical areas [6,16,41,42,44,45]. Specifically, seven populations were created: the EASTMED-CAU population (Algeria, Cyprus, Georgia, Israel, Lebanon, Tunisia, and Turkey), the BP population (Bosnia and Herzegovina, Bulgaria, Croatia, Greece, Serbia, Slovenia, and Montenegro), the ITA population (Italy), the FRA-CEU population (Austria, France, Germany, Hungary, and Switzerland), the IP population (Spain and Portugal), and the IC and HI populations. Figure S6a shows the neighbour-joining circular dendrogram of the 308 individuals of the world population grouped in populations corresponding to the 7 defined geographical areas. Once the 7 populations containing the 308 varieties were grouped, an assignment test was performed. A 60% goodness of fit was found. In Figure S6a,b, it can be seen how three main branches are formed. The first one is where the IC and HI varieties are placed, together with a few IP varieties. In the second one, almost all the IP varieties are placed, and the third branch holds the rest of the world populations. It can also be observed how the first branch is subdivided and, in the third subdivision, gives rise to a differentiation into two sub-branches, one containing exclusively IC varieties, and the other containing all the HI varieties, some IC admixed, and some peninsular varieties (pure (all unknown assigned to IP) and admixed (Mollar cano and Molinera)). In Figure S6a, it can also be seen how in El Hierro population, there are two admixed varieties (Tesoro blanco and Seis de Carlos) and the admixed varieties in the IC population are Vijariego blanco de la granja, Diego chinija, Sabro, Burra chinija, and Breval negro. Another aspect to take into account is the location of the three admixed varieties of IC in the third large branch together with other PI admixed types. These are Albillo forastero, Albillo del monte Lentiscal, and Bienmesabe tinto. Finally, another fact to note is regarding the variety from La Palma Island, Albillo criollo (pure), that is displaced to the FRA-CE population, and is located with a group of admixed grapes from this same cluster. The composition of each of these seven groups can also be seen in detail in Figure S5.

The low goodness of fit of the assignment test forced the misplaced varieties (which were mostly the result of admixture) to be discarded so that the world population was reduced to 184 varieties. A further assignment test was carried out to check the goodness of fit of the new distribution. With these new conditions, the goodness of fit reached 92%. Figure 8a,b shows a much sharper and more precise distribution than the previous one, where the first of the three large branches corresponded to the EASTMED-CAU and BP populations, the second branch contained the ITA and FRA-CE populations, and the last branch contained the IP, IC, and HI populations.

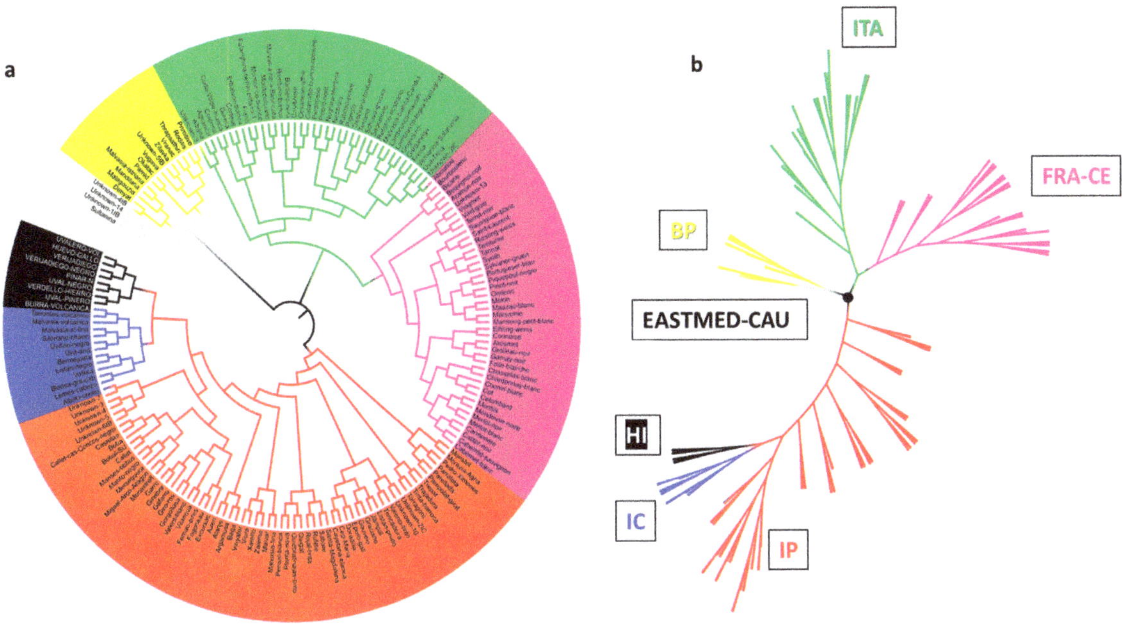

Figure 8. World population (184 individuals) distributed in populations corresponding to seven geographical areas. (**a**) Circular neighbour-joining dendrogram of the 184 pure individuals of the world population, highlighting the location of the varieties from El Hierro. (**b**) Phylogenetic tree of the distribution of these seven populations with all their individuals.

The PCoA representation of these populations is shown in Figure 9. Figure 9a.1 shows how IC and HI both overlap with IP, but there is differentiation between the first two. Moreover, it is coordinate 1 (with a goodness of fit of 6.19%) that separates these populations of Spanish origin from the ITA and FRA-CE populations. In this image, the EASTMED-CAU and BP populations are blurred in the centre, occupying all the quadrants. The discrimination made by coordinate 2 (with a goodness of fit of 4.07%) in this figure is not clear. The population representation itself (Figure 9a.2), is much more decisive. In this way, coordinate 1 again differentiates (with a goodness of fit of 46.36%) the Spanish populations from the rest, and coordinate 2 (with a goodness of fit of 17.6%) separates the EASTMED-CAU, IP, and HI populations from the rest. Thus, only IC is located in the upper right quadrant, IP and HI are in the lower right quadrant are IP and HI, and in the upper left quadrant, very close to coordinate 2, are ITA and FRA-CE, and BP is a little further away. Finally, EASTMED-CAU is in the lower left quadrant and very far from the axis. These positions are confirmed by the values of the Fst statistic shown in Figure 9a.3. In Figure 9b, there is a slight differentiation between the IC and HI populations with respect to IP, as the third dimension raises most of the IC and HI individuals above the position of most of the IP varieties. The behaviour of the rest of the populations is similar to that shown in the two-dimensional plot.

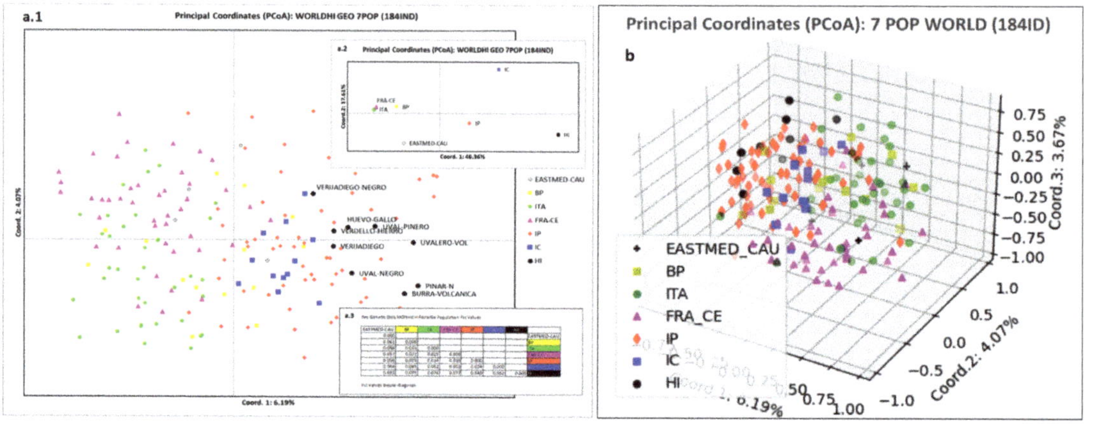

Figure 9. World grapevine varieties population PCoA representation (184 individuals) according to the geographical criterion. (**a.1**) Two-dimensional representation of the seven populations per individuals; (**a.2**) two-dimensional representation of the seven populations per population; (**a.3**) values of the Fst statistic for each population. (**b**) Three-dimensional representation of the seven populations per individuals.

4. Discussion

4.1. SSR Polymorphism

The analysis of grapevine varieties from El Hierro island produced interesting results; however, it is first necessary to assess the goodness of the 20 SSRs that have been used. These microsatellites have been used in all of TECNENOL's work [6,16,41,42] and, in this way, this research group has been creating its own database, which allows for the exhaustive and precise comparison of new MP-SSRs. So far, this SSR kit has proven to be efficient and effective. Table S4 shows a summary of the results of the main statistical parameters obtained for this study. It should be taken into account that the comparison with other studies may be very approximate, as the number of SSRs used, the number of samples analysed, and the closeness of the population samples to be analysed may substantially vary the final results of the studies. [46,47]. The total Na obtained from El Hierro population was 185 alleles. These results are slightly lower than those obtained in the study on the island of Lanzarote [42], because more than twice as many samples (223 vs. 87) were analysed in Lanzarote for the same SSRs. In the case of the Balearic and Canary Archipelagos prospection, approximately the same number of samples with the same SSRs were analysed [6,16], but significantly higher values were also found with respect to HI. Obviously, this was because it was a group of islands with different and more numerous local varieties than a single island. The several studies carried out on the mainland [48–50], always give higher results due to the greater diversity of their samples. On the other hand, the mean Na value was not significantly different from studies on varieties from Turkey [48] and Croatia [49], but lower than studies on the Balearic and Canary Islands [6,16]. The mean genetic diversity index, or He, was 0.737, within the range of most of the studies consulted and slightly higher than the one found in Lanzarote, as its population was more uniform [42]. Fourteen SSRs were found with an F of less than 0.01, meaning that they have a small excess of heterozygosity, reaffirming the consistency of the homozygous individuals. The accumulative PI was also within the expected range (9.4×10^{-23}) for such a study, indicating that this SSR kit was able to guarantee that two individuals with the same MP-SSR at all *loci* were the same individual (except for the "sport" loci).

The best SSR for this population were VVS2 (Na: 13; Ne: 6.23; He: 0.840; F: −0.165; PI: 4.5×10^{-2}), VVMD5 (Na: 12; Ne: 8.42; He: 0.881; F: −0.034; PI: $2{,}6 \times 10^{-2}$) and y el

VVZAG79 (Na: 12; Ne: 6.17; He: 0.838; F: −0.087; PI: 4.3 × 10^{-2}). The least informative were VVS3 (Na: 3; Ne: 2.00; He: 0.500; F: −0.154; PI: 3.6 × 10^{-1}), VVS29 (Na: 5; Ne: 1.17; He: 0.148; F: −0.054; PI: 7.3 × 10^{-1}), and VVMD6 (Na: 5; Ne:3.59; He: 0.721; F: 0.076; PI: 1.2 × 10^{-1}). At this point, it can be concluded that this SSR kit continues to be suitable for characterising and identifying samples from El Hierro island.

4.2. Grapevine Varieties Analysis

Regarding grapevine population analysis, sample uniformity has to be emphasised. In the first data standardisation, 47% of the individuals were discarded because they were identical to other samples of the population. The 46 individuals from El Hierro with unique MP-SSRs were in turn reduced to 28 varieties (28 individuals). This fact indicated that 43% of the population of unique MP-SSRs (20 individuals, as there are 2 mutated individuals that are taken as representatives of the variety, as this collection does not have the generic MP-SSR) were mutations and, therefore, variability within a vine variety (intra-varietal variability) was detected. When working with 20 SSRs, 40 alleles are being compared. If a variation in a MP-SSR is detected in one allele, it is considered a mutated individual with a similarity of 97.5% with respect to the most widespread MP-SSR; when the variations reach two alleles, it is considered a mutated individual with a similarity of 95%; when the variation is three alleles, the individual has a similarity of 92.5%, and the similarity will be 90% when the MP-SSR varies in four alleles or 87.5% when the mutation reaches five alleles. From six variations onwards, it is already considered a new variety. This arbitrary delimitation is based on the works of Ibañez et al. [51] (SSR (2 alelos/26) 92%)), Vélez [52] (SSR (2 alelos/18) 89%)) y Cabezas et al. [53] (SNP, 90%). In Tables S1 and S2, all this intra-varietal variability is detailed, which not only reaches the typical numerical variation with respect to one of the two alleles of an SSR, but in this population, two cases of tri-allelism have also been described. These are the accession entered under the name Mierda de gallina (whose prime name (PN) in the VIVC is Isabella) and the sample entry under the name Diego de El Hierro (PN: Vijariego blanco). Since the beginning of the century, several authors have described the appearance of a third (or even a fourth) allele in a given SSR, indicating cases of hybridisation or chimerism [54,55]. For the 20 cases of individuals showing variations in their MP-SSR, 2 match a molecular profile already described in the Lanzarote prospection [42]. For this reason, they already have a name proposed to be included in VIVC in the corresponding publication; however, for the remaining 18 individuals, 18 particular names are proposed to name them and to be included in the world database. This proposal is made because VIVC provides the names of Pinot meunier for a mutation of Pinot noir, Chasselas cioutat for a mutation of Chasselas blanc, and Bastardo blanco for a colour mutation (sport) of Trousseau noir. In addition, 11 accessions have been identified and entered with the name "unknown". Fifteen errors have been detected (very possibly due to the vine grower's lack of knowledge), six new varieties have been characterised and identified for the first time (Uval piñero, Uvalero volcánico, Pinar negro, Seis de Carlos, Tesoro blanco, and Uval negro), and a new colour mutation for the Bermejuela variety (Bermejuela rosada) that already possessed a sport, Bermejuela tinta, has been identified [18]. It is also proposed to incorporate five new synonymous names (Bermajuelo, Bermajuelo rosada, Mulata rosada, Bermajuelo mulato de El Llano y Negramuelle), plus the term Baboso blanco as a new synonym of the Portuguese grapevine variety Samarrinho. Finally, the detection of a homonym for the Uval blanco grapevine variety, which would be one of the new synonyms used on El Hierro to name the varieties Verijadiego and Malvasia fina, is proposed to incorporate for both varieties.

4.3. El Hierro Grapevine Population Genetic Structure

Representatives of 28 varieties have been found on El Hierro island, of which 8 correspond to local grapevine varieties (Tabla S2). Of these eight varieties, six have MP-SSR described for the first time, while the remaining two have been described before [17,18]. El Hierro grapevine varieties have been distributed among three clusters (Figures 4 and 5 and Table S5), with four pure and four admixed placed as follows: four in POP1 (two pure vs. two admixed), two in POP2 (one to one), and two in POP3 (one to one). In POP1, mainly Spanish and Canary Islands grapevine varieties were grouped together, with the exception of the Portuguese admixed, Malvasia fina, a cross between the Heben variety (very widespread in the northern half of the Iberian Peninsula) and the Portuguese Alfrocheiro (located in POP2). POP2 (green colour) is a Portuguese group, with the exception of the French Trousseau noir (very widespread in Portugal and known as Bastardo negro). All the known components of POP2 have their pedigrees described, and in all of them the variety of Central European origin (unknown) Savaning blanc appears as one of its progenitors. It would also form part of the second generation of progenitors of the admixed varieties and one from POP1, namely Albillo forastero (Palomino fino × Verdelho branco) and Malvasia fina (Heben × Alfrocheiro), respectively, with a large green area in its genome, as can be seen in Figure 3. Finally, POP3 groups together other varieties, such as Muscat of Alexandria or Malvasia Dubrovacka, which means that the group will have a strong influence from the Eastern Mediterranean area. In addition, an American DPH is observed (Isabella variety) a cross between a *Vitis labrusca* and the Meslier petit vinifera (Heunisch weiss × Savaning blanc). The Savaning blanc variety, so common in Portuguese varieties, seems to have entered the Iberian Peninsula via the "Camino de Santiago". [56], leaving a remarkable progeny in addition to the Central European trace in the peninsular genetic profiles. In this study, the presence of this variety represents the genesis of a grouping (POP2).

Figure 4 shows how the Isabella variety moves away from the POP3 group and from the *vinifera* in general, due to the influence of *Vitis labrusca*, but remains close to POP2, due to the presence of the Savaning blanc variety in the second generation of parents. In short, this sample of varieties from El Hierro represents the history of the introduction of the domesticated vine on this island from the Spanish colonisation to the successive incorporations of Portuguese settlers from the archipelagos of the Azores and Madeira [17,57]. Thus, the most genuinely Spanish-influenced varieties are Uval negro and Pinar negro, the admixed Seis de Carlos and Uvalero volcanico, as they show a marked influence from the Eastern Mediterranean and Central Europe (via Portugal), respectively. El Hierro grapevine varieties present in POP2 and, therefore, with a strong Portuguese (Central European) influence are the Uval piñero (pure) and the admixed Verdello de El Hierro with a strong Eastern Mediterranean influence. The pure POP3, Tesoro blanco, and the admixed Verijadiego (with a strong peninsular and marked Central European influence) show in their MP-SSR traces of the influence of the Eastern Mediterranean on the peninsular varieties, which later, as can be seen, was transferred to the Canary Islands and the island of El Hierro.

4.4. El Hierro Grapevine Population Genetic Structure with Respect to the World Population

The objective of this section is to show the uniqueness of El Hierro population, compared to the world population. As has already been mentioned in other articles the introduction of vines in the Canary Islands dates back to the 15th century, so the uninterrupted evolution of the vine in this new island ecosystem is older than 500 years old [57]. Thus, the goal is to check whether adaptation to the island's soil and climatic conditions and both natural and anthropogenic selection have left an identifying mark on these varieties in their MP-SSRs. The first strategy adopted was to group the varieties by genetic proximity. The 11 varieties from El Hierro (new: Uval piñero, Uvalero volcánico, Pinar negro, Seis de Carlos, Tesoro blanco, Uval negro; already published: Burra volcánica, Huevo de gallo, Verdello de El Hierro, Verijadiego and Verijadiego negro) were compared with the remaining 297 individuals from the TECNENOL database from 22 countries. After Structure 2.3. allowed us to detect and eliminate the admixed varieties ($q < 85\%$) of the 2 ancestral groups formed, the 258 pure individuals of the world population were represented by a neighbour-joining circular dendrogram (Figure 6b). It can be seen how all the population from el Hierro was located in the fourth arm of POP2, and it can also be seen how the varieties Tesoro blanco and Uval piñero were located at a distance from the rest of the group. In this figure, the IC, HI, and POP2 populations were in the same grouping and highlighted in green, which was logical since the IC and HI populations had been extracted from POP2 (a grouping made up mainly of Spanish varieties). In the individual representation by PCoA (Figure 6a.1 and Figure 7b) a slight differentiation between CI and HI is shown, while when the representation is visualised by populations (Figure 6a.2) a clear differentiation between POP2, CI, and HI is seen, with the latter occupying a quadrant by itself. This result is validated in Figure 6a.3, where results are presented for the Fst statistic, which takes into account the identity of alleles in the infinite allele model (IAM) when populations are compared pairwise (AMOVA). This is a method used to estimate the differentiation between populations from molecular data. Therefore, it can be concluded that the HI population has its own uniqueness and that there is, above all, a special distinction between the Tesoro blanco variety and the rest of the grapevine varieties from El Hierro.

Another aspect to consider is the low percentage of the goodness of fit of the graphical representations by individuals using PCoA [4,58]. This is because by using 20 SSRs and having 2 alleles each, we are using 40 numerical data to represent an individual on a graph. It would then have to be possible to represent a graph with 40 dimensions, and this is not possible. The reduction to two dimensions, or at best to three, is what reduces the reliability of the coordinates of the graph. It then becomes necessary to talk about trends.

In order to support the thesis on the uniqueness of El Hierro grapevine varieties, another strategy was developed. The next step was to look at the behaviour of the population studied under a geographical component. The 308 varieties were, thus, grouped into 7 groups corresponding to 7 different geographical areas, according to the country of origin in the VIVC database. The corresponding assignment test was carried out, and a goodness of fit of 60% was achieved. Figures S5 and S6 show the distribution of the resulting populations and which varieties were well assigned and which were not (admixed). Once the varieties misassigned by admixture were removed, the world population was reduced to 184 individuals with a goodness of assignment of 92%. It should be noted that under these conditions the variety Tesoro blanco was not well assigned and was consequently removed, which reinforces the idea that it is a markedly admixed variety, while the variety Uval piñero was retained within the group of varieties from El Hierro. Figure 8, both in the circular diagram and in the phylogenetic tree, confirms the uniqueness of both the Island of El Hierro and the Canary Archipelago, occupying a branch by themselves that divides, defining the two populations. This behaviour is also observed in PCoA representations, whether performed by individuals (Figure 9a.1) or by populations (Figure 9a.2), either in two (Figure 9a.1) or three dimensions (Figure 9b). In all cases, the trend of the El Hierro grapevine population is to differentiate itself not only from the Canarian archipelago population, but also from the other populations. Once again, Fst (Figure 9a.3) confirmed the uniqueness of the population of varieties from El Hierro, showing the highest values of the matrix compared to the rest of the populations. The notable exception will be that it is closer to the IP population than to the IC population, which indicates its greater influence from the Iberian Peninsula compared to the Canary Islands.

5. Conclusions

After carrying out the prospection, obtention, and analysis of data, the main conclusions obtained are as follows. On the one hand, the SSR kit that TECNENOL has been using proved to be suitable to continue with studies of this nature, and also allowed the presentation of six new varieties: Uval piñero, Uvalero volcánico, Pinar negro, Seis de Carlos, Tesoro blanco, Uval negro, and a new rose "sport" of Bermejuela (W). Two other interesting aspects are the fact that all the entries of individuals with the name Pedro Ximenez corresponded to the variety Albillo forastero and that all the accessions registered with the name Baboso blanco turned out to be the minority Portuguese variety Samarrinho. Fifteen errors were detected in total. Eleven varieties were identified that were unknown to the vine growers, and twenty individuals with variations (mutations) were found, of which two had already been described in a previous prospection in Lanzarote Island.

It is proposed to incorporate the 7 new names of the new identified varieties and the "sport" into the VIVC database, in addition to the 18 names of the detected mutations (Airen del pinar, Forastera de la isla Redonda, Baboso negro de Frontera, Bermajuelo del Echedo, Bermajuelo del puerto, Bermajuelo rosado del tesoro, Mierda de gallina, Listan negro del tesoro, Listan prieto chijo, Listan prieto herreño, Malvasia fina gabetera, Molar tintilla, Molar herreño, Verijadiego blanco de Frontrea, Vijariego blanco de El Hierro, Eusebia, and Diego de El Hierro y Diego de Frontera), 5 new synonyms (Bermajuelo, Bermajuelo rosado de El Llano, Bermajuelo rosado, and Mulata rosada y Negra muelle), the new synonym for Samarrinho (Baboso blanco), and, additionally, the term Uval blanco for the grapevine variety Malvasia fina y Verijadiego, which has turned out to be the only case of homonymy.

The El Hierro grapevine population has a marked influence from the islands in its MP-SSR, and basically has three major sources of influence: the Iberian Peninsula through the Spanish varieties, Central Europe through the Portuguese varieties (Savaning blanc) and the east part of the Mediterranean Sea (Muscat of Alexandria and Malvasia Dubrovacka). Finally, it is necessary to highlight the case of the Tesoro blanco variety. This variety is significantly different from the rest of the El Hierro population, presenting a marked influence from the Eastern Mediterranean. For all of the above, El Hierro appears to have a unique population, not only worldwide, but even among the rest of the varieties of the Canary Islands archipelago. These results once again support the proposal that the Canary Islands and the El Hierro island are one of the few centres of biodiversity on our planet.

Supplementary Materials: The following supporting information can be downloaded at: https://www.mdpi.com/article/10.3390/horticulturae9121297/s1, Table S1: Original and conclusive information on 87 accessions from El Hierro. Similarity to the nearest genome (TECNENOL database); Table S2: List of the 46 unique profiles belonging to the population of the island of El Hierro. Values of the 7 international SSRs; Table S3: SSR groups by annealing Temperature (Ta); Table S4: Characterization of the twenty microsatellite markers used in this study; Figure S1: The 4 steps of the graphical method of Evanno et al. (2005) allow the estimation of the true number of ancestral K groups for a population of 28 varieties from the El Hierro collection; Table S5: Genetic structure of the El Hierro population. Distribution K = 3 (individuals belonging to each group or population); Figure S2: The 4 steps of the graphical method of Evanno et al. (2005), allow the estimation of the true number of ancestral K groups for a population of 308 varieties from the TECNENOL database; Figure S3: Genetic structure of the world population. Distribution K = 2 (Individuals belonging to each group or population). Detail of the proportion of pure and admixed individuals as a function of of q value. Nationalities that make up each group. El Hierro grapevine varieties are highlighted in red; Figure S4: World population (308 individuals) distributed in 2 populations. (a) Graphical representation of K = 2 according to the Structure 2.3. program. (b) Neighbour-joining circular dendrogram of the 308 individuals of the world population, highlighting the pure and admixed individuals, as well as the location of the El Hierro grapevine varieties (indicator arrows); Figure S5: Genetic structure of the world population. Distribution in 7 geographical areas. Detail of the proportion of well-assigned (pure) and misassigned (admixed) individuals. Nationalities comprising each of the groups: EASTMED-CAU (Algeria, Cyprus, Georgia, Israel, Lebanon, Tunisia, and Turkey), BP (Bosnia and Herzegovina, Bulgaria, Croatia, Greece, Serbia, Slovenia, and Montenegro), ITA (Italy), FRA-CEU (Austria, France, Germany, Hungary, and Switzerland), IP (Spain and Portugal), IC (Canary Archipelago), and HI (El Hierro Island); Figure S6: World population (308 individuals) distributed in populations corresponding to 7 geographic areas. (a) Circular Neighbour-joining dendrogram of the 308 individuals of the world population, differentiating the well-placed individuals (pure) from the admixed (poorly placed), and highlighting the location of El Hierro grapevine varieties (indicator arrow); (b) phylogenetic tree of the distribution of these 7 populations with all their individuals.

Author Contributions: Conceptualization, F.F. and Q.L.-Y. methodology, F.F., Q.L.-Y. and L.R.S.-A.; software, P.S.-G., Q.L.-Y. and L.R.S.-A.; validation, F.F. and Q.L.-Y. formal analysis, F.F., Q.L.-Y. and L.R.S.-A.; investigation, F.F., Q.L.-Y. and L.R.S.-A.; resources, F.F.; data curation, F.F. and Q.L.-Y. writing—original draft preparation, F.F. and P.S.-G.; writing—review and editing, F.F. and P.S.-G.;

visualization, J.M.C. and F.Z.; supervision, F.F., Q.L.-Y. and L.R.S.-A.; P.S.-G., J.M.C. and F.Z. All authors have read and agreed to the published version of the manuscript.

Funding: This project has been funded by the Cabildo Insular de la Isla de El Hierro, through the Regulatory Council of the Denominación de Origen El Hierro.

Data Availability Statement: Data not available because it is confidential.

Acknowledgments: The authors are grateful to Javier Ibañez, and the viticulture specialists Alejandro Déniz, Alfredo Hernández and Andrés Acosta, for their valuable support. We would also like to thank Gemma Marsal, Isabel Araque, Rosa Pastor, Braulio Esteve-Zarzoso, Laia Fañanás, and Santiago Moreno for their support in the laboratory.

Conflicts of Interest: The authors declare no conflict of interest.

References

1. This, P.; Lacombe, T.; Thomas, M.R. Historical origins and genetic diversity of wine grapes. *Trends Genet.* **2006**, *22*, 511–519. [CrossRef]
2. Marsal, G. Caracterización e Identificación de 449 Accesiones de Vitis vinifera L. Procedentes de dos Colecciones Ampelográficas. Ph.D. Thesis, Universidad Rovira i Virgili, Tarragona, Spain, 2015.
3. Wan, Y.; Schwaninger, H.; Baldo, A.M.; Labate, J.A.; Zhong, G.Y.; Simon, C.J. A phylogenetic analysis of the grape genus (*Vitis* L.) reveals broad reticulation and concurrent diversification during neogene and quaternary climate change. *BMC Evol. Biol.* **2013**, *13*, 141. [CrossRef]
4. Dong, Y.; Duan, S.; Xia, Q.; Liang, Z.; Dong, X.; Margaryan, K.; Musayev, M.; Goryslavets, S.; Zdunić, G.; Bert, P.-F.; et al. Dual domestications and origin of traits in grapevine evolution. *Science* **2023**, *379*, 892–901. [CrossRef]
5. Wolkovich, E.M.; García de Cortázar-Atauri, I.; Morales-Castilla, I.; Nicholas, K.A.; Lacombe, T. From Pinot to Xinomavro in the world's future wine-growing regions. *Nat. Clim. Chang.* **2018**, *8*, 29–37. [CrossRef]
6. Marsal, G.; Bota, J.; Martorell, A.; Canals, J.M.; Zamora, F.; Fort, F. Local cultivars of *Vitis vinifera* L. in Spanish Islands: Balearic Archipelago. *Sci. Hortic.* **2017**, *226*, 122–132. [CrossRef]
7. Spence, C. Multisensory experiential wine marketing. *Food Qual. Prefer.* **2019**, *71*, 106–116. [CrossRef]
8. Sancho-Galán, P.; Amores-Arrocha, A.; Palacios, V.; Jiménez-Cantizano, A. Identification and Characterization of White Grape Varieties Autochthonous of a Warm Climate Region (Andalusia, Spain). *Agronomy* **2020**, *10*, 205. [CrossRef]
9. GEVIC (Gran Enciclopedia Virtual Islas Canarias). 2007. Available online: https://www.gevic.net/info/contenidos/mostrar_contenidos.php?idcat=22&idcap=91&idcon=526 (accessed on 7 July 2023).
10. Wikimedia Commons. 2014. Available online: https://commons.wikimedia.org/w/index.php?curid=33566121#file (accessed on 7 July 2023).
11. NASA. 2011. Available online: https://www.flickr.com/photos/gsfc/6630087415/in/photostream/ (accessed on 7 July 2023).
12. Macías, A.M. Colonización y viticultura. El caso de las Canarias, 1350–1550. *Douro Estud. Doc.* **2002**, *VII*, 285–296. Available online: https://ojs.letras.up.pt/index.php/dou/article/view/12586 (accessed on 7 July 2023).
13. Hidalgo, J.; Hidalgo, L. La Filoxera. In *Tratado de Viticultura*, 2nd ed.; Mundi-Prensa: Madrid, Spain, 2019; Volume 1.
14. NASA World Wind. 2006. Available online: https://es.m.wikipedia.org/wiki/Canarias#/media/Archivo%253ASanta_Cruz_de_Tenerife_SPOT_1320.jpg (accessed on 7 July 2023).
15. DOP Vinos de El Hierro. Available online: http://doelhierro.es/ (accessed on 7 July 2023).
16. Marsal, G.; Mendez, J.J.; Mateo-Sanz, J.M.; Ferrer, S.; Canals, J.M.; Zamora, F.; Fort, F. Molecular characterization of *Vitis vinifera* L. local cultivars from volcanic areas (the Canary Islands and Madeira) using SSR markers. *Oeno One* **2019**, *4*, 667–680. [CrossRef]
17. Zerolo, J.; Cabello, F.; Espino, A.; Borrego, J.; Ibañez, J.; Rodriguez-Torres, I.; Muñoz-Organero, G.; Rubio, C.; Hernández, M. *Variedades de Vid de Cultivo Tradicional en Canarias*; Instituto Canario de Calidad Agroalimentaria. Gobierno de Canarias: Santa Cruz de Tenerife, Spain, 2006.
18. Rodriguez-Torres, I. *Variedades de vid Cultivadas en Canarias. Descriptores Morfológicos. Caracterización Morfológica, Molecular, Agronómica y Enológica*; Instituto Canario de Investigaciones Agrarias. Gobierno de Canarias: Santa Cruz de Tenerife, Spain, 2018.
19. Marsal, G.; Baiges, I.; Canals, J.M.; Zamora, F.; Fort, F. A fast, efficient method for extracting DNA from leaves, stems, and seeds of *Vitis vinifera* L. *Am. J. Enol. Vitic.* **2011**, *62*, 376–381. [CrossRef]
20. Marsal, G.; Boronat, N.; Canals, J.M.; Zamora, F.; Fort, F. Comparison of the efficiency of some of the most usual DNA extraction methods for woody plants in different tissues of *Vitis vinifera* L. *J. Int. Sci. Vigne Vin* **2013**, *47*, 227–237. [CrossRef]
21. Fort, F.; Hayoun, L.; Valls, J.; Canals, J.M.; Arola, L.; Zamora, F. A new and simple method for rapid extraction and isolation of high-quality RNA from grape (*Vitis vinifera*) berries. *J. Sci. Food Agric.* **2008**, *88*, 179–184. [CrossRef]
22. Thomas, M.R.; Scott, N.S. Microsatellite repeats in grapevine reveal DNA polymorphisms when analyzed as sequence-tagged sites (STSs). *Theor. Appl. Genet.* **1993**, *86*, 985–990. [CrossRef] [PubMed]
23. Bowers, J.E.; Dangl, G.S.; Vignani, R.; Meredith, C.P. Isolation and characterization of new polymorphic simple sequence repeat loci in grape (*Vitis vinifera* L.). *Genome* **1996**, *39*, 628–633. [CrossRef] [PubMed]

24. Bowers, J.E.; Dangl, G.S.; Meredith, C.P. Development and characterization of additional microsatellite DNA markers for grape. *Am. J. Enol. Vitic.* **1999**, *50*, 243–246. [CrossRef]
25. Sefc, K.M.; Regner, F.; Turetschek, E.; Glössl, J.; Steinkellner, H. Identification of microsatellite sequences in *Vitis riparia* and their applicability for genotyping of different *Vitis* species. *Genome* **1999**, *42*, 367–373. [CrossRef] [PubMed]
26. Scott, K.D.; Eggler, P.; Seaton, G.; Rosseto, M.; Abblet, E.M.; Lee, L.S.; Henry, R.J. Analysis of SSRs derived from grape ESTs. *Theor. Appl. Genet.* **2000**, *100*, 723–726. [CrossRef]
27. Lefort, F.; Kyvelos, C.; Zervou, M.; Edwards, K.; Roubelakis-Angelakis, K. Characterization of new microsatellite loci from *Vitis vinifera* and their conservation in some *Vitis* species and hybrids. *Mol. Ecol. Resour.* **2002**, *2*, 20–21. [CrossRef]
28. Cipriani, G.; Spadotto, A.; Jurman, I.; Di Gaspero, G.; Crespan, M.; Meneghetti, S.; Frare, E.; Vignani, R.; Cresti, M.; Morgante, M.; et al. The SSR-based molecular profile of 1005 grapevine (*Vitis vinifera* L.) accessions uncovers new synonymy and parentages and reveals a large admixture amongst varieties of different geographic origins. *Theor. Appl. Genet.* **2010**, *121*, 1569–1585. [CrossRef]
29. Dalbó, M.A.; Ye, G.N.; Weeden, N.F.; Steinkellner, H.; Sefc, K.M.; Reisch, B.I. A gene-controlling sex in grapevines is placed on a molecular marker-based genetic map. *Genome* **2000**, *43*, 333–340. [CrossRef]
30. Maul, E.; Röckel, R. *Vitis* International Variety Catalogue (VIVC): A cultivar database referenced by genetic profiles and morphology. *BIO Web Conf.* **2015**, *5*, 01009. [CrossRef]
31. This, P.; Jung, A.; Boccacci, P.; Borrego, J.; Botta, R.; Costantini, L.; Crespan, M.; Dangl, G.S.; Eisenheld, C.; Ferreira-Monteiro, F.; et al. Development of a standard set of microsatellite reference alleles for the identification of grape cultivars. *Theor. Appl. Genet.* **2004**, *109*, 1448–1458. [CrossRef] [PubMed]
32. Paetkau, D.; Calvert, W.; Stirling, I.; Strobeck, C. Microsatellite analysis of population structure in Canadian polar bears. *Mol. Ecol.* **1995**, *4*, 347–354. [CrossRef] [PubMed]
33. Peakall, R.; Smouse, P.E. GenAlEx 6.5: Genetic analysis in Excel. Population genetic software for teaching and research-an update. *Bioinformatics* **2012**, *28*, 2537–2539. [CrossRef] [PubMed]
34. Paetkau, D.; Slade, R.; Burden, M.; Estoup, A. Genetic assignment methods for the direct, real-time estimation of migration rate: A simulation-based exploration of accuracy and power. *Mol. Ecol.* **2004**, *13*, 55–65. [CrossRef] [PubMed]
35. Pritchard, J.K.; Stephens, M.; Donnelly, P. Inference of population structure using multilocus genotype data. *Genetics* **2000**, *155*, 945–959. [CrossRef] [PubMed]
36. Falush, D.; Stephens, M.; Pritchard, J.K. Inference of population structure using multilocus genotype data: Linked loci and correlated allele frequencies. *Genetics* **2003**, *164*, 1567–1587. [CrossRef] [PubMed]
37. Evanno, G.; Regnaut, S.; Goudet, J. Detecting the number of clusters of individuals using the software structure: A simulation study. *Mol. Ecol.* **2005**, *14*, 2611–2620. [CrossRef]
38. Saitou, N.; Nei, M. The neighbor-joining method: A new method for reconstructing phylogenetic threes. *Mol. Biol. Evol.* **1987**, *4*, 406–425. [CrossRef]
39. Kumar, S.; Stecher, G.; Tamura, K. MEGA7: Molecular Evolutionary Genetics Analysis version 7.0 for bigger datasets. *Mol. Biol. Evol.* **2016**, *33*, 1870–1874. [CrossRef]
40. Three-Dimensional Plotting in Matplotlib (Python Data Science Handbook). Available online: https://jakevdp.github.io/PythonDataScienceHandbook/04.12-three-dimensional-plotting.html (accessed on 8 August 2023).
41. Marsal, G.; Mateo, J.M.; Canals, J.M.; Zamora, F.; Fort, F. SSR analysis of 338 accessions planted in Penedes (Spain) reveals 28 unreported molecular profiles of *Vitis vinifera* L. *Am. J. Enol. Vitic.* **2016**, *67*, 466–470. [CrossRef]
42. Fort, F.; Marsal, G.; Mateo-Sanz, J.M.; Pena, V.; Canals, J.M.; Zamora, F. Molecular characterisation of the current cultivars of *Vitis vinifera* L. in Lanzarote (Canary Islands, Spain) reveals nine individuals which correspond to eight new varieties and two new sports. *Oeno One* **2022**, *56*, 281–295. [CrossRef]
43. Maul, E.; Röckel, F. Vitis International Variety Catalogue. 2015. Available online: http://www.vivc.de (accessed on 7 July 2023).
44. Bacilieri, R.; Lacombe, T.; Cunff, L.L.; Di Vecchi-Staraz, M.; Laucou, V.; Genna, B.; Perós, J.P.; This, P.; Boursiquot, J.M. Genetic structure in cultivated grapevines is linked to geography and human selection. *BMC Plant Biol.* **2013**, *13*, 25. [CrossRef] [PubMed]
45. Arroyo-García, R.; Ruiz-García, L.; Bolling, L.; Ocete, R.; López, M.A.; Arnold, C.; Ergul, A.; Söylemezo˝lu, G.; Uzun, H.I.; Cabello, F.; et al. Multiple origins of cultivated grapevine (*Vitis vinifera* L. ssp. *sativa) based on chloroplast DNA polymorphisms. Mol. Ecol.* **2006**, *15*, 3707–3714. [CrossRef] [PubMed]
46. Aliquo, G.; Torres, R.; Lacombe, T.; Boursiquot, J.M.; Laucou, V.; Gualpa, J.; Fanzone, M.; Sari, S.; Pérez-Peña, J.; Prieto, J.A. Identity and parentage of some South American grapevine cultivars present in Argentina. *Aust. J. Grape Wine Res.* **2017**, *23*, 452–460. [CrossRef]
47. Moita, A.; Santos, R.; Catarina, A. Unraveling the origin of *Vitis vinifera* L. Verdelho. *Aust. J. Grape Wine Res.* **2018**, *24*, 450–460. [CrossRef]
48. Arslan, N.; Yılmaz Baydu, F.; Hazrati, N.; Yüksel Özmen, C.; Ergönül, O.; Uysal, T.; Yaşasın, A.S.; Özer, C.; Boz, Y.; Kuleyin, Y.S.; et al. Genetic Diversity and Population Structure Analysis of Anatolian Kara Grapevine (*Vitis vinifera* L.) Germplasm Using Simple Sequence Repeats. *Horticulturae* **2023**, *9*, 743. [CrossRef]
49. Žulj Mihaljević, M.; Maletić, E.; Preiner, D.; Zdunić, G.; Bubola, M.; Zyprian, E.; Pejić, I. Genetic Diversity, Population Structure, and Parentage Analysis of Croatian Grapevine Germplasm. *Genes* **2020**, *11*, 737. [CrossRef]
50. Jiménez-Cantizano, A.; Puig-Pujol, A.; Arroyo-García, R. Identification of *Vitis vinifera* L. Local Cultivars Recovered in Andalusia (Spain) by Using Microsatellite Markers. *Horticulturae* **2023**, *9*, 316. [CrossRef]

51. Ibañez, J.; De Andrés, M.T.; Molino, A.; Borrego, J. Genetic study of key Spanish grapevine varieties using microsatellite analysis. *Am. J. Enol. Vitic.* **2003**, *54*, 22–30. [CrossRef]
52. Vélez, M.D.; Ibáñez, J. Evaluation of the uniformity and stability of Microsatellite markers in grapevine. *Acta Hortic.* **2009**, *827*, 163–168. [CrossRef]
53. Cabezas, A.; Ibañez, J.; Lijavetzky, D.; Vélez, D.; Bravo, G.; Rodríguez, V.; Carreño, I.; Jermakow, A.M.; Carreño, J.; Ruiz-García, L.; et al. A 48 SNP set for grapevine cultivar identification. *BMC Plant Biol.* **2011**, *11*, 153. [CrossRef] [PubMed]
54. Riaz, S.; Garrison, K.E.; Dangl, G.S.; Boursiquot, J.M.; Meredith, C.P. Genetic divergence and chimerism within ancient asexually propagated wine grape cultivars. *J. Am. Soc. Hort. Sci.* **2002**, *127*, 508–514. [CrossRef]
55. Grigoriou, A.; Tsaniklidis, G.; Hagidimitriou, M.; Nikoloudakis, N. The Cypriot Indigenous Grapevine Germplasm Is a Multi-Clonal Varietal Mixture. *Plants* **2020**, *9*, 1034. [CrossRef]
56. Casanova, J.; Mozas, P.; Ortiz, J.M. Ampelography and microsatellite DNA analysis of autochthonous and endangered grapevine cultivars in the province of Huesca (Spain). *Span. J. Agric. Res.* **2011**, *9*, 790–800. [CrossRef]
57. Macías, A. El paisaje vitícola de Canarias. Cinco siglos de historia. *Ería* **2005**, *68*, 351–364.
58. Emanuelli, F.; Lorenzi, S.; Grzeskowiak, L.; Catalano, V.; Stefanini, M.; Troggio, M.; Myles, S.; Martínez-Zapater, J.M.; Zyprian, E.; Moreira, F.M.; et al. Genetic diversity and population structure assessed by SSR and SNP markers in large germplasm collection of grape. *BMC Plant Biol.* **2013**, *13*, 39. [CrossRef]

horticulturae

MDPI

Article

Exploring Diversity among Grapevines Varieties (*Vitis vinifera* L.) in Ibiza and Formentera (Balearic Islands, Spain) Using Microsatellite Markers, Ampelographic Methods and an Ethnobotanical Approach

Raquel González [1,*], Alba María Vargas [2], Teresa Garnatje [3,4,*], Joan Vallès [1,5] and María Teresa de Andrés [2]

1 Laboratori de Botànica (UB), Unitat Associada al CSIC, Facultat de Farmàcia i Ciències de l'Alimentació—Institut de Recerca de la Biodiversitat (IRBio), Universitat de Barcelona, Av. Joan XXIII 27-31, 08028 Barcelona, Catalonia, Spain; joanvalles@ub.edu
2 Instituto Madrileño de Investigación y Desarrollo Rural, Agrario y Alimentario, Finca El Encín, Autovía A2, km 38.200, 28805 Alcalá de Henares, Madrid, Spain; alba.vargas@madrid.org (A.M.V.); maite.deandres@madrid.org (M.T.d.A.)
3 Institut Botànic de Barcelona (IBB), CSIC-CMCNB, Passeig del Migdia s/n, 08038 Barcelona, Catalonia, Spain
4 Jardí Botànic Marimurtra-Fundació Carl Faust, Passeig Carl Faust, 9, 17300 Blanes, Catalonia, Spain
5 Institut d'Estudis Catalans (IEC), Carrer del Carme, 47, 08001 Barcelona, Catalonia, Spain
* Correspondence: rgonzalezdunford@educaib.eu (R.G.); tgarnatje@ibb.csic.es (T.G.)

Abstract: *Vitis vinifera* L. has been present in Ibiza and Formentera, two islands of the Balearic Islands (Spain), since the 7th century BC. In the past few years, there have been several studies and investigations on the Balearic Islands. These have focused mainly on Mallorca and Menorca with a small representation of Ibiza and none that take into account Formentera. This research aims to contribute to the knowledge of *Vitis* cultivars cultivated on those islands and to investigate whether there are local cultivars still being grown. To do this, using an ethnobotanical approach, 15 persons were interviewed to gather information about local grapevines, and 36 accessions from 12 plots were characterized using ampelographic descriptors and identified using SSR markers. Relationships of the accessions studied with other cultivars were also assessed. The results show 21 different genotypes profiles, where six were new genotypes: 'Colló de gall', 'Grec', 'Maçanet', 'VIEIV015-Maçanet', 'Morzacà', and 'Vermelleta'. Ten new synonyms and three homonyms have been proposed. Additionally, we suggest three new relationships for the 'Hebén' cultivar, one new relationship for the 'Llora' cultivar and one new relationship for the 'Beba' cultivar. These results show the first reported information for Ibiza and Formentera on *Vitis*.

Keywords: ethnobotany; grapevine; local cultivar; somatic variant; SSR markers; parentage

Citation: González, R.; Vargas, A.M.; Garnatje, T.; Vallès, J.; de Andrés, M.T. Exploring Diversity among Grapevines Varieties (*Vitis vinifera* L.) in Ibiza and Formentera (Balearic Islands, Spain) Using Microsatellite Markers, Ampelographic Methods and an Ethnobotanical Approach. *Horticulturae* **2023**, *9*, 1307. https://doi.org/10.3390/horticulturae9121307

Academic Editors: Antonio Amores-Arrocha and Ana Jiménez-Cantizano

Received: 31 October 2023
Revised: 27 November 2023
Accepted: 29 November 2023
Published: 5 December 2023

1. Introduction

Ibiza (also known by its Catalan name, Eivissa) and Formentera are the two smallest inhabited islands of the Balearic Islands (Spain), which are together known as the Pityusic Islands. These islands have a strong tradition of growing grapevines and winemaking. It is thought that *Vitis vinifera* L. (*Vitaceae*) was brought to Ibiza and Formentera by the Punics in the 7th century BC, who even created structures purposely to cultivate grapevines [1]. During the Middle Ages, other authors, such as Al-Zuhri or Al-Himyari, pointed out the production of sultanas in Ibiza and that the trade with wine continued [2].

In 1851, grape production in the Balearic Islands was affected by powdery mildew (*Uncinula necator*), which substantially reduced production [3]. Then, in 1862, when phylloxera (*Daktulosphaira vitifoliae*) expanded in France, Ibiza and Formentera's production increased to supply the French market [4]. The presence of this disease was detected in Mallorca in 1891 and later in Menorca in 1892 [5]. Although, at that time, phylloxera had not been detected in Ibiza and Formentera, present day evidence indicates symptoms in various areas of the islands [6].

Today, grapevines continue to be some of the most important plants when considered in socio-economic terms. Spain is the global leader in terms of the area dedicated to this crop cultivation, with 964 kha [7] in the Balearic Islands [8], whose surface is 2336 ha. From this total, Ibiza Island cultivates 58.6 ha and Formentera cultivates 14.3 ha [4]. Despite their relatively small surface, each of these islands has its own wine appellation. Cultivars allowed in these appellations are mainly well-known cultivars and include only a small number of minority cultivars, such as 'Monestrell' or 'Malvasia' [9,10]. This could be one of the main reasons that in Ibiza, 'Monestrell' and 'Garrut' together represent 76.45% of the registered vineyard surface, while in Formentera, 57.5% of the surface comprises 'Monestrell', 'Garrut', and 'Garnatxa tinta' [11].

Grapevines are not only restricted to vineyards for the winemaking industry; they are also a common Mediterranean crop that is present in every farmhouse in the Pityusic Islands as well as other parts of the Balearic Islands [12,13].

Organic agriculture is promoting a revival of the recovery and identification of local germplasm to fight against diseases and climate change in grapevine cultivation [14,15]. In the last 20 years, there has been an interest in the Balearic Islands to study, protect and reintroduce to the market local minor grapevine cultivars in order to differentiate wines produced in this region [16–21]. Most of the investigations undertaken have been in the islands of Mallorca and Menorca, usually on vineyards belonging to wineries. In those works, samples are studied from a molecular point of view in order to assess the enological properties [22–25]. Two other studies have analyzed the local cultivars growing in the Balearic Islands in order to establish their origin and phylogeny [26,27].

Although SSR markers (simple sequence repeats or microsatellites) have been used largely to identify the diversity present in vineyards and to discard any homonyms and synonyms [26–29], in the works concerning the Balearic Islands, Ibiza and Formentera are scarcely represented. Additionally, none of the studies that consider Ibiza samples conducted ampelographic descriptions or ethnobotanical interviews. The only exception is Grup d'Acció Local per al Desenvolupament Rural i Pesquer d'Eivissa i Formentera, who prospected and conducted an ampelographic study between 2012 and 2014 in these territories [13]. That study constitutes the starting point for this research.

Although numerous characterizations, through several molecular markers, have been carried out [30–32], ampelography and other morphological methods [33,34] still remain the first steps when identifying cultivars, and these are useful when distinguishing somaclonal variants or mislabels [35,36].

Despite the fact that classic ampelographic identification is usually difficult, because ampelographic traits can be influenced by cultivar age [37], viruses [38], crop management [39] or even environmental characteristics [35,40], an ampelographic description is needed to inscribe any cultivar in the register of commercial cultivars or the variety catalogue [41,42].

Ethnobotanical interviews have rarely been used in grapevine studies to register information related to the samples considered. Through information given by farmers during the investigation of local cultivars, it is often found that these cultivars have not been registered [43]. These interviews are able to gather important information describing the cultivars, how to grow them and the different particularities that have been transmitted over the generations.

Although there have been some studies in recent years [13,28] that have studied grape samples from the Pityusic Islands, no research work that specifically focuses on grapevine landraces in Ibiza or in Formentera and that takes into account microsatellite identification, ampelographic study and ethnobotanical interviews altogether has been carried out.

Given the above-described current state of the art of *Vitis* research in both of the considered islands, the main goals of this study were to (1) identify vine diversity growing in the Pityusic Islands using microsatellite markers, ampelography, ethnobotanical interviews and bibliography, (2) characterize the cultivars and landraces found in these islands from genetic and ampelographic points of view, and (3) analyze relationships of the accessions studied with other cultivars and their possible origin.

2. Materials and Methods

2.1. Plant Material and Origin

Between 2012 and 2014, a project was conducted to prospect local grapevine germplasm in the Pityusic Islands on behalf of Grup d'Acció Local d'Eivissa i Formentera [13]. In this work, five farmers had been interviewed, and six plots were prospected. Between 2017 and 2023, adopting an ethnobotanical approach, another 10 persons were interviewed, and another four plots were prospected and added to the study. The plots included in this study involve individuals that allowed us to conduct ampelography and genetic analysis on the cultivars sown, and they also shared valuable knowledge about their vineyard. In total, 12 plots were included in this study, 10 in Ibiza and 2 in the Formentera Islands. The Ibiza samples were mostly located in the Sant Josep de sa Talaia municipality (Figure 1). Most of the plots were orchards in which *Vitis* was one of the products, and as the cultivated extension was less than 1000 m^2, those did not have to be compulsorily registered to an official body [44].

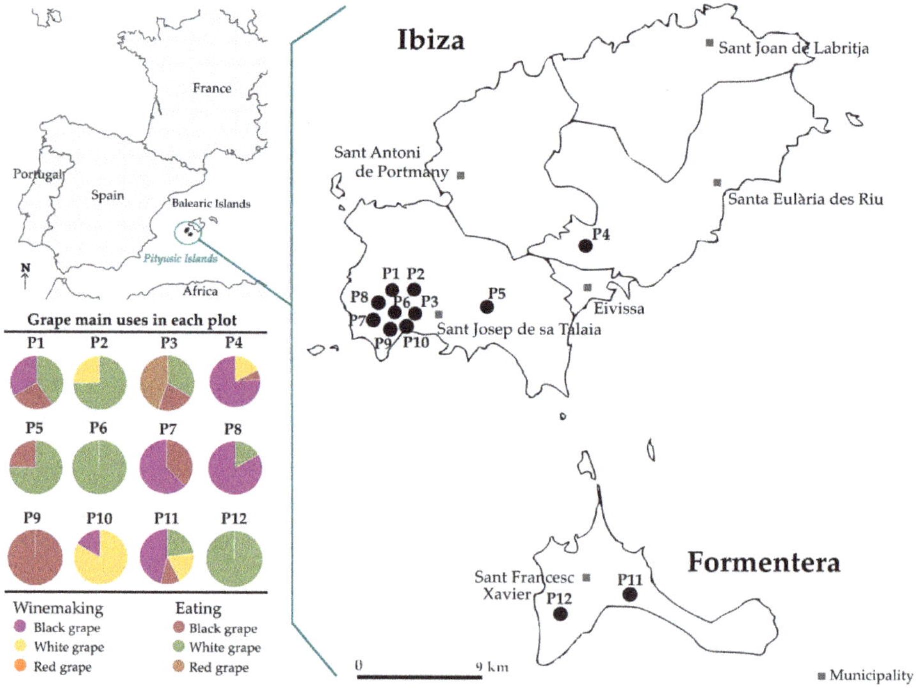

Figure 1. Plot localization and main grape uses in each plot studied in Ibiza and Formentera Islands.

Vineyards were all over 10,000 m^2, and grapes were sold to elaborate the Appellation of Control Origen wines of the islands; in these cases, local cultivars constituted less than 10% of all the grapevines. The plots' characteristics are detailed in Table S1 and summarized in Table 1.

Table 1. Plots and number of plants studied in each plot.

Plot Number	Municipalities, Island	No. of Plants Studied	Rootstock Used	Soil Type	Plant Management	Plant Age	Irrigation	Plot Type
P1	Sant Josep de sa Talaia, Ibiza	15	None	Silt soil	Double cordon	>40 years	No	Plot
P2	Sant Josep de sa Talaia, Ibiza	8	None	Silt soil	Double cordon	>40 years	No	Plot
P3	Sant Josep de sa Talaia, Ibiza	9	American rootstock	Silt soil	Double cordon	>40 years	No	Plot
P4	Santa Eulària des Riu, Ibiza	17	American rootstock	Clay soil	Goblet	>40 years	No	Vineyard
P5	Sant Josep de sa Talaia, Ibiza	4	None	Silt soil	Goblet	<10 years	Yes	Plot
P6	Sant Josep de sa Talaia, Ibiza	4	None	Silt soil	Goblet	>40 years	No	Plot
P7	Sant Josep de sa Talaia, Ibiza	8	None	Silt soil	Goblet	> 40 years	No	Plot
P8	Sant Josep de sa Talaia, Ibiza	6	None	Silt soil	Goblet	>40 years	No	Plot
P9	Sant Josep de sa Talaia, Ibiza	2	None	Silt soil	Goblet	>40 years	No	Vineyard
P10	Sant Josep de sa Talaia, Ibiza	6	None	Silt soil	Goblet	>40 years	No	Plot
P11	Formentera, Formentera	26	None	Silt soil	Double cordon	>40 years	No	Vineyard
P12	Formentera, Formentera	4	None	Silt soil	Goblet	>40 years	No	Vineyard

The accessions selected for this study were those that farmers had identified as local landraces during the prospections, or they had doubts about their origin. The accession named 'Monestrell' was present in all plots, so it was used as a control cultivar to determine the phenology on the other accessions, although samples were not collected on all plots, as it was identified by farmers as the same accession.

The sampling comprised a total of 109 plants grouped into 36 accessions. Those accessions were cultivated in 12 plots, 10 in Ibiza and 2 in Formentera, with different numbers of individuals investigated in each of the *Vitis vinifera* accessions, 27 of which were from Ibiza and 9 of which were from Formentera. Codes VIEIV001 to VIEIV028 corresponded to accessions from Ibiza Island, while code VIEIV004 was removed from this study because it had a doubtful origin and name. Accessions coded VIFOR029 to VIFOR037 were collected from Formentera Island.

The taxon code, local name, number of plants studied, and voucher specimens' number, place and date of collection are shown in Table S2. Voucher specimens for each accession are deposited in the herbarium BC (Botanical Institute of Barcelona).

2.2. Semi-Structured Interviews

A total of 15 semi-structured interviews were performed during the 2012–2014 and 2017–2018 periods. Interviews were conducted with farmers, most of whom were men [45]. Additionally, interviews were held with other farmers following the snowball methodology [46] to gather different information about traditional vines cultivated on the island. In those interviews, the main subjects discussed were cultivars planted or known, plant age, rootstock used, crop management and winemaking processes.

This information is used, together with ampelographic characterization and SSR markers, to identify and clarify synonyms, homonyms and misnaming in the samples here studied.

2.3. Ampelographic Characterization

The morphological characterization has been carried out during two consecutive years, 2017 and 2018, by a single ampelographer, using 35 descriptors from the International

Organization of Vine and Wine (OIV) [42]. For every descriptor, the minimum number of observations established in the OIV protocol [42] has been made. Where this could not be achieved, this has been stated (Table S3). Data, including the registered number for each accession, are available in Table S4.

Metric descriptors such as bunch and grape weight, width and length have been first measured and then converted to category following the descriptors' indications.

The mode of all qualitative descriptors for both years of study has been calculated. If the mode fell between two categories, the most observed one, based on ampelographer experience on field and with those cultivars, has been selected.

2.4. DNA Extraction and Microsatellite Analysis

DNA extractions were carried out from herbarium material using the Qiagen DNeasy 96 Plant Kit (Hilden, Germany) with minor modifications. Twenty-six nuclear microsatellites were simultaneously amplified in two multiplex PCR. The SSR markers included in this study are VVIP60, VVIB01, VVIQ52, VVIH54, VVIN73, VVIP31 [47], VRZAG83, VRZAG79, VRZAG62, VRZAG112 [48], VVMD7, VVMD25, VVMD24 [49,50], VMC1B11 [51] and VVS2 [52]); set B: VRZAG29, VRZAG67 [48], VVMD28, VVMD32, VVMD27, VVMD21, VVMD5 [49,50], VVIV37, VVIV67, VVIN16 and VMC4F3-1 [51]. These 26 microsatellites [53] were chosen because they presented clear profiles that were easy to score, carrying a higher number of alleles, and they are evenly distributed along grapevine chromosomes.

Multiplex PCRs were performed in a final volume of 20 µL containing 1× Multiplex PCR Master Mix (Qiagen, Hilden, Germany) and 5 ng of DNA template. The thermocycler conditions, primers concentration, and primers fluorochrome labeling were those used by Zarouri [54] for Mx01 and Mx02. PCR products were analyzed in an ABI 3130 Genetic Analyzer, and the fragments were sized with GeneMapper 5.0 using GeneScan™-600 LIZ™ Size Standard as the internal marker [54].

2.5. Data Analysis

Standard measures of genetic variation including number of alleles per locus (Na), number of effective alleles per locus (Ne), Shannon information index (I), the observed (*Ho*) and expected (He) heterozygosity, and the probability of identity (PI) were calculated using GenAlEx 6.5 software [55]. The polymorphism information content (*PIC*) was calculated employing Cervus 3.0 software [56]. To determine the genetic uniqueness of each accession, the Excel add-in Microsatellite Toolkit 2001 [57] was used.

The microsatellite genotypes obtained were compared with SSR profiles stored in the Instituto Madrileño de Investigación y Desarrollo Rural, Agrario y Alimentario (IMIDRA, hereafter) database [58] with around 8000 unique SSR profiles. These genotypes include 2061 unique profiles that correspond to accessions maintained in the *Vitis* Germplasm Bank "Finca El Encín" (IMIDRA, Alcalá de Henares, Spain). The rest of the genotypes, not present in this collection, came from 209 cultivars included or pending inclusion in the Grapevine Spanish Catalogue [59], 1541 genotypes from information exchange with other European germplasm banks, and 4182 from other IMIDRA investigation works and/or scientific papers. All have been compared after being standardized following a similar method described in [30].

Parentage analysis was performed based on 26 SSR profiles, using Cervus 3.0 software [56]. The analysis was performed with 4208 accessions of the IMIDRA database [58].

Genetic structure was analyzed using a panel diversity comprising 140 cultivars from five different countries, including the studied samples (Table S7). The selection of these cultivars considered the historical connection of Ibiza and Formentera with other countries. Twenty-six non-linked microsatellite mentioned before were used in both analyses. Only unique SSR profiles were used; cultivars with somatic mutations are not represented.

Firstly, the genetic structure was conducted by a Bayesian analysis using the software Structure 2.3.4 [60]. A model, with a putative number between one and ten populations and correlated allele frequencies [61], was assumed. A Monte Carlo Markov Chain run-length period of 100,000, with 100,000 burn-in steps, and 10 iterations for each number of putative populations were used. The Evanno criterion was used to decide the population number [62]. The membership coefficient threshold defined for individual assignment to a given cluster was Q = 0.80.

Secondly, GeneAlex 6.5 software [55,63] was used to conduct principal coordinate analysis (PCoA), based on standardized covariance of the genetic distances previously calculated for codominant markers in this software.

All statistical analyses were performed with Microsoft Excel [64] and SPSS [65].

3. Results

3.1. Material Prospection, Ampelographic and Molecular Characterization

From this sample group, 285 alleles have been detected for the 26 SSRs studied. The most polymorphic markers for this group of cultivars have been loci VMC4F3-1 and VVMD28, which showed 11 alleles each, whereas VVIQ52, VVIN73, and ZAG29 markers only had three different alleles. The highest informative marker, using the allele effective value for measurement, is the VVMD5 locus with a value of 7.1 that presents 14 different genotypic combinations in this sample group. The 21 different genotypes found in this study can be differentiated using exclusively the genotypes obtained in VMC4F3-1 and VVMD5 loci. A mean value of unbiased expected heterozygosity (He) of 72% and a Shannon's Information Index (I) of 1.5 has been obtained (Table S5).

3.2. Identifications, Synonyms, and Homonyms

One sample from each of the 36 accessions was selected for the microsatellite marker study based on the interviews conducted and considering that all plants in the same accession group had come from the same original plant.

These analyzed samples resulted in 21 different genetic profiles between both islands (Table S6). From the 27 accessions studied in Ibiza Island, there were 21 genotype profiles detected that correspond to 22 cultivars. In Formentera Island, from nine prospected accessions, eight belong to different genotype profiles. Redundant genotypes were due to the existence of somatic mutations such as 'Beba' cultivar (local name 'Palop') that had white and pink berry skin color for accessions VIEIV001, VIEIV007 (local name 'Palop blanc') and VIEIV026 (local name 'Palop vermell'), and on the other hand to misnaming errors and homonyms. Results obtained from comparing genetic profiles with the ones in IMIDRA's database are included in Table 2.

Table 2. Synonyms, homonyms, and misnaming.

Genotype Code	Accession Code	Island	Local Name	Prime Name *	Observation	VIVC Proposed Name
GEN_0074	VIEIV001	Ibiza	Palop blanc	Beba	New synonym	Palop blanc
GEN_0105	VIEIV002	Ibiza	Primerenc	Valencí tinto	New synonym	Primerenc
GEN_0204	VIEIV003	Ibiza	Moscatell	Moscatel de Alejandría	Described synonym	-
MEXT_0504	VIEIV005	Ibiza	Monestrell	Llora	New synonym	Monestrell
GEN_0208	VIEIV006	Ibiza	Mamella de vaca	Ahmeur bou Ahmeur	New synonym/ new homonym	Mamella de vaca
GEN_0074	VIEIV007	Ibiza	Palop blanc	Beba	New synonym	Palop blanc
MEXT_3954	**VIEIV008**	Ibiza	Morzacà	-	New found genotype	Morzacà
MEXT_3940	**VIEIV009**	Ibiza	Grec	-	Homonym/New found genotype	Grec
MEXT_3955	**VIEIV010**	Ibiza	Maçanet	-	New found genotype	Maçanet
GEN_0022	VIEIV011	Ibiza	Sultanita	Sultanina	New synonym	Sultanina
GEN_0881	VIEIV012	Ibiza	Fresa	Agawam	Misnaming; *Vitis* interspecific crossing	-
GEN_0003	VIEIV013	Ibiza	Sant Jaume	Santa Magdalena	New synonym	Sant Jaume

Table 2. *Cont.*

Genotype Code	Accession Code	Island	Local Name	Prime Name *	Observation	VIVC Proposed Name
GEN_0575	VIEIV014	Ibiza	Ferrana	Danugue	New synonym	Ferrana
MEXT_3957	**VIEIV015**	Ibiza	Maçanet	-	New found genotype	VIEIV015-Maçanet
GEN_1259	VIEIV016	Ibiza	Giró	Callet negrella	Misnaming	-
GEN_0104	VIEIV017	Ibiza	Fogoneu	Fogoneu		-
MEXT_3958	**VIEIV018**	Ibiza	Vermelleta	-	New found genotype	Vermelleta
MEXT_3940	**VIEIV019**	Ibiza	Grec	-	Homonym/New found genotype	Grec
GEN_0105	VIEIV020	Ibiza	Palop negre	Valencí tinto	New synonym	Palop negre
GEN_0003	VIEIV021	Ibiza	Santa Margalida	Santa Magdalena	New synonym	Santa Margalida
MEXT_0504	VIEIV022	Ibiza	Monestrell de xingló	Llora	New synonym	Monestrell
MEXT_0504	VIEIV023	Ibiza	Monestrell d'Alger	Llora	New synonym	Monestrell
MEXT_3955	**VIEIV024**	Ibiza	Blanqueta	-	Misnaming/New found genotype	Maçanet
MEXT_3959	**VIEIV025**	Ibiza	Colló de gall	-	New found genotype	Colló de gall
GEN_0074	VIEIV026	Ibiza	Palop vermell	Beba roja	Somatic mutation	-
GEN_0575	VIEIV027	Ibiza	Ferrana	Danugue	New synonym	Ferrana
MEXT_3955	**VIEIV028**	Ibiza	Maçanet	-	New found genotype	Maçanet
MEXT_3955	**VIFOR001**	Formentera	Moscatel	-	Misnaming	Maçanet
GEN_0391	VIFOR002	Formentera	Batista	Mansès de Tibbús	Misnaming	-
GEN_0018	VIFOR003	Formentera	Garnatxa blanca	Garnacha tinta	Somatic mutation	-
GEN_0062	VIFOR004	Formentera	Fogoneu	Tinto velasco	Misnaming	-
MEXT_0504	VIFOR005	Formentera	Monestrell	Llora	New synonym	Monestrell
GEN_0575	VIFOR006	Formentera	Palop negre	Danugue	Misnaming	-
MEXT_3940	**VIFOR007**	Formentera	Palop blanc	-	Misnaming/New found genotype	Grec
MEXT_3955	**VIFOR008**	Formentera	Mancet	-	Misnaming	Maçanet
GEN_0111	VIFOR009	Formentera	Grec	Quigat	Misnaming	-

Bold: new genotypes found in this study. * Name registered in VIVC catalogue.

Those 21 identified cultivars are mostly cultivars used traditionally in the Spanish winemaking industry; only five cultivars came from other countries ('Ahmeur bou Ahmeur', 'Agawam', 'Danugue', 'Moscatel de Alejandría', and 'Sultanina'), which either have double use or are exclusively used as table grapes.

The most interesting result about the genetic profiles is that six genotypes, corresponding to 12 accessions, could not be identified either using the extensive IMIDRA's database [59] or the *Vitis* International Variety Catalogue (VIVC) [66]. Those accessions correspond to 'Colló de gall' (mext_3959), 'Grec' (mext_3940), 'Maçanet' (mext_3955), 'VIEIV015_Maçanet' (mext_3957), 'Morzacà' (mext_3954) and 'Vermelleta' (mext_3958) (Table 2).

On the other hand, all 'Monestrell' accessions studied in this work (VIEIV005, VIEIV022, VIEIV013, VIFOR005) have been identified as 'Llora' (mext_0504). These accessions have the same genetic profile as an accession prospected in 2012 in Menorca Island by Institut de Recerca i Formació Agrària i Pesquera (IRFAP) of a minority cultivar named 'Llora' [67], which has not been published in VIVC but is currently in the process of being included in the Spanish vine cultivar catalogue [68].

In total, 2 somatic mutants, 18 cultivars and an interspecific cross were verified.

Unidentified genotype cultivars were named with the name given by the farmer during the interviews. If more than one name had been given for different accessions resulting in the same genotype, the most common cited name was chosen [46]. Those names were 'Maçanet', 'Grec', 'Morzacà', 'Colló de gall' and 'Vermelleta'. Within those new genotypes, there are three detected homonyms. One is for accession VIEIV015 that was named 'Maçanet' by the farmer, but the genotype is not the same as the other accessions identified as 'Maçanet'. The other homonym would be the 'Grec' cultivar name, as it is a registered synonym for the 'Alcanon' cultivar registered in VIVC [66] (Table 2). On the other hand, the 'Grec' cultivar found in the Formentera sample has been identified as 'Quigat', which would be considered to be a misnaming for 'Grec' until more research has been made (see Table 2).

Ten accession names have been identified as new synonyms. The four 'Monestrell' accessions, three from Ibiza and one from Formentera Island, have all been identified as the same genotypes for the cultivar 'Llora'. 'Sant Jaume' and 'Santa Margalida' are synonyms for the 'Santa Magdalena' cultivar. 'Primerenc' is a synonym for 'Valencí tinto', 'Palop negre' is a synonym for 'Valencí tinto', 'Palop blanc' is a synonym for 'Beba' and 'Sultanita' is a synonym for 'Sultanina' (Table 2).

Also, two other synonyms have been detected during this study. One is found for accessions VIEIV014 and VIEIV027, which is named 'Ferrana' for 'Danugue'. This would be considered a homonym, as 'Ferrana' is registered in VIVC [66] as a synonym for 'Planta fina', and in this case study, two farmers have cited this cultivar with the same name [45]. The last homonym is the accession VIEIV006, identified as 'Ahmeur bou Ahmeur' and locally named 'Mamella de vaca', which is a name that is registered for another genotype profile in VIVC [66].

Two somatic mutations were detected: one for the 'Beba' cultivar of accessions VIEIV001 and VIEIV007 which are named 'Palop blanc' and accession VIEIV026 named 'Palop vermell'. When contrasted with the morphological characterization (Table S3), it is revealed that VIEIV026 has a red grape color, which could be due to a somatic mutation in the cultivar as García-Muñoz [26] detected in her study, too. The other possible somatic mutation that has been detected in previous works is for the 'Garnacha tinta' cultivar, as the local accession is named 'Garnatxa blanca' [69].

There are eight misnaming cases considered in this study for the following cultivars: 'Batista' for 'Mansès de Tibbús', 'Fogoneu' instead of 'Tinto Velasco', 'Giró' for 'Callet negrella', 'Grec' for 'Quigat', 'Blanqueta', 'Moscatell' and 'Mancet' for 'Maçanet', as 'Mancet' was cited by the farmers as a red grape cultivar, and this one has white grapes [46], and 'Palop blanc' for 'Grec' (Table 2).

3.3. Putative Parentage Relationships

The search for compatible trios (parents and offspring) and duos (parent–offspring) was performed based on 26 SSRs loci. From the analysis of the parentage, we established 4 compatible trios (Table 3) and 71 duos (Table 4). The trio analysis results confirm what García-Muñoz [26] had already established for 'Fogoneu' and 'Callet negrella', syn. 'Mansès de cabdell' (Table 3).

Table 3. Compatible trios in Ibiza and Formentera samples.

Offspring ID	First Candidate ID	Pair LOD Score	Second Candidate ID	Pair LOD Score	Trio Loci Mismatching	Trio LOD Score	Published
Fogoneu	Mansès de cabdell	2.26×10^{15}	Excursac	2.40×10^{15}	0	4.89×10^{15}	[26]
Callet negrella	Beba	2.00×10^{15}	Giró sardo	2.10×10^{15}	0	4.44×10^{15}	[26]
Maçanet	Hebén	2.05×10^{15}	Cigenera	2.33×10^{15}	0	5.34×10^{15}	Not published
Colló de gall	Hebén	2.21×10^{15}	Breval negro	2.19×10^{15}	0	5.10×10^{15}	Not published

From the primary analysis conducted, we have a main group with 'Hebén' being the parent of six cultivars. From this, five relationships had already been reported in Cipriani [70], García-Muñoz [26] and D'Onofrio [71]. Three new first-grade relationships have been found ('Maçanet', 'Morzacà' and 'Colló de gall') and one new second grade relationship with 'Beba', a descendent of 'Hebén' and an unknown cultivar, with another undetermined cultivar obtaining the newfound genotype named 'Grec'. The other five new first-grade relationship have been detected within cultivars: 'Mansès de Tibbús' with 'Mansès de cabdell'; 'Callet negrella' with 'Valencí tinto'; 'Grec' with 'Valencí tinto'; 'VIEIV015-Maçanet' with 'Llora', although it could not be determined which would be the parent and which would be the offspring in this last one; and 'Danugue' with 'Albaranzeuli blanco' (Table 4).

Table 4. Possible parent–offspring relationships for the accessions studied.

Progeny	Parent ID	Consistent Loci	Pair LOD Score	Published
Santa Magdalena	Planta Fina de Pedralba	26/26	8.42×10^{14}	[71]
Beba	Hebén	26/26	5.44×10^{14}	[70]
Valencí tinto	Beba	26/26	1.17×10^{15}	[71]
Valencí tinto	Callet negrella	26/26	9.93×10^{14}	**Not published**
Quigat	Hebén	26/26	5.60×10^{14}	[71]
Moscatel de Alejandría (Moscatel de Málaga/Moscatel de Setúbal/Moscatel Graúdo)	Moscatel de grano menudo (Moscatel Morisco/ Sárga muskotály)	26/26	2.13×10^{15}	[70]
Mansès de Tibbús	Mansès de cabdell	26/26	1.51×10^{15}	**Not published**
Callet negrella	Valencí tinto/Grumiere	26/26	9.93×10^{14}	**Not published**
Grec *	Beba	26/26	6.02×10^{14}	**Not published**
Grec *	Valencí tinto/Grumiere	26/26	6.01×10^{14}	**Not published**
Morzacà *	Hebén	26/26	7.73×10^{14}	**Not published**
Maçanet *	Hebén	26/26	1.06×10^{15}	**Not published**
VIEIV015-Maçanet *	Llora	26/26	1.68×10^{15}	**Not published**
Colló de gall *	Hebén	26/26	9.51×10^{14}	**Not published**
Danugue	Albaranzeuli bianco	22/22	5.42×10^{14}	**Not published**

Bold: new parentage relations found in the study. * New genotypes found.

3.4. Genetic Structure

To analyze the genetic structure, 140 unique genotypes were considered, including the 20 obtained from the Pityusic Islands prospection and a *Vitis* cultivar diversity panel. The 'Agawam' genetic profile (interspecific cross) has not been considered because it could distort results. Also, two different approaches were used, Structure and PCoA, to infer the population structure and geographical assignment of the accessions studied in this work.

From structural analysis, three main grouping levels were identified: K = 2, K = 3 and K = 5 (Figure S1). The membership coefficient threshold defined for individual assignment to a given cluster was Q = 0.80.

K = 2 showed a group with Spanish-origin cultivars with 52.38% of the studied samples grouped showing a possible relationship with Spanish cultivars and another group that included 30% of the cultivars analyzed whose origin was diverse (Maghreb, Italy, France, Greece, etc.). Cultivars assigned to group 2 represented 30% of the cultivars analyzed from Ibiza and Formentera. There were three cultivars that could not be assigned to a certain group as probabilities were under 0.8 (Table S8), which were then considered to be admixed genotypes. Those were 'Santa Magdalena', 'Quigat' and 'VIEIV015-Maçanet' (Figure S2).

Three groups were observed in K = 3, group 1, which had French-related cultivars, (indicated by red-colored bars); group 2 with Maghrebi, Italian and Greek origin (indicated by green-colored bars); and group 3, which had Spanish-related cultivars (indicated by blue-colored bars) (Figure 2).

In this case, as in the previous K = 2, more than half of the samples (55%) from Ibiza and Formentera were grouped with the cultivars that had Spanish origin (Table S9). Spanish cultivars in this analysis have been organized into two groups: cultivated in Balearic Islands, according to García-Muñoz [26], and the rest were from the other important or ancient Spanish cultivars.

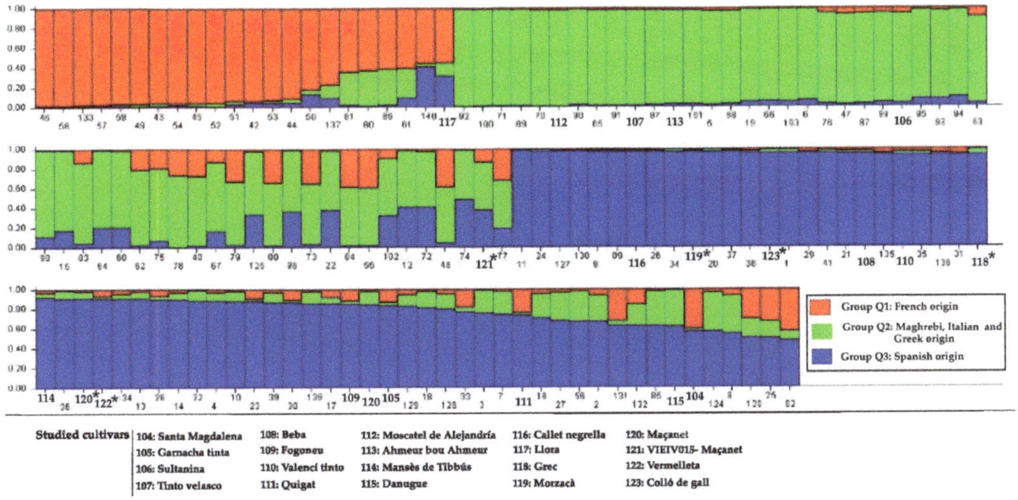

Figure 2. Representation of the 20 genotypes found in Ibiza and Formentera Island using Structure software 2.3.4. version for K = 3. In **bold** are the samples studied, samples with * are new found genotypes [60].

Samples that were grouped into the Spanish-origin group included the following: 'Garnacha tinta' (q = 0.842), 'Fogoneu' (q = 0.857), 'Vermelleta' (q = 0.911), 'Maçanet' (q = 0.917), 'Mansès de Tibbús' (q = 0.923), 'Grec' (q = 0.938), 'Valencí tinto' (q= 0.942), 'Beba' (q = 0.950), 'Colló de gall' (q = 0.960), 'Morzacà' (q = 0.975) and 'Callet negrella' (q = 0.977). This aggregation, except the 'Danugue' cultivar, was the same as that obtained in K = 2.

Overall, 20% of the samples have relations with cultivars that have Maghrebi, Italian and Greek origin. These are 'Ahmeur bou Ahmeur' (q = 0.952), 'Sultanina' (q = 0.906), 'Tinto Velasco' (q = 0.963) and 'Moscatel de Alejandría' (q = 0.970).

Meanwhile, 25% of the cultivars studied that could not be assigned to a certain group.

In K = 5 aggrupation, five groups can be observed, and 55% of the cultivars studied are located in group 3, where Spanish origin cultivars are located. These cultivars are 'Callet negrella' (q = 0.968), 'Morzacà' (q = 0.939), 'Vermelleta' (q = 0.903), 'Beba' (q = 0.893), 'Valencí tinto' (q = 0.869), 'Grec' (q = 0.866), 'Fogoneu' (q = 0.850), 'Colló de gall' (q = 0.837), 'Mansès de Tibbús' (q = 0.837), 'Maçanet' (q = 0.8200) and 'Garnacha tinta' (q = 0.814). In group 2, which has cultivars with Maghrebi origin, 10% of the cultivars are found; those are 'Ahmeur bou Ahmeur' (q = 0.870) and 'Tinto velasco' (q = 0.813). In group 5, there is only the 'Sultanina' cultivar (q = 0.810). The admixed origin accounted for 25% of the studied cultivars (Figure 3) (Table S10).

Genetic analysis by principal coordinates (PCoA, Figure 4) was carried out, and the results obtained are consistent with the results obtained with Structure software 2.3.4 version. In this analysis, prospected cultivars are located in three of the five main origin groups. Group 1 was formed by Spanish origin cultivars, where 'Hebén' is represented at the far end of the group, which means that it is the least mixed cultivar in this group. This matches with the Structure results, where the cultivar has the largest value for Q within its group. In this group, the following cultivars are included: 'Danugue', 'Maçanet', 'Grec', 'Colló de gall', 'Valencí tinto', 'Santa Magdalena', 'Garnacha tinta', 'Quigat', 'Vermelleta', 'Fogoneu', 'Mansès de Tibbús' and 'Callet negrella'. Group 2 is mostly grouping cultivars originated in Maghreb, and 'Ahmeur bou Ahmeur', 'Tinto velasco', 'Sultanina' and 'VIEIV015-Maçanet' appear to be related. The last group, group 3, has cultivars with French origin, and it is where cultivar 'Llora' is: between French, Italian and Greek groups in this analysis.

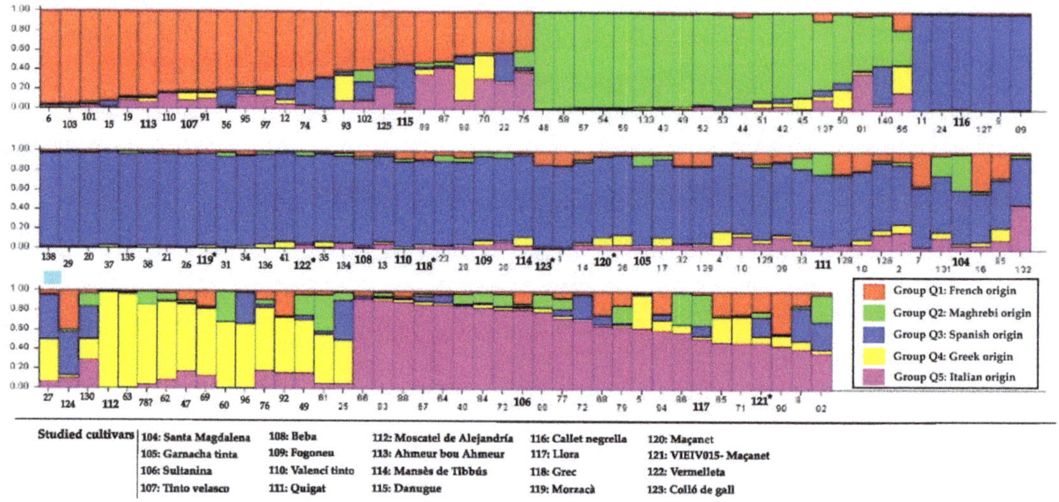

Figure 3. Representation of the 20 genotypes found in Ibiza and Formentera Island using Structure Software 2.3.4 version for K = 5. In **bold** are the samples studied, samples with * are new found genotypes [60].

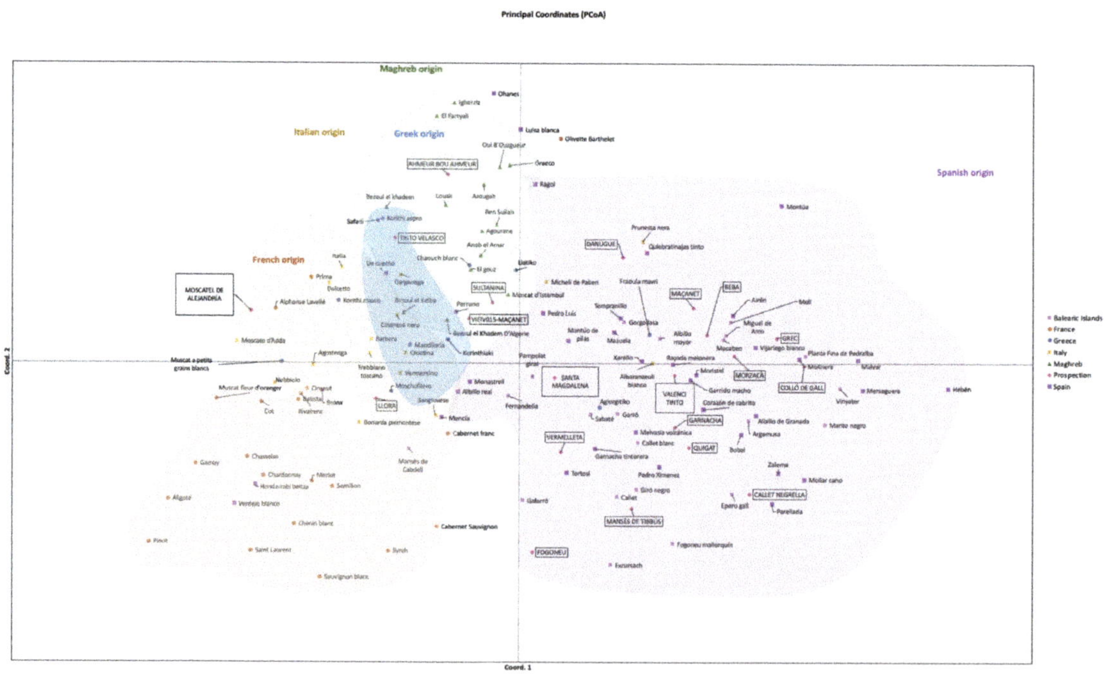

Figure 4. Representation of individuals of the genetic analyses by principal coordinates based on standardized covariance of the genetic distance calculated for 26 codominant markers with GeneAlex software 6.5, using 140 cultivars that include cultivars from five countries as well as the studied samples [55].

Considering only traditional Spanish and Balearic cultivars (Figure 5), some differentiation can be observed between Balearic cultivars, 'Hebén' relatives, and cultivars related to those with Oriental or European origin.

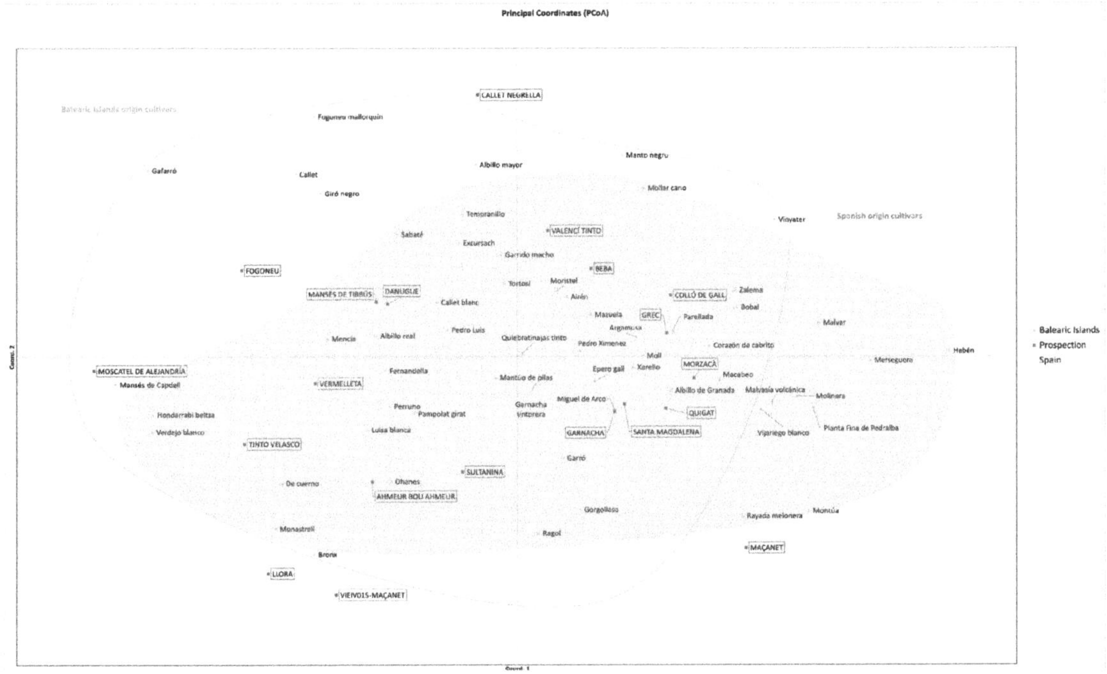

Figure 5. Representation of individuals of the genetic analyses by principal coordinates based on standardized covariance of the genetic distance calculated for 26 codominant markers with GeneAlex software 6.5, considering 78 traditional Spanish and Balearic Islands cultivars, prospection samples included [55].

4. Discussion

4.1. Possible Origin of the Studied Samples

Structure and PCoA results show that most accessions studied here were grouped according to their geographical regions of origin. The highest value of ΔK (Evanno Criterion [62]) was obtained for K = 2 (Figure S1), dividing accessions in a group made up by Spanish cultivars, against another group constituted by cultivars from the rest of Mediterranean origins (Greek, French, Italian and Maghrebi). This clustering is inconsistent with other authors whose results reflect Negrul [72] eco-geographic classification [73,74], since *pontica* and *occidentalis* proles are grouped together, which could indicate a bias in the analysis because of a higher number of Spanish cultivars, many of which are related to 'Hebén'. Regarding K = 3, those previous results are corroborated, since proles *pontica* and *occidentalis* are differentiated, being the third group the Spanish cultivars. Regarding K = 5, the groups correspond to *pontica* (Q5), *occidentalis* (Q3), and *orientalis* subproles *capsica* or Muscat (Q4), according to Emanuelli (2010), who grouped Spanish cultivars with Sultanina descendants in *proles orientalis* subproles *antiasiatica,* using SNPs to differentiate between both groups. A new independent group is obtained in this work shaped by Maghrebi cultivars (Q2) and located in PCoA between *pontica* and the Spanish cultivars group, which in this study, together with Spanish group, constitutes the group proles *orientalis* subproles *antiasiatica*.

According to the results obtained in this work with Structure software 2.3.4 version and PCoA, most of the new or traditional cultivars found in the prospection are related to the Spanish cultivars group, except for 'VIEIV015-Maçanet' and 'Llora', which show a higher component of Greek origin. 'Llora' would be mixed with prole *occidentalis* while 'VIEIV015-Maçanet' would be mixed with prole *orientalis-antiasiatica*. This could correspond with the Phoenician commercial routes around the Mediterranean Sea, where they would have interchanged grapevines between different islands (Cerdà [75] in Piqueras [76]).

Analyzing only Spanish and Balearic cultivars by PCoA, it is observed that new genotypes 'Colló de gall', 'Grec', 'Maçanet' and 'Morzacà' have a strong relationship with traditional Spanish cultivars, as they are all related to the 'Hebén' cultivar, as it is also shown by parentage analysis for some of them. The prospection cultivars 'Fogoneu', 'Danugue', 'Mansès de Tibbús' and 'Callet negrella' could show a stronger relationship with traditional Balearic cultivars as 'Callet', 'Gafarró', 'Fogoneu mallorquí' or 'Giró negre'.

4.2. *Cultivar Analysis and Their Possible Origin*

Recent works on prospection of grapevine cultivars in the Balearic Islands cite some of the cultivars prospected in this study (Table 5).

Table 5. Prospected cultivars in Ibiza and Formentera with same genotypes as others cited in recent works.

Cultivar	Cited in
Beba	[19–21,26,27,67,77]
Callet negrella	[19,21,26]
Fresa	[19]
Fogoneu	[19–21,26,27,77,78]
Quigat	[19–21,26,27,77]
Llora	[28,67,77]
Moscatel de Alejandría	[19,27,77]
Mansès de Tibbús	[19,21,27]
Santa Magdalena	[21,26–28,77]
Sultanina	[19,27]
Tinto velasco	[27,77]
Valencí tinto	[19,21,27,77]

4.2.1. Traditional Balearic Cultivars

Nowadays, the 'Monestrell' cultivar is authorized in wine appellation that exists in both Ibiza and Formentera, and two other traditional cultivars are authorized in Formentera's wine appellation, which are 'Fogoneu' and 'Premsal' [9,10]. There is another wine appellation that can be used by all the winemakers in the Balearic Islands, Vi de les Illes Balears, that has three Balearic cultivars authorized: 'Moll', 'Callet' and 'Manto negro' [79].

Some of the cultivars, like 'Beba', cited as 'Calops', 'Fogoneu', 'Quigat' quoted as 'Cagat', probably because of its pronunciation in the Catalan spoken in the Balearic Islands has been spelt in two different ways, 'Llora', 'Maçanet', 'Monestrell' or 'Moscatell' have been reported since the end of the 19th century [80,81] in the Balearic Islands, which suggests that these cultivars are traditionally cultivated in this geographic area, although citations about the presence of specific cultivars in Ibiza and Formentera Islands are scarce (Table S11). 'Callet negrella', 'Fogoneu', and 'Quigat' are minority cultivars from the Balearic Islands, as García-Muñoz [26] and Marsal [27] confirm with their work.

'Beba', a well-known eating cultivar distributed in different parts of Spain [82], has been reported in Ibiza by Pérez-Cabrero [83] as 'Palop', which is a synonym registered in VIVC for this cultivar. Ludwig Salvator [81] describes for Ibiza and Formentera a white grape cultivar that could also be found producing red-colored grapes, although he does not cite the name of these grapes: it could be 'Beba roja', a somatic mutation with red grapes that García-Muñoz [26] detected, which we have also found when interviewing farmers cited as 'Palop vermell' and confirmed with the SSR markers results (Table S6). Locally, farmers use 'Palop blanc' to refer to the cultivar that has white grapes, as *blanc* means white in Catalan. In the work of García-Muñoz [26], the names used are 'Calop blanc' and 'Calop roig' or 'Calop vermell', as both vermell and *roig* mean red in Catalan. The 'Palop blanc' name for 'Beba' is not registered in VIVC and would be proposed as a synonym for this cultivar.

When comparing ampelographic characterizations [58,84] with ours, it can be observed that leaves from somatic mutations are quite similar, although 'Palop blanc' accessions have shorter teeth in the leaves than 'Palop vermell' (Table S3). Differences in genome size have also been found [6] that are probably caused by somatic mutations when the management and environmental conditions of the crops are the same.

'Beba' is an offspring of 'Hebén' (syn. 'Gibi'), a female Spanish cultivar that is a parent of many cultivars in Spain [85] and also many Balearic cultivars [26]. Our results are coincident with the ones obtained in those previous works.

An offspring of 'Beba' and 'Giró sardo', as García-Muñoz [26] first published, and we confirm with our results (Table 3), is 'Callet negrella', which has been recorded in this study as 'Giró'. This cultivar has been reported for the Balearic Islands in Mestre [86]. It is not registered in the VIVC database [66], but it has been registered in the Spanish variety catalogue [59], and it appears in the Balearic Islands local cultivars catalogue [84]. In this case, the name given by the farmer would be considered a case of misnaming for this cultivar.

Another traditional Balearic cultivar is 'Fogoneu' [26]. Although there are no antique bibliographical references for its presence in Ibiza and Formentera, it has been cited by farmers as a local cultivar and is used in winemaking to improve wine color [45], and it is also authorized in Formentera's wine appellation [10]. During parentage analysis, our results confirm that 'Fogoneu' is an offspring of 'Excursac' and 'Mansès de cabdell' (syn. 'Giro nero'), as García-Muñoz [26] had published.

'Mansès de Tibbús' is also considered to be traditionally grown in the Balearic Islands, as it has been reported by García de los Salmones [87] and García-Muñoz [26]. Although it has not been specifically cited for Ibiza and Formentera, it has been discovered in one of the studied plots. This cultivar that had not been located in any previous prospection on Balearic Islands *Vitis* cultivars and until now was only conserved in IMIDRA's [88] and IRFAP's collections [27].

'Llora' is considered a local minor cultivar cited in Menorca in the end of the 19th century [80,89]. There are also citations [87] for the 'Lloras' cultivar in Menorca and 'Lloreta' in Mallorca that could refer to this cultivar due to the name similarities. In recent times, a sample of this cultivar was detected in Menorca in 2012 [67] but was not located before any other prospection in the Balearic Islands. It also appears in Bota's report [28] (Table 5).

While the 'Llora' cultivar was not specifically mentioned for Ibiza and Formentera, 'Monestrell' was cited [87], which is the name provided by the farmers during the interviews for 'Llora' accessions [45]. 'Monestrell' is also the main black grapevine grown in the Pityusic Islands [11]. It is among the authorized cultivars in Formentera and Ibiza wine appellations [9,10] and was the most cited cultivar for winemaking during the interviews [45]. García de los Salmones [87] also cites 'Mostrell' for Ibiza Island, which phonetically resembles 'Monestrell'; this could either be a misspelling or two cultivars.

The use of 'Monestrell' as a synonym of 'Llora' could be supposed to be widely spread on the Pityusic island as the four 'Monestrell' accessions studied in this work, three samples located in Ibiza and one in Formentera, have all been identified as the 'Llora' cultivar.

Comparing the results on ampelographic descriptors from IMIDRA's database for 'Monestrell' cultivars and ours for 'Llora' cultivars, ampelographic traits are similar, so this might explain the use of 'Monestrell' as a synonym.

The four Monestrell accessions studied have shown different results in the ampelographic descriptors (Table S3). Descriptor OIV065 for leaf blade size is bigger for both 'Monestrell' named accessions, while it is smaller for 'Monestrell d'Alger' and 'Monestrell de xingló' (Figure 6). From the genome size studies, 'Monestrell d'Alger' had a higher 2C

DNA content than the other accessions within the same group [6]. There are even differences in the naming of the accession, where two of them are cited as 'Monestrell d'Alger', because a farmer's relative brought the original plant from Algier [90] and 'Monestrell de xingló' because the plant produces double bunches.

Figure 6. 'Monestrell' cultivars studied in Ibiza and Formentera: (**a**,**b**) 'Monestrell' leaf and bunch accession VIEIV005 (Ibiza); (**c**,**d**) 'Monestrell' leaf and bunch accession VIFOR005 (Formentera); (**e**,**f**) 'Monestrell de xingló' leaf and bunch accession VIEIV022 (Ibiza); (**g**,**h**) 'Monestrell d'Alger' leaf and bunch accession VIEIV023 (Ibiza).

During the interviews and the prospection prior to the ampelographic characterization, farmers cited 'Monestrell fort' and 'Monestrell ros' accessions that could not be located. One of the farmers interviewed had stated that the 'Monestrell' cultivar found in Ibiza Island was different from those that he had seen on the mainland [45]. No references could be found in the bibliography consulted for these different names given by farmers.

From the origin point of view, on one hand, the PCoA results show that this cultivar could have an Italian origin. Italian-origin cultivars in PCoA are also related to Greek and French origin, as these cultivars are located in between them (Figure 4). On the other hand, Structure results showed that this cultivar could be related to French or Greek origin cultivars. From K = 5, 'Llora', although that is an admixed cultivar, it has a stronger component of Greek origin, and the K = 3 result shows that the origin is nearest to the French cultivars, which corroborates the results obtained in PCoA.

A bibliographic reference could confirm this French origin as the 'Llora' cultivar is also cited as 'Marselleres' for Menorca Island [80]. 'Marselleres' could refer to Marseille in France. Piqueras [76] found a citation from the Phoenicians culture about a wine that was named *Lora*, but no other information was provided.

Taking into account the above results and considering that the Phoenician spread vines and viniculture across the Mediterranean territories, such as Marseille in France (Brun [91] in Terral [92]), Ibiza [1] or Maghreb (Rivera [93], Buxó [94] cited in Terral [92]), it could be possible that 'Llora' is a Greek-origin cultivar and is also related to French-origin cultivars because it might be an ancestor of one of them. Further studies should be conducted to confirm this hypothesis and to explore the presence of this cultivar in other places such as Marseille or Algiers.

On the other hand, while reviewing the literature, the names 'Monastrell' and 'Monestrell' have been used to refer to the same cultivar; this could lead to misinterpretation as two different cultivars. Because of this, we suggest that from now on, the name 'Monestrell' should be written as in this paper, and not *Monastrell*, to be able to trace future works and have a spelling homogeneity criterion. The origin of name is the Latin word *monesteriellu*, diminutive of *monasterium*, abbey (*monestir* in Catalan), as it is thought to be a cultivar sown mostly in abbeys [95].

After the aforementioned findings, 'Monestrell' would be proposed to be included in the VIVC catalogue as a synonym of 'Llora'. Further studies should be made to verify if the other 'Monestrell' names detected as 'Monestrell d'Alger' or 'Monestrell de xingló' are found in other plots and also identified as the 'Llora' cultivar to be able to propose those as synonyms. Also, other plots in the Pityusic Islands should be prospected as there are other citations for 'Monestrell' such as 'Monestrell ros' or 'Monestrell fort' [45] that had not been located yet.

'Quigat', a traditional Balearic cultivar [26], has been misnamed in this study for the newly found genotype 'Grec'. However, more samples should be studied in the islands to find any more use cases of this to better determine if it is a misnaming or a new synonym.

4.2.2. Spanish Cultivars

A traditional Spanish cultivar, such as 'Garnacha tinta', has been detected as a somatic mutation known as 'Garnatxa blanca', which was also cited with this name by the farmer. This somatic mutation is published in VIVC [66]. In García de los Salmones [87], there is only a reference for the 'Garnatxa negra' cultivar in Ibiza but not the white cultivar that has been found in this study.

Another Spanish traditional cultivar found in this study is 'Ahmeur bou Ahmeur', although it has been named locally as 'Mamella de vaca', which means cow's udder in Catalan. This name would considered a synonym for 'Ahmeur bou Ahmeur', as this cultivar has a registered synonym in VIVC, 'Teta de vaca', which is the Spanish translation of 'Mamella de vaca'. There is no reference of this cultivar in the Balearic Islands in the works consulted; only Marsal [27] had a sample with the same name, but this was identified as 'Afus ali', and they proposed 'Mamella de vaca' as a new synonym for that cultivar.

'Santa Magdalena' and 'Valencí tinto' cultivars were also identified in this study. The 'Santa Magdalena' cultivar has been found locally named by farmers as 'Sant Jaume' and 'Santa Margalida' (Table 4). Favà [96] cited 'Sant Jaume' and 'Santa Magdalena' as different cultivars in the Ibiza islands, but he did not conduct SSR makers identification. 'Santa Magdalena' has been cited as a synonym for 'Jaumillo' [77], which is a diminutive for the name Jaume. No references have been found for 'Santa Margalida' apart from the connection between the farmers reporting that this cultivar is an early ripening one [45], and Saint Margaret's day which is in early July in the Spanish catholic calendar, so this could explain the relation with the name. Comparing the ampelographic traits of 'Santa Magdalena' cultivars found (Table S3; [58,84]), the leaves from the Balearic Grapevine Catalogue and ours are very similar, while the ones in IMIDRA's descriptors are slightly different. As a result, for the 'Santa Magdalena' cultivar, 'Sant Jaume' and 'Santa Margalida' would be proposed as new synonyms to be included in VIVC.

The 'Valencí tinto' cultivar has been identified in two samples in this study: one was cited as 'Palop negre', which has also been cited for 'Valencí tinto' in Mallorca [77], and another one was cited as 'Primerenc'. Both names would be proposed to be registered in VIVC as new synonyms for the 'Valencí tinto' cultivar in the Balearic Islands, although further studies are needed to validate these propositions.

4.2.3. French Cultivars

'Danugue' is also a French cultivar that had not been identified until now in any of the recent works (Table 5). It has been identified in three samples, two of which were named 'Ferrana' by two different farmers and one of which was cited as 'Palop negre'. García de los Salmones [87] cites a 'Ferrana negra' cultivar for Ibiza. Nowadays, 'Ferrana' has been used as a synonym for 'Planta fina' and as a homonym for the Moroccan cultivar 'Bourboulenc' [66]. In this case, we propose that 'Ferrana' would also be registered in the VIVC database as a synonym for 'Danugue', and we consider 'Palop negre' a misnaming for the 'Ferrana' cultivar.

In the VIVC database, 'Danugue' has French origin and is related to 'Breval negro' [71,96]. From this study, we have found that 'Danugue' is also related to 'Albarazeuli bianco', which is an Italian cultivar offspring of 'Hebén'. D'Onofrio [71] has cited other local cultivars in the island of Sardinia, which were also related to 'Hebén', showing that there was a cultivar movement from Spain to Italy. Also, the Structure and the PCoA results show that this cultivar seems to have a Spanish origin, so these results with the above mention might suggest that these cultivar was spread through the Mediterranean in the past.

4.2.4. New Identified Genotypes

Six new genotypes have been discovered, which do not appear in any of the databases consulted [58,66]. The proposed names, 'Colló de gall', 'Grec', and 'Maçanet', the latter with its homonyms 'VIEIV05-Mançanet', 'Morzacà and 'Vermelleta', are the ones local farmers reported, as these have not been found for these genotypes.

'Colló de gall' is one of the newfound genotypes. There are no citations of this cultivar for Ibiza Island before Favà's [95] work on *Vitis* names in the Catalan language. In Bota [28], there is a sample cited as a newfound genotype named 'Colló de gall', but no ampelographic description or microsatellite results have been published yet. This cultivar name means rooster's testicle, translated from Catalan (being *gall*, rooster, and *colló*, testicle). He also cites other synonyms for this cultivar, such as 'Botó de gall', 'Colló de gat' or 'Botó de gat (where *gat*, means cat in Catalan). He suggests that *botó* is used as a synonym of *colló* in order not have to use *colló* with its sexual connotation.

Favà's [95] descriptions for the 'Colló de gall' cultivar does not match with our ampelographic characterization (Table S3). In his work, this cultivar is described to have large bunches and to be very productive, while the studied plants in this work are not very productive, and bunches tend to be small to medium. In any case, it must be considered that the plants studied are very old [37], not grafted on American rootstock, and are not watered, and all this could alter the ampelographic descriptions [35].

From the parentage analysis, this cultivar might be an offspring of Spanish cultivar 'Hebén', which makes sense with the results obtained in Structure analyses, where this cultivar is grouped with Spanish-origin cultivars in all groupings obtained.

The new identified genotype locally named 'Grec' has been one of the most reported cultivar names during the interviews [45]. It is cited in García de los Salmones [87] in Ibiza Island as 'Grech' (an old Catalan spelling for *Grec*, meaning Greek), although with red grapes, and with white grapes in Menorca. This last reference might refer to our cultivar as it has white grapes (Table S3). The name *Grec* was also used to refer to a type of wine [76,95]. Favà [95] and the farmers [45] cite also 'Grec' as a synonym of 'Malvasia'. 'Malvasia' is one of the authorized cultivars in winemaking appellation regulations in Ibiza and Formentera [9,10]. Taking this synonym of 'Malvasia' for 'Grec', these cultivars are being used to produce wines within the appellation in Ibiza Island. In Bota [28], there is a 'Grec' accession from Ibiza, but this report does not specify any other information. Also, there is a recent ampelographic description of a 'Grec' cultivar [84] from Ibiza that has similar results to our ampelographic descriptions (Table S3). Consulting the VIVC catalogue, the 'Grec' name appears as a synonym of 'Alcanon' and 'Greco', this last one was considered by Favà [95] as a translation to Italian of the name 'Grec', and it is also registered in the VIVC catalogue as 'Malvasia'. So, in this case, 'Grec' would be proposed as a homonym for the new genotype.

From our Structure analysis results, 'Grec' might have Spanish origin, as in all groupings, it falls within the Spanish-origin cultivars. Parentage analysis suggests this, as 'Grec' is related to 'Beba', which is an offspring of 'Hebén' [26]: an ancient Spanish cultivar that has originated many other cultivars [85]. The results also suggest it could be a sibling of 'Valencí tinto', as this is confirmed to be 'Beba' offspring by VIVC (Table 3).

'Maçanet' cultivar is another of the new genotypes discovered in this study. It has been cited for Menorca Island [80] and for Ibiza [83,87], although García de los Salmones [87] describes this cultivar as a black grape color cultivar, while ours has white grapes (Table S3). In Favà [95], he registered 'Mançaneta' as a synonym of 'Maçanet' as being originally from outside Ibiza island, but all farmers interviewed have cited 'Maçanet' as a traditional cultivar of Ibiza [45]. He proposes that the ancient cultivar 'Masanel' from the south of Spain could be related to 'Maçanet and also to 'Manzanilla' cultivars, as both have similarities to the names *mançana* and *manzana*, meaning apple in Catalan and Spanish, respectively.

The above-mentioned information suggests a Spanish origin for this cultivar, which the Structure results obtained seem to back up, as this cultivar falls in the Spanish origin group either in K = 2 and K = 3. From the parentage results, this cultivar is the offspring of 'Hebén' and 'Cigenera', which are both Spanish-origin cultivars [66] (Table 3). Despite this, the relationship with the 'Manzanilla' cultivar [95] has been checked, but there is a mismatch of 10 loci from the 26 compared, so there is not a significant relation between these two cultivars that might only be related etymologically.

For the 'Maçanet' cultivar, a homonym has been detected, as accession 'VIEIV015-Maçanet' has resulted in a different genotype from the 'Maçanet' cultivar that had not been detected in either IMIDRA's database [58] or VIVC [66]. From the parentage analysis (Table 4), there is a significant relationship between 'VIEIV015-Maçanet' and 'Llora', but it is not possible to determine which is the parent and which is offspring without chloroplast DNA, which has not been analyzed in this study.

One of the accessions that was identified as 'Maçanet' was locally named 'Blanqueta'. The name 'Blanqueta' has been cited as a synonym of 'Merseguera', which is a cultivar that has been cultivated in the Valencian community since 1800 [95]. More samples must be studied to be able to determine if this is a synonym of the 'Maçanet' cultivar, although farmers reported them as different cultivars. One of them also stated that the 'Blanqueta' cultivar was very hard to graft, which he did not state for the 'Maçanet' cultivar [45]. Bota [28] names a 'Maçanet' sample as a non-registered cultivar in her report but does not provide any other information.

Another unregistered genotype found in this study is the 'Morzacà' cultivar. This cultivar has only been found cited by a farmer in Ibiza Island in Favà's work [95]. Etymologically, this name could be derived from the word *morzar*, meaning to have breakfast in Catalan, and *ca*, which means dog in Catalan, suggesting that this cultivar was not good to eat, as it was dog's food [95]. In the interviews conducted in this study, this cultivar was cited for winemaking, and a farmer also explained that it was not a very aromatic grape [45], which could make sense with Favà's [95] etymology of the name. From Structure analysis, it is grouped with the Spanish-origin cultivars, and from parentage analysis, it appears that it could be an offspring of 'Hebén', which aligns with the Structure results. 'Morzacà' would be proposed as a prime name to be included in VIVC for this genotype and as a local cultivar from Ibiza and Formentera.

'Vermelleta' is the last cultivar in this study that has been found to have a unique genotype profile that has not been detected before. The name of this cultivar might be related to the color of the grapes as it has red, small grapes (Table S3) and 'Vermelleta' is a feminine diminutive of red in Catalan. Ludwig Salvator [80] cites the 'D'en Vermell' cultivar in Menorca Island and Bota [28] cites a cultivar named 'Vermellet'. Both are related to the name red in Catalan, although 'D'en Vermell' refers to be property of a man named *Vermell*, and the name cited in Bota seems to refer to the color of the grape as a masculine diminutive. In the case of Bota's study [28], no other information has been published yet, so it cannot be known if they are the same cultivars.

In the case of 'Morzacà' and 'Vermelleta', there is also no earlier reference before 2000s [95] which might confirm that those are old minor cultivars.

5. Conclusions

This study to assess and identify *Vitis* cultivars in the Pityusic islands has been conducted from a different point of view than most similar works conducted in Europe, as it uses ethnobotanical interviews to select samples in plots with the idea to recover cultivars that are not officially documented and also to gather information that could help to identify the cultivars and their origin.

With this approach, it has been possible to have an image of the current state of cultivars maintained in plots and vineyards in parts of Ibiza and Formentera islands, discovering six new genotypes that had not been published before, one that was only maintained in a conservation collection, and ten new synonyms, which have been detected and proposed to be included in the VIVC catalogue. Most of these cultivars belong to local traditions, especially the 'Monestrell' (syn. 'Llora') cultivar, which is present in the 100% of the studied plots.

As the plots in this study were concentrated mostly in the Sant Josep de sa Talaia municipality, in Ibiza, further studies should be made with the approach used in this study to cover all Ibiza and Formentera Islands, as more minor cultivars could still be found.

Supplementary Materials: The following supporting information can be downloaded at https://www.mdpi.com/article/10.3390/horticulturae9121307/s1, Figure S1. ΔK graphic., Figure S2. Graphical representation of the 20 genotypes found in Ibiza and Formentera Island by Structure software for K = 2 (Stanford University, 2.3.4 version, 2012); Table S1. Plot information; Table S2. Samples codes and vouchers numbers; Table S3. Ampelographic descriptors mode values 2017 and 2018 for the samples accessions studied; Table S4. Observation number for each accession, descriptor and year; Table S5: SSR markers diversity indexes; Table S6. SSR markers accession analyses results; Table S7. Cultivars considered Structure analysis; Table S8. Sample codes and probabilities for K = 2; Table S9. Sample codes and probabilities for K = 3; Table S10. Sample codes and probabilities for K = 5; Table S11. Bibliographic references for research cultivars in the Balearic Islands and Ibiza and Formentera.

Author Contributions: Conceptualization, methodology, and investigation, R.G., A.M.V., T.G., J.V. and M.A; data curation, resources, software and validation, R.G., M.A and A.M.V.; formal analysis, R.G., M.A and A.M.V.; writing—original draft preparation and writing—review and editing, R.G., A.M.V., T.G., J.V. and M.T.d.A. All authors have read and agreed to the published version of the manuscript.

Funding: This study was partially funded by project 1.323.119 by Grup d'Acció Local d'Eivissa i Formentera through the European Agricultural Fund for Rural Development (EAFRD) and by project 2021SGR00315 of the Catalan government.

Data Availability Statement: Data are contained within the article and supplementary materials.

Acknowledgments: We thank all farmers who allowed us to collect and study plants from their plots as well as providing valuable information; Airy Gras (Laboratori de Botànica, Facultat de Farmàcia i Ciències de l'Alimentació, Universitat de Barcelona) and Roi Rodríguez (Institut Botànic de Barcelona (IBB), CSIC-Ajuntament de Barcelona) for their assistance in processing the samples; and Iván Pérez-Lorenzo (Institut Botànic de Barcelona (IBB), CSIC-Ajuntament de Barcelona) for his assistance in vouchers preparation for the herbarium of the Botanical Institute of Barcelona.

Conflicts of Interest: The authors declare no conflict of interest.

References

1. López Garí, J.M.; Marlasca Martin, R.; Yll, R. El Paissatge Fòssil d'Eivissa Antiga. Els Retalls Pel Conreu a Època Púnica i Romana. *Fites* **2009**, *9*, 7–18.
2. Ferrer Abárzuza, A. *El Llibre Dels Mostassaf d'Eivissa. La Vila d'Eivissa a La Baixa Edat Mitjana*, 1st ed.; Costa, M., Ed.; Editorial Mediterrània: Ibiza, Spain, 2002; pp. 118–129.
3. Oliver Moragues, M. Vitivinicultura Balear Extremos de Su Historia. In *Actas del I Encuentro de Historiadores de la Vitivinicultura Española*; Ribas, C., Ed.; Imprenta Politécnica: Palma de Mallorca, Spain, 2001.
4. Denominations and Marks from Balearic Islands. Available online: http://www.illesbalearsqualitat.es/iquafront/denominacio/llistat/331 (accessed on 10 June 2023).

5. Ballester, A. *La Invasión Filoxérica En España y Estado en 1909 de La Reconstrucción Del Viñedo;* de MG Hernández, H., Ed.; Ministerio de Fomento: Madrid, Spain, 1911.
6. González, R.; Vallès, J.; Garnatje, T. Genome Size Variation Assessment in *Vitis vinifera* L. Landraces in Ibiza and Formentera (Balearic Islands). *Plants* **2022**, *11*, 1892. [CrossRef] [PubMed]
7. International Organisation of Vine and Wine. *Balance de La OIV Sobre La Situación Vitivinícola Mundial;* International Organisation of Vine and Wine: Paris, France, 2019; pp. 1–5.
8. Escudero, D.; Mancheño, S.; López, J.J. Capítulo 7: Superficie y producciones de cultivos. In *Anuario de Estadística. Avance 2021;* Ministerio de Agricultura, Pesca, y Alimentación: Madrid, Spain, 2022; pp. 878–882. Available online: https://www.mapa.gob.es/estadistica/pags/anuario/2021-Avance/AVANCE_ANUARIO/AvAE21.pdf (accessed on 5 July 2022).
9. Order 18 February 2013 in Which Specification for the Winemaking Appellation Ibiza Is Approved. BOIB N. 24. 19 February 2013; pp. 7946–7962. Available online: https://www.caib.es/eboibfront/eli/es-ib/o/2013/02/18/(1)/dof/spa/html (accessed on 10 October 2023).
10. Order 11 March 2013 in Which Specification for the Winemaking Appellation Formentera Is Approved. BOIB N. 35. 14 March 2013; pp. 11744–11755. Available online: https://www.caib.es/eboibfront/eli/es-ib/o/2013/03/11/(1)/dof/spa/html (accessed on 10 October 2023).
11. Registre Vitivinícola de les Illes Balears. Available online: http://registreviticola.caib.es/viticolafront/ (accessed on 10 January 2023).
12. Prats, E.; Ramón, F.; Vallès, R. *Geografia de les Illes Pitiüses. Les Activitats Agràries;* Institut D'Estudis Eivissencs: Eivissa, Spain, 1984; pp. 20–29.
13. González, R. Prospecció, Localització i Ampelografia de Varietats Locals de Vinya En Eivissa i Formentera. Grup d'Acció Local per al Desenvolupament Rural d'Eivissa i Formentera: Ibiza, Spain, 2014; *Unpublished work*.
14. Gutiérrez-Gordillo, S.; Pérez-González, J.M.; Sancho-Galán, P.; Amores-Arrocha, A.; Palacios, V.; Jiménez-Cantizano, A. Physiological, Enological and Agronomic Characterization of Pedro Ximénez Grapevine Cultivar under Organic Farming in a Warm Climate Zone. *Agronomy* **2023**, *13*, 1732. [CrossRef]
15. Ruiz-García, L.; Romero, P.; Tornel, M.; Menéndez, C.; Cabello, F.; Martínez-Cutillas, A. La viticultura frente al cambio climático: Adaptación y estrategias de mejora. In *Influencia del Cambio Climático en la Mejora Genética de Plantas*, 1st ed.; García, J., Pérez, O., Cos, J., Ruiz-García, L., Sánchez, E., Eds.; Comunidad Autónoma de la Región de Murcia, Consejería de Agua, Agricultura, Ganadería y Pesca, Dirección General de Agricultura, Ganadería, Pesca y Acuicultura, Servicio de Formación y Transferencia Tecnológica: Murcia, Spain, 2018; pp. 165–198.
16. Carámbula, C.; Bota, J.; Medrano, H. Recuperación, Saneamiento y Evaluación En Condiciones de Campo de La "Malvasia de Banyalbufar". In Proceedings of the III International Symposium Malvasias, La Palma, Spain, 26 May 2009; pp. 1–6.
17. Escalona, J.; March, J.; Luna, J.; Martorell, A. Recuperación de variedades minoritarias de Baleares. In Proceedings of the 5° Foro Mundial del Vino, Logroño, Spain, 4 November 2005.
18. Cretazzo, E. Selección Clonal de Variedades de Vid Autóctonas de Mallorca: Aspectos Sanitarios, Genéticos y Agronómicos. Ph.D. Thesis, Universitat de les Illes Balears, Palma de Mallorca, Spain, 2010.
19. Martorell, A. Història i Evolució de Les Varietats de Raïm de Les Illes Balears. Ph.D. Thesis, Universitat de les Illes Balears, Palma de Mallorca, Spain, 2011.
20. Bota, J. *Estado Actual de las Variedades Minoritarias de la Islas Baleares;* Gesevid: Madrid, Spain, 2012.
21. García Muñoz, S. Estudio de Variedades Minoritarias de Vid (*Vitis vinifera* L.): Descripción, Caracterización Agronómica y Enológica de Material Procedente de las Islas Baleares. Ph.D. Thesis, Universidad de Valladolid, Valladolid, Spain, 2011.
22. Carámbula, C.; Tomás, M.; Medrano, H.; Cifre, J. Selección clonal, saneamiento y conservación de material vegetal vitícola de Baleares. *Tecnol. Vino Trat. Equipos Vitic. Enol.* **2005**, *22*, 79–84.
23. Bota, J.; Escalona, J.M.; Luna, J.M.; Mulet, D.; Martorell, A. Variedades Minoritarias de Las Baleares: Sabater, Caracterización Ampelográfica, Evaluación Agronómica y Enológica. In Proceedings of the XXIII Reuniones del Grupo de Trabajo de Experimentación en Viticultura y Enología, Madrid, Spain, 6–7 February 2009.
24. Carámbula, C.; Moreno, M.T.; Riera, D.; Cretazzo, E.; Tomás, M.; Pou, A.; Escalona, J.M.; Martorell, A.; Medrano, H.; Cifre, J. Selección clonal de las principales variedades autóctonas de Baleares. In Proceedings of the V Fórum Mundial del Vino, Logroño, Spain, 28–30 March 2006; Ministerio de Agricultura, Pesca y Alimentación: Madrid, Spain, 2006.
25. Escalona, J.; March, J.; Luna, J.M.; Nicolau, A.M.; Rallo, J. Caracterización y Aptitud Enológica de Variedades Minoritarias de Vid de Baleares: Perfil Aromático. In Proceedings of the XXIX Congreso Mundial de la Viña y del Vino, Logroño, Spain, 25–30 June 2006; Ministerio de Agricultura, Pesca y Alimentación: Madrid, Spain, 2006.
26. García-Muñoz, S.; Lacombe, T.; de Andrés, M.T.; Gaforio, L.; Muñoz-Organero, G.; Laucou, V.; This, P.; Cabello, F. Grape Varieties (*Vitis vinifera* L.) from the Balearic Islands: Genetic Characterization and Relationship with Iberian Peninsula and Mediterranean Basin. *Genet. Resour. Crop Evol.* **2012**, *59*, 589–605. [CrossRef]
27. Marsal, G.; Bota, J.; Martorell, A.; Canals, J.M.; Zamora, F.; Fort, F. Local Cultivars of *Vitis vinifera* L. in Spanish Islands: Balearic Archipelago. *Sci. Hortic.* **2017**, *226*, 122–132. [CrossRef]
28. Bota, J. Justificació Tècnica Ajuda BIA05/07. Recuperació Sanitària i Valoració Enològica de Varietats Minoritàries de Vinya de les Illes Balears. Universitat de les Illes Balears: Palma de Mallorca, Spain, 2019; Unpublished work.

29. Jiménez-Cantizano, A.; Muñoz-Martín, A.; Amores-Arrocha, A.; Sancho-Galán, P.; Palacios, V. Identification of Red Grapevine Cultivars (*Vitis vinifera* L.) Preserved in Ancient Vineyards in Axarquia (Andalusia, Spain). *Plants* **2020**, *9*, 1572. [CrossRef]
30. This, P.; Jung, A.; Boccacci, P.; Borrego, J.; Botta, R.; Costantini, L.; Crespan, M.; Dangl, G.S.; Eisenheld, C.; Ferreira-Monteiro, F.; et al. Development of a Standard Set of Microsatellite Reference Alleles for Identification of Grape Cultivars. *Theor. Appl. Genet.* **2004**, *109*, 1448–1458. [CrossRef]
31. Theocharis, A.; Hand, P.; Pole, J.; Cevik, V.; Fisarakis, I.; Henderson, J. Study of Genetic Diversity among Inter-Intraspesific Hybrids and Original Grapevine Varieties Using AFLP Molecular Markers. *Aust. J. Crop Sci.* **2010**, *4*, 1–8.
32. Dallakyan, M.; Esoyan, S.; Gasparyan, B.; Smith, A.; Hovhannisyan, N. Genetic Characterization of Grape Varieties in Armenia. *VITIS—J. Grapevine Res.* **2015**, *54*, 23–26.
33. Galet, P. *A Practical Ampelography: Grapevine Identification*; Cornell University: London, UK, 1979; p. 226.
34. Swanepoel, J.J.; de Villiers, C.E. A Numerical-Taxonomic classification of *Vitis* spp. and Cultivars Based on Leaf Characteristics. *S. Afr. J. Enol. Vitic.* **1987**, *8*, 31–35. [CrossRef]
35. Mena Morales, A. Recuperación, Caraterización y Conservación de Variedades de Vid (*Vitis vinifera* L.) Minoritarias de Castilla-La Mancha. Ph.D. Thesis, Universidad de Castilla-La Mancha, Castilla-La Mancha, Spain, 2013.
36. Garcia-Muñoz, S.; Muñoz-Organero, G.; Andrés, M.T.D.; Cabello, F. Ampelography—An Old Technique with Future Uses: The Case of Minor Varieties of *Vitis vinifera* L. from the Balearic Islands. *J. Int. Sci. Vigne Vin* **2011**, *45*, 125–137. [CrossRef]
37. Crespan, M.; Cabello, F.; Giannetto, S.; Ibáñez Marcos, J.; Karoglan Kontic, J.; Maletic, E. Malvasia Delle Lipari; Malvasia Di Sardegna; Greco Di Gerace; Malvasia de Sitges and Malvasia Dubrovačka—Synonyms of an Old and Famous Grape Cultivar. *VITIS—J. Grapevine Res.* **2006**, *45*, 69–74.
38. El Aou-ouad, H.; Montero, R.; Baraza, E.; Bota, J. Recovering Ancient Grapevine Cultivars in the Balearic Islands: Sanitary Status Evaluation and Virus Elimination. *Plants* **2022**, *11*, 1754. [CrossRef] [PubMed]
39. Poni, S.; Gatti, M.; Palliotti, A.; Dai, Z.; Duchêne, E.; Truong, T.-T.; Ferrara, G.; Matarrese, A.M.S.; Gallotta, A.; Bellincontro, A.; et al. Grapevine Quality: A Multiple Choice Issue. *Sci. Hortic.* **2018**, *234*, 445–462. [CrossRef]
40. Nicolosi, E.; Sicilia, A.; Ferlito, F.; Bonfante, A.; Monaco, E.; Lo Piero, A.R. Phenotypic Plasticity in Bud Fruitfulness Expressed in Two Distinct Wine Grape Cultivars Grown under Three Different Pedoclimatic Conditions. *Agriculture* **2022**, *12*, 1660. [CrossRef]
41. Chomé, P.M.; Sotés, V.; Benayas, F.; Cayuela, M.; Hernández, M.; Cabello, F.; Ortiz, J.M.; Rodríguez-Torres, I.; Chaves, J. *Variedades de vid. Registro de Variedades Comerciales*; Ministerio de Agricultura, Pesca y Alimentación: Madrid, Spain, 2003.
42. UPOV (International Union for the Protection of New Varieties of Plants). Grapevine. UPOV Code: VITIS. *Vitis* L. Guidelines for the Conduct of Tests for Distinctness, Uniformity and Stability. 9 April 2008. Available online: https://www.upov.int/edocs/tgdocs/en/tg050.pdf (accessed on 5 February 2023).
43. Dallakyan, M.; Esoyan, S.; Gasparyan, B.; Smith, A.; Hovhannisyan, N. Genetic Diversity and Traditional Uses of Aboriginal Grape (*Vitis vinifera* L.) Varieties from the Main Viticultural Regions of Armenia. *Genet. Resour. Crop Evol.* **2020**, *67*, 999–1024. [CrossRef]
44. Decree 31/2014 of 18th of July in Which It Stablish General Principles in Planting and Replanting Grapevines, Vine Registration and Compulsory Declarations in Winemaking Scheme. BOCAIB N 97. 19 July 2014; pp. 33145–33177. Available online: https://www.caib.es/eboibfront/eli/es-ib/d/2014/07/18/31/dof/spa/html (accessed on 20 October 2023).
45. González, R.; Vallès, J.; Garnatje, T. Laboratori de Botànica-Unitat associada al CSIC, Facultat de Farmàcia i Ciències de l'Alimentació-Institut de la Biodiversitat IRBio, Universitat de Barcelona: Barcelona, Spain. Manuscript in preparation. 2023.
46. Sabaté, A.; Perdomo, A.; Afonso, V. *Las Fuentes Orales en los Estudios de Agroecología*; CCBAT: Tenerife, Spain, 2008; pp. 65–68.
47. Merdinoglu, D.; Butterlin, G.; Bevilacqua, L.; Chiquet, V.; Adam-Blondon, A.-F.; Decroocq, S. Development and Characterization of a Large Set of Microsatellite Markers in Grapevine (*Vitis vinifera* L.) Suitable for Multiplex PCR. *Mol. Breed.* **2005**, *15*, 349–366. [CrossRef]
48. Sefc, K.M.; Regner, F.; Turetschek, E.; Glössl, J. Identification of Microsatellite Sequences in Vitis Riparia and Their Applicability for Genotyping of Different Vitis Species. *Genome* **1999**, *42*, 367–373. [CrossRef]
49. Bowers, J.E.; Dangl, G.S.; Vignani, R.; Meredith, C.P. Isolation and Characterization of New Polymorphic Simple Sequence Repeat Loci in Grape (*Vitis vinifera* L.). *Genome* **1996**, *39*, 628–633. [CrossRef]
50. Bowers, J.E.; Boursiquot, J.-M.; This, P.; Chu, K.; Johansson, H.; Meredith, C. Historical Genetics: The Parentage of Chardonnay, Gamay, and Other Wine Grapes of Northeastern France. *Science* **1999**, *285*, 1562–1565. [CrossRef]
51. Adam-Blondon, A.F.; Roux, C.; Claux, D.; Butterlin, G.; Merdinoglu, D.; This, P. Mapping 245 SSR Markers on the *Vitis vinifera* Genome: A Tool for Grape Genetics. *Theor. Appl. Genet.* **2004**, *109*, 1017–1027. [CrossRef] [PubMed]
52. Thomas, M.R.; Scott, N.S. Microsatellite Repeats in Grapevine Reveal DNA Polymorphisms When Analysed as Sequence-Tagged Sites (STSs). *Theor. Appl. Genet.* **1993**, *86*, 985–990. [CrossRef] [PubMed]
53. Zarouri, B.; Vargas, A.M.; Gaforio, L.; Aller, M.; de Andrés, M.T.; Cabezas, J.A. Whole-Genome Genotyping of Grape Using a Panel of Microsatellite Multiplex PCRs. *Tree Genet. Genomes* **2015**, *11*, 17. [CrossRef]
54. Applied Biosystems Inc. *GeneMapper* ®, version 5.0; Applied Biosystems Inc.: Foster City, CA, USA, 2012.
55. Peakall, R.; Smouse, P.E. Genalex 6: Genetic Analysis in Excel. Population Genetic Software for Teaching and Research. *Mol. Ecol. Notes* **2006**, *6*, 288–295. [CrossRef]
56. Kalinowski, S.T.; Taper, M.L.; Marshall, T.C. Revising How the Computer Program CERVUS Accommodates Genotyping Error Increases Success in Paternity Assignment. *Mol. Ecol.* **2007**, *16*, 1099–1106. [CrossRef] [PubMed]

57. Park, S.J. *Microsatellite Toolkit for Excel*; Smurfit Institue of Genetics, Trinity College, University of Dublin: Dublin, Ireland, 2001.
58. Instituto Madrileño de Investigación y Desarrollo Rural y Agrario. Partially Published. Available online: http://www.madrid.org/ (accessed on 7 July 2023).
59. Ministerio de Agricultura, Pesca y Alimentación. Grapevine Spanish Catalogue. Available online: https://www.mapa.gob.es/app/regVar/ResBusVariedades.aspx?nOrden=1&id=es&TxtEspecie=VID&IDEspecie=119&nPag=4 (accessed on 7 July 2023).
60. Pritchard, J.K.; Stephens, M.; Donnelly, P. Inference of Population Structure Using Multilocus Genotype Data. *Genetics* **2000**, *155*, 945–959. [CrossRef] [PubMed]
61. Falush, D.; Stephens, M.; Pritchard, J.K. Inference of Population Structure Using Multilocus Genotype Data: Linked Loci and Correlated Allele Frequencies. *Genetics* **2003**, *164*, 1567–1587. [CrossRef] [PubMed]
62. Evanno, G.; Regnaut, S.; Goudet, J. Detecting the Number of Clusters of Individuals Using the Software Structure: A Simulation Study. *Mol. Ecol.* **2005**, *14*, 2611–2620. [CrossRef]
63. Peakall, R.; Smouse, P.E. GenAlEx 6.5: Genetic Analysis in Excel. Population Genetic Software for Teaching and Research—An Update. *Bioinformatics* **2012**, *28*, 2537–2539. [CrossRef]
64. Microsoft Corporation. *Microsoft Excel*, version 16.60; Microsoft Corporation: Redmond, WA, USA, 2020.
65. IBM Corp. *IIBM SPSS Statistics for Macintosh*, version 26.0; IBM Corp.: Armonk, NY, USA, 2019.
66. Vitis International Variety Catalogue. Available online: https://www.vivc.de/ (accessed on 20 October 2023).
67. Mayrata, T. Caracterització Morfològica i Molecular de Vinya de L'illa de Menorca. Bachelor's Thesis, Universitat de les Illes Balears, Palma de Mallorca, Spain, 2013.
68. Bota, J.; (Universitat de les Illes Balears, Palma de Mallorca, Spain). Personal Communication, 2023.
69. Galet, P. *Cépages et Vignobles de France*; Le Pay san du Midi: Montpellier, France, 1957; p. 3500.
70. Cipriani, G.; Spadotto, A.; Jurman, I.; Di Gaspero, G.; Crespan, M.; Meneghetti, S.; Frare, E.; Vignani, R.; Cresti, M.; Morgante, M.; et al. The SSR-Based Molecular Profile of 1005 Grapevine (*Vitis vinifera* L.) Accessions Uncovers New Synonymy and Parentages, and Reveals a Large Admixture amongst Varieties of Different Geographic Origin. *Theor. Appl. Genet.* **2010**, *121*, 1569–1585. [CrossRef]
71. D'Onofrio, C.; Tumino, G.; Gardiman, M.; Crespan, M.; Bignami, C.; de Palma, L.; Barbagallo, M.G.; Muganu, M.; Morcia, C.; Novello, V.; et al. Parentage Atlas of Italian Grapevine Cultivars as Inferred from SNP Genotyping. *Front. Plant Sci.* **2021**, *11*, 605934. [CrossRef] [PubMed]
72. Negrul, A.M. *Evolution of Cultivated Forms of Grapes*; Comptes Rendus (Doklady) de l'Académie des Sciences de l'URSS; Librairie Agricole: Paris, France, 1938; Volume 18, pp. 585–588.
73. Emanuelli, F.; Lorenzi, S.; Grzeskowiak, L.; Catalano, V.; Stefanini, M.; Troggio, M.; Myles, S.; Martinez-Zapater, J.M.; Zyprian, E.; Moreira, F.M.; et al. Genetic Diversity and Population Structure Assessed by SSR and SNP Markers in a Large Germplasm Collection of Grape. *BMC Plant Biol.* **2013**, *13*, 39. [CrossRef] [PubMed]
74. Cretazzo, E.; Moreno Sanz, P.; Lorenzi, S.; Benítez, M.L.; Velasco, L.; Emanuelli, F. Genetic Characterization by SSR Markers of a Comprehensive Wine Grape Collection Conserved at Rancho de La Merced (Andalusia, Spain). *Plants* **2022**, *11*, 1088. [CrossRef] [PubMed]
75. Cerdà, D. *El Ví En l'ager Pollentinus i en el Seu Entorn*; Col·lecció la Deixa; Consell de Mallorca: Palma de Mallorca, Spain, 1999.
76. Piqueras, J. *Historia de La Vid y el Vino en España. Edades Antigua y Media*; Universidad de Valencia: Valencia, Spain, 2014.
77. Martorell, A.; Escalona, L.; Luna, J.M.; Bota, J.; Garau, C. *Varietats de Vinya Autòctones de Les Illes Balears*; Conselleria de Medi Ambient, Agricultura i Pesca: Palma de Mallorca, 2016.
78. Medrano, H.; Balda, P. Informe de Justificación Técnica. Projecte AIA03/15. Les Varietats Minoritàries de Vinya de Les Illes Balears: Recuperació i Sanejament de Varietats, Identificación Morfológica i Molecular, Autorització i Certificació de Material Vegetal de Vinya. 2016. Available online: https://docplayer.es/118022139-Informe-de-justificacion-tecnica-projecte-aia03-15.html (accessed on 3 June 2023).
79. Order 29 January 2013 in Which Specification for the Winmeaking Appellation Illes Balears Is Approved. BOIB N. 20. 9 February 2013; pp. 6441–6451. Available online: https://www.caib.es/eboibfront/eli/es-ib/o/2013/01/29/(1)/dof/cat/html (accessed on 10 October 2023).
80. Ludwig Salvator, Archduke of Austria. *Las Baleares Descritas por la Palabra y el Grado (Die Balearen)*; Ajuntament de Palma: Palma de Mallorca, Spain, 1981.
81. Ludwig Salvator, Archduke of Austria. *Les Antigues Pitiüses. Les Balears Descrites per Paraule i la Imatge*; Promallorca Edicions S.L.: Palma de Mallorca, Spain, 1984.
82. Cabello, F.; Rodriguez-Torres, I.; Muñoz-Organero, G.; Rubio, C.; Benito, A.; García-Beneytez, S. *La Colección de Variedades de Vid de "El Encín"*; Consejería de Economía e Innovación Tecnológica: Madrid, Spain, 2003.
83. Pérez-Cabrero Tur, A. *Ibiza Arqueológica*; Establecimiento Gráfico Thomas: Barcelona, Spain, 1911.
84. Institut de Recerca i Formació Agroalimentaràia i Pesquera. Varietats Locals de les Illes Balears. Available online: https://varietatslocalsib.com/ca/ (accessed on 25 August 2023).
85. Zinelabidine, L.H.; Cunha, J.; Eiras-Dias, J.E.; Cabello, F.; Martínez-Zapater, J.M.; Ibáñez, J. Pedigree Analysis of the Spanish Grapevine Cultivar "Hebén". *Vitis—J. Grapevine Res.* **2015**, *54*, 81–86.

86. Mestre, A.; Forteza, B. *Apuntaciones de Viticultura y Enología*; Estación de Viticultura y Enología de Felanitx: Palma de Mallorca, Spain, 1933.
87. De los Salmones, G. Estudio general de conjunto de las viníferas españolas. In *Congreso Nacional de Viticultura: Memoria General de las Sesiones del Congreso y Ponencias Presentadas*; Ezquerro, J., Ed.; Imprenta Provincial: Pamplona, Spain, 1914; pp. 512–534.
88. Consejería de Agricultura, Pesca y Alimentación. Agricultura Amplía con 15 Variedades Autóctonas la Colección de Viña de la Finca Pública de sa Granja. Available online: https://www.caib.es/pidip2front/jsp/es/ficha-convocatoria/9451290 (accessed on 1 September 2023).
89. Junta Consultiva Agronómica. *Avance Estadístico Sobre el Cultivo y la Producción de la vid en España en 1889*; Peant e Hijos, L., Ed.; Ministerio de Fomento, Dirección General de Agricultura, Industria y Comercio: Madrid, Spain, 1891.
90. Marí, V.; (Ibiza, Spain). Personal Communication, 2020.
91. Brun, J.P. *Le Vin et L'huile Dans la Méditerranée Antique. Viticulture, Oléiculture et Procédés de Fabrication*; Errance: Paris, France, 2003; p. 204.
92. Terral, J.-F.; Tabard, E.; Bouby, L.; Ivorra, S.; Pastor, T.; Figueiral, I.; Picq, S.; Chevance, J.-B.; Jung, C.; Fabre, L.; et al. Evolution and History of Grapevine (*Vitis vinifera*) under Domestication: New Morphometric Perspectives to Understand Seed Domestication Syndrome and Reveal Origins of Ancient European Cultivars. *Ann. Bot.* **2010**, *105*, 443–455. [CrossRef]
93. Rivera Nuñez, D.; Walker, M.J. A review of palaeobotanical findings of early *Vitis* in the Mediterranean and the origins of cultivated grape-vines, with special reference to new pointers to prehistoric exploitation in the Western Mediterranean. *Rev. Palaeobot. Palynol.* **1989**, *61*, 205–237. [CrossRef]
94. Buxó, R. The Agricultural Consequences of Colonial Contacts on the Iberian Peninsula in the First Millennium BC. *Veg. Hist. Archaeobotany* **2008**, *17*, 145–154. [CrossRef]
95. Favà i Agud, X. *Diccionari Dels Noms de Ceps i Raïms. L'ampelonímica Catalana*, 1st ed.; Institut d'Estuids Catalans: Barcelona, Spain, 2001; Volume 1, pp. 33–404.
96. Laucou, V.; Launay, A.; Bacilieri, R.; Lacombe, T.; Adam-Blondon, A.F.; Bérard, A.; Chauveau, A.; de Andrés, M.T.; Hausmann, L.; Ibáñez, J.; et al. Extended Diversity Analysis of Cultivated Grapevine *Vitis vinifera* with 10K Genome-Wide SNPs. *PLoS ONE* **2013**, *2*, e0192540. [CrossRef]

horticulturae

Article

Wild Grapevine (*Vitis vinifera* L. subsp. *sylvestris* (C.C. Gmelin) Hegi)—Novel Species to the Israeli Flora

Oshrit Rahimi [1], Jotham Ziffer Berger [2], Ilana Shtein [3,4], Mafatlal M. Kher [1], Suembikya Frumin [5], Sariel Hübner [6], Ehud Weiss [5] and Elyashiv Drori [1,3,*]

1 Chemical Engineering Department, Ariel University, Ariel 40700, Israel; oshritra@ariel.ac.il (O.R.); mafatlalmkher@gmail.com (M.M.K.)
2 Department of Biology, Levinsky-Wingate Academic College, Tel Aviv 6937808, Israel; jotham.z@gmail.com
3 Eastern Regional R&D Center, Ariel 40700, Israel; ilanash@ariel.ac.il
4 Department of Molecular Biology, Ariel University, Ariel 40700, Israel
5 Archaeobotany Lab, Martin (Szusz) Department of Land of Israel Studies and Archaeology, Bar-Ilan University, Ramat-Gan 590002, Israel; suembikya.frumin@biu.ac.il (S.F.); ehud.weiss@biu.ac.il (E.W.)
6 Galilee Research Institute (MIGAL), Tel Hai Academic College, Upper Galilee 1220800, Israel; sarielh@migal.org.il
* Correspondence: droris@ariel.ac.il; Tel.: +972-54-323-168

Abstract: The wild grapevine, *Vitis vinifera* subsp. *sylvestris*, grows naturally throughout the northern hemisphere, including the Mediterranean region. Wild grapevines have also been observed sporadically across the southern Levant and are considered a non-native feral plant. Nevertheless, no formal characterization has been conducted for wild grapevines in this region; thus, its taxonomical assignment remains elusive. Previously, we have shown that the wild grapevine populations growing in northern Israel are genetically separated from the feral domesticated forms. This work aimed to comprehensively describe the morphological, anatomical, and ecological traits of wild grapevines naturally thriving in two distinct habitats in Israel. The dioicous nature of the wild grapevine, the flower and pollen morphology, and the characteristic Sylvestris fruit and seed morphology, in addition to the occurrence of the natural germination of seeds in close vicinity of the mother plant, have all led to the conclusion that these plants belong to *Vitis vinifera* subsp. *sylvestris* and should be included in the Flora Palaestina. These findings, combined with the recently published genetic evidence for these populations, significantly advance our understanding of the species' ecology and the importance of its preservation.

Keywords: Israeli flora; morphological characterization; Sylvestris; *Vitis*; *Vitis vinifera* subsp. *sylvestris*; wild grapevine species persistence

Citation: Rahimi, O.; Berger, J.Z.; Shtein, I.; Kher, M.M.; Frumin, S.; Hübner, S.; Weiss, E.; Drori, E. Wild Grapevine (*Vitis vinifera* L. subsp. *sylvestris* (C.C. Gmelin) Hegi)—Novel Species to the Israeli Flora. *Horticulturae* **2023**, *9*, 998. https://doi.org/10.3390/horticulturae9090998

Academic Editors: Ana Jiménez-Cantizano and Antonio Amores-Arrocha

Received: 31 July 2023
Revised: 29 August 2023
Accepted: 31 August 2023
Published: 4 September 2023

1. Introduction

The *Vitaceae* family mostly includes shrubs and woody lianas that climb using leaf-opposed tendrils. Most grape cultivars belong to the *Vitis* genus, which consists of 83 subordinate taxa [1] that primarily prevail in the northern Hemisphere, including North America and East Asia. The Eurasian species of common grapevine, *Vitis vinifera* subsp. *sylvestris* (C.C. Gmelin) Hegi [2] (hereafter, Sylvestris), is the best-known species, as it is the ancestor of most of the grapes cultivated today [3–5]. The cultivated form, *V. vinifera* subsp. *sativa* (hereafter, Sativa), is one of the most notable perennial crops, cultivated across 7.3 million hectares worldwide [6]. The distinction between these two subspecies is mainly based on the morphological differences in their reproductive organs, as, while the wild grapevine is dioecious, Sativa is a hermaphrodite [5,7,8].

Wild grapevine (Sylvestris) grew abundantly in natural habitats in Europe until the mid-nineteenth century, when the invasion of pests and pathogens from North America, including phylloxera and powdery and downy mildews, caused a decrease in their populations [8]. Later, accelerating urbanization processes and extensive anthropogenic land use dramatically damaged the natural habitats of the wild grapevine (Sylvestris), reducing its distribution range and endangering its persistence [9,10]. While the Sylvestris populations shrank, the cultivated grapevine (Sativa) flourished throughout Europe and the eastern Mediterranean region, where, by the end of the 19th century, according to Post's seminal work, it was "cultivated everywhere in numerous varieties, but nowhere strictly spontaneous" [11].

The southern Levant was considered a region beyond the distribution range of the wild grapevine; thus, early studies of Israeli vegetation during the 20th century considered all grapevine plants as Sativa, i.e., feral populations. Indeed, feral grapevines may emerge due to vineyard abandonment or neglect, wherein the previously cultivated grapevines begin to proliferate without human intervention. Over time, the seeds originating from these plants may germinate and be mistakenly observed as wild grapevines, adopting similar climbing growth habits. Because of the mixture of both feral and real wild subspecies in the same habitat, and the false assumption that Sylvestris does not grow in such a way down south, the sylvestris species was not considered to be native to the land. The first indication of wild grapevine in the southern Levant region was in 1994, discovered in surveys in the Upper Galilee region, along the banks of the Jordan River [12,13]. Unfortunately, the available records of these surveys lack the necessary description and exact location of the observed plants. For these reasons, until 2004, the species was not included in wild flora records in this region [14]. Over the years, more evidence of indigenous southern Levantine Sylvestris has accumulated. Fossil pips, pollen, and wood of wild grapevines were discovered in archaeological sites from the Lower Paleolithic Gesher Benot Ya'aqov (780,000 BP) through the Upper Paleolithic Ohalo II (23,000 BP) [15–19]. These sites are located around the area of the Hula Lake, the Sea of Galilee, and the Jordan River, in high geographical proximity to the populations observed in the 1994 survey.

Recently, a dedicated comprehensive survey of grapevines in Israel uncovered new populations of hermaphrodite and dioecious plants [20]. Their genetic analysis, with SSR markers and morphological characterization, identified a marked separation between the hermaphrodite accessions and the dioecious groups, with the latter displaying unequivocal wild grapevine characteristics (leaf, bunch, and berry shape, and growth habits). Moreover, the dioecious wild grapevine (Sylvestris) accessions show a genetic split between two distinct subgroups, in accordance with ecogeographic divergence. The dioecious populations, occurring primarily along the main streams in the Upper Galilee region and around the Sea of Galilee, marked the southern edge of the distribution range of the wild grapevine [21].

An in-depth analysis of whole-genome sequencing data from these accessions has further supported the previous observations that wild grapevine grew naturally in the southern Levant for millennia [22]. In fact, these wild grapevine accessions were identified as the progenitors of the domesticated indigenous varieties from the Levant, which genetically contributed to some of the European varieties [22,23]. These conclusions were further supported in a recent comprehensive genomic study of more than 3500 accessions, which provided unequivocal evidence for the contribution of southern Levantine Sylvestris populations to domesticated grapevine species around the world [24]. The Sylvestris populations from Israel (E1) were identified as the genetic source of the table-grape group (CG1) [24]. This group later genetically contributed to the most well-known wine grape varieties used in modern times worldwide.

Despite the number of studies and amount of evidence for thriving wild grapevine populations in Israel, there remains a considerable knowledge gap with regards to the evolution, geographic distribution, ecology, and domestication of this species. This work provides a systematic and comprehensive morpho-anatomical characterization of the wild grapevine population that spread between the two different sites in Israel, including a thorough characterization of the male and female flowers, the fruit and seed morphology, the distribution dynamics of the male and female individuals in each group, and natural regeneration by spontaneous germination. The findings clearly affirm that the inspected wild grapevine populations were accurately assigned to the protologue and type specimens of *Vitis vinifera* subsp. *sylvestris*.

2. Materials and Methods

Plant material and research area: The sampling was carried out during the spring (May 2022), when the plants were in full blossom, which enabled the identification of developing pistils and stamens in the dioecious (unisexual plants) or monoecious plants. The samples were collected in northern Israel, where stable populations of wild grapevines, denoted here as Sylvestris, had been previously observed [21]. Thirty-two wild grapevine accessions were collected from the Beit Tsaida site, located next to the Sea of Galilee (32°53′09.2″ N 35°38′34.6″ E), and fourteen accessions were collected from the Banias River site in the Upper Galilee region (33°12′12.3″ N 35°38′27.6″ E), from an area of about 8000 m^2 in each location (Figure 1). The geographical positions of the wild grapevine accessions were georeferenced using the free application Google Maps (Google, Inc., Mountain View, CA, USA). Branches with young and mature leaves, as well as inflorescences, were collected from each accession. A subsample of each accession was dried and prepared for Herbaria deposition, and the remaining parts were fixed and stored in FAA solution (formaldehyde: acetic acid: 70% ethanol, 1:1:20) for histomorphological inspection (samples were stored at Ariel University). The exsiccata of all accessions were stored at Tel Aviv University (voucher specimen numbers from TELA4443 to TELA4450).

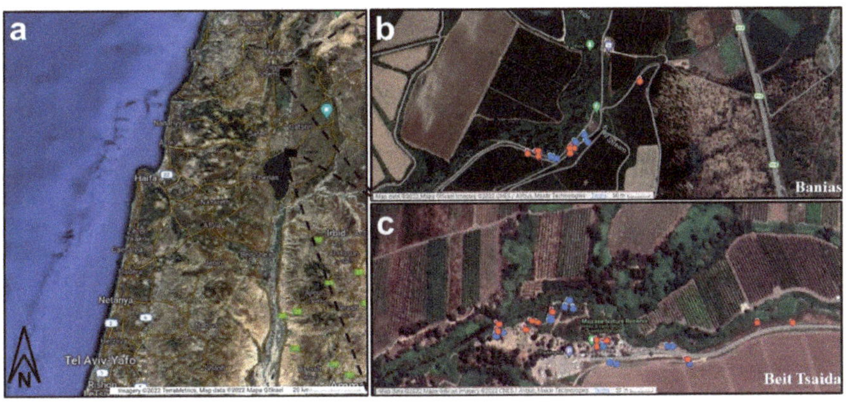

Figure 1. Study sampling sites. (**a**) Satellite map of the northern part of Israel. The black squares indicate the location of the two study areas. (**b**) An aerial photo of the Banias area and (**c**) an aerial photo of the Beit Tsaida area. Blue and red dots represent male and female wild grapevine accessions, respectively. The map was created with Google Maps App. (Mapa GISrael).

Analysis of growth variation between male and female individuals: The growth variation between male and female populations was examined by assessing the variation in internode length and diameter, using a digital caliper and a normal scale. This investigation encompassed three male and three female accessions, measuring three branches from each plant and ten internodes from each branch. The plants were collected from Ariel University's Grapevine Germplasm collection, where plants are maintained under uniform growing conditions (common garden) to ensure uniform irrigation, soil, climate, and irrigation conditions [25].

2.1. Morphological Characterization

Leaf: The leaf morphology of the dry herbarium specimens was examined. The length of the petiole and the length and width of the leaves in their greatest extension were measured. The ratio between leaf length and width and the length ratios of the blade to the petiole were then calculated. The shade differences between the abaxial and adaxial sides of the leaf lamina were recorded in an effort to determine whether the leaves were concolorous or discolorous. Leaf form was described according to the glossary in Kafkafi [26].

Seed and berry: To obtain healthy berries and seeds of optimal size for characterization, plants that were cultivated under ideal conditions at the Ariel University's Grapevine Germplasm collection, located at Ariel University, were studied. Approximately 60 seeds and 100 berries from each site were used for morphological characterization. Berry diameter and seed length and width were measured using a Sparkfun electronics digital caliper (0–15 mm), and the mean and standard deviation are presented herein (Table 1). In addition, the morphological descriptors of the International Organization of Vine and Wine (OIV) [27] were used to describe the morphological traits of the seeds and berries.

Table 1. Comparison of the main morphological features of wild grapevines collected at Beit Tsaida and Banias, Israel.

	Beit Tsaida	Banias
Abaxial/adaxial leaf color contrast	Concolor	Discolor
Blade/petiole length ratio	1.72 ± 0.92 *	1.09 ± 0.3 *
Blade shape	Not constant	Trilobate
Blade size ratio	0.8 ± 0.11 ^{ns}	0.76 ± 0.093 ^{ns}
Color of berry skin (OIV 225)	Blue black (6) #	Blue black (6) #
Berry shape (OIV 223)	Globose (2) #	Globose (2) #
Berry diameter (mm)	10.04 ± 0.90 *	11.34 ± 0.97 *
Seed shape	Ellipsoid	Ellipsoid
Seed length (OIV 242)	Very short (1) #	Very short (1) #
Seed length (mm)	5.38 ± 0.45 *	5.10 ± 0.42 *
Seed width (mm)	3.66 ± 0.29 ^{ns}	3.68 ± 0.23 ^{ns}
Observed ♀/♂ratio	0.88	1

Measured values are mean ± SD; # numbers indicate OIV description parameters [27]; * indicates significant differences between the two locations (p-value < 0.01); ns: non-significant.

Flower: Male and female flowers were placed on a glass microscope slide, with or without black paper wrap, and illuminated with a white LED light. Images were captured with a Nikon SMZ25 stereomicroscope (Nikon Ltd., Tokyo, Japan) equipped with a camera (Nikon DS-Ri2). Sixty digital photomicrographs (resolution: 4908 × 3264 pixels) were taken at different focal planes, at ~50 μm intervals, and compiled into a single image using ND2-NIS elements software version 5.02.02, with an extended depth of focus (EDF) patch (Nikon Instruments, Tokyo, Japan). The images were then transformed into a single high-quality focused image using dedicated software.

Pollen: Dried anthers with pollen grains were coated with gold using a sputter machine (Quorum Q-150T ES). The prepared samples were then imaged with a Field Emission Scanning Electron Microscope (FE-SEM) (Tescan Ultra-High-Resolution MAIA 3), with a beam voltage of 1.0 kV and an SE detector.

Tissue histology: Tissue samples of male and female flowers were fixed in FAA, as described above, embedded in paraplast, sectioned to a 12-μm thickness with a rotary microtome (SLEE medical GmbH, Nieder-Olm, Germany) and stained with safranin-alcian blue [28]. The slides were photographed under an Olympus SZX7 stereomicroscope (Tokyo, Japan) equipped with a camera (Pixelink USB 3.0 version PL-D795CU, Ottawa, ON, Canada) using the PixeLINK Capture program.

2.2. Statistical Analysis

A statistical analysis was carried out on each dataset to compute the mean, standard deviation, and minimum and maximum values. The normality of distribution was evaluated for each variable. In cases where a normal distribution was ascertained, a t-test was executed. Cases of non-normal distribution were studied by applying a Wilcoxon 2-sample test. A significant difference, defined by a p-value < 0.05, served as the basis for inferring equal variances. To explore the morphological attributes of the leaf and petiole measurements, a principal component analysis (PCA) was performed on the correlation matrix. All statistical analyses were performed using JMP® Pro 16.0.0 software [29].

3. Results and Discussion

Sylvestris growth habits in wild habitats: This study focused on the two sites where stable and large grapevine populations had been previously reported [21]: the northern site at the Banias River and the southern site at Beit Tsaida (Figure 1). Both of the sites are located on alluvial soils within protected natural reserves (Banias and Majrase). The sampling was performed randomly along the water streams where grapevines grow. Though the sampling areas were of equal size (circa 8000 m^2), more individuals were found at the Banias site than at the Beit Tsaida site. In total, forty-six accessions were collected, with thirty-two wild grapevine accessions originating from Beit Tsaida and fourteen accessions from the Banias River. The variation in the population size of the wild grapevine accessions between the Beit Tsaida and Banias River areas can likely be attributed to factors such as differing riverbank sizes, micro-climatic conditions, wildfire frequencies and intensity, nutrient availability, ecological characteristics, fluctuation in the water level of the Sea of Galilee, and varying levels of site disturbance due to anthropogenic activities [21]. Notably, the Banias River area exhibited a reduction in species diversity due to pronounced anthropogenic activities [30].

All *Vitis* plants at both of the sampling sites were dioicous, supporting their identification as true Sylvestris. The ratio between the male and female plants was close to 1, with a slight dominance of male accessions (Table 1). This finding strengthens the assumption that the examined populations are wild, with no bias toward the fructiferous female form [31].

The growth habits varied between the male and female grapevines at both locations. The male grapevines were taller and tended to climb to the tops of trees, while the female grapevines were generally shorter and tended to prostrate in a tangle of low vegetation (Figure 2a,b). The male grapevines were characterized by multiple and densely packed flower clusters, usually located near the top of the vine (Figure 2b), while the females produced fewer and smaller flower clusters (Figure 2c), which were located lower along the stem. The male climbing habits can be attributed to the wind-pollination strategy [9], and the low stature of the females may be required to structurally support the heavy bunches of fruits. The sexual dimorphism observed in the plant height and inflorescence position seem to be adaptive to the dioecy and pollination by wind and insects, i.e., to a cross-pollination reproductive system. The taller growth of the male population aligned with their significantly longer internode length and wider stem diameter, as compared to the female accessions grown in a common garden (Figure 3).

In the current survey, the grapevines grew in deep uncultivated soils very close to flowing sweet water streams, as was previously reported [21]. The male and female plants were spread randomly. At both of the sites, the wild grapevines grew in proximity to fig trees (*Ficus carica*), plane trees (*Platanus orientalis*), and holy raspberries (*Rubus sanguineus*), a plant community typical to water-rich habitats along the Mediterranean basin. Interestingly, the observed Sylvestris female plants tended to grow between the spiny holy raspberry plants, which, apparently, provide protection during flowering against herbivores that are abundant in these regions, including wild boar (*Sus scrofa*) [32] and mountain gazelle (*Gazella gazella*) [33], which commonly feed on grapevine shoots and leaves in cultivated vineyards. Spiny vegetation has been suggested to have played a vital role in protecting wild vines from both wild and domestic animals [34,35]. Overall, the observed populations at both sites had a distinctive appearance, with a significant polymorphism in the leaves, and clusters of small, greenish-yellow flowers, which developed into black grape berries. Their growth habits and woody stems make them hardy plants, providing cover, and serving as a habitat for various animals sharing the same natural environment [36,37].

Figure 2. Growth habits of Sylvestris plants in northern Israel. (**a**) Male grapevine climbing on a tree in Beit Tsaida. (**b**) Female grapevine by the side of the road in Banias. (**c**) Male inflorescence and (**d**) female inflorescence from Beit Tsaida.

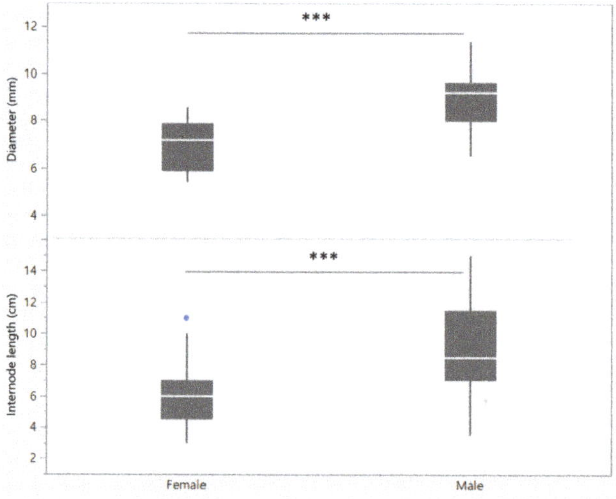

Figure 3. Differences in internode length and diameter between male and female wild grapevines. Box plot comparison of the internode length (cm) and diameter (mm) of three male and female wild grapevine accessions from a common garden. The horizontal white lines in the graph represent the median values. The boxes indicate the range between the 25th and 75th percentile of each group's distribution of values. Observation that falls outside of this range is denoted by a blue dot. *** p value: < 0.001.

4. Morphology

Leaf morphology: The leaf shape and morphology were highly polymorphic, ranging from reniform, with weak lobation, to cordate with pronounced lobation. No significant correlation was observed between the blade (length—width) ratios (p-value > 0.05) and the sampling location (Table 1), presumably due to the wide variation in leaf length. The blade—petiole-length ratio of the wild grapevine population was higher in Beit Tsaida (p-value < 0.001) than in Banias. The dorsal surface of the wild grapevine leaves from Banias was hairy, in contrast to the grapevine leaves from Beit Tsaida [21]. This resulted in a shade contrast, with the leaves in the Banias populations being discolored, while the Beit Tsaida populations had concolor leaves. The PCA of the leaf and petiole data revealed a very weak separation between the two populations, where the first two principal components explained 89.2% of the variation in the analysis (Supplementary Figure S1). The variation in the PCA was explained mainly by the leaf length (43%), followed by the leaf—petiole length ratio (23%), which further support the above-mentioned statistical analysis and corroborate the slight separation of the populations derived from a greater number of samples and other OIV attributes in our prior study [21]. Taken together, it is reasonable to conclude that the sampled populations do not represent discrete entities, but rather constitute a single population that underwent adaptive changes in response to diverse environmental conditions across the different geographic regions, thereby giving rise to specific morphological characteristics.

Berry and seed morphology: The growing conditions have a dramatic effect on the number and size of grape berries and seeds. To obtain healthy, optimal- sized berries and seeds for inspection and characterization, cuttings were sampled from plants at both of the sites and were grown under optimal conditions at the Ariel University's Grapevine Germplasm collection, at Ariel University [25]. When they were cultivated at the university, the fruit bunches were well-grown and dense (Figure 4a), while the bunches were sporadic when grown in the wild (Figure 4b). A broad range of polymorphism in the cluster shapes was observed among the samples, yet most of them had sparse clusters of tiny, typically black, berries, each typically with 2–3 seeds (Figure 4c). The berry diameter was significantly different (p-value < 0.001) between the populations, with Banias Sylvestris berries being larger than those of Beit Tsaida (Table 1). In both populations, the berry skin was thin, and the pulp was juicy and sweet, with high acid levels, which were, however, lower than those reported for European Sylvestris grapes [20]. The seed length differed significantly between the sampling sites (t-test, p-value < 0.001), and was longer in Beit Tsaida (mean = 5.38, sd = 0.45) (Table 1). These seed values correspond to the previously recorded Sylvestris varieties [38,39].

Flower morphology: It is well established that flower development, and, more specifically, that of the reproductive organs, is the main feature that distinguishes the differences between wild and domesticated grapevines [8]. The female flower includes an ovary and reflexed rudimentary/atrophied stamens that angle downwards, while the male flower displays upright stamens and a reduced pistil, without a style or stigma, and a rudimentary/atrophied ovary at the base (Figure 5). These features are clearly distinctive to Sylvestris plants, while the Sativa forms, also found occasionally growing feral, all have a hermaphrodite phenotype [20]. Grapevine flowering in the studied areas occurred in the month of May, giving rise to fruit in the female individuals in the summer (August). In the present analysis, the flowers of the wild grapevines were small, greenish-yellow, and arranged in panicles. The individual flowers had a diameter of around 5 mm and contained five petals fused at the base. *Vitis vinifera* subsp. *sylvestris* is the wild ancestor of the cultivated grapevine varieties of *Vitis vinifera* [24]. In the case of Sylvestris, the presence of both male and female plants in wild populations requires crossbreeding for the formation of fruit, while the domesticated types are hermaphroditic and can self-pollinate [40]. The hermaphroditic nature of the Sativa types was an essential step of the domestication process, leading to stable yields and full bunches, due to better pollination and the possibility of planting only reproductive plants without the need for male pollinator plants [8].

Figure 4. Morphological traits of wild grapevine berries and seeds collected from northern Israel. Female grapevine bunches (**a**) from Ariel University's Grapevine Germplasm collection, at Ariel University, and (**b**) from the wild (Beit Tsaida). (**c**) A whole grape berry next to a sectioned berry (Accession number 22, from Ariel University's Grapevine Germplasm collection). (**d**) Ventral (right) and dorsal (left) sides of the seeds (Accession number 189, from Ariel University's Grapevine Germplasm collection) on graph paper with 1-mm squares.

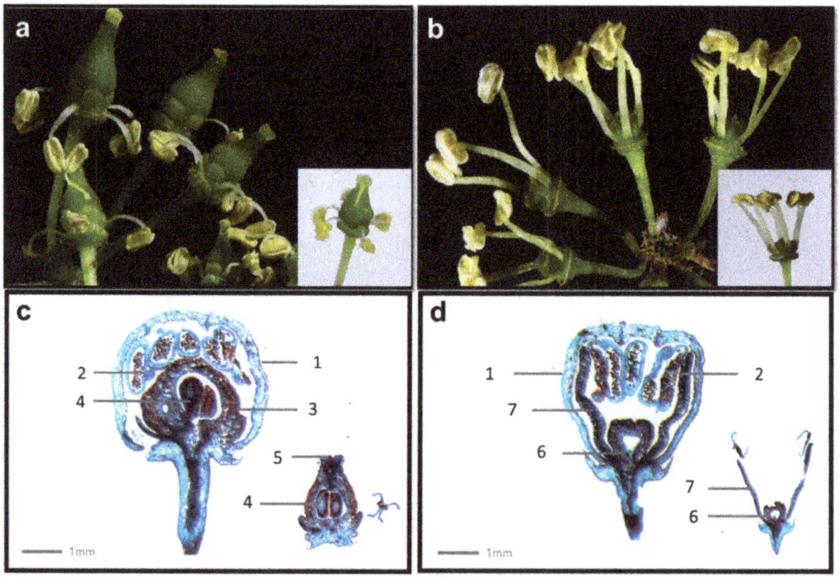

Figure 5. Female and male flowers of wild grapevines and their histological sections (**a**) Female and (**b**) male wild grapevine specimens at flowering. Histological sections of closed and open (lower right) (**c**) female and (**d**) male flowers. Parts of the flower: 1—Petal, 2—Anther, 3—Ovary, 4—Ovules, 5—Stigma, 6—Atrophied ovary remnant, and 7—Anther filament.

The histological sections of the female grapevine flower showed a single ovary (Figure 5c), a style, and a stigma, which comprised the pistil. The ovary was located at the base of the flower and contained ovules that eventually developed into seeds, if fertilized with pollen (Figure 5c). The structure of the male flower was distinctly different from that of the female flower (Figure 5). The male grapevine flower, sampled at the same location, consisted of an atrophied ovary and stamens that were arranged in a tight cluster at the base of the flower. The stamens produced and released pollen grains that were transferred by wind and/or insects to the female flowers, allowing for fertilization and fruit production [41].

Pollen grain morphology: A SEM scan was utilized to image the pollen of the Sylvestris accessions sampled at the experimental vineyard (Figure 6a,b). In the male flowers, the pollen grains exhibited tricolpate morphology (with three furrows) and were ellipsoidal in shape (Figure 6b). In the female flowers, the pollen grains were inaperturate and spheroidal in shape (Figure 6a). Additionally, the pollen grains found on the anthers in the female flowers appeared collapsed or exhausted, while being potent in the male flowers, further supporting dioecy. The morphology of the grapevine pollen grains was consistent with previous findings in studies comparing sterile and fertile pollen grains in Sylvestris [42,43].

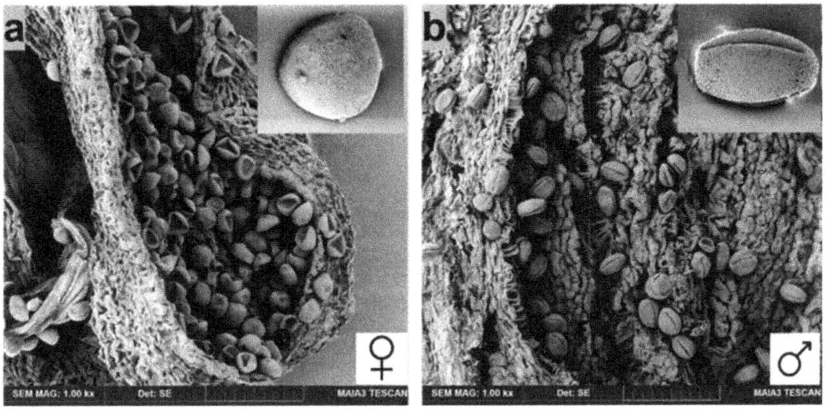

Figure 6. Scanning electron microscope images of pollen grains produced by (a) female and (b) male flowers of wild grapevine (*Vitis vinifera* subsp. *sylvestris*) inside the anther. The scale bar is 50 µm. Enlarged single pollen grains are presented on the top right of each image.

Natural seed germination: A thorough survey of the Sylvestris habitats in the wild was conducted in search of the natural germination of seedlings. In the Banias area, five seedlings were found, all beneath female plants (Figure 7a). This germination habit was abundant, indicating its success under the ecological conditions of this specific natural habitat. The young plantlets were carefully removed from the soil, including the remnants of their outer integuments (Figure 7c,d). The grapevine seeds were clearly identifiable, despite being slightly damaged and soiled. To the best of our knowledge, this is the first time that the natural germination of Sylvestris in its natural habitat has been recorded. The occurrence of natural Sylvestris germination in a natural habitat provides strong evidence for the spontaneous nature of the population and its persistent strength as a stable population—as an indigenous plant in Israel. Due to the growth habits of Sylvestris inside of a dense bush of spiny raspberry plants, young seedlings were not identified in the Beit Tsaida area.

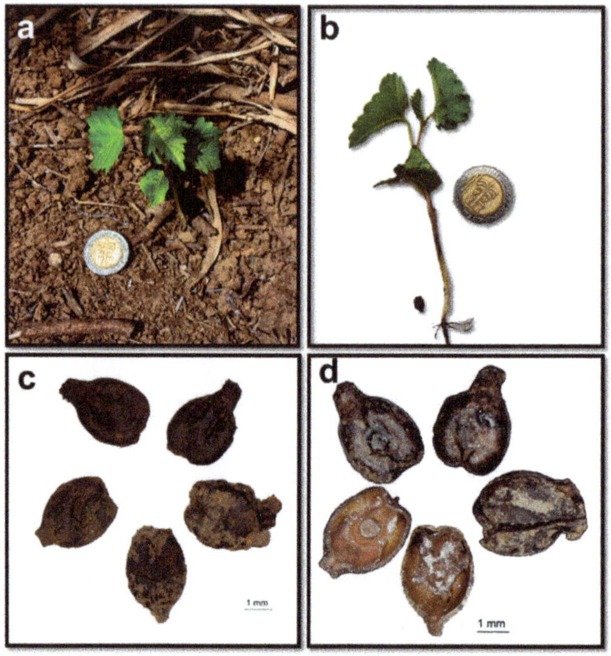

Figure 7. ***Vitis vinifera*** **subsp.** ***sylvestris*** **germination in the field.** (**a**) Naturally germinated grapevine seedling found in the Banias area of north Israel under female grapevines (coin diameter 23 mm), and (**b**) the seedling following its extraction from the soil, including the remnant integument (coin diameter 23 mm). (**c**,**d**) Wild grape seeds (c—ventral side, d—dorsal side), which were carefully dug out and removed from the ground.

To summarize, previously identified grapevines growing in the wild in the northern part of Israel were considered to be feral Sativa plants, and there was no confidence as to the occurrence of *Vitis vinifera* subsp. *sylvestris* in Israel. This was likely due to the failure to carry out a comprehensive survey, and the minimal description of this population by the surveyors, who provided only brief notes on the species [12,13]. The present work has systematically described the habitats, growth habits, morphology, and anatomy of widely spread wild grapevine populations growing in two distinct habitats in north Israel. All of the findings, including the dioecious nature of the wild grapevine plants, the sexual dimorphism between the male and female plants, the characteristic traits of their flower development, pollen, berry size and morphology, seed structure, and the spontaneous regeneration of the population from seeds, together with our previous genomic findings, showing a clear separation of these populations from feral Sativa accessions [20,22], all lead to the conclusion that wild grapevine populations grow naturally in Israel. The results indicate that wild grapevines occur in natural habitats located within the region of their ancient area of appearance during the Pleistocene, as corroborated by archaeological findings [24].

This botanical description clears up the previous uncertainty as to the definition of these populations, and is of particular significance in light of the emerging notion that these wild populations are representatives of the core population from which the cultivated grapevine was first domesticated, circa 11,000 years ago [24]. These emerging new data emphasize the significance of the conservation of the environmental conditions and biodiversity of the Sea of Galilee, the prevention of drastic water level fluctuations, and the preservation of the stream banks and forest habitats of the Upper Galilee region, as the main habitats of this important population.

5. Conclusions

The research presented here supports the persistence of the wild grape species in the Israeli flora, extending the southern edge of its global distribution. As a result, Israel can be confidently added to the map of the native distribution of wild *Vitis vinifera*, and this plant can be included in the Flora Palaestina.

Supplementary Materials: The following supporting information can be downloaded at: https://www.mdpi.com/article/10.3390/horticulturae9090998/s1, Figure S1: PCA (principal component analysis) biplot analysis performed on 46 Sylvestris accessions, using the leaves and petioles measured data.

Author Contributions: Conceptualization, O.R., E.D., S.F. and E.W.; methodology, O.R., E.D., I.S. and J.Z.B.; validation, O.R., E.D., I.S. and J.Z.B.; formal analysis, O.R. and J.Z.B.; investigation, O.R., S.F. and E.D.; resources, E.D.; data curation, O.R. and J.Z.B.; writing—original draft preparation, O.R., M.M.K. and E.D.; writing—review and editing, O.R., J.Z.B., I.S., M.M.K., S.F., S.H., E.W. and E.D.; visualization, O.R., M.M.K., E.D. and S.F.; supervision, E.D. and S.H.; project administration, E.D.; funding acquisition, E.D. All authors have read and agreed to the published version of the manuscript.

Funding: This work was supported by the Israeli Ministry of Innovation, Science, and Technology (MOST) and the Eastern Regional R&D Center.

Acknowledgments: We thank Olga Krichevski for her great help with the SEM imaging and INPA (Israel Nature and Parks Authority), especially to Margareta Walczak for help with data for the Nature Reserves.

Conflicts of Interest: The authors declare no conflict of interest.

References

1. WFO The World Flora Online. Available online: http://www.worldfloraonline.org/taxon/wfo-4000040377 (accessed on 30 July 2023).
2. Ferrer-Gallego, P.P.; Ferrer-Gallego, R.; Laguna, E.; Pipia, I. (2682) Proposal to Conserve the Name *Vitis sylvestris* C.C. Gmel. (Vitaceae) against *V. sylvestris* W. Bartram. *Taxon* **2019**, *68*, 409–410. [CrossRef]
3. Hegi, G. *Vitis* L. In *Illustrierte Flora von Mittel Europa (5,1 ([1925]) Dicotyledones: (III. Teil)*; Karl Hans Verlag: Munich, Germany, 1925; pp. 359–361. (In German)
4. Heywood, V.H.; Zohary, D. A Catalogue of the Wild Relatives of Cultivated Plants Native to Europe. *Flora Medit* **1995**, *5*, 375–415.
5. Zohary, D.; Spiegel-Roy, P. Beginnings of Fruit Growing in the Old World. *Science.* **1975**, *187*, 319–327. [CrossRef] [PubMed]
6. International Organisation of Vine and Wine (OIV). *State of the World Vine and Wine Sector*; Roca, P., Ed.; International Organisation of Vine and Wine Intergovernmental Organisation: Dijon, France, 2023; p. 4.
7. Caporali, E.; Spada, A.; Marziani, G.; Failla, O.; Scienza, A. The Arrest of Development of Useless Reproductive Organs in the Unisexual Flower of *Vitis vinifera* Ssp *silvestris*. In *Acta Horticulturae*; ISHS: Kecskemet, Hungary, 2003; Volume 603, pp. 225–228.
8. This, P.; Lacombe, T.; Thomas, M.R. Historical Origins and Genetic Diversity of Wine Grapes. *Trends Genet.* **2006**, *22*, 511–519. [CrossRef] [PubMed]
9. Di Vecchi-Staraz, M.; Laucou, V.; Bruno, G.; Lacombe, T.; Gerber, S.; Bourse, T.; Boselli, M.; This, P. Low Level of Pollen-Mediated Gene Flow from Cultivated to Wild Grapevine: Consequences for the Evolution of the Endangered Subspecies *Vitis vinifera* L. Subsp. *silvestris*. *J. Hered.* **2009**, *100*, 66–75. [CrossRef] [PubMed]
10. Ocete, R.; López, M.Á.; Gallardo, A.; Arnold, C. Comparative Analysis of Wild and Cultivated Grapevine (*Vitis vinifera*) in the Basque Region of Spain and France. *Agric. Ecosyst. Environ.* **2008**, *123*, 95–98. [CrossRef]
11. Post, G.E. *Flora of Syria, Palestine, and Sinai, from the Taurus to Ras Muhammad, and from the Mediterranean Sea to the Syrian Desert*; Syrian Protestant College: Beirut, Syria, 1896.
12. Zohary, D.; Hopf, M. *Domestication of Plants in the Old World*, 2nd ed.; Cambridge University Press: Cambridge, UK, 1994; Volume 30.
13. Rottenberg, A. Sex Ratio and Gender Stability in the Dioecious Plants of Israel. *Bot. J. Linn. Soc.* **1998**, *128*, 137–148. [CrossRef]
14. Danin, A. *Distribution Atlas of Plants in the Flora Palaestina Area*; Israel Academy of Sciences and Humanities: Jerusalem, Israel, 2004.
15. Weinstein-Evron, M. The Paleoecology of the Early Wurm in the Hula Basin, Israel. *Paléorient* **1983**, *9*, 5–19. [CrossRef]
16. Kislev, M.E.; Nadel, D.; Carmi, I. Epipalaeolithic (19,000 BP) Cereal and Fruit Diet at Ohalo II, Sea of Galilee, Israel. *Rev. Palaeobot. Palynol.* **1992**, *73*, 161–166. [CrossRef]
17. Weiss, E.; Kislev, M.; Simchoni, O.; Nadel, D. Morphogenetics of Dicots and Large- and Small-Grained Wild Grasses from the Paleolithic Era (Old Stone Age) Ohalo II., Israel (23,000 Bp). In *Plant Archaeogenetics*; Gyulai, G., Ed.; Nova Science Publishers: New York, NY, USA, 2011; pp. 23–30.

18. Goren-Inbar, N.; Alperson, N.; Kislev, M.E.; Simchoni, O.; Melamed, Y.; Ben-Nun, A.; Werker, E. Evidence of Hominin Control of Fire at Gesher Benot Ya'aqov, Israel. *Science* **2004**, *304*, 725–727. [CrossRef]
19. Melamed, Y.; Kislev, M.E.; Geffen, E.; Lev-Yadun, S.; Goren-Inbar, N. The Plant Component of an Acheulian Diet at Gesher Benot Ya'aqov, Israel. *Proc. Natl. Acad. Sci. USA* **2016**, *113*, 14674–14679. [CrossRef]
20. Drori, E.; Rahimi, O.; Marrano, A.; Henig, Y.; Brauner, H.; Salmon-Divon, M.; Netzer, Y.; Prazzoli, M.L.; Stanevsky, M.; Failla, O.; et al. Collection and Characterization of Grapevine Genetic Resources (*Vitis vinifera*) in the Holy Land, towards the Renewal of Ancient Winemaking Practices. *Sci. Rep.* **2017**, *7*, 44463. [CrossRef]
21. Rahimi, O.; Ohana-Levi, N.; Brauner, H.; Inbar, N.; Hübner, S.; Drori, E. Demographic and Ecogeographic Factors Limit Wild Grapevine Spread at the Southern Edge of Its Distribution Range. *Ecol. Evol.* **2021**, *11*, 6657–6671. [CrossRef]
22. Sivan, A.; Rahimi, O.; Lavi, B.; Salmon-Divon, M.; Weiss, E.; Drori, E.; Hübner, S. Genomic Evidence Supports an Independent History of Levantine and Eurasian Grapevines. *Plants People Planet* **2021**, *3*, 414–427. [CrossRef]
23. Drori, E.; Levy, D.; Smirin-Yosef, P.; Rahimi, O.; Salmon-Divon, M. CircosVCF: Circos Visualization of Whole-Genome Sequence Variations Stored in VCF Files. *Bioinformatics* **2017**, *33*, 1392–1393. [CrossRef]
24. Dong, Y.; Duan, S.; Xia, Q.; Liang, Z.; Dong, X.; Margaryan, K.; Musayev, M.; Goryslavets, S.; Zdunić, G.; Bert, P.-F.; et al. Dual Domestications and Origin of Traits in Grapevine Evolution. *Science* **2023**, *379*, 892–901. [CrossRef]
25. Shecori, S.; Kher, M.M.; Tyagi, K.; Lerno, L.; Netzer, Y.; Lichter, A.; Ebeler, S.E.; Drori, E. A Field Collection of Indigenous Grapevines as a Valuable Repository for Applied Research. *Plants* **2022**, *11*, 2563. [CrossRef]
26. Kafkafi, I. *Biological Source Book for Scientific Names*; Hakibbutz Hameuchad: Bnei Brak, Israel, 1988; p. 383. (In Hebrew)
27. Organisation Internationale de la Vigne et du Vin (OIV) Second Edition of the OIV Descriptor List for Grape Varieties and Vitis Species. Available online: https://www.oiv.int/public/medias/2274/code-2e-edition-finale.pdf (accessed on 10 August 2023).
28. Wolberg, S.; Haim, M.; Shtein, I. Simple Differential Staining Method of Paraffin-Embedded Plant Sections with Safranin-Alcian Blue. *IAWA J.* **2023**, *44*, 170–175. [CrossRef]
29. MP Prof, SAS Institute Inc. JMP®Pro *Version* 16. Available online: https://www.jmp.com/en_gb/home.html (accessed on 10 August 2023).
30. Koner, L.; Avisar, A.; Yanay, Z. *Nahal Hermon and the Springs of the Banias—A Case Study about Drought and Climate Change (Report no. 7)*; DSHA Institute: Tel Aviv, Israel, 1987; pp. 1–54. Available online: https://deshe.org.il/ (accessed on 10 August 2023). (In Hebrew)
31. Fisher, R.A. *The General Theory of Natural Selection*; Clarendon Press: Oxford, UK, 1930; p. 272.
32. Ditchkoff, S.S.; Mayer, J.J. Biology of Wild Pigs: Wild Pig Food Habits. In *Wild Pigs: Biology, Damage, Control Techniques and Management*; Mayer, J.J., Brisbin, I.L., Eds.; Savannah River National Laboratory: Aiken, SC, USA, 2009; pp. 105–143. ISBN 9780080248219.
33. Mauri, L.; Masin, R.; Tarolli, P. Wildlife Impact on Cultivated Lands: A Multi-Temporal Spatial Analysis. *Agric. Syst.* **2020**, *184*, 102890. [CrossRef]
34. Ocete Rubio, R.; Ocete Rubio, E.; Ocete Pérez, C.; Ángeles Pérez Izquierdo, M.; Rustioni, L.; Failla, O.; Chipashvili, R.; Maghradze, D. Ecological and Sanitary Characteristics of the Eurasian Wild Grapevine (*Vitis vinifera* L. Ssp. *sylvestris* (Gmelin) Hegi) in Georgia (Caucasian Region). *Plant Genet. Resour.* **2012**, *10*, 155–162. [CrossRef]
35. Ronel, M.; Lev-Yadun, S. The Spiny, Thorny and Prickly Plants in the Flora of Israel. *Bot. J. Linn. Soc.* **2012**, *168*, 344–352. [CrossRef]
36. Arnold, C.; Schnitzler, A.; Douard, A.; Peter, R.; Gillet, F. Is There a Future for Wild Grapevine (*Vitis vinifera* Subsp. silvestris) in the Rhine Valley? *Biodivers. Conserv.* **2005**, *14*, 1507–1523. [CrossRef]
37. Zito, P.; Serraino, F.; Carimi, F.; Tavella, F.; Sajeva, M. Inflorescence-Visiting Insects of a Functionally Dioecious Wild Grapevine (*Vitis vinifera* Subsp. *sylvestris*). *Genet. Resour. Crop Evol.* **2018**, *65*, 1329–1335. [CrossRef]
38. Cervantes, E.; Martín-Gómez, J.J.; Espinosa-Roldán, F.E.; Muñoz-Organero, G.; Tocino, Á.; de Santamaría, F.C.S. Seed Morphology in Key Spanish Grapevine Cultivars. *Agronomy* **2021**, *11*, 734. [CrossRef]
39. Susaj, L.; Susaj, E.; Ferraj, B.; Dragusha, B. Identification of the Main Characters and Accompanying Plants of Wild Type Grapevine [*Vitis vinifera* L. Ssp *sylvestris* (Gmelin) Hegi], through Shkrelis Valley, Malësia e Madhe. In Proceedings of the Vth Annual International Conference "Albanians and Europa 2020—Statement and Prospects", Prishtina, Kosovo, 1–3 October 2010; p. 111.
40. Cunha, J.; Baleiras-Couto, M.; Cunha, J.P.; Banza, J.; Soveral, A.; Carneiro, L.C.; Eiras-Dias, J.E. Characterization of Portuguese Populations of *Vitis vinifera* L. Ssp. sylvestris (Gmelin) Hegi. *Genet. Resour. Crop Evol.* **2007**, *54*, 981–988. [CrossRef]
41. Zito, P.; Scrima, A.; Sajeva, M.; Carimi, F.; Dötterl, S. Dimorphism in Inflorescence Scent of Dioecious Wild Grapevine. *Biochem. Syst. Ecol.* **2016**, *66*, 58–62. [CrossRef]
42. Lukšić, K.; Zdunić, G.; Mucalo, A.; Marinov, L.; Ranković-Vasić, Z.; Ivanović, J.; Nikolić, D. Microstructure of Croatian Wild Grapevine (*Vitis vinifera* Subsp. sylvestris Gmel Hegi) Pollen Grains Revealed by Scanning Electron Microscopy. *Plants* **2022**, *11*, 1479. [CrossRef] [PubMed]
43. Jovanovic-Cvetkovic, T.; Micic, N.; Djuric, G.; Cvetkovic, M. Pollen Morphology and Germination of Indigenous Grapevine Cultivars Zilavka and Blatina (*Vitis vinifera* L.). *Agrolife Sci. J.* **2016**, *5*, 105–109.

horticulturae

Article

Grapevine in the Ancient Upper Euphrates: Horticultural Implications of a Bayesian Morphometric Study of Archaeological Seeds

Javier Valera [1,*], Gonzalo Matilla-Seiquer [1], Concepción Obón [2], Francisco Alcaraz [3] and Diego Rivera [3,*]

[1] Departamento de Prehistoria, Arqueología, Historia Antigua, Historia Medieval y Ciencias y Técnicas Historiográficas, Facultad de Letras, Campus de la Merced, Universidad de Murcia, C/Santo Cristo, 1, 30001 Murcia, Spain; gmatilla@um.es

[2] Departamento de Biología Aplicada, EPSO, Universidad Miguel Hernández de Elche, Carretera de Beniel Km 3.2, 03312 Orihuela, Spain; cobon@umh.es

[3] Departamento de Biología Vegetal, Facultad de Biología, Universidad de Murcia, 30100 Murcia, Spain; falcaraz@um.es

* Correspondence: javier.valera@um.es (J.V.); drivera@um.es (D.R.); Tel.: +34-670973968 (J.V.); +34-868884994 (D.R.)

Abstract: The origins of the main cultivar groups of *Vitis vinifera*, their relationships with wild grapevine populations, and the use of other *Vitaceae* are relevant issues for the improvement and conservation of *Vitis* diversity. Morphometric studies, domestication indices, multivariate analyses, and Bayesian hypothesis testing have been used. Eight different seed types have been identified in the 24 samples analyzed from materials from the Upper Euphrates sites of Tell Khâmis and Tell Qara Quzaq (Early Bronze Age to Hellenistic), ranging from highly domesticated to purely wild. We have been able to establish the predominance among the domesticated of *Proles orientalis* Negrul (three samples, Domestication Index = 1), the existence of and extinct *Proles euphratica* (six samples, Domestication Index = 0.67–0.83) and numerous intermediates and hybrids (eight samples). We have determined the continued presence throughout the period studied of wild grapevines related to *Vitis sylvestris* C.C.Gmelin and *V. caucasica* Vavilov (5 samples, with Domestication Indices = 0.17–0.5). The existence of *Ampelopsis* seeds was established for three samples. We determined that the oldest *Vitaceae* seed linked to human presence, in the Acheulense (780 myr), also belongs to *Ampelopsis*. Finally, "stenosperms" appear associated with *Ampelopsis* seeds (three samples), suggesting anomalies in seed formation due to intergeneric cross-pollination. Moreover, if isolated, they suggest the presence of "stenospermocarpic" *Vitis vinifera* raisins of the *Sultanina* type. Finally, we must reflect on the role that *Ampelopsis* species may have played and their possible cultivation and domestication almost 4000 years ago.

Keywords: ampelography; archaeobotany; oenology; viticulture; plant genetic resources

Citation: Valera, J.; Matilla-Seiquer, G.; Obón, C.; Alcaraz, F.; Rivera, D. Grapevine in the Ancient Upper Euphrates: Horticultural Implications of a Bayesian Morphometric Study of Archaeological Seeds. *Horticulturae* **2023**, *9*, 803. https://doi.org/10.3390/horticulturae9070803

Academic Editors: Antonio Amores-Arrocha and Ana Jiménez-Cantizano

Received: 19 June 2023
Revised: 9 July 2023
Accepted: 11 July 2023
Published: 13 July 2023

1. Introduction

1.1. Grapevine Relevance and Diversity

The grapevine, olive, date palm, fig, and pomegranate constitute the core of domesticated fruit species in Western Asia and the Mediterranean [1]. The fruits of the grapevine, *Vitis vinifera* L. (*Vitaceae*), can be consumed directly as table grapes, dried as raisins, or pressed into a must that can be fermented into wine, which contains 12–17% alcohol. The consumption of alcoholic beverages, and not only grape wine, was an important element of the nutrition, ritual and economy of early societies in Mesopotamia, Egypt, Syria, and the Levant [2–4].

Domesticated grapevine belongs to the Genus *Vitis* (*Vitaceae*), which comprises two subgenera and over 60 species. Grapevine (*Vitis vinifera* L.) is widely cultivated, especially in Mediterranean-type climates. More than 40.000 grapevine cultivar names exist worldwide, corresponding to a little more than 15.000 grapevine genotypes [5–7]. Grapevine is a glycophyte, so with low salt tolerance (ClNa up to 40 mM, EC close to 4 dS/m), concentrations of 80 mM (EC close to 8 dS/m) produce significant damage [8].

Wild (*Vitis sylvestris* C.C.Gmelin) and cultivated (*V. vinifera*) grapevines mainly differ in their reproductive biology. Wild grapevines are dioecious, with males producing great quantities of pollen; on the other hand, most cultivated grapevines are self-pollinated hermaphrodites, producing small pollen amounts [9]. Negrul [10,11] argued in 1946 that hermaphrodite-cultivated grapes result from the selection of hermaphrodite branches accidentally appearing in male *V. sylvestris*. According to Sosnovszky [12,13], the ancestors of *Vitis* had bisexual flowers, and unisexual development is the result of reduction through evolution. Some cultivars, such as *Ohanes* and *Bicane*, are functionally female and may require assisted pollination.

Four main theories on the origin of cultivated grapevine have been published [14] with their variants. They are summarized as follows:

1. Monophyletic and Monospecific: Local populations of cultivated grapevine descend from local wild populations. Both are conspecific. This theory was proposed between 1882 and 1946 by De Candolle [15], Hegi [16], Planchon [17], Baranov et al. [18], and Negrul [10] and has been clearly supported by Levadoux [19], who also refers to some Pliocene *Vitis* fossils from Europe, known as *V. parasylvestris* Kirch., *V. tokayensis* St., or *V. ausoniae* Gaud. et Str, as conspecific.
2. Monophyletic and Bispecific: Cultivated grapevine descend from an extinct ancestor that is also presumably an ancestor of wild grapevine, being both two distinct separate species. Occasional hybridization may have produced some cultivars or cultivar groups [20]. Sosnovszky [12,13] stated that the Eurasian cultivated grapevine did not directly derive from *V. sylvestris*, which is morphologically well distinct from *V. vinifera* and extremely polymorphic, with its own history, geographical area, and natural habitat. This author [12,13] suggests that *V. sylvestris* and *V. vinifera* developed independently from a bisexual extinct ancestor who gave place to diverse types of cultivated grapevines; it is quite possible that the cultivated grapevine consists of an anthropogenic hybrid swarm involving crossing with *V. sylvestris* of several extinct *Vitis*.
3. Polyphyletic and Multispecific: Regional populations of cultivated grapevine descend from different wild ancestors extinct or not. Cultivated grapevine is divided into species with their corresponding wild relatives. The primary species hybridized, producing new cultivar groups. In 1925, Andrasovzsky [21] recognized five fundamental species, organized geographically, and the offspring of bispecific crosses between them, as well as pedigrees involving three species.
4. Hybrid Hypothesis: Cultivated grapevine descend through hybridization from wild European and Asiatic grapevines. Terpó [22] attributes the origin of cultivated grapevines to the domestication and crossing among populations of at least two species: *Vitis sylvestris* Gmel. (dioecious) and *Vitis nuristanica* Vassilcz. (hermaphrodite).

In 1807, Clemente [23–26] proposed the first systematic approach to grapevine diversity. Kolenati [27] first discussed in 1846 the origins of cultivated grapevines and proposed a classification of grapevines, wild and cultivated, in Georgia. Different authors followed Clemente's point of view; however, it was not possible to acquire a better view of grapevine diversity patterns until the beginning of the 20th century, when Russian agronomists carried out an in-depth study on the wild and cultivated grapevines of Western and Central Asia, especially in the *Ampelographia USSR* [18,28]. In this framework, the Russian agronomist Negrul proposed the recognition of three groups of cultivars, or *Proles*, namely: *Occidentalis*, *Pontica*, and *Orientalis* [10,19].

1.2. Grapevine in the near East Origins and Domestication

The Near East includes the eastern Mediterranean regions, the territories along the Euphrates and Tigris rivers, and the nearby regions of Central Asia, with the boundary to the north in the southern Caucasus and to the south in the Arabian and Sahara deserts. *Vitis* traces from the area are derived from pollen, wine residues, grapes (especially seeds), and wood remains [1,9,29]. The archaeobotanical remains that provide the most information are the seeds, which have been preserved by being charred, dried, or waterlogged. The carbonization process, with many variables (exposure time, temperature, humidity, and chemical composition), or conservation in an aquatic environment, can affect the morphology of grape seeds and hinder their taxonomic identification, i.e., their ascription to wild or domesticated populations [29–32].

The pollen record from cores in the present range of wild grape within this area shows low but consistent *Vitis* counts, at least from the beginning of the Holocene, e.g., Ghab Valley (Syria), Lake Van (Turkey), and Lake Urmia (Iran) [33].

The oldest wild grape (*Vitis sylvestris*) seeds (8400 B.C.) associated with human activity, about 3 mm long, were excavated in Turkey at Nevali Çori, near the city of Urfa, on the slope of a tributary valley of the Euphrates (Hilvan province, Turkey). Domestication and cultivation of the grapevine seems to have occurred between the seventh and fourth millennium B.C., and between the Black Sea and Iran, including the Caucasus and the Upper Euphrates [9,34]. Slightly to the east of Lake Urmia, Lake Zeribar (Zagros Mountains, Iran), *Vitis* pollen first occurred in the core just before c. 4300 cal BC. This evidence was interpreted as grape cultivation spreading to the south-east, but there is no indication of substantial plantings. At present, the earliest evidence for grape used in wine production comes from the sixth millennium BC (Neolithic) site of Hajji Firuz Tepe (Lake Urmia basin, Iran) in the form of a tartaric acid residue [33]. The first convincing evidence of grapevine (*Vitis vinifera*) seeds, with indications of grape cultivation, was uncovered in Turkey at Kurban Höyük (5700–5200 B.P. non-calibrated radiocarbon date) [35].

Grapevine cultivation seems to have spread westward from western Asia. In Crete and Greece, the beginning of grapevine cultivation may have started around the fifth millennium B.C. [36]; however, archaeobotany in Greece suggests that there was a transitional period when grapevine seeds were neither domesticated nor wild [37]. This could have a connection between the seeds and the wine pressings found at Dikili Tash, suggesting that the use of grapes to produce wine may have begun independently of the domestication process [38]. In Spain, Phoenician influence during the first millennium B.C. seems to have played an important role in the development of viticulture and wine production, although the grapevine was exploited by local populations in the Neolithic before contact with Mediterranean cultures [29,39]. This would support the theory of an independent domestication center in Western Europe [40].

The analysis of archaeological grapevine seed remains from West Asia and nearby areas and their comparison with modern cultivars may provide interesting data to reconstruct the history of grapevine domestication and cultivation [41]. In West Asia, numerous archaeological grape seeds have been recovered, notably from Chalcolithic and Bronze Age levels, and are attributed to cultivated grapevine [42,43].

The objectives of the present study are, therefore:

- To study the morphology of archaeological and modern seeds from the Upper Euphrates.
- To compare them in order to establish a classification that will allow us to distinguish wild from domesticated seeds in the archaeological repertoire.
- To determine, as far as possible, the major groups or *Proles* of *Vitis vinifera* present in the archaeological repertoire and in the modern cultivars analyzed.
- To evaluate to what extent archaeobotanical data can contribute to the understanding of the origin of cultivated grapevines.

2. Materials and Methods

2.1. Archaeological Sites Sampled

In the present paper, we study grapevine seed samples from two sites that our group has excavated, both located in Syria near the Euphrates, Tell Qara Qûzâq [44,45] and Tell

Khamîs [46–50] (Figure 1), and compare them with a wide range of published archaeological seeds from the area and with local modern cultivars.

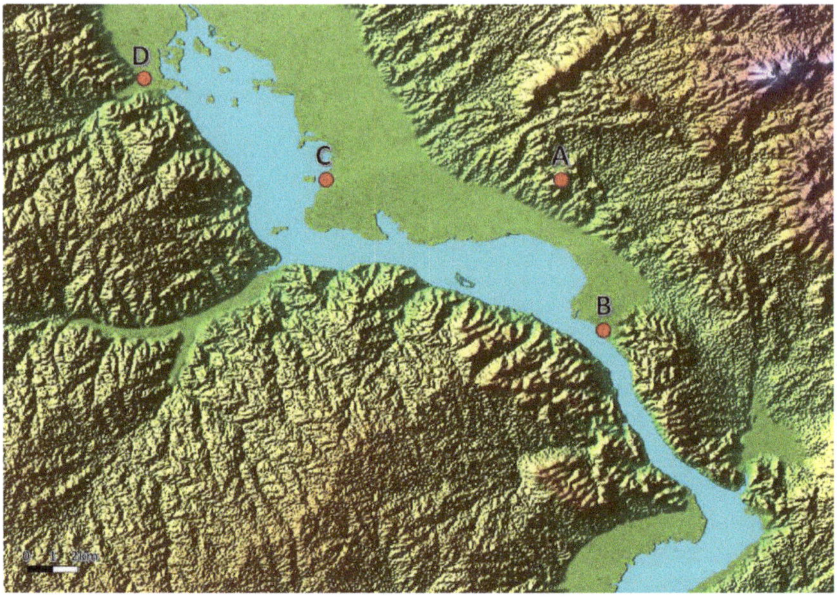

Figure 1. Geographical situation of the sites analyzed in the Syrian Upper Euphrates. A: Tell Khâmis (Ar Raqqah governorate). B Tell Quara Quzaq (Ar Raqqah governorate). Other relevant sites: C, Tell Ahmar (Ar Raqqah governorate), and D, Tell Amarna (Aleppo governorate). Note: the Euphrates waters fill here the Lake Assad since the construction of Tabqa dam.

2.1.1. Tell Qara Qûzâq

Tell Qara Qûzâq is located about 30 km from the Turkish border (36°37′57.80″ N, 38°12′52.92″ W) on the banks of the Euphrates River and about 325 m above sea level. Excavations continued uninterrupted from 1989 to 1999, when the waters of the reservoir rose and flooded the village of Qara Qûzâq, turning the site into an island. Five archaeological levels of occupation have been defined from the Early Bronze Age II (ca. 2080 BC) to the Roman period, with a very marked temporal hiatus between the Middle Bronze Age and the Roman level, signifying the inoccupation of the Tell for about 2000 years [51]. Several archaeobotanical studies have been carried out with the samples collected from the site in a systematic way [44]. The archaeological samples of grape seeds analyzed here from this site are six, for a total of eleven seeds.

2.1.2. Tell Khamîs

Tell Khamîs is located in the Upper Jazira region of Syria (36°43′56.36″ N, 38°07′09.81″ W), 3 km from the eastern edge of the Euphrates River, and at 330 m above sea level. Excavations at this small site began in 1992 and were completed in 2000 by the Institute of the Near East and Antiquity (IPOA/University of Murcia). The chronological period covered by the site dates back to the first half of the third millennium BC. After several temporary hiatuses over 11 archaeological levels, Tell Khamîs was abandoned in the middle of the 2nd century BC [46], with a long period of time in which there was no further occupation until the conversion of Syria into Islamic territory, when the site became a regular burial place [48]. A large number of seeds in a carbonized state belonging to different chronological horizons have been identified, highlighting those of barley and grapevine for their number and preservation [50]. The archaeological samples of grape seeds analyzed here from this site are 18, for a total of 22 seeds.

2.2. Seed Samples

The study involves the analysis of 782 grape seed samples, including the 24 archaeological samples from the Euphrates Valley that we intend to identify within a range of probability. Comparison seed samples were provided by Erika Maul (The Institute for Grapevine Breeding Geilweilerhof, Siebeldingen, Germany) (cultivars from the Near East), the U.S. National Plant Germplasm System (United States Department of Agriculture, Agricultural Research Service), Rancho Santa Ana, Istituto ed Orto Botanico di Palermo, Botanisches Garten Johanes Gutenberg—Universität Mainz, Berlin Botanisches Garten, Smithsonian Institution, and Giardino Botanico di Padova (seeds of American and Asian species of *Vitis* and related genera). Seeds of other cultivars came from the vine collections of Rioja (Spain) in Mendavia, the La Casa de las Vides nursery in Agullent (Valencia, Spain), the Rojas Clemente vine collection of the Real Jardín Botánico de Madrid (Spain), and the Istituto Agrario di San Michele all'Adige (Trentino, Italy). Wild populations were sampled by the authors, Rafael Ocete, Emilio Laguna, and Encarna Carreño. Anna Nebish and David Maghradze supplied seeds of wild, feral, and cultivated vines from the southern Caucasus. Seeds from other archaeological sites were also included in the main analysis.

The primary raw data matrix used for this study consists of 4028 single rows of analyzed seeds, belonging to 782 seed samples, and 20 columns with observed variables, 11 quantitative, 6 allometric indices and 3 qualitative. The modern reference material used aims to give a global image of the diversity within *Vitis* species and cultivars using a wide range of samples that represent a large number of cultivars, wild and subspontaneous grapevine, in order to compare the morphological differences between them. Of the total number of seeds analyzed, 3483 are modern (481 samples), 398 are archaeological (194 samples), where part of the material is carbonized (251 seeds, 87 samples), and 147 are dry or waterlogged (107 samples).

2.3. Characters

Each seed was individually described according to 20 characters. Of these, 11 are quantitative: total length, breadth, and thickness of the seed, breadth of the beak at the junction with the body and at the seed base, length of the beak in dorsal and in ventral view, thickness of beak at the junction, length and breadth of the chalaza scutellum, and distance from the chalaza apex to the seed apex (Figure 2) [52]. Six are allometric: width/length index, width/thickness index, prism volume index, beak length/seed length index, beak width/beak length index, and chalaza width/length index. The qualitative characters are three: Contour type (assessment of shape), with five states (ovoid, quadrangular, triangular, rounded, and pentagonal), arrangement of the fossettes, with four states (parallel, furcate, convergent and divergent) (Figure 2) and presence/absence of radial furrows.

The quantitative and qualitative characters were measured and analyzed using digitally scaled images. In total, 10 seeds of each sample were individually placed, except when the number of seeds available was inferior, on a plasticine support with a built-in scale to be photographed in dorsal, ventral, and lateral views with a Samsung A40 camera and measured using open-source Fiji software [53]. All photographs were taken under the same zoom conditions. Additionally, scale images of fossilized and archaeological seeds from specialized literature were used for measurements. The characters were recorded in an Excel spreadsheet, where the allometric relationships were automatically calculated using algorithms.

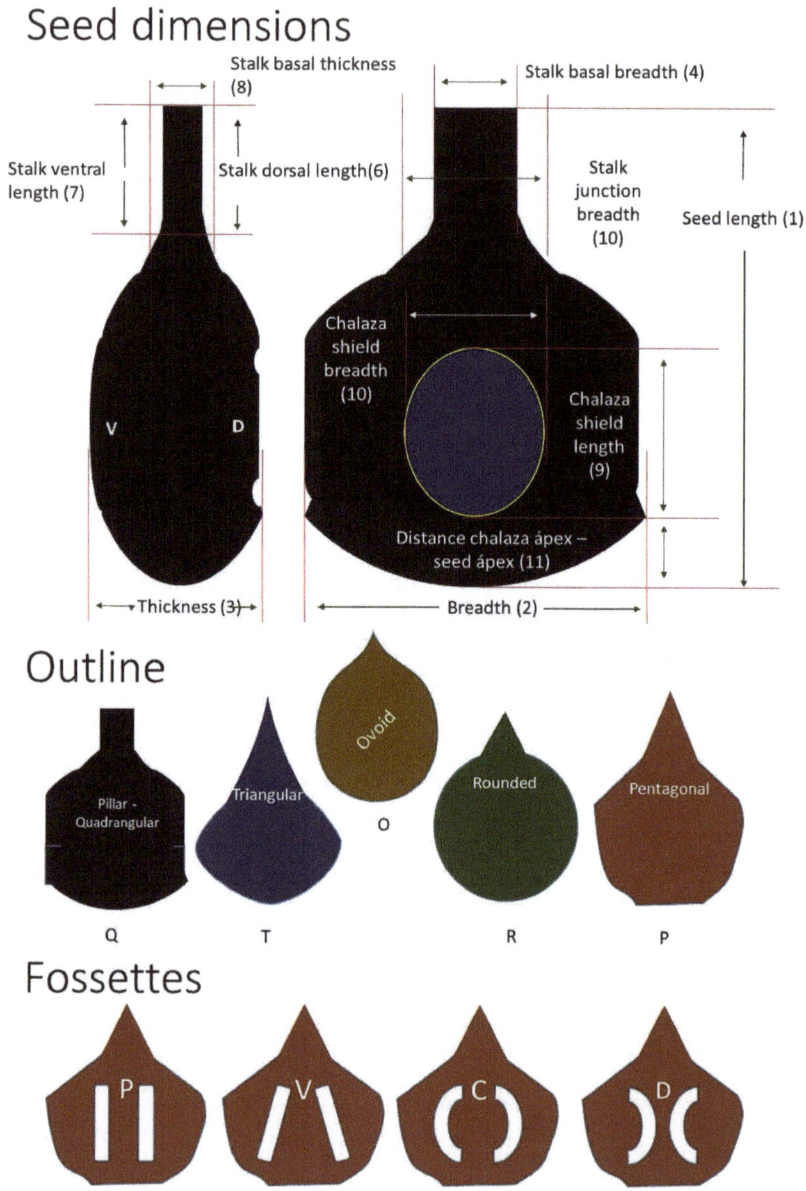

Figure 2. Qualitative and quantitative characters analyzed in the grapevine seeds. Abbreviations for Fossettes types: P, parallel. V, furcate. C, convergent and D, divergent.

The SEM images were made at the Scientific and Technical Research Area of the University of Murcia. The microstructure of the archaeological seeds was investigated by means of field emission scanning electron microscopy (FE-SEM) (ApreoS Lovac IML, Thermofisher, Waltham, MA, USA). Specimens were mounted on aluminum stubs and platinum sputter-coated with a 5.0 nm thin layer (Leica EM ACE 600). Samples were examined using a FE-SEM (ApreoS Lovac IML, Thermofisher, Waltham, MA, USA) with a selected voltage of 10 kV and 0.20 nA for imaging.

2.4. Morphometric Indices

2.4.1. Stummer's Index

In 1911, Stummer [54] proposed an index based on the allometric relationship between seed width and seed length. This index makes it possible to quite effectively differentiate the extreme forms, but intermediate values are found in both wild and cultivated populations (Table 1). Stummer's index values ranging from 0.44 to 0.53 would be exclusive to cultivars, while 0.76 to 0.83 would be unique to Austrian wild/ferals. Values between 0.53 and 0.76 were found in both cultivars and wild vines. In 1956, Levadoux [19] demonstrated that this index has limited validity and is not useful for distinguishing wild vines from cultivated vines.

Table 1. Stummer's index for wild and domesticated grapevine seeds [1].

Range of Values	Taxa
44–53	*V. vinifera*
54–75	Intermediate or hybrids
76–83	*V. sylvestris*

[1] Formula: B/L × 100.

2.4.2. Facsar–Perret's Index

In 1997, Perret [55] proposed a new index based on the allometric relationship between the length of the beak or column and the total length of the seed. Apparently, this index makes it possible to quite effectively differentiate between wild and cultivated populations, with the border situated between 18 and 19 (Table 2). Although it was attributed to Perret, this index was previously proposed by Facsar in 1970 and Facsar and Jerem in 1986 [56,57].

Table 2. Facsar–Perret's index for wild and domesticated grapevine seeds [2].

Range of Values	Taxa
12–18	*V. sylvestris*
19–30 (35)	*V. vinifera*

[2] Formula: LS/L × 100. LS: stalk lengt, L: seed length.

2.4.3. Mangafa and Kotsaki's Indices

The formulae proposed by Mangafa and Kotsakis in 1996 [58] were successfully applied to local Greek samples of both modern seeds and archaeological remains. The four formulae (Table 3) are based on the combined use of relationships and constants involving variables such as seed length (L), stalk length (LS), and chalaza position (PCH).

Table 3. Mangafa and Kotsakis's indices for wild and domesticated grapevine seeds [3].

Range of Values	Taxonomic Information
Range of values (Formula (1))	Seed classification
<−0.2	Wild grapes
−0.2 < x < 0.2	Wild grapes (64.7% probability to be wild)
0.2 < x < 0.8	Domesticated grapes (76.2% probability to be cultivated)
>0.8	Domesticated grapes
Range of values (Formula (2))	Seed classification
<−0.2	Wild grapes
−0.2 < x < 0.4	Wild grapes (64.7% probability to be wild)
0.4 < x < 0.9	Domesticated grapes (76.2% probability to be cultivated)
>0.9	Domesticated grapes

Table 3. *Cont.*

Range of Values	Taxonomic Information
Range of values (Formula (3))	Seed classification
<0	Wild grapes
0 < x < 0.5	Wild grapes (90.1% probability to be wild)
0.5 < x < 0.9	Domesticated grapes (63.3% probability to be cultivated)
>0.9	Domesticated grapes
Range of values (Formula (4))	Seed classification
<−0.9	Wild grapes
−0.9 < x < 0.2	Wild grapes (90.1% probability to be wild)
0.2 < x < 1.4	Domesticated grapes (63.3% probability to be cultivated)
>1.4	Domesticated grapes

[3] Formula 1: −0.3801 + (−30.2 LS/L) + 0.4564 PCH − 1.386 L + 2.88 PCH/L + 9.4239 LS). Formula 2: 0.2951 + (−12.64 PCH/L − 1.6416 L + 4.5131 PCH + 9.63 LS/L). Formula 3: −7.491 + (1.7715 PCH + 0.49 PCH/L + 9.56 LS/L). Formula 4: 0.7509 + (−1.5748 L + 5.297 PCH − 14.47 PCH/L). LS, stalk length; L, seed length; PCH, chalaza position.

2.4.4. Domestication Index

Although the above indices serve the same purpose, to separate wild from domesticated forms, their results differ from case to case. The combined use of the six indices may be able to better discriminate seeds from wild or cultivated grapevines. The combined domestication/wild index is calculated individually for each seed using the following Formula: (1), where *NIT* means indices exceeding, above or below, the threshold value, and *NI* means the indices considered:

$$DW_i = \sum_{i=1}^{n} NIT_i / \sum_{i=1}^{n} NI_i \quad (1)$$

Threshold values for recognizing a seed as wild: Stummer > 75, Facsar-Perret < 19, Mangafa and Kotsakis F1 < −0.2, Mangafa and Kotsakis F2 < −0.2, Mangafa and Kotsakis F3 < 0 and Mangafa and Kotsakis F4 < −0.9.

The sum of the wild index (*WI*) and the domestication index (*DI*) values, which is complementary to the previous one, will always be equal to one.

Values of the *DI* (domestication index) range from 0 to 1, with intermediate values, based on six indices: 0.17, 0.33, 0.5, 0.67, and 0.83. Seeds with index values between 0.67 and 1, both included, would undoubtedly be domesticated seeds and present the "domestication syndrome", and those with values between 0 and 0.33 would be truly wild and present the "wild syndrome". In the present work, the value 0.5 is tentatively interpreted as wild.

We usually work with samples consisting of several seeds, which, in the case of modern populations, wild or cultivated, usually come from the same cluster, although not always. There are three relevant parameters when inferring from the results of individual seeds the characteristics of the whole sample, both for *DI* and *WI*:

1. The mean of the *WI*, wild index values for individual seeds, ranging from 0 to 1.
2. The standard deviation of the *WI* index values.
3. The proportion of seeds within each sample exceeding the wild threshold, *PW*, proportion wild, ranging from 0 to 1.

2.4.5. Hybridization Index

The standard deviation of the *WI* index values has shown to be useful to distinguish hybrids and hybrid swarms from pure wild and pure domesticated individual populations. Values above 0.2 of the standard deviation of the WI index points to the hybrid or mixed nature of the sample.

2.5. Multivariate Analysis

2.5.1. Variables

The data matrix consists of 782 samples (rows) and 227 columns of variables resulting from the segmentation in mutually excluding states or classes of the 20 primary variables above described, in the form of a spectrum of frequencies expressed in percentages with the following structure, from left to right: length (25 classes), width (21), thickness (9), width/length ratio (29), width/thickness ratio (10), volume (12), beak length in dorsal view (9) and in ventral view (9), beak length/seed length ratio (16), beak width at base (11), beak width (11) and beak thickness at junction with body (6), beak width/length ratio (9), chalaza shield length (18), chalaza shield width (6), chalaza width/chalaza length ratio (9), chalaza apex to seed apex distance (10), outline (5), dorsal radial grooves (2).

2.5.2. Data Analyses

The chi square dissimilarity index was calculated based on the above data matrix using the tool Darwin 6.0 [59,60]. This measure expresses a value x_{ik} as its contribution to the sum x_i on all variables and is a comparison of unit profiles (2).

$$d_{ij}^2 = \sum_{k=1}^{K} \left(\frac{x_{ik}}{x_i} - \frac{x_{jk}}{x_j} \right)^2 \left(\frac{x}{x_k} \right) \tag{2}$$

For $j \neq i$.

where d_{ij}: dissimilarity between units i and j; $i, j = 1, 2, \ldots \ldots, N$ (samples, rows), $N = 782$; $k = 1, 2, \ldots \ldots, K$ (variables, columns).

where $d_{ij} = 1$ means varieties i and j differ in all variables, and $d_{ij} = 0$ means varieties i and j are identical.

These pairwise dissimilarities can be represented in a multidimensional space, but in order to obtain a meaningful graphic representation of these relationships in a two-dimensional plane, we used cluster analysis.

Cluster analysis is a term used to name a set of numerical techniques whose main purpose is to divide the objects of study into discrete groups. These groups are based on the characteristics of the objects. We used minimum variance clustering (Ward's method), which focuses on determining how much variation is within each cluster. In this way, the clusters will tend to be as distinct as possible since the criterion for clustering is to have the least amount of variation [61]. Ward's method produces a single tree. For the graphic representation, we opted for the software Figtree version 1.4.4 [62].

The use of distance-based trees to allocate archaeological seed samples is not new; in 2015, Pagnoux et al. [63] assigned archaeological grape seeds to the groups defined by UPGMA cluster analysis; their tree is based on Mahalanobis distances among comparison grapevine wild individuals and cultivars. Rivera et al. in 2014 [64] tentatively allocated archaeological *Phoenix* seed samples using a method based on the minimum variance Ward's principle.

2.6. Allocation of Archaeological Samples to Categories and Taxa

2.6.1. Bayes–Laplace Theorem

For the interpretation of archaeological seed samples, we adopted a Bayesian approach. We try to answer the question: What is the conditional probability that an archaeological seed or seed sample belongs to a determined *Vitis* taxon Θ_i given that it presents the domestication index value x_j and/or it belongs to the cluster y_j? The framework is based on the knowledge provided by hundreds of comparison samples (c. 700), whose taxonomic identity we "a priori" know in each case not only from the morphology of the seeds but also from the study of the grapevine plant from which the sample was collected. Identification is based on ampelographic characters. So far, the most relevant ampelographic data included in the Vine Descriptors (IPGRI-UPOV-OIV 1997) [65] have been collected, especially those related to the hairs covering the leaves and the characteristics of the grape berry, either recorded directly in the field, since more than two hundred of these vines were grown on a farm in Molina de Segura (Spain), or from the databases (FNGR 2023; VIVC 2023) [66].

This allows us to construct a discrete joint probability function $p(X,\Theta)$ that assigns a posterior probability value to each particular combination of a *Vitis* taxon and a domestication index value, or of a *Vitis* taxon and a Ward's tree cluster.

The Bayes' rule (3) makes it possible to approximate the answer:

$$p(\theta|x) = p(x|\theta)p(\theta)/p(x) \tag{3}$$

where $p(\theta \mid x)$ is the posterior probability distribution for the parameter θ given a single observed value of the variable $X = x_j$, in our case the degree of domestication, which is represented by the domestication index value, which ranges from 0 (clearly wild) to 1 (cultivar with fully domesticated traits).

When considering the Bayes' rule in terms of individual probabilities, (3) can be read as (4).

$$Posterior\ probability = \frac{likelihood \times prior\ probability}{marginal\ likelihood} \tag{4}$$

Given a value for the data, for instance $X = x_4$, and a specific value for the parameter θ (*Vitis* taxa), such as, $theta = \theta_3$, we can obtain (5)

$$p(\theta_3|x_4) = p(x_4|\theta_3)p(\theta_3)/p(x_4) \tag{5}$$

In (5), both likelihood $p(x_4|\theta_3)$ and marginal likelihood $p(x_4)$ are values that can be calculated on the basis of the joint distribution generated from the comparison samples. The prior probability $p(\theta_3)$ can also be calculated as the sum of probabilities for this taxon given the distribution of all x values on the sample data alone. However, the very nature of the prior allows the inclusion of data on the regional prevalence of the different taxa from other well-established sources of evidence (catalogues of local varieties, germplasm collections, field studies). In this study, we have paid attention, in the case of domesticated vines, to the geographical variation in the proportions of the different *Vitis vinifera "Proles"* and, in the case of the rest, to the ratio, *V. sylvestris*/*V. caucasica*. *Vitis caucasica* Vavilov is still unclear as a taxon, but we use this to name the ensemble of wild grapevines in the Southern Caucasus that present a very low domestication index value of <0.5. Furthermore, we pay attention to the low relevance of the fossil grapevine species, introduced as an outgroup and treated as such.

2.6.2. Application to the near East Seeds Question

Following the Bayesian approach, we can advance in our evaluation of the probabilities of the different hypotheses by considering a set of the available "a priori" evidence on the relative frequencies of the hypotheses, especially considering time range and geographic constraints. For example, it is much less likely to find seeds of an American grapevine species in a European Neolithic site than those of *Vitis sylvestris*. A careful elaboration of the "a priori" distribution of probabilities based on solid and logically coherent evidence is as fundamental as a clear definition of the different hypotheses and of the variable(s) to be considered.

Among the cultivars, we assume "a priori" a proportion of *Proles* and *Subproles* in the sense of Negrul (1946) [11], similar to the present one in West and Central Asia [66–68], which implies: *Orientalis antasiatica* ($p = 0.43$), *Orientalis caspica* ($p = 0.25$), *Pontica* ($p = 0.17$), *Occidentalis* ($p = 0.07$), although most of the stages of introduction and translocation of varieties had not yet taken place. For hybrids, using the same sources ($p = 0.07$), this leads us to the following "a priori" scenario (Table 4). The "a priori" or "prior" probabilities used are also based on the assumption that at that time, 3rd mill. BC, 30% of the vines in the territory were wild and 70% cultivated. If we were to assume different proportions, the following probabilities would differ.

Table 4. Alternative hypotheses and their respective priors and likelihoods [1].

Groups	Prior	L1	L2
Vitis vinifera Cultivars			
Oriental cultivars. *Proles orientalis* Negrul *Subproles antasiatica.*	0.303	0.178	0.042
Oriental cultivars. *Proles orientalis* Negrul *Subproles caspica.*	0.177	0.086	0.041
Western and Mediterranean cultivars in a broad sense. *Proles Pontica* Negrul.	0.118	0.103	0.035
Western and Mediterranean cultivars in a broad sense. *Proles Occidentalis* Negrul.	0.051	0.059	0.056
Varieties with intermediate characteristics resulting from hybridization between the previous groups.	0.049	0.082	0.077
Wild grapevines in natural habitats			
Vitis sylvestris, we include the variability inherent to western wild vines that do not descend from cultivated plants.	0.024	0.070	0.104
Feral grapevines, which descended from cultivated plants and although they show partial reversion to ancestral characters, they conserve traits derived from domestication.	0.028	0.037	0.068
Vines related to wild vines from the Caucasus (*Vitis caucasica* Vavilov *sensu auct.*) or other eastern regions, their probability is small but we do not rule them out. They are divided into:			
• Direct hybrids of wild Caucasian grapevines with cultivars.	0.060	0.099	0.038
• Purely Caucasian feral.	0.060	0.052	0.041
• Wild Caucasian grapevines.	0.043	0.001	0.068
Unlikely hypotheses			
American grapevine species: we should rule out the possibility that an American vine could have been present in Western Asia in such early times, but what we do assume is a very low probability.	0.0005	0.076	0.074
Eastern Asian grapevine species: These are unlikely, but given the ancient connection facilitated by the Silk Road, their presence is not impossible.	0.084	0.156	0.101
Finally, fossils: These are extremely unlikely, but we do not rule out the survival of a living fossil.	2.5×10^{-5}	6×10^{-5}	0.255

[1] L1, Posterior probability combining prior and the Likelihood based on DI values. L2, Normalized Marginal Likelihood considering the allocation to clusters 1 to 13 as in Figure 3.

In the Bayesian method, we can combine the evidence resulting from the study of several variables by concatenating results in which the posterior probability distribution function of the first analysis will be used as the a priori distribution for the second and so on. In the present study, we combine the results of the morphometric indices summarized in the domestication index for each of the seeds as a starting analysis and then use the results of the multivariate analysis in terms of the assignment of each of the samples to one or another of the 13 clusters.

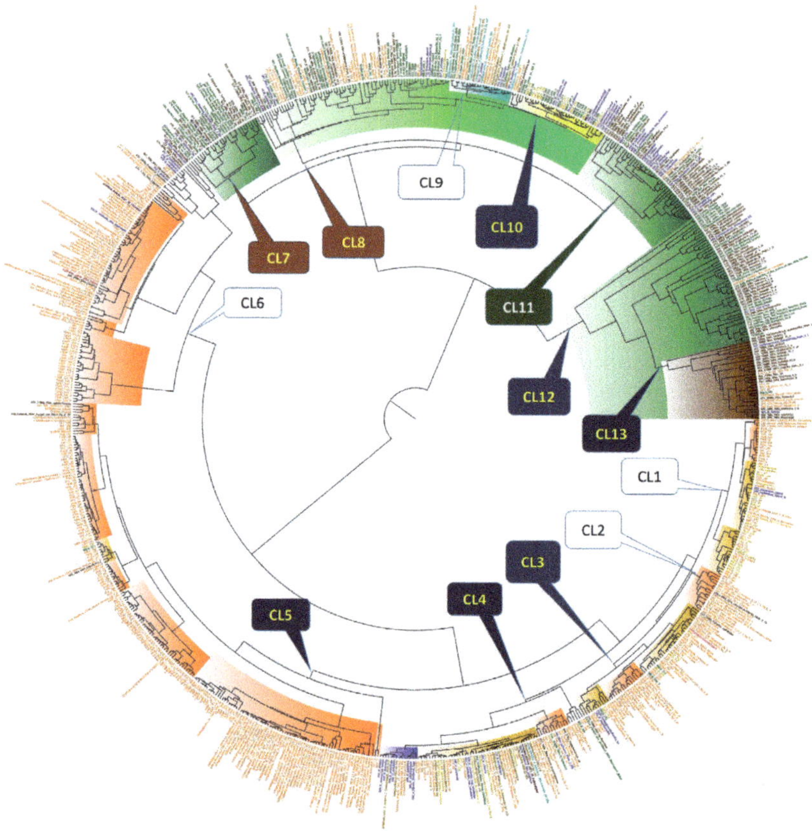

Figure 3. Minimum variance Ward's tree. Cluster labels: Deep blue filled and yellow characters for clusters with Tell Khâmis seed samples. Dark green filled and white characters for clusters with those from Tell Quara Quzaq. Dark brown for those with both. White filled and black characters for the rest. Color codes (RGB system) for sample labels: *Vitis caucasica*: 0-153-153; *V. sylvestris*: 0-102-0; *V. vinifera*: 255-102-0; *V. vinifera* × *V. sylvestris*: 204-153-0; other wild *Vitis* species: 0-0-204; *Vitis* seeds fossil: 102-51-0; *V. vinifera* × *V. caucasica*: 153-102-0; *V. vinifera* × *V. amurensis*: 102-0-0.

3. Results

3.1. Probability-Based Allocation of Grapevine Seeds to Major Types

The Bayesian sequential combination of domestication indices and cluster allocation (Figure 3) has allowed us to assign the archaeological samples with varying degrees of probability (Supplementary Table S1) to the different groups of wild, domesticated, and hybrid grapevines. It should be emphasized that the method followed gives priority in its results to the information obtained from the multivariate analysis of the morphology of the seeds and the assignment of the seeds to each of the 13 clusters identified, previous data on the characteristics of the cultivated and wild vines in the area, and, finally, the information provided by the six domestication indices combined into a single one that we abbreviate as DI and whose values range from 0 to 1 for each of the seeds.

3.1.1. Domesticated Grapevines of the *Proles orientalis* Negrul

Domesticated grapevines are related to modern cultivars belonging to *Proles orientalis Subproles antasiatica* (probability, p = 0.55–0.87) and associated with high domestication values ($DI = 1$). Additionally, they are related to a lesser extent with *Proles orientalis Subproles caspica* ($p = 0–0.15$), *Proles Pontica* ($p = 0–0.16$), and *Proles Occidentalis* ($p = 0.1–0.12$), but clearly unrelated to wild grapevines. Three samples of Tell Khâmis presented this high domestication syndrome. The group is otherwise highly heterogeneous as the seeds have been assigned separately to clusters 4, 5, and 13 (Figure 3).

- Middle Bronze Age: 268 TK bm 68 11;
- Assyrian: 264 TK as 95 6B;
- Persian–Hellenistic: 565 TK 011 2017.

Modern seeds that presented similar profiles in the study include numerous Eastern cultivars, such as an Uzbekistani cultivar, several Afghan raisin cultivars, Chaouch Blanc, Chaouch Rose, Shiradzouli violet, Beylerce, Konya Bozkir; Rasheh, Besni, Hasandede Beyazi, Rumi Ahmar, Razakisi Antep, Haci Tespihi, Zonguldak, Dabouki, Mecka, Nehelescol, and Chasselas.

Similar seeds were found in the Urartian levels of Yoncatepe (Van, Eastern Turkey) [69].

3.1.2. Domesticated Grapevines Belonging to a Particular Euphrates *Proles*

Domesticated grapevine cultivars are unrelated to modern cultivars but related to Caucasian and Asian wild grapevines. Numerous groups of samples present the paradox of having high values for the domestication index ($DI = 0.67–0.83$) but a very low probability of belonging to any of the known domesticated *Proles* (at least those included in our collection of comparison samples) and at the same time a high probability ($p = 0.36$ to 0.42), of being an Asian *Vitaceae* species, or, to a lesser extent, being *V. caucasica* ($p = 0.26$ to 0.32), or a hybrid ($p = 0.21$ to 0.22). Furthermore, all of them cluster together in cluster 8 of the multivariate analysis (Figure 3). This suggests the existence in the ancient Upper Euphrates of a peculiar set of domesticated varieties, possibly derived from hybridization between Asian and South Caucasian vines. At present, these varieties seem to have disappeared, although it would be very useful to prospect the tributary valleys of the Euphrates in Syria and Iraq in search of them, either cultivated or feral. It should be noted that no relationship has been found between this group of archaeological samples and the numerous current varieties from Turkey, Georgia, or Armenia that have been introduced to the comparison matrix. It would also be very interesting to be able to include in further studies the wild and feral vines recently studied in Israel.

- Early Bronze Age: 180 AR QQ 1 14-92.
- Middle Bronze Age: 580AR QQ MBA W78b.
- Assyrian: 263 TK as 95 6; 266 TK as 108 1, (Figure 4E).
- Aramean: 268 TK bm 68 12, (Figure 4B).
- Persian-Hellenistic: 559 TK 028 2017.

No single modern seed was found close to this group, and only one archaeological seed sample was allocated to this group from the coastal plain of Sidon, dated from the Late Iron Age at Phoenician Tell el-Burak (Lebanon) [70].

3.1.3. Domesticated Grapevines Related with Asian Wild Grapevines

Unrelated to modern cultivars but related to Asian *Vitaceae* species ($p = 0.5–1$) and to a lesser extent with European wild ($p = 0–0.37$) and hybrid ($p = 0–0.29$) grapevines, this group also presents the paradox of associating high values of the domestication index (DI = (0.5) 0.67 to 1) with a very low probability of belonging to any known domesticated grapevine *Proles* and, at the same time, a high probability of being an Asian wild species. This group could be related to the previous one and draws our attention, once again, to the influence of the Asiatic species of *Vitis* or *Ampelopsis* on the ancient grapevines of the Euphrates and the Near East. All of them were allocated together in clusters 7, 12, and 13 of the multivariate analysis (Figure 3).

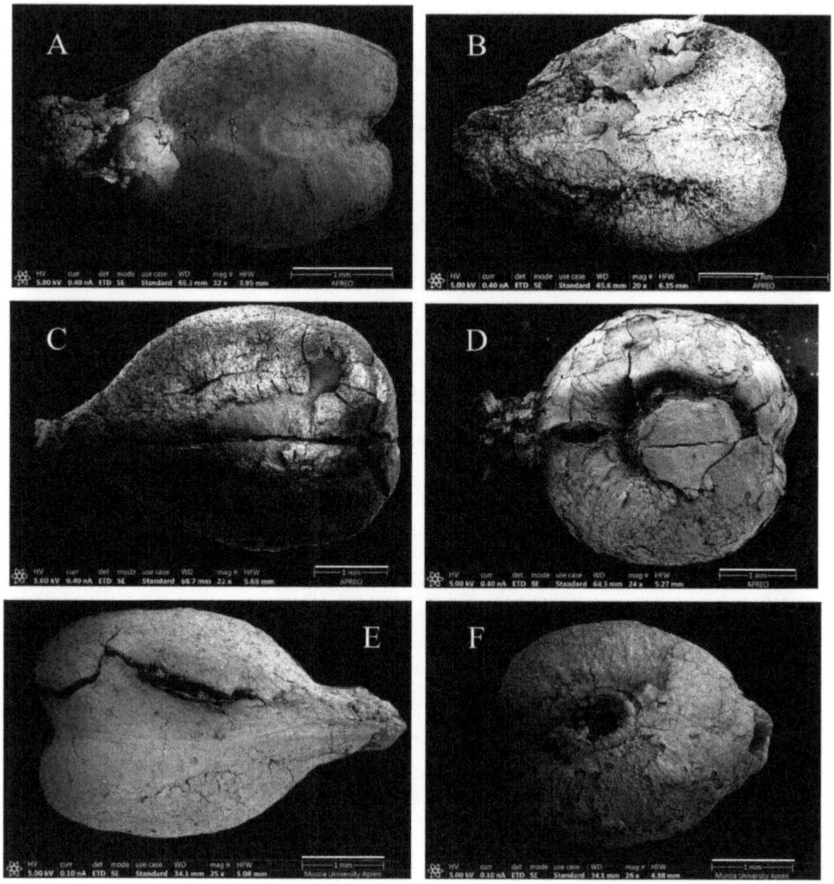

Figure 4. Archaeological *Vitis* seeds. SEM images. (**A**) Domesticated grapevines related with Asian wild grapevines, Assyrian: 264 TK as 95 6A. (**B**) Domesticated grapevines belonging to a particular Euphrates *Proles*, Aramean: 268 TK bm 68 12. (**C**) Domesticated grapevines related with Asian wild grapevines. Early Bronze Age: 180AR QQ 2 14-92. (**D**) Wild Eurasian-Caucasian grapevines, Middle Bronze Age: 185AR QQ 1 137-92. (**E**) Domesticated grapevines belonging to a particular Euphrates *Proles*, Assyrian: 266 TK as 108 1. (**F**) Domesticated grapevines related with Asian wild grapevines. Assyrian: Assyrian: 264 TK as 95 6A; 262 TK as 85 5 (the beak was broken during the handling for SEM). Scale bars: above 1 mm, below 1 mm, except B, where is 2 mm.

- Early Bronze Age: 180AR QQ 2 14-92, (Figure 4C).
- Assyrian: 264 TK as 95 6A, (Figure 4A); 262 TK as 85 5, (Figure 4F).

Archeological seeds with similar features were found in Bronze Age levels of Konar Sandal, Halil Rud basin, southeastern Iran (but see below) [71] and in Late Bronze levels of Tall al-'Umayri, Jordan [72]. Finally, from the Petra Garden and Pool Complex in Jordan, a Hellenistic/Roman pleasure garden [73], and in medieval Tashbulak, on the Silk Road [74].

With a similar pattern of likely relationships but with lower domestication index values (*DI* = 0.17–0.33), seeds of the wild type were identified from Tuzusai, Central Asian mountains, Kazakhstan (410–150 BC) [75], in the Euphrates zone of Urartian levels in Yoncatepe (Van), eastern Turkey, mixed with other *Vitaceae* seed types [76], and in Bronze Age levels (2480–2290 calibrated BC) of Konar Sandal, Halil Rud basin, southeastern Iran [71] mixed with the domesticated type. In addition, also related to the wild syndrome are two seed samples from Nepal from the Late Pleistocene Besigaon section of the Gokarna Formation, Kathmandu Valley, in central Nepal: 53 kyr [77] and 45 kyr [78].

The few grape seeds found at Tuzusai (Kazakhstan, 410–150 BC). These are assumed to be cultivated and likely domesticated [75] however our results may question this view. Miller [33] states that northern Central Asia lies outside the range of wild *Vitis sylvestris* indicating that the pips probably came from domesticated grapes. These pips may represent exotic exchange goods or a locally grown horticultural product. The presence of a small number of grape pips does not prove viticulture occurred at Tuzusai. Grapes may have been imported from other regions in the form of raisins [75].

3.1.4. Eurasian Hybrid Wild Grapevines

Hybrid wild (DI = 0.17–0.5) grapevines are related to Eurasian hybrid modern grapevines (p = 0.19–0.64) (Supplementary Table S1) and to European wild (p = 0.30–0.81) grapevines, but are unrelated to Caucasian or Asian wild grapevines. Only two samples follow this anomalous pattern. The samples are allocated to clusters 3 and 10 (Figure 3).

- Assyrian: 265 TK as 100 1.
- Persian-Hellenistic: 560 TK 009 2017 A.

No modern samples of wild or cultivated vines from the Near East have been found that could be included in the group. The single archaeological grapevine seed sample that approximates this group is that found at Petra, Jordan; 150 B.C.–A.D 40 by Jacquat and Martinoli [79].

3.1.5. Eurasian Domesticated Hybrid and Feral Grapevines

Domesticated grapevines (DI = 0.67–0.83) are related to Eurasian feral modern grapevines (p = 0.42–0.57) and, to a similar extent, to European hybrid grapevines (p = 0.43–0.55), but are unrelated to wild grapevines (Supplementary Table S1). All the samples are allocated to cluster 3 (Figure 3). Three samples of Tell Khâmis present this clearly transitional syndrome:

- Middle Bronze Age: 186 AR TK.
- Aramean: 268 TK bm 68 13.
- Persian-Hellenistic: 560 TK 009 2017 B.

Similar grapevine seeds were found in Middle Bronze Age levels of Mezraa Höyük, Upper Euphrates, Turkey [69], and mixed with hybrid wild types in the same sample of charred grape pips from the Nabataean period, Petra, Jordan, 50 B.C–A.D 100 [79], also at the Early Bronze Age settlement of Ḫirbet ez-Zeraqōn in Northern Jordan [80] and in Dayr al-Barshā Middle Egypt during the late Antique/early Islamic period [81].

3.1.6. Wild Eurasian—Caucasian Grapevines

Wild grapevines are unrelated to modern cultivars, directly related to Caucasian wild grapevine with a medium probability (p = 0.3 to 0.5), and, to a lesser extent, are related to Asian wild grapevines (p = 0.23 to 0.33) and Eurasian wild grapevines (*V. sylvestris*) (p = 0.13–0.3). These are typical wild grapevines with low domestication index values (p = 0.17 to 0.33 (0.5)). Furthermore, all of them cluster together in cluster 8 of the multivariate analysis and with the samples allocated above to the *Proles Euphratica* (Figure 3), from which they differ in their relatively low domestication index (DI = 0.17–0.5).

- Early Bronze Age: 577AR QQ EBA H84; 578AR QQ EBA W80.
- Middle Bronze Age: 579AR QQ MBA W27; 185AR QQ 1 137-92, (Figure 2D).
- Persian-Hellenistic: 562 TK 029 2017. S

Similar grapevine seeds of the wild type were found in pigeon dung samples from a Byzantine (6th century AD) dovecote near Shivta, Israel [82]. This suggests the presence of this type of wild grapevine in the Eastern Mediterranean, whose seeds were transported by doves and other birds. Relatively similar to this group but with an even lesser relation to Eurasian wild and a relatively high relation to Caucasian hybrids were found in samples from the Euphrates and Levant areas: in Middle Bronze Age levels of Mezraa Höyük, Upper Euphrates, Turkey [69] and Early Bronze Age levels of Tell Lachish, Israel [83].

More striking is the similarity with the analyzed samples of *Vitis heyneana* Schult, a species that currently ranges from Afghanistan to China and Japan [84].

3.1.7. Wild Asian Grapevines Allocated to Genus Ampelopsis

Wild grapevines are unrelated to Caucasian wild grapevines and related to Asian wild grapevines belonging to the genus *Ampelopsis* with an extremely high probability ($p = 0.99$). These are typical wild grapevines with low domestication index values ($DI = 0$ to 0.5) and a very low probability of belonging to any of the known domesticated *Proles*. The multivariate analysis allocates them all to cluster 12 (Figure 3), except for the QaraQuzaq Early Bronze Age sample, which is allocated to cluster 11.

- Early Bronze Age: 185AR QQ 2.
- Assyrian: 267 TK as 108 1.
- Persian-Hellenistic: 566 TK 040 2017 (Figure 5B).

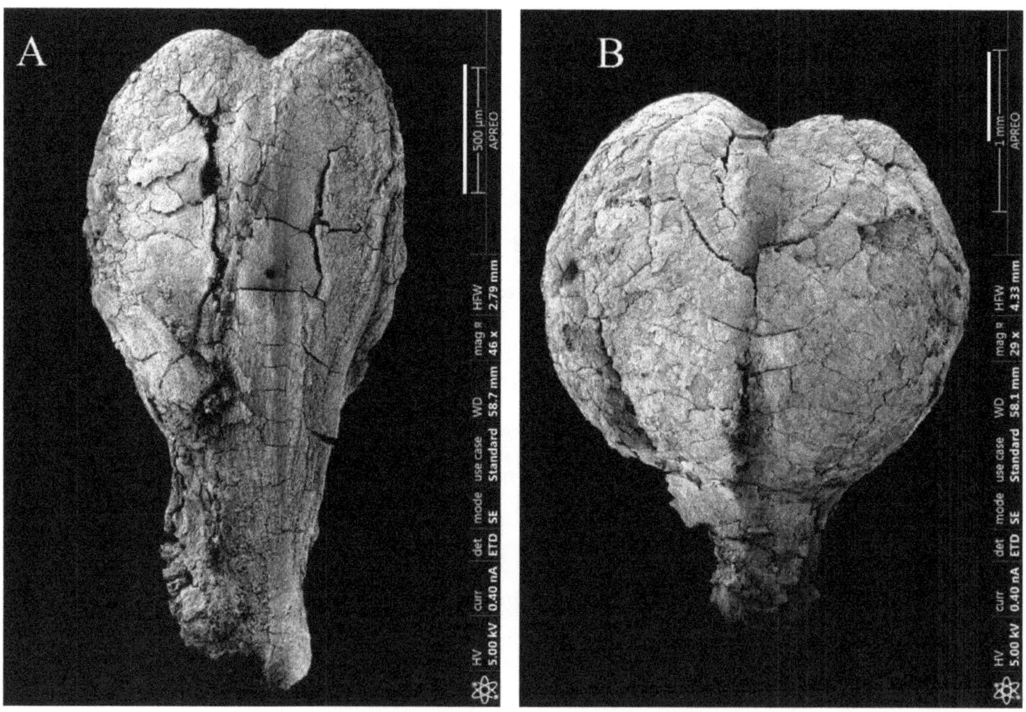

Figure 5. Archaeological *Vitaceae* seeds of the Upper Euphrates. Raisin type seeds with low domestication syndrome remotely linked to genus *Ampelopsis*, SEM images. (**A**) "Stenosperm" type. Wild Asian grapevines allocated to genus *Ampelopsis*; Persian-Hellenistic: 561 TK 064 2017. (**B**) *Ampelopsis* seed type, Persian-Hellenistic: 566 TK 040 2017. Scale bars: 0.5 mm.

Modern specimens of wild *Vitaceae* with similar characteristics have been identified (Supplementary Table S1) as: *Ampelopsis orientalis* (Lam.) Planch., which currently grows in Syria and Turkey [85]; and *Ampelopsis vitifolia* (Boiss.) Planch., which currently extends through Central Asia, from Iran to the Himalayas. Along with the above are included modern seeds determined as *Vitis* sp. by the US gene bank (S518_DVIT_1445) from Afghanistan, which are possibly *Ampelopsis*, and a fossil seed, 45 myr, from Late Pleistocene plant macro-fossils from the Gokarna Formation of the Kathmandu Valley, Central Nepal [78], determined as *Vitis* sp., but which may possibly also be ascribed to the genus *Ampelopsis*. The seed from Acheulian levels at Gesher Benot Ya'aqov, Israel [86], very rounded, is allocated to this group. Carbonized remains from Iron Age Raja-Nal-ka-Tila (1st millennium BC) of Ganga Plain, India [87] are included in this group. Kohneh Pasgah Tepesi (eastern Azerbaijan, Iran) in levels of the Late Chalcolithic and the Early Bronze Age [88] yielded *Vitaceae* seeds that also fall within this group and in 7th to 5th cent. BC levels of Deir'Alla ,Jordan. In this period, Deir 'Alla was a temple tell, as in Late Bronze Age times. A small group of priests would have lived there permanently, while seminomadic tribes stayed in the vicinity of the Tell only during the winter months [89].

3.1.8. Raisin Type Seeds with Low Domestication Syndrome Remotely Linked to Genus Ampelopsis

Thin sterile-like types with intermediate domestication index values $DI = 0.5–0.67$. Three seeds were found conforming to this type in Tell Khâmis.

- Persian-Hellenistic: 561 TK 064 2017; 563 TK 059 2017; 564 TK 038 2017 (Figure 5A).

It is relevant to mention that this group includes apparently sterile, tiny, elongated seed samples, such as those from Early Bronze Age Ras an-Numayra, Jordan [90]. This abnormal "stenosperm" seed type was relatively common in the study of modern grapevine tables and raisin seedless-like varieties of the "Sultanina" type from Afghanistan [67] and in some of Afghan comparison samples within this study (Figure 6).

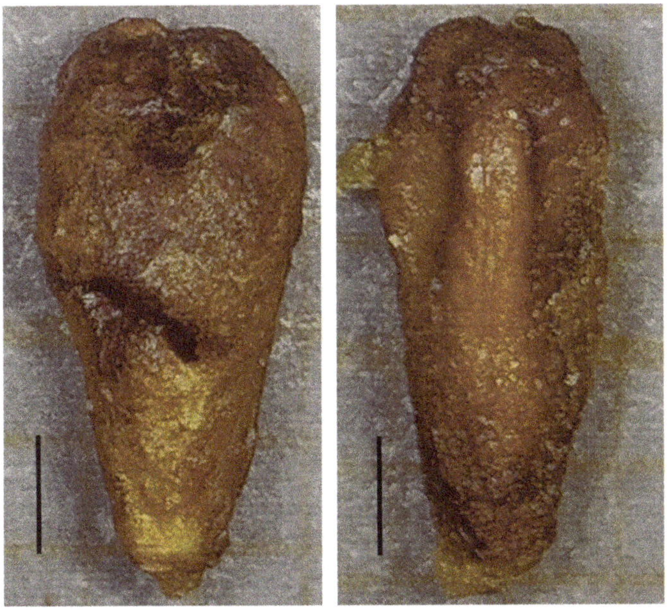

Figure 6. Modern *Vitis* seeds of a modern Afghan Raisin cultivar "Stenosperm" type. Digital microscopy. Scale bars: 1 mm.

4. Discussion

4.1. The Horticultural Relevance of of the Seed Types Identified

4.1.1. Domesticated Grapevines of the *Proles orientalis* Negrul

A group of seeds presents clearly domesticated characteristics and is assigned in the present work to the set of the three *Proles* of Negrul [11] (Table 5) with a probability = 1. Among the recognized *Proles*, assignment to the *Proles orientalis* is in the majority, with values of $p = 0.7–0.87$, predominating in all cases the *Subproles antasiatica*. The results confirm our "prior" hypothesis about the existence of this set of table and raisin varieties, but also for wine, known as *Proles orientalis*, in the Near East, at least since the Middle Bronze Age.

Table 5. Negrul's *Proles* of *Vitis vinifera* [5].

Taxa	Leaf	Cluster Size and Shape	Berry	Area and Typical Cultivars
Pontica Negr.	Mature leaf below with mixed hairy cover: cobwebby and hairy, irregularly curved leaf edge	Medium, compact, table, raisins and wine	Spherical, rarely ovoid, bittersweet, medium size, acidity 0.6–1%; seeds small to large. some seedless cultivars	West Caucasus, Balkans, Anatolia, Mediterranean. Frost hardy (Saperavi, Rkatsiteli, Cinuri).
Occidentalis Negr.	Mature leaf below with hairy cover: cobwebby, leaf edge folded downward	Small, compact, raisins and wine	Spherical; rarely ovate, small or medium, acidity 0.6–1%; seeds small shortly stalked, no seedless cultivars	Western and Central Europe. Frost hardy (Pinot, Gamay, Semillon, Riesling, Mourvedre, Muscat of Alexandria).
Orientalis Negr	Leaf above green, glabrescent, below glabrous to hairy-pubescent, leaf edge folded upwards	Large, conical, often branched, table	Ovate or oblong, medium to large; acidity 0.3–0.6%, seeds medium to large, long stalked, frequently seedless	Central Asia, East Caucasus. West Asia, frost tender (Baian Shirey, Muscat, Chasselas, and Kismisci).

[5] Data elaborated and synthesized by the authors based on different bibliographic sources, particularly [11,91].

De Lorenzis et al. [92], by genotyping germplasm from Central Europe, Armenia, Azerbaijan, Georgia, and Moldova by SSR markers, investigated the genetic relationships among samples along an East-to-West gradient. The identification of three different groups was explained based on geographical origin and human uses and it was in agreement with the *Proles* classification proposed by Negrul [11]:

1. Wine varieties from the West (Central European Cultivars); cf. *Proles occidentalis* (Table 5).
2. Wine varieties from the East (Armenia, Georgia, Moldova); cf. *Proles pontica* (Table 5).
3. Table varieties from the East (Azerbaijan). cf. *Proles orientalis* (Table 5).

Ampelographic studies of selected Syrian grapevine cultivars show their relationships with *Proles orientalis*, especially with regard to five highly discriminating traits (shoot internode length, berry weight, berry elongation, 100-seed weight, and titratable juice acidity) [93], and ampelographic analysis of Israeli grapevine shows clear differences between the *V. vinifera* and *V. sylvestris* groups in terms of flower, leaf and bunch parameters, and that most of *V. vinifera* in the area belongs to *Proles orientalis*, which is consistent with our results based on archaeological seed samples [94].

We must highlight the relevance of the *Proles orientalis*, in front of the *Proles pontica*, at least 4000 years ago, which extended the original area of the former from Central Asia (Negrul 1946) to Western Asia.

Among the grapevine cultivars described by Galet in 1970 [67] from Afghanistan, 29 belong to *Proles orientalis Subproles caspica*, 16 to *Proles orientalis Subproles antasiatica*, 6 to *Proles pontica*, and only two to *Proles occidentalis*. It is unclear to what extent the *Proles occidentalis* and *Pontica* varieties were recently introduced into Afghanistan or are due to a hypothetical introduction by the Greeks in the 4th century B.C. as a consequence of the campaigns of Alexander the Great. One of the open questions, especially concerning the ancient varieties of *Proles antasiatica*, is what the prioritized destination of their fruits was: table grapes or raisins? The available evidence deserves to be analyzed. In the stated

Near Eastern sites, it is common to interpret grape remains as raisins in preference to table grapes. While their use for winemaking is interpreted only when specialized containers and contexts are also found, along with the presence of marker chemical substances such as tartrates and characteristic anthocyanins, the fragmentary nature of our Upper Euphrates archaeological samples prevents us from going further in determining whether the grapes were actually consumed fresh or in the form of raisins. However, their preservation in the archaeological register of raisins' pips seems more likely.

4.1.2. Domesticated Grapevines Belonging to a Particular Euphrates *Proles*

While there is clearly an existence of a group of vines with domesticated features (DI = 0.67–0.83) in their seeds of hybrid origin, the fact they would have intervened with wild vines of the Caucasus and Asia leads us to propose the existence of a fourth singular *Proles euphratica* with more rounded seeds than in *Proles orientalis* and with Stummer [54] index values from 0.73 to 0.82 instead of 0.49 to 0.7, similar to the characteristics of wild vines. *Proles euphratica* is the name we propose here for this group of domesticated grapevines with characteristic seeds (4.7–5.5 $\times$ 3.5–4.1 mm) found in Middle Bronze, Assyrian, Aramaic, Persian, and Hellenistic levels from Upper Euphrates sites. These varieties appear to be unrelated to modern cultivars from the area or elsewhere but are related to Caucasian and Asian wild vines. This suggests that this peculiar set of domesticated varieties probably derived from hybridization between Asian and South Caucasian vines. The non-existence of similar grape vines in Turkey or other neighboring areas today may be explained by the abandonment of their cultivation due to their low productivity or the fact that they were exclusively wine grapes in a context where, since the 8th century, wine consumption was excluded for religious reasons.

To verify the persistence, or, if applicable, the date of extinction and further ampelographic characteristics of the here proposed *Proles euphratica*, it would be necessary to extend the ampelographic sampling of wild and feral vines, especially in the rivers and streams tributaries of the Euphrates in northern Syria and Iraqi Kurdistan, to expand the study of their seeds and archaeological remains of vines from the Near East with the methodology proposed here.

However, similar DI values were found in our study in samples also allocated to cluster 8 (Figure 2) from archaeological Chalcolithic to Medieval sites of Spain, Italy, Armenia, and the Balkans, and modern samples of wild and feral grapevines from Georgia, and Spain, and *Vitis piasezkii* Maxim. from China.

Vitis vinifera was probably cultivated in Emar (Syria). Apart from a fragment of wood from a Late Bronze Age stratum, many grape pips have been identified in levels of this period [95]. Carbonized seeds and charcoal of *Vitis sylvestris* and *V. vinifera* were repeatedly recovered from mid-third millennium BC contexts at Tilbeshar, Horum, and Jerablus Tahtani in the Middle Euphrates Valley; tartaric acid associated with gypsum basins found in houses at Titris Höyük suggests that grape processing was widespread there. The area continued to be a recognized center of wine production and export into the early second millennium [96]. Vine seeds, fruits, and pedicels have been recovered from archaeobotanical samples from the Middle Bronze Age site of Tell Tweini, Lattakia (Syria), probably the ancient city of Gibala [97]. It would be particularly useful to be able to study samples from these deposits in detail in order to establish the possible extent of *Proles euphratica* and the other types detected in this work.

4.1.3. Domesticated Grapevines Related with Asian Wild Grapevines

This group establishes a bridge between the cultivated vines of the *Proles euphratica* with various Central Asian vines and, possibly, with *Ampelopsis* species.

A stem was discovered in the Yanghai Tombs, Turpan District in Xinjiang, China. Anatomical features showed it to be of grape (*Vitis vinifera* L.). Radiocarbon dating indicates it to be nearly 2300 years old, which would suggest that there was grape cultivation at least from that time. To date, this is the earliest physical evidence of *V. vinifera* cultivation in China [98]. Two grape (*V. vinifera*) pips were discovered in the gut contents (665–770 cal. years AD) of a person unearthed at tomb 75TAM601 in Astana Cemetery, the public graveyards of the ancient Gaochang people in Turpan, Xinjiang. Grapevine (*V. vinifera*) cultivation was introduced into Turpan c.300 BCE, and Xinjiang is considered to be the earliest place to cultivate grapevine in China [99]. We cannot rule out the existence of other *Vitis* species or even *Ampelopsis* in these *Vitaceae* materials.

4.1.4. Eurasian Hybrid Wild Grapevines

By hybrids we mean individuals and populations that present, on average, intermediate characteristics between the domesticated vine and the wild vine and, at the same time, a great variability in the diagnostic parameters. With similar DI values, we found in clusters 3 and 10 (Figure 3) samples from hybrid populations (European wild vine x domesticated vine or American vines x domesticated grapevine) in natural habitats in Spain as well as archaeological remains from the Levant and the Balkans.

The above suggests the existence in the Upper Euphrates of hybridization processes in wild grapevines populations produced by the domesticated ones, which may have been due to the presence of vineyards in proximity to the wild populations. The analysis of genetic diversity among grapevine (*Vitis vinifera* L.) cultivars in Tartous province (Syria) using microsatellite markers detected high levels of polymorphism, which may be due to continuous seed propagation by birds, natural hybridization between native and introduced plants, and human selection, and also found that vines classified as table grapes with white to raspberry-colored fruit were included in the same cluster despite their genetic variation [100].

4.1.5. Eurasian Domesticated Hybrid and Feral Grapevines

This group is closely related to the previous one but with a significant degree of domestication and is assigned to the high DI values part of cluster 3 (Figure 3), together with feral grapevines from natural habitats in Spain and France, and archaeological materials from Spain, Italy, Hungary, the Balkans, Egypt, the Near East, and the Caucasus. The above also suggests the existence in the Upper Euphrates and abroad, since at least the Bronze Age, of hybridization processes between wild and domesticated grapevine populations, which could have been intentional or casual, which modified the morphology of cultivated grapevines. This would indicate that hybridization is a phenomenon that may have been relevant in the origin of many modern grapevine varieties.

Drori et al. [94] show the existence in Israel of two distinct populations that cluster closely together, suggesting a common genetic origin, one mostly of *V. vinifera* (together with Central Asian cultivars and some from *Proles orientalis Subproles caspica*), another cluster with the *V. sylvestris* from Northern Israel, and a third mixed. This suggests the possibility of wide hybridization between domesticated and wild grapevines in the Near East, which, according to our study of archaeological seeds, may have occurred over a few centuries.

4.1.6. Wild Eurasian—Caucasian Grapevines

These seeds are typical of wild vines and have similarities with the various types of wild vines that we have recognized "a priori" that witnesses the continued presence of clearly wild vines in the high Euphrates throughout all the periods studied, living, for some of these, with the presence of domesticated grapevines.

The group of vines whose seeds present low DI values allocated to cluster 8 includes Asian vines, such as *Vitis ficifolia* Bunge, which is widespread in China, Korea, and Japan, *Vitis amurensis* Rupr. from China, Korea, Siberia, and Japan, and *Vitis wilsoniae* H.J.Veitch; numerous wild grapevine populations (*Vitis sylvestris*) in Spain, Italy, and the Caucasus, American wild vines, fossil vines, and archaeological materials from Hungary, the Balkans, Spain, and the Near East.

Wan et al. [101] place the origin of the *V. heyneana*/*V. vinifera*/*V. sylvestris* clade at around 6.3 Mya and report an unexpected *V. sylvestris*-derived position that conflicts with *V. sylvestris* being the progenitor of *V. vinifera*, as the phylogenetic position suggests that *V. sylvestris* is derived from *V. vinifera*. They interpret this discrepancy in terms of erroneous inference due to the effects of clonal propagation in the grapevine cultivars analyzed [101]. We detected high variability in the domestication values of *V. heyneana*, *DI* = 0–0.67.

The phylogenetic structure within the genus *Vitis* was analyzed by Aradhya et al. [102] using simple sequence repeat (SSR) and amplified fragment length polymorphism (AFLP) markers, resulting in fourteen clusters, most of which contained a moderate frequency of mixed genotypes, suggesting interspecific gene flow within the subgenus *Vitis*. The AFLP-based tree clearly separated two clusters within *V. sylvestris*, assigned *V. vinifera* cultivars predominantly to one of them, and supported the close relatedness of *V. heyneana*. The SSR-based analysis was less conclusive.

The nuclear microsatellite-based study by Doulati-Baneh et al. [103] of wild grapevine populations in the Zagros Mountains of Iran grouped the accessions into three clusters corresponding to the geographical distribution of the populations. Most of the populations had an unbalanced sex ratio, with a modest number of female individuals in most of the Zagros Mountains, reducing the species seed dispersal ability. This was also confirmed by the field assessment during the plant sampling activity; a large proportion of individuals were older than 25–30 years, and only a few young seedlings were detected in only one population. Thus, Iranian wild grape populations have a reduced regeneration capacity, probably due to modest seed production and/or environmental disturbances leading to an inhospitable habitat for young seedlings.

All grape pips analyzed in Greece by Pagnoux et al. [104] from the Late Neolithic are morphologically wild. The change from the wild to the domesticated form occurred during the Middle Bronze Age (1900–1700 BC). The picture we get from the Upper Euphrates area is far more complex because wild types persisted until later periods coexisting with domesticated types.

4.1.7. Wild Asian Grapevines Allocated to Genus Ampelopsis

The presence of endemic *Ampelopsis* species in the area even today makes us think that, logically, they were also present during the periods analyzed. It is pending identification whether that presence is merely accidental from wild populations or if they were cultivated for any purpose. The finding of similar seeds in a garden purely dedicated to enjoyment at Hellenistic levels of Petra in Jordan is particularly informative.

Likely, the wild grapevine, which was a plant component of an Acheulian diet at Gesher Benot Ya'aqov, Israel, in the southern Hula Valley and assigned to the Lower–Middle Pleistocene, dated c. 780,000 BC [86], belongs to the genus *Ampelopsis* [78]. With the DI = 0.5, this seed is neither typically wild nor domesticated; however, we are not aware that this index remains useful for *Ampelopsis*.

Most of the wild vine species, both American and Asian, as well as the fossils studied, were ascribed to cluster 11.

4.1.8. Raisin Type Seeds with Low Domestication Syndrome Remotely Linked to Genus Ampelopsis

Abortive seeds alone do not explain this strange morphology since they are often morphologically similar to normal seeds and only differ in that they are empty instead of containing an embryo and endosperm [105].

Stenospermocarpic grapes, such as the "Sultanina" variety, are heritable, and these embryos can be obtained via normal pollination and fertilization, but seed development ends prematurely, leading to embryo abortion occurring at 2–4 weeks after blooming. Together with parthenocarpic grapes, these conform to two groups of seedless cultivars [106]. *Keshmesh*, meaning sultana, is a very common name for grapes used to make sultanas in Iran. In this case, similar names imply identical usage but do not necessarily indicate genetic similarity between the cultivars [107].

Proles antasiatica vines are now often used for raisin production and show anomalies in seed formation. Among the 55 Iranian grapevine cultivars investigated by Abiri et al. [108], 20 cultivars formed seeds in a rudimentary form, while seeds were well developed in 34 cultivars and one cultivar was seedless. On average, the seed number per berry is two, with a maximum of four. As expected, larger berries were found in less dense bunches.

Levchenko et al. [109] have shown the close genetic relationship within the *Vitaceae* family between the genera *Vitis* and *Ampelopsis*, which are closer in terms of the evolution of the grape culture as a whole. They obtained artificial intergeneric hybrids that produced c. 25% of plump seeds and 10% of viable embryos. Thus, 75% of the seeds were abnormally thin. It is relevant to mention here that when we found these abnormal types of seeds together with those that were *Ampelopsis*-like, we suspected that this anomaly could be due to hybridization or merely intergeneric cross-pollination.

5. Conclusions

The development of a new domestication index for the seeds of *Vitis vinifera–V. sylvestris*, based on the combination of six pre-existing indices, together with the multivariate analysis of the morphometry of the seed, the hierarchical classification of the samples, based on a very comprehensive collection of comparisons, and the integration of data from both sources using the Bayes–Laplace method, constitutes a very sensitive tool that allows one to successfully analyze alternative hypotheses about the identity of the archaeological seeds of *Vitis*.

As a consequence, in the study of the archaeological seeds of two sites of the Upper Euphrates, we have been able to establish the predominance, among the clearly domesticated ancient varieties, of *Proles orientalis* Negrul, together with the existence of some domesticated vines with peculiar characteristics, which we include in a possible extinct *Proles* that we call *Euphratica*.

Along with the above, it was determined that wild vines related to *Vitis sylvestris* C.C.Gmelin and *V. caucasica* Vavilov continued to be present throughout the period studied, from Early Bronze to Hellenistic.

In parallel, seeds with hybrid characteristics have been recorded, and much more significantly, the existence of seeds of the *Ampelopsis* type has been documented, among which is the oldest vitaceous seed linked to human presence in the Acheulean (780 myr).

Finally, the presence of "stenosperms" that appear to be associated with fully developed seeds of the *Ampelopsis* type has been identified, which suggests the existence of anomalies in the formation of the seeds due to intergeneric cross-pollination. Additionally, in the cases where they appear isolated, they suggest the presence of "stenospermocarpic" *Vitis vinifera* raisins of the Sultana type.

If we consider the relevance of this study for the future of the grapevine, specifically in terms of adaptation to climatic changes, resistance to pests, and new varieties with new characteristics, we can affirm that what is most relevant in our opinion is the detection of the early existence of "stenosperms" and therefore of seedless grapes of the "sultanina" type about four thousand years ago, which indicates the capacity of adaptation of this type of grape as well as those of the *Proles orientalis* Negrul to climatic and salinity changes along millennia. The possible extinction of *Proles euphratica* would suggest that not all local vines were able to overcome climatic crises, changes in use, or the appearance of eventual pests and would point to the value of ancient local varieties for their resilience. Finally, the study of history can help the farmer improve his vineyard through the idea of not despising local cultivars and local wild populations that have survived environmental adversities for centuries and whose communities are also repositories of feral grapevines. Furthermore, farmers can appreciate them as sources of genetic resources for the improvement of their vineyards, with preference for the introduction of exotic varieties whose possibilities of adaptation to local conditions and climate change are uncertain.

Supplementary Materials: The following supporting information can be downloaded at: https://www.mdpi.com/article/10.3390/horticulturae9070803/s1; Table S1: Summary of relevant parameters and samples for the eight seed main types recognized.

Author Contributions: This paper is part of the PhD thesis of one of the authors, J.V. Data curation, J.V.; Formal analysis, D.R.; Investigation, J.V.; Methodology, C.O. and D.R.; Project administration, G.M.-S.; Resources, G.M.-S.; Software, F.A.; Validation, D.R.; Writing—original draft, J.V.; Writing—review & editing, G.M.-S., C.O., F.A. and D.R. All authors have read and agreed to the published version of the manuscript.

Funding: This research received no external funding.

Data Availability Statement: Not applicable.

Acknowledgments: This article is part of the PhD thesis research of one of the authors, Javier Valera. We deeply acknowledge Lucía Cazorla Blázquez for the translation and linguistic revision of early versions of this document, to Maria-Teresa Coronado for the realization of carbonized seed images from FE-SEM, and David Arce for the help with the map of Syria. We thanks the Embassy of Spain in Syria for the support provided while we were excavating; to the General Directorate of Antiquities and Museums of the Arab Republic of Syria both for having allowed us to excavate in Tell Khamîs and for the support provided to the botanical expedition carried out in 1999; to Dina Bahkur, who accompanied us in the different excavation campaigns; to Abbas al-Hamza and his entire extended family from Qara Qûzâq; and to the members of the Archaeological Mission of the University of Murcia in Qara Qûzâq (Syria).

Conflicts of Interest: The authors declare no conflict of interest.

References

1. Zohary, D.; Spiegel-Roy, P. Beginnings of fruit growing in the Old World. *Science* **1975**, *187*, 319–326. [CrossRef]
2. Joffe, A.H. Alcohol and social complexity in ancient western Asia. *Curr. Anthropol.* **1998**, *39*, 297–322. [CrossRef]
3. Daler, S.; Cangi, R. Characterization of grapevine (*V. vinifera* L.) varieties grown in Yozgat province (Turkey) by simple sequence repeat (SSR) markers. *Turk. J. Agric. For.* **2022**, *46*, 38–48.
4. Taskesenlioglu, M.Y.; Ercisli, S.; Kupe, M.; Ercisli, N. History of Grape in Anatolia and Historical Sustainable Grape Production in Erzincan Agroecological Conditions in Turkey. *Sustainability* **2022**, *14*, 1496. [CrossRef]
5. Alleweldt, G.; Detweiler-Münch, E. *The Genetic Resources of Vitis. Genetic and Geographic Origin of Grapevine Cultivars, Their Prime Names and Synonyms*, 3rd ed.; Institut für Rebenzüchtung: Siebeldingen, Germany, 1992.
6. Sefc, K.M.; Steinkellner, H.; Lefort, F.; Botta, R.; Machado, A.D.C.; Borrego, J.; Maletić, E.; Glössl, J. Evaluation of the genetic contribution of local wild vines to European grapevine cultivars. *Am. J. Enol. Vitic.* **2003**, *54*, 15–21. [CrossRef]
7. This, P.; Jung, A.; Boccacci, P.; Borrego, J.; Botta, R.; Costantini, L.; Crespan, M.; Dangl, G.S.; Eisenheld, C.; Ferreira-Monteiro, F.; et al. Development of a common set of standard varieties and standardized method of scoring microsatellites markers for the analysis of grapevine genetic resources. *Theor. Appl. Genet.* **2004**, *109*, 1448–1458. [CrossRef]
8. Charbaji, T.; Ayyoubi, Z. Differential growth of some grapevine varieties in Syria in response to salt in vitro. *Vitr. Cell. Dev. Biol.-Plant* **2004**, *40*, 221–224. [CrossRef]
9. Zohary, D.; Hopf, M. *Domestication of Plants in the Old World*; Oxford University Press: New York, NY, USA, 2000.
10. Negrul, A.M. Family *Vitaceae* Lindley (Ampelideae Kunth.). In *Ampelografiia SSSR*; Baranov, A., Kai, Y.F., Lazarevski, M.A., Negrul, A.M., Palibin, T.V., Prosmoserdov, N.N., Eds.; Pischepromizdat: Moscow, Russia, 1946; Volume 1, pp. 45–133. (In Russian)
11. Negrul, A.M. Origin of the cultivated grapevine and its classification. In *Ampelografiia SSSR*; Baranov, A., Kai, Y.F., Lazarevski, M.A., Negrul, A.M., Palibin, T.V., Prosmoserdov, N.N., Eds.; Pischepromizdat: Moscow, Russia, 1946; Volume 1, pp. 159–216. (In Russian)
12. Sosnovszky, D.I. *Vitaceae* . In *Flora SSSR*; Komarov, V.L., Ed.; Academy Sci. S.S.S.R.: Moscow, Russia, 1949; Volume 14, pp. 674–710.
13. Sosnovszky, D.I. Vitaceae. In *Flora SSSR*; Komarov, V.L., Ed.; Israel Program of Scientific Translations: Jerusalem, Israel, 1974.
14. Obón, C.; Rivera, D.; Carreño, E.; Alcaraz, F.; Palazón, J.A. Seed morphology of *Vitis vinifera* and its relationship to ecogeographical groups and chlorotypes. *Acta Hortic.* **2008**, *799*, 51–55. [CrossRef]
15. De Candolle, A. *Origine des Plantes Cultivées*; Librairie Germer Baillière et Cie.: Paris, France, 1882.
16. Hegi, G. (Ed.) *Illustrierte Flora von Mittel-Europa*; Julius Friedrich Lehmann: Munich, Germany, 1925; Volume 5, pp. 350–426.
17. Planchon, J.E. Monographie des Ampélidées vrais. In *Monographiae Phanaerogamarum*; De Candolle, A.P., de Candolle, A.C.P., Eds.; Treuttel et Würtz: Paris, France, 1887; Volume 5, pp. 305–654.
18. Baranov, A.; Kai, Y.F.; Lazarevski, M.A.; Negrul, A.M.; Palibin, T.V.; Prosmoserdov, N.N. *Ampelografiia SSSR*; Pischepromizdat: Moscow, Russia, 1946; Volume 1. (In Russian)
19. Levadoux, L. Les populations sauvages et cultivées de *Vitis vinifera* L. *Ann. L'amel. Plantes* **1956**, *1*, 59–118.

20. Khorszhinzkii, S.I. *Ampelography of Crimea*; Glavnogo Upravleniya Udelov Yalta: Yalta, Russia, 1910; Volume I. (In Russian)
21. Andrasovszky, D. *Vitis* . In *Magyar Flora, 2*; Javorka, S., Ed.; Studium: Budapest, Hungary, 1925; pp. 701–708.
22. Terpó, A. Origin and distribution of *Vitis sylvestris* Gmel. In Proceedings of the II International Symposium on the Problems of Balkan Flora and Vegetation, Istanbul, Turkey, 3–10 July 1978.
23. Clemente, S.R. *Ensayo Sobre las Variedades de la Vid Común que Vegetan en Andalucía*; Imprenta de Villalpando: Madrid, Spain, 1807.
24. Clemente, S.R. *Essai sur les Variétés de la Vigne qui Végétent en Andalousie*; Poulet: Paris, France, 1814.
25. Clemente, S.R. En que se trata que tierras, aires y sitios son Buenos para las viñas. In *Agricultura General de Gabriel Alonso de Herrera*; Imprenta Real: Madrid, Spain, 1818; Volume 1, pp. 314–544.
26. Clemente, S.R. *Ensayo Sobre las Variedades de la Vid Común que Vegetan en Andalucía*; Estereotipia Perojo: Madrid, Spain, 1879.
27. Kolenati, F.A. Versuch einer systematischen Anordnung der in Grusinien einheimischen Reben nebst einem oekonomisch-technischen Anhange. *Bull. Soc. Sci. Nat. Moscou* **1846**, *19*, 283–371.
28. Frolov-Bagreev, A.M.; Negrul, A.M.; Blagonravov, P.P. *Ampelogafiya SSSR (Ampelography of SSSR)*; Pishchepromizdat: Moscow, Russia, 1953–1956; Volume 2–6. (In Russian)
29. Rivera, D.; Walker, M. A review of palaeobotanical findings of early *Vitis* in the Mediterranean and of the origins of cultivated grape-vines, with special reference to new pointers to prehistoric exploitation in the western Mediterranean. *Rev. Palaeobot. Palynol.* **1989**, *61*, 205–237.
30. Smith, H.; Jones, G. Experiments on the effects of charring on cultivated grape seeds. *J. Archaeol. Sci.* **1990**, *17*, 317–327. [CrossRef]
31. Bouby, L.; Bonhomme, V.; Ivorra, S.; Pastor, T.; Rovira, N.; Tillier, M.; Pagnoux, C.; Terral, J.F. Back from burn out: Are experimentally charred grapevine pips too distorted to be characterized using morphometrics? *Archaeol. Anthropol. Sci.* **2018**, *10*, 943–954. [CrossRef]
32. Van der Veen, M. Formation processes of desiccated and carbonised plant remains—The identification of routine practice. *J. Archaeol. Sci.* **2007**, *34*, 968–990. [CrossRef]
33. Miller, N. Sweeter Than Wine? The Use of the Grape in Early Western Asia. *Antiquity* **2008**, *82*, 937–946. [CrossRef]
34. Châtaignier, C. La Transcaucasie au Néolithique et au Chalcolithique. *Br. Archaeol. Ser.* **1995**, *624*, 1–240.
35. Mravcsik, Z.; Gyulai, F.; Vinogradov, S.; Emődi, A.; Rovner, I.; Gyulai, G. Digital seed morphometry for genotype identification–Case study of seeds of excavated (15th century Hungary) and current vinegrape (*Vitis* v. *vinifera*) varieties. *Acta Bot. Hung.* **2015**, *57*, 169–182. [CrossRef]
36. Valamoti, S.M.; Mangafa, M.; Koukouli-Chrysanthaki, C.; Malamidou, D. Grape-pressings from Northern Greece: The earliest wine in the Aegean? *Antiquity* **2007**, *81*, 54–61. [CrossRef]
37. Logothetis, V. Contribution of the grape vine and wine in Greek and Eastern Mediterranean civilization. *Sci. Ann. Univ. Thessalon.* **1974**, *17*, 1–286.
38. McGovern, P.; Fleming, S.; Katz, S. *The Origins and Ancient History of Wine*; Gordon & Breach: New York, NY, USA, 1995.
39. Buxó, R. The agricultural consequences of colonial contacts on the Iberian Peninsula in the first millennium BC. *Veg. Hist. Archaeobot.* **2008**, *17*, 145–154. [CrossRef]
40. Arroyo-Garcia, R.; Ruiz-Garcia, L.; Bolling, L.; Ocete, R.; Lopez, M.A.; Arnold, C.; Ergul, A.; Söylemezoğlu, G.; Uzun, H.I.; Cabello, F.; et al. Multiple origins of cultivated grapevine (*Vitis vinifera* L. ssp. *sativa*) based on chloroplast DNA polymorphisms. *Mol. Ecol.* **2006**, *12*, 3707–3714. [CrossRef] [PubMed]
41. Willcox, G. The distribution, natural habitats and availability of wild cereals in relation to their domestication in Near East: Multiple events, multiple centres. *Veg. Hist. Archaeobot.* **2005**, *14*, 534–541. [CrossRef]
42. Zohary, D. The domestication of the grapevine *Vitis vinifera* L. In *Near East, in The Origins and Ancient History of Wine*; McGovern, P.E., Fleming, S.J., Katz, S.H., Eds.; Gordon & Breach: New York, NY, USA, 1995; pp. 23–30.
43. Rivera, D.; Matilla, G.; Obón, C.; Alcaraz, F. *Plants and Humans in the Near East and the Caucasus*; Editum-Plants and Humans: Murcia, Spain, 2012; Volumes 1 and 2.
44. Matilla, G.; Rivera, D.; Obon, C. Estudio Paleoetnobotânico de Tell Qara Qûzâq, 1. Campañas I–III. In *Tell Qara Qûzâq I*; del Olmo, G., Ed.; AUSA: Sabadell, Spain, 1994; pp. 151–182.
45. Matilla, G.; Garcia-Cano, J.M.; Gallardo, J.; Page, V. *Tell Qara Qûzâq-III, Campañas VII–XI (1995–1999)*; AUSA: Sabadell, Spain, 2012.
46. Matilla, G. Las tres primeras campañas de excavaciones en Tell Jamis (Siria). *Aula Orient* **1996**, *14*, 191–247.
47. Matilla, G. Tell Khamîs. In *International Symposium on the Archaeology of the Upper Syrian Euphrates (Tishrin Dam Area)*; Olmo, G., Ed.; AUSA: Sabadell, Spain, 1997; pp. 205–225.
48. Matilla, G. Arqueología y Antropología en el Alto Éufrates Sirio: Tell Jamis. Ph.D. Thesis, Universidad de Murcia, Murcia, Spain, 1999.
49. Matilla, G. El complejo religioso de Tell Khamîs (norte de Siria) durante el bronce medio. In *Nomina in Aqua Scripta: Homenaje a Joaquín María Córdoba Zoilo*; del Cerro, C., Ed.; Universidad Autónoma de Madrid España: Madrid, Spain, 2021; pp. 703–718.
50. Valera, J.; Matilla-Seiquer, G.; Obón, C.; Rivera, D. Archaeobotanical Study of Tell Khamîs (Syria). *Heritage* **2022**, *5*, 1687–1718. [CrossRef]
51. Matilla, G.; Gallardo, J. Tell Qara Quzaq. *Rev. Arqueol.* **1997**, *18*, 24–35.
52. Rivera, D.; Miralles, B.; Obón, C.; Carreño, E.; Palazón, J.A. Multivariate analysis of *Vitis* subgenus *Vitis* seed morphology. *Vitis* **2007**, *46*, 158–167.

53. Schindelin, J.; Arganda-Carreras, I.; Frise, E.; Kaynig, V.; Longair, M.; Pietzsch, T.; Preibisch, S.; Rueden, C.; Saalfeld, S.; Schmid, B.; et al. Fiji: An open-source platform for biological-image analysis. *Nat. Methods* **2012**, *9*, 676–682. [CrossRef]
54. Stummer, A. Zur Urgeschichte der Rebe und des Weinbaues. *Mitteilungen Anthropol. Ges. Wien* **1911**, *61*, 283–296.
55. Perret, M. *Caractérisation et Evaluation du Polymorphisme des Génotypes Sauvages et Cultivés de Vitis vinifera L. à L'aide de Marqueurs RAPD et de Certains Traits Morphologiques*; Travail de diplôme, Université de Neuchâtel: Neuchâtel, Switzerland, 1997.
56. Facsar, G. Összehasonlitö morfolögiai vizsgälatok kerti szölöfajtäk magjain I (Vergleichende morphologische Untersuchungen der Samen von Gartenrebensorten. I). *Bot. Közlemenyek* **1970**, *57*, 221–231.
57. Facsar, G.; Jerem, E. Zum urgeschichtlichen Weinbau in Mitteleuropa. Rebkernfunde von *Vitis vinifera* L. aus der urnenfelder-, hallstatt-und latenezeitlichen Siedlung Sopron-Krautacker. *Wiss. Arb. Aus Dem Burgenland* **1985**, *71*, 121–144.
58. Mangafa, M.; Kotsakis, K. A new method for the identification of wild and cultivated charred grape seeds. *J. Archaeol. Sci.* **1996**, *23*, 409–418. [CrossRef]
59. Perrier, X.; Flori, A.; Bonnot, F. Data analysis methods. In *Genetic Diversity of Cultivated Tropical Plants*; Hamon, P., Seguin, M., Perrier, X., Glaszmann, J.C., Eds.; Enfield: Montpellier, France, 2003; pp. 43–76.
60. Perrier, X.; Jacquemoud-Collet, J.P. DARwin Software. Available online: http://darwin.cirad.fr/ (accessed on 23 January 2016).
61. Kovach, M. MVSP–A MultiVariate Statistical Package for Windows. Available online: https://www.kovcomp.co.uk/mvsp/ (accessed on 11 May 2022).
62. Rambaut, A. Figtree. 2018. Available online: https://github.com/rambaut/figtree/releases (accessed on 1 January 2023).
63. Pagnoux, C.; Bouby, L.; Ivorra, S.; Petit, C.; Valamoti, S.-M.; Pastor, T.; Picq, S.; Terral, J.-F. Inferring the agrobiodiversity of *Vitis vinifera* L. (grapevine) in ancient Greece by comparative shape analysis of archaeological and modern seeds. *Veg. Hist. Archaeobot.* **2015**, *24*, 75–84.
64. Rivera, D.; Obón, C.; García-Arteaga, J.; Egea, T.; Alcaraz, F.; Laguna, E.; Carreño, E.; Johnson, D.; Krueger, R.; Delgadillo, J.; et al. Carpological analysis of Phoenix (Arecaceae): Contributions to the taxonomy and evolutionary history of the genus. *Bot. J. Linn. Soc.* **2014**, *175*, 74–122. [CrossRef]
65. IPGRI-UPOV-OIV. Descriptors for Grapevine (*Vitis* spp.). International Union for the Protection of New Varieties of Plants, Geneva, Switzerland/Office International de la Vigne et du Vin, Paris, France/International Plant Genetic Resources Institute, Rome, Italy. 1997. Available online: http://archive-ecpgr.cgiar.org/fileadmin/bioversity/publications/pdfs/393_Descriptors_for_grapevine__Vitis_spp._-cache=1415188920.pdf (accessed on 14 May 2023).
66. FNGR. French Network of Grapevine Repositories, Version 4 © 2017–2019 INRA-IFV-Montpellier SupAgro, Database of the Collections. 2023. Available online: https://bioweb.supagro.inra.fr/collections_vigne/SearchS.php?l=EN (accessed on 1 April 2023).
67. Galet, P. Rapport sur la viticulture en Afghanistan. *Vitis* **1970**, *9*, 15–46.
68. VIVC. *Vitis* International Variety Catalogue VIVC. 2023. Available online: https://www.vivc.de/index.php?r=www-europ-catalogue%2Findex (accessed on 1 April 2023).
69. Dönmez, E.O. Prehistoric and Medieval plant remains from two sites on the Euphrates, south-eastern Turkey. *Turk. J. Bot.* **2006**, *30*, 11–38.
70. Orendi, A.; Deckers, K. Agricultural resources on the coastal plain of Sidon during the Late Iron Age: Archaeobotanical investigations at Phoenician Tell el-Burak, Lebanon. *Veg. Hist. Archaeobot.* **2018**, *27*, 717–736. [CrossRef]
71. Mashkour, M.; Tengberg, M.; Shirazi, Z.; Madjidzadeh, Y. Bio-archaeological studies at Konar Sandal, Halil Rud basin, southeastern Iran. *Environ. Archaeol.* **2013**, *18*, 222–246. [CrossRef]
72. Ramsay, J.; Mueller, N. Telling Seeds: Archaeobotanical Investigations at Tall al-'Umayri, Jordan. In *The Archaeology of Agro-pastoralist Economies in Jordan*; McGeough, K.M., Ed.; The American Schools of Oriental Research: Boston, MA, USA, 2016; pp. 1–26.
73. Ramsay, J.; Bedal, L.A. Garden variety seeds? Botanical remains from the Petra Garden and Pool Complex. *Veg. Hist. Archaeobot.* **2015**, *24*, 621–634. [CrossRef]
74. Spengler, R.N.; Maksudov, F.; Bullion, E.; Merkle, A.; Hermes, T.; Frachetti, M. Arboreal crops on the medieval Silk Road: Archaeobotanical studies at Tashbulak. *PLoS ONE* **2018**, *13*, e0201409. [CrossRef]
75. Spengler, R.N.; Chang, C.; Tourtellotte, P.A. Agricultural production in the Central Asian mountains: Tuzusai, Kazakhstan (410–150 bc). *J. Field Archaeol.* **2013**, *38*, 68–85. [CrossRef]
76. Dönmez, E.O.; Belli, O. Urartian plant cultivation at Yoncatepe (Van), eastern Turkey. *Econ. Bot.* **2006**, *61*, 290–298. [CrossRef]
77. Paudyal, K.N.; Bhandari, S.; Momohara, A. Late Quaternary plant macrofossil assemblages from the Besigaon section of the Gokarna Formation, Kathmandu Valley, central Nepal. *J. Nepal Geol. Soc.* **2011**, *42*, 1–12. [CrossRef]
78. Bhandari, S.; Momohara, A.; Paudayal, K.N. Late Pleistocene plant macro-fossils from the Gokarna Formation of the Kathmandu Valley, central Nepal. *Bull. Dep. Geol.* **2009**, *12*, 75–88. [CrossRef]
79. Jacquat, C.; Martinoli, D. *Vitis vinifera* L.: Wild or cultivated? Study of the grape pips found at Petra, Jordan; 150 B.C.–A.D. 40. *Veg. Hist. Archaeobot.* **1999**, *8*, 25–30. [CrossRef]
80. Riehl, S. Archaeobotany at the early Bronze Age settlement of Ḫirbet ez-Zeraqōn: A preliminary report. *Z. Des Dtsch. Palästina-Ver.* **2004**, *120*, 101–122.

81. Marinova, E.; van Loon, G.J.; De Meyer, M.; Willems, H. Plant economy and land use in Middle Egypt during the late Antique/early islamic period: Archaeobotanical analysis of mud bricks and mud plasters from the area of Dayr al-Barshā. In *Reports in African Archaeology, 3, Windows on the African Past, Current Approaches to African Archaeobotany*; Fahmy, A., Kahlheber, S., D'Andrea, C., Eds.; Africa Magna Verlag: Frankfurt, Germany, 2012; pp. 120–136.
82. Ramsay, J.; Tepper, Y. Signs from a green desert: A preliminary examination of the archaeobotanical remains from a Byzantine dovecote near Shivta, Israel. *Veg. Hist. Archaeobot.* **2010**, *19*, 235–242. [CrossRef]
83. Goor, A. The history of the grape-vine in the Holy Land. *Econ. Bot.* **1966**, *20*, 46–64. [CrossRef]
84. POWO. *Vitis heyneana* in Plants of the World Online. Available online: https://powo.science.kew.org/taxon/urn:lsid:ipni.org:names:68768-1 (accessed on 1 May 2023).
85. POWO. *Ampelopsis orientalis* in Plants of the World Online. Available online: https://powo.science.kew.org/taxon/urn:lsid:ipni.org:names:67448-1 (accessed on 1 May 2023).
86. Melamed, Y.; Kislev, M.E.; Geffen, E.; Lev-Yadun, S.; Goren-Inbar, N. The plant component of an Acheulian diet at Gesher Benot Ya'aqov, Israel. *Proc. Natl. Acad. Sci. USA* **2016**, *113*, 14674–14679. [CrossRef] [PubMed]
87. Pokharia, A.K.; Sharma, S.; Tripathi, D.; Mishra, N.; Pal, J.N.; Vinay, R.; Srivastava, A. Neolithic– Early historic (2500–200 BC) plant use: The archaeobotany of Ganga Plain, India. *Quat. Int.* **2017**, *443*, 223–237. [CrossRef]
88. Decaix, A.; Mohaseb, F.A.; Maziar, S.; Mashkour, M.; Tengberg, M. Subsistence economy in Kohneh Pasgah Tepesi (eastern Azerbaijan, Iran) during the Late Chalcolithic and the Early Bronze Age based on the faunal and botanical remains. In *The Iranian Plateau during the Bronze Age. Development of Urbanisation, Production and Trade*; Meyer, J.W., Vila, E., Mashkour, M., Casanova, M., Vallet, R., Eds.; Maison de l'Orient et de la Méditerranée: Lyon, France, 2019; pp. 75–88.
89. Van Zeist, W.; Heeres, J.A. Paleobotanical studies of Deir' Alla, Jordan. *Paléorient* **1973**, *1*, 21–37. [CrossRef]
90. White, C.E.; Chesson, M.S.; Schaub, R.T. A recipe for disaster: Emerging urbanism and unsustainable plant economies at Early Bronze Age Ras an-Numayra, Jordan. *Antiquity* **2014**, *88*, 363–377. [CrossRef]
91. Dalmasso, G.; Tyndalo, V. *Viticoltura e ampelografia dell'U.R.S.S*; Accad Ital della Vite e del Vino: Florence, Italy, 1957; Volume IX, pp. 446–548.
92. De Lorenzis, G.; Maghradze, G.; Biagini, B.; Di Lorenzo, G.S.; Melyan, G.; Musayev, M.; Savin, G.; Salimov, V.; Chipashvili, R.; Failla, O. Molecular investigation of Caucasian and Eastern European grapevine cultivars (*V. vinifera* L.) by microsatellites. *Vitis-J. Grapevine Res.* **2015**, *54*, 13–16.
93. Khalil, S.; Tello, J.; Hamed, F.; Forneck, A. A multivariate approach for the ampelographic discrimination of grapevine (*Vitis vinifera*) cultivars: Application to local Syrian genetic resources. *Genet. Resour. Crop Evol.* **2017**, *64*, 1841–1851. [CrossRef]
94. Drori, E.; Rahimi, O.; Marrano, A.; Henig, Y.; Brauner, H.; Salmon-Divon, M.; Netzer, Y.; Prazzoli, M.L.; Stanevsky, M.; Failla, O.; et al. Collection and characterization of grapevine genetic resources (*Vitis vinifera*) in the Holy Land, towards the renewal of ancient winemaking practices. *Sci. Rep.* **2017**, *7*, 44463. [CrossRef]
95. Deckers, K. Anthracological research at the archaeological site of Emar on the Middle Euphrates, Syria. *Paléorient* **2005**, *31*, 153–167. [CrossRef]
96. Peltenburg, E. Conflict and Exclusivity in Early Bronze Age Societies of the Middle Euphrates Valley. *J. Near East. Stud.* **2013**, *72*, 233–252. [CrossRef]
97. Marinova, E.; Linseele, V.; Vandorpe, P.; Van der Valk, J. Middle Bronze Age Ritual, Subsistence and Environment at Tell Tweini-Infered from Bioarchaeological Evidence. In *The Ancient Near East, A Life!* Boiy, T., Bretschneider, J., Goddeeris, A., Hameeuw, H., Jans, G., Tavernier, J., Eds.; Festschrift Karel Van Lerberghe: Leuven, Belgium, 2012; pp. 345–364.
98. Jiang, H.E.; Zhang, Y.B.; Li, X.; Yao, Y.F.; Ferguson, D.K.; Lu, E.G.; Li, C.S. Evidence for early viticulture in China: Proof of a grapevine (*Vitis vinifera* L., *Vitaceae*) in the Yanghai tombs, Xinjiang. *J. Archaeol. Sci.* **2009**, *36*, 1458–1465. [CrossRef]
99. Chen, T.; Wang, B.; Mai, H.; Jiang, H. Last meals inferred from the possible gut contents of a mummy: A case study from Astana Cemetery, Xinjiang, China. *Archaeometry* **2020**, *62*, 847–862. [CrossRef]
100. Dauob, R.; Makhoul, G.; Mahfoud, H. Genetic Diversity among Grapevine (*Vitis vinifera* L.) Cultivars of Tartous Province (Syria) using Microsatellite Markers. *Int. J. Agric. Environ. Sci.* **2018**, *5*, 54–58.
101. Wan, Y.; Schwaninger, H.R.; Baldo, A.M.; Labate, J.A.; Zhong, G.Y.; Simon, C.J. A phylogenetic analysis of the grape genus (*Vitis* L.) reveals broad reticulation and concurrent diversification during Neogene and quaternary climate change. *BMC Evol. Biol.* **2013**, *13*, 1–20.
102. Aradhya, M.; Wang, Y.; Walker, M.A.; Prins, B.H.; Koehmstedt, A.M.; Velasco, D.; Gerrath, J.M.; Dangl, G.S.; Preece, J.E. Genetic diversity, structure, and patterns of differentiation in the genus *Vitis*. *Plant Syst. Evol.* **2013**, *299*, 317–330. [CrossRef]
103. Doulati-Baneh, H.; Mohammadi, S.A.; Labra, M.; De Mattia, F.; Bruni, I.; Mezzasalma, V.; Abdollahi, R. Genetic characterization of some wild grape populations (*Vitis vinifera* subsp. *sylvestris*) of Zagros Mountains (Iran) to identify a conservation strategy. *Plant Genet. Resour.* **2015**, *13*, 27–35. [CrossRef]
104. Pagnoux, C.; Bouby, L.; Valamoti, S.M.; Bonhomme, V.; Ivorra, S.; Gkatzogia, E.; Karathanou, A.; Kotsachristou, D.; Kroll, H.; Terral, J.F. Local domestication or diffusion? Insights into viticulture in Greece from Neolithic to Archaic times, using geometric morphometric analyses of archaeological grape seeds. *J. Archaeol. Sci.* **2021**, *125*, 105263. [CrossRef]
105. Ebadi, A.; Sedgley, M.; May, P.; Coombe, B.G. Seed development and abortion in *Vitis vinifera* L., cv. Chardonnay. *Int. J. Plant Sci.* **1996**, *157*, 703–712. [CrossRef]

106. Li, S.; Liu, K.; Yu, S.; Jia, S.; Chen, S.; Fu, Y.; Sun, F.; Luo, Q.; Wang, Y. The process of embryo abortion of stenospermocarpic grape and it develops into plantlet in vitro using embryo rescue. *Plant Cell Tissue Organ Cult.* **2020**, *143*, 389–409. [CrossRef]
107. Fatahi, R.; Ebadi, A.; Bassil, N.; Mehlenbacher, S.A.; Zamani, Z. Characterization of Iranian grapevine cultivars using microsatellite markers. *Vitis* **2003**, *42*, 185–192.
108. Abiri, K.; Rezaei, M.; Tahanian, H.; Heidari, P.; Khadivi, A. Morphological and pomological variability of a grape (*Vitis vinifera* L.) germplasm collection. *Sci. Hortic.* **2020**, *266*, 109285. [CrossRef]
109. Levchenko, S.V.; Volynkin, V.A.; Zlenko, V.A.; Polulyakh, A.A.; Likhovskoi, V.V.; Vasylyk, I.A. Experimental Variability of Grapes in *Vitaceae* Family Based on the Use of Allotetraploidy and in vitro embryo culture. *Acta Hortic.* **2018**, *1324*, 179–185. [CrossRef]

horticulturae

Article

Analysis of the Aroma Volatile Profile of Muscadine Grape Germplasm by Headspace Solid-Phase Microextraction Coupled with Gas Chromatography-Mass Spectrometry

Gaurab Bhattarai [1], Orestis Giannopoulos [2], Ramsey Nathanal Corn [2], Camille E. E. McAvoy [2], Angelos Deltsidis [2], Margaret L. Worthington [3] and Patrick J. Conner [2,*]

1 Institute of Plant Breeding, Genetics & Genomics, University of Georgia, Athens, GA 30602, USA; gbhattarai@uga.edu
2 Department of Horticulture, University of Georgia-Tifton Campus, Tifton, GA 31793, USA; orestis.g@uga.edu (O.G.); ramsey.corn@uga.edu (R.N.C.); ceemcavoy@uga.edu (C.E.E.M.); adeltsidis@uga.edu (A.D.)
3 Department of Horticulture, University of Arkansas, Fayetteville, AR 72701, USA; mlworthi@uark.edu
* Correspondence: pconner@uga.edu

Abstract: Muscadine grapes (*Vitis rotundifolia*) are native to the southeastern U.S., where they are valued for their unique flavor and fruity aroma. Despite having a diverse aroma profile, muscadine germplasm is virtually unexplored in terms of its aroma volatile content and composition, which is crucial in determining the value of its products. The aim of this research was to characterize 24 muscadine genotypes with distinct uses and origin for their aroma-related volatile profiles using the headspace solid-phase microextraction method coupled with gas-chromatography mass spectrometry. In total, 63 volatile compounds were detected, and genotypes significantly differed for 43 of the volatile compounds. We also profiled the aroma volatile content and composition of the commercially cultivated muscadine cultivar Carlos at various stages of berry ripeness. Characteristic differences were observed in the composition of the volatile compounds as ripening progressed. This is the first study to have evaluated the aroma volatile composition of a wide variety of muscadine germplasms, including juice and fresh fruit cultivars, as well as the related species *Vitis popenoei* and its complex hybrids between *V. rotundifolia* and *Vitis vinifera*. The results obtained from this study will help identify muscadine genotypes and better design crosses to produce fresh fruit and wine selections with the desired aroma profiles. This knowledge will lead to the development of new muscadine cultivars and significantly contribute to the expansion of muscadine use in the future.

Keywords: *Vitis rotundifolia*; *Muscadinia*; flavor; quality; GC-MS; HS-SPME

Citation: Bhattarai, G.; Giannopoulos, O.; Corn, R.N.; McAvoy, C.E.E.; Deltsidis, A.; Worthington, M.L.; Conner, P.J. Analysis of the Aroma Volatile Profile of Muscadine Grape Germplasm by Headspace Solid-Phase Microextraction Coupled with Gas Chromatography-Mass Spectrometry. *Horticulturae* **2023**, *9*, 1054. https://doi.org/10.3390/horticulturae9091054

Academic Editors: Antonio Amores-Arrocha and Ana Jiménez-Cantizano

Received: 25 August 2023
Revised: 15 September 2023
Accepted: 18 September 2023
Published: 20 September 2023

1. Introduction

Muscadine grapes (*V. rotundifolia*) are a native North American fruit found commonly throughout the southeastern U.S. with excellent adaptation to hot, humid summers and warm winters. It belongs to the *Muscadinia* subgenera in the genus of *Vitis* [1–4]. The subgenus *Muscadinia* has only two species: *V. rotundifolia* Michx. and *V. popenoei* J.H. Fennel [5]. Muscadines are a regional fruit with commercial production concentrated in the southern piedmont of North Carolina, the eastern coastal plain of North Carolina and South Carolina, and the piedmont and coastal plain of Georgia [6]. Production also extends to other states, especially Florida and Arkansas, but in lesser amounts. This fruit crop possesses tremendous potential for sustainable fruit systems in this region as it is much more productive than bunch grapes when grown in low-input systems [7]. Muscadines are primarily produced for fresh-market sales and processed products such as wine and juice. Muscadines have a high antioxidant capacity similar to that of blueberries and blackberries, leading to them often being described as a "superfruit" [8,9].

Muscadines are prized throughout their native range for their unique flavor. The basic tastes (sweetness, sourness, and bitterness) impacting fruit flavor are perceived by the taste receptors on the tongue, while volatile compounds are responsible for typical aromas (smells) and aromatic flavors (perceived while in the mouth). The combination of taste and olfaction provide the sensation of flavor [10,11]. Muscadines have very pronounced aromas that are often described as "fruity", "foxy", or "candy-like". The aroma of fresh muscadines is very desirable to those who know and love the fruit, but it can be overwhelming to those who are only familiar with *V. vinifera* grapes. Desired flavors vary depending on whether the fruit is to be used for fresh market, juice, or wine. Muscadine grape germplasm is highly diverse for appealing fruit aromas, likely in part to facilitate the location and consumption of dehiscent berries by an array of frugivore seed dispersers [12]; however, muscadine germplasm is virtually unexplored for its aroma-related volatile composition and content.

Baek et al. [13] performed a gas chromatography-mass spectrometry (GC-MS) study of 'Carlos' muscadine grape juice and identified 33 volatile compounds related to aroma. Among them, 2,5-Dimethyl-4-hydroxy-3(2H)-furanone (furaneol) was the most abundant, which, along with o-aminoacetophenone, was thought to give the candy and foxy-like aroma notes to muscadine juice. At low concentrations, furaneol has a pineapple- or strawberry-like aroma. It has also been found in other North American *Vitis* species, including *Vitis labrusca*. However, at higher concentrations, the same compound gives off an undesirable caramel-like aroma, which is often avoided in wine made from bunch grapes (*V. vinifera*). This study also found several other predominant compounds and their aromas such as 2,3-butanedione (buttery/cream cheese), ethyl butanoate (bubble gum/fruity), ethyl 2-methylbutanoate (green apple/fruity), and 2-phenylethanol (rosy). In a separate study of ripe 'Cowart' muscadine grapes by Lee et al. [14], volatile esters associated with fruity, floral, and pleasant odors were detected. Wine samples of 'Welder' and 'Noble' muscadine grapes produced relatively high concentrations of alcohols and esters of fatty acids [15]. Recently, Deng et al. [16] investigated the aroma-related volatile profile of five commercially grown muscadine cultivars. This study was able to identify 44 compounds, including esters, aldehydes, alcohols, fatty acids, terpenes, ketones, and furan via the solid-phase microextraction (SPME) GC-MS procedure. They found that geraniol and cinnamyl alcohol were the key volatile components that distinguished the Alachua cultivar from the rest. This study also showed that (Z)-3-hexenal, and (E)-2-hexenol were more prominent in the Fry and Granny Val cultivars, respectively. Although these studies indicate the presence of a wide variety of aroma-related volatile compounds in muscadine, focusing on only a few pure *V. rotundifolia* muscadine cultivars was not sufficient to investigate and exploit this genetic resource.

Fruit ripeness is the key attribute in muscadines that determines the appropriate time of harvesting fruits for both fresh market and juice production. This is especially important for fruits like muscadines with a relatively short shelf-life (about four weeks) and harvesting window (late summer to early autumn). The composition and abundance of aroma volatiles are linked with the fruit maturity and level of ripeness. Studies have shown differences in volatile composition in the different ripening stages of bunch grapes [17], raspberries [18], bananas [19], lulos [20], figs [21], and muskmelons [22]. However, no research has been conducted to understand the effect of berry ripeness on the aroma volatile composition of muscadines.

In this research, we investigated the aroma-related volatile profile of a wide variety of muscadine germplasms, including closely related species and hybrids that have been an important gene pool in muscadine breeding. Additionally, we identified volatile compounds and their abundance pattern at various ripening stages in a commercial muscadine wine cultivar, Carlos. The systematic characterization of the volatile compound profile of key germplasms will aid muscadine breeding by giving breeders the ability to select and combine appropriate parents for the desired flavor profile of the product. This eventually helps fulfill consumers' demand for specific muscadine flavors in the market, and it will ultimately boost muscadine utilization. Similarly, understanding the pattern of volatile abundance at various stages of fruit ripeness could provide information for the muscadine industry in understanding how berry ripening stage will influence the flavor profile of muscadine products.

2. Materials and Methods

2.1. Plant Materials and Sampling

The muscadine vines for germplasm characterization were grown at the University of Georgia (UGA) breeding program's experimental vineyards located in Tifton, GA, USA (lat. 31°28′39.81″ N, long. 83°31′39.61″ W). A list of these genotypes, their utility, and derivation is presented in Table 1. Vines were planted 3.04 m apart within the row and trained to two cordons on a 1.5 m high wire. Vines were irrigated and received commercial level care, which includes fertilization and fungicide applications as recommended by Poling et al. [23]. For the experiment to study the aroma volatiles at different stages of ripening, the fruit samples collected from the Carlos cultivar were planted at a commercial vineyard (Paulk Vineyards) in Wray, GA, USA.

Table 1. Muscadine germplasm used in aroma-volatile characterization.

Genotype	Female Parent	Male Parent	Berry Color [z]	Reference	Notes
AM-195	AM-19	AM-1	Black	M.L.W. breeding records	Fresh-market selection.
AM-77	Carlos	NC. 67A015-26	Black	M.L.W. breeding records	Red wine and juice breeding selection.
Carlos	Howard	NC. 11-173	Bronze	[24]	Leading white wine muscadine cultivar.
Cowart	Higgins	Ga. 28	Black	[24]	Fresh-market cultivar with a pronounced aroma.
Fennel's 3-way hybrid	Fennell's 2-way hybrid	*V. popenoei*	Black	[25]	Fennel's 2-way hybrid is 'Scuppernong' × *V. rotundifolia* var. *munsoniana*.
Fry	Ga. 19-13	Ga. 19-11	Bronze	[24]	Leading bronze fresh-market cultivar.
Ga. 13-3-36	Ga. 6-9-91	Ga. 6-1-217	Bronze	P.J.C. breeding records	Ga. 18-5 is a grandparent, Ga. 13-3-36 10.9% *V. vinifera* and 89.1% *V. rotundifolia*.
Ga. 1-6-14	Scarlet	Tara	Bronze	P.J.C. breeding records	Fresh-market muscadine selection with a "honey" flavor.
Ga. 18-5	Ga. 14-32	Ga. 12-2-1	Black	P.J.C. breeding records	Interspecific hybrid that is 43.75% *V. vinifera* and 56.25% *V. rotundifolia*.
Golden Isles	Fry	Ga. 19-6	Bronze	[24]	White wine muscadine cultivar released for more neutral flavored wine.
Hall	Fry	Tara	Bronze	[24]	Fresh-market muscadine cultivar.
Lane	Supreme	Tara	Black	[24]	Fresh-market muscadine cultivar.
Magnolia	Unnamed seedling	Topsail × Tarheel	Bronze	[24]	White wine muscadine cultivar.
Magoon	Thomas	Burgaw	Black	[24]	Black fresh-market muscadine cultivar.
Noble	Thomas	Tarheel	Black	[24]	Leading red wine muscadine cultivar.
Oh My!	JB99-1-4-15	JB03-20-1-21	Bronze	[24,26]	Stenospermocarpic seedless quasi-BC2 hybrid with an 86.9% *V. rotundifolia* background.
Paulk	Supreme	Tara	Black	[24]	Fresh-market muscadine cultivar.
Pineapple	Fry	Senoia	Bronze	[24]	Fresh-market muscadine cultivar with a "pineapple" flavor.
Ruby Crisp	Supreme	Tara	Red	[24]	Fresh-market muscadine cultivar with a red color and mild flavor.
Scuppernong	Unknown	Unknown	Bronze	[24]	Bronze colored native selection.
Southern Home	Summit	Fla. P9-15	Black	[25]	Interspecific hybrid of *V. rotundifolia*, *V. popenoei*, and *V. vinifera*.
Supreme	Black Fry	Dixieland	Black	[24]	Leading black fresh-market muscadine cultivar.
Tarheel	Luola	V68 R14 B2	Black	[24]	Older red wine muscadine cultivar.
V. popenoei DVIT 2970	Unknown	Unknown	Purple		Native selection of *V. popenoei*.

[z] Underlined genotypes were used in the study of difference in the aroma volatile composition between black and bronze muscadines.

Muscadine berries ripen asynchronously, and berries are harvested individually rather than in clusters. Therefore, most muscadine cultivars require several pickings to remove all the ripe fruit. For germplasm characterization, berries judged to be at optimal commercial ripeness (fully colored and with some softness), as well as free from defects, were harvested. Fruits were harvested at different times depending upon the harvest period of the genotype, but all fruits were harvested in the month of August. Four berry samples were collected separately from the two vines of each genotype on two different days (two samples each day and a sample from each vine). Depending on berry size, a sample of at least five berries were combined for each replicate (more berries were required per replicate for small-berry-sized genotypes). For the ripening stage study, fruit samples of the Carlos cultivar were collected on the same day from five separate vines with berries at various stages of ripening. Fruits were collected ranging from green, firm, and unripe to dark bronze, soft, and overripe. Fruits were harvested, brought into the lab, sorted as described below, and processed for GC-MS, as well as a fruit quality study, within 24 h. In addition to the analysis of aroma volatiles, several fruit-quality- and flavor-related traits were measured for the muscadine samples. A sample of 10 berries were separated from the same samples harvested for GC-MS study as a biological replicate; in addition, color, firmness, total soluble solids (TSS), and titratable acidity (TA) were measured.

2.2. Sample Preparation for Quality and Volatile Analysis

Fruits were washed with distilled water and dried with paper towels. Berries were then cut open into halves and the seeds were removed. For aroma volatile analysis, the halves of at least five berries were then collected for each sample in a 50 mL centrifuge tube and homogenized for 20 s using Power Gen 500 (Thermo Fisher Scientific, Waltham, MA, USA). Five grams of the homogenized samples were immediately pipetted into the 20 mL amber glass vials containing 5 g of a saturated salt (NaCl) solution and vortexed for homogenization. The addition of salt to the homogenized sample lowers the partitioning coefficient (K) for some volatiles and thus increases their concentration in the headspace. A total of 10 µL of 1000 ppm 1-Heptanol (Sigma-Aldrich Co., Saint Louis, MO, USA) was added as an internal standard (IS) in the vials for the relative quantification of volatiles in a homogenized sample. The vials were then stored at −20 °C until analysis.

For the ripening stage study, 'Carlos' berries were first density sorted by floating berries in sodium chloride brine solutions of 8%, 9%, 10%, and 11% [27] to determine the four grades of the berries (Stages 1–4) (Figure 1). Berries of increasing ripeness sunk in progressively denser brine. Sorted berries were then immediately rinsed with distilled water and processed as outlined above.

Figure 1. Four ripening stages in 'Carlos' muscadine berries (S1: unripe, S2: slightly ripe, S3: fully ripe, and S4: overripe).

Skin color was measured using a Chroma Meter CR-400 (Konica Minolta Sensing Americas Inc., Wayne, NJ, USA). Five different color components (L*: Lightness, a*: Red/Green value, b*: Blue/Yellow value, C*: Chroma, and h: Hue) were measured. Fruit firmness was measured using a FirmTech 2 Automatic Fruit Firmness tester (BioWorks, Inc., Wamego, KS, USA). Total soluble solids (TSS) and titratable acidity (TA) are commonly used indicators of fruit maturity, ripening, and flavor. For both TA and TSS measurement, the same protocol of homogenization was followed as used for the volatile analysis; in addition, the homogenized muscadine samples in 20 mL tubes were centrifuged at 4000 rpm for 30 min at 4 °C. The solid portion was separated by filtering with the cheesecloth and the supernatant flow, which was collected and stored immediately at −20 °C until analysis. TSS was measured using a digital refractometer (PAL-1; Minato-ku City, Tokyo, Japan). Six grams of a filtered sample was used for TA analysis. A 0.1 N NaOH solution was used as the titrant, and the percent TA was measured using a Mettler Toledo DL15 titrator (Greifensee, Switzerland).

2.3. Sample Incubation and GC-MS Conditions

The homogenized fruit samples were incubated for 30 min at 40 °C with continuous agitation at a speed of 250 rpm. After equilibration, the volatile compounds were collected using a 1 cm SPME-fiber-assembly Divinylbenzene/Carboxen/Polydimethylsiloxane (DVB/CAR/PDMS) (Supelco Inc., Bellefonte, PA, USA), and this was achieved by exposing the fiber to the headspace for 30 min under the same temperature. The fibers were activated before sampling according to the instructions of the manufacturer. After the incubation, the SPME fiber was inserted directly into the injection port for desorption (4 min at 250 °C) in a spitless mode. An ultra-inert liner of straight geometry and a 0.75 mm inner diameter (Agilent Technologies Inc., Santa Clara, CA, USA) was used. Aroma volatiles were analyzed using an Agilent 7890A gas chromatography system that was connected with a 5977B mass spectrometer (Agilent Technologies Inc., Santa Clara, CA, USA). The sample preparation was fully automated and carried out by a Gerstel MultiPurpose Sampler (MPS) (GERSTEL GmbH & Co. KG, Mülheim, Germany), which was coupled with GC-MS. Helium gas with a 99.9% purity was used as a carrier gas. A back-inlet purge flow rate was maintained at 3 mL min^{-1}, and a constant gas flow rate of 1.2 mL min^{-1} was utilized through the column. Volatiles were separated using an Agilent HP-5MS (30 m × 250 µm ×0.25 µm) (Agilent Technologies Inc., Santa Clara, CA, USA) column. The GC-MS methodology for this experiment was optimized based on the literature of previous research on muscadine volatile analysis, and it was modified according to the equipment and the needs of the specific cultivars used in this project. The oven temperature was programmed at an initial temperature of 35 °C for 7 min. The oven temperature was increased to 120 °C at 8 °C min^{-1}, held for 5 min, ramped to 150 °C at 4 °C min^{-1}, and then held for 2 min. The post run temperature was set at 280 °C for 5 min before returning to 35 °C. The thermal Aux 2 MSD transfer line, ion source, and quadrupole mass detector temperature values were set to 250 °C, 230 °C, and 150 °C, respectively. The solvent delay time was set to 2 min in order to avoid detection of the unnecessary carbon dioxide peak in the chromatogram. The fragmentation data from a mass spectrometer were collected in scan mode from m/z 25 to 300. The mass spectra in the electron impact ionization (ME-EI) mode were recorded at an ionization energy of 70 eV. MS data were analyzed in Agilent MSD software ChemStation F.01.03 (Santa Clara, CA, USA), and volatile compounds were identified by comparing the mass spectral data with the NIST 2.0 reference library (National Institute of Standards and Technology, Gaithersburg, MA, USA). The peak area of volatiles was normalized against the peak area of the internal standard. Furthermore, their relative concentration in a sample (ng/g of fresh fruit) were calculated with respect to the known concentration of the internal standard.

2.4. Statistical Analysis

Statistical analysis was performed in R statistical software, version 4.2.1 [28]. One-way ANOVA (analysis of variance) and Tukey's HSD (honestly significant difference) test was performed with four biological replicates for germplasm characterization study and five biological replicates for the maturity study. Principal component analysis (PCA) was carried out, and genotype and variable biplots were generated to visualize the difference between muscadine genotypes, as well as to identify the correlated volatile compounds or classes for such variation. Heatmaps were generated for both germplasm and ripeness studies using the 'pheatmap' package in R [29]. Euclidian distance between samples or volatile compounds were calculated and a complete method of clustering was applied to generate heatmaps. Pure *V. rotundifolia* genotypes were divided into black or bronze categories, and a two-sample t-test was performed to identify the volatile compounds that significantly differ between two color classes. To correct the type I error due to multiple testing, a Bonferroni correction was applied to the *p*-values obtained.

3. Results and Discussion

3.1. Role of Berry Maturity on Volatile Composition

Ripening 'Carlos' berries were density sorted into four grades from the least mature (stage 1) to most mature (stage 4). Stage 1 berries were very firm, partially green, and clearly unripe. Stage 4 berries were dark bronze in color, very soft, obviously overripe, and would likely be rejected for fresh fruit sales. Stages 2 and 3 were more similar to each other, with stage 3 berries being darker in color (Figure 1, Table S1). Both stages would likely be considered acceptable for fresh fruit sales, but experienced pickers would recognize stage 3 by its darker color and increased softness; as such, this stage would be targeted for harvest. The berry color darkened, the firmness and titratable acidity decreased, and the sugar content increased with increasing maturity classification (Table S1).

The 'Carlos' berries produced 18 different volatile compounds, 15 of which varied significantly among the four fruit maturity stages (Table S2). A heatmap of the volatile compounds in ripening fruit indicated two distinct clusters: compounds that increase in maturing fruit and compounds that decrease with maturity (Figure 2). A cluster of four volatiles, i.e., Eucalyptol, 2-hexenal, (E)-, Hexanal, and 2,4-hexadienal, (E, E)-, decreased with maturity, especially from stage two to stage three. These volatiles impart fresh mint-like, fresh green, freshly cut grass, unripe, and citrus odors (Table S2). These volatiles were among the most dominant aroma volatiles at ripening stages 1 and 2. Three out of four of the volatiles were from the aldehyde class, and one (Eucalyptol) was a monoterpene. The second cluster of volatiles, which increased with berry maturity, primarily belonged to the ester class. Among the 14 volatile compounds, 10 were from the ester, 1 from the sesquiterpene, 2 from the alcohol, and 1 from the aldehyde chemical class. These volatiles produced floral, fruity, warm, peppery, sweet, rose, and mango-like aromas. The increased abundance of these volatile compounds in ripening stages 3 and 4 marked the onset and progression, respectively, of the berry maturation in the muscadine.

A previous study conducted on ripe berries from five muscadine cultivars has shown similar categories of volatile compounds (ethyl hexanoate, ethyl octanoate, benzene acetaldehyde, 1-octanol, and phenylethyl alcohol) being abundant in 'Carlos' [16]. Similarly, among the 21 positively identified volatile compounds in 'Carlos' juice samples, ethyl acetate, butanoic acid, ethyl ester, hexanal, hexanoic acid, ethyl ester, benzeneacetaldehyde (phenyl acetaldehyde), and 2-phenylethyl alcohol were detected with concentrations ranging from 1.3 to 51 folds higher compared to the detectable aroma threshold (in ppb) [13]. Baek et al. [13] found furaneol to be amongst the most important aromatic compounds in ripe 'Carlos' juice samples by following a liquid–liquid continuous method of extraction (LLCE). However, it was not detected in our study likely due to the difference in extraction methods between the two studies. Future studies can implement extraction methods that allow for the better detection of this unstable and polar aromatic volatile compound [30,31]. A proteome analysis of ripening 'Carlos' berries indicated 55 proteins with a change of $\pm$1.5-fold, and these were recognized to be associated with flavor and aroma components [32]. The enzymes associated with terpenes, benzenoids, fatty acid degradation, and phenylpropanoid pathways were all detected during the ripening of the 'Carlos' berries.

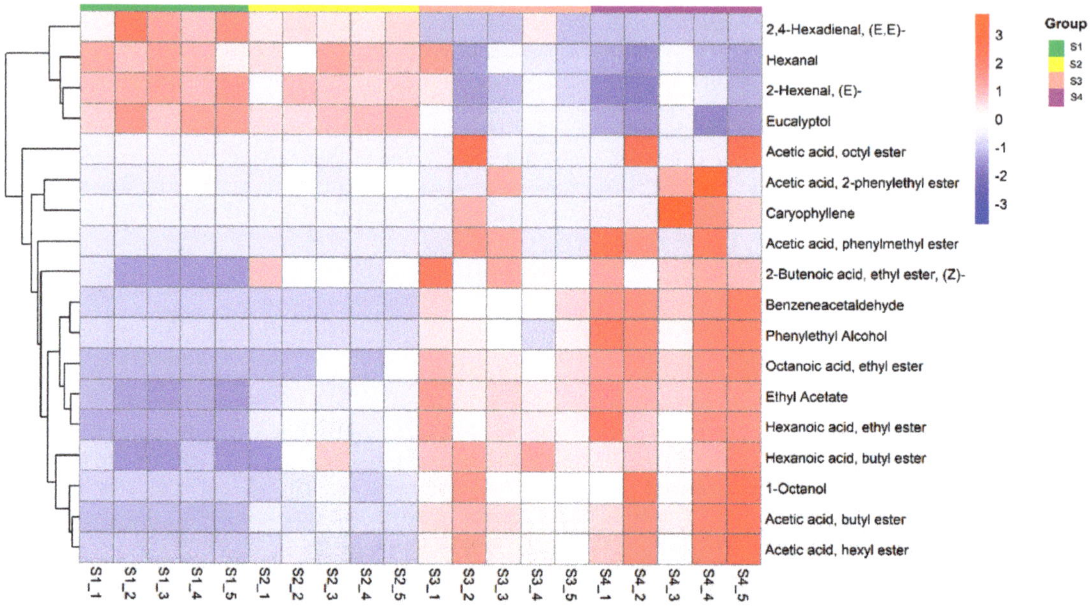

Figure 2. Heatmap showing the cluster of volatile compounds at various ripening stages of the 'Carlos' muscadine cultivar. (S1–S4: the four different stages of increasing berry ripeness).

The volatile profile we obtained in this study can also be compared with the profile obtained in bunch grapes (*V. vinifera*). Gu et al. [17] monitored four red wine grape varieties (*V. vinifera* cvs. Cabernet Sauvignon, Cabernet Gernischet, Cabernet Franc, and Merlot) near harvest time for their aroma volatile composition and found a very similar pattern of volatile composition during berry ripening. As the ripening progressed, the content of favorable bound aroma compounds such as free alcohols, esters, and terpenes increased, and the content of C-6 aldehydes such as 2-hexenal, (E)-, and hexanal decreased in most cultivars. Similarly, Yang [33] found most esters tended to accumulate during and after maturation, while C-6 volatiles increased until early maturation and then decreased.

3.2. *Diversity of Aroma Volatiles in Muscadine Germplasm*

In total, 63 aroma-related volatile compounds were detected in 24 different muscadine genotypes. These compounds can be compared to the 45 [34], 60 [35], and 94 [36] volatiles detected in other various groups of the *Vitis* germplasm. The abundance of the 43 volatile compounds were significantly different in at least one cultivar (Table S3). Based on the functional group, these volatile compounds represented seven different chemical classes: aldehyde, alcohol, ester, furan, monoterpene, aromatic hydrocarbon, and sesquiterpene. In the principal component (PC) analysis performed, each of the 43 volatile compounds were considered as a separate variable. Furthermore, the first (PC1) and second (PC2) components explained 23.9% and 12.4% of total variation in the aroma profile due to the effect of the genotypes (Figure 3). The cumulative contribution of the first ten principal components explained 84.4% of the total variation (Figure 4, Table S4), indicating that there is large variation among the genotypes for their content and composition of aroma volatiles, thus resulting in the scattered contribution of several PCs. The clustering of genotypes and volatile compounds clearly shows the variation between genotypes for aroma volatiles (Figure 5). The PCA loading plot indicated that the separation of genotypes in the first principal component was strongly correlated with the volatile compounds that belong to the aldehyde, furan, ester, and alcohol classes (Figure 6, Table S5). The ester and alcohol class volatile compounds had a positive correlation, while the aldehyde and furan class compounds were negatively correlated with the first principal component. The second PC was positively affected by the monoterpene content of the genotypes. *V. popenoei* DVIT 2970 had both the lowest PC1 and PC2 score, and it was slightly separated from the other genotypes. *V. popenoei* is one of two species of the *Muscadinia* subgenera. It is a native of southern Mexico [37] with a very thick skin, and this accession lacks the typical fruity aroma of *V. rotundifolia*. *V. popenoei*, which has not been used in muscadine breeding to develop new cultivars due to the exception of its appearance in the pedigree of the complex interspecific hybrid cultivar Southern Home [25]. The volatile composition showed that DVIT 2970 had relatively large amounts of C-6 aldehydes [hexanal (47.1%); 2,4-hexadienal, (E, E)- (3.9%); 2-hexenal, (E)- (43.6%)], furan (Furan, 2-pentyl- (0.56%)), as well as low ester and alcohol class volatiles. Another genotype that has a lower PC1 and PC2 score is Fennel's 3-way hybrid, which is the early genotype developed from a complex cross between *V. rotundifolia* var. *rotundifolia*, *V. rotundifolia* var. *munsoniana*, and *V. popenoei*. *V. popenoei* is the immediate parent of Fennel's 3-way hybrid, although DVIT 2970 is not the accession of *V. popenoei* that was used as the parent [38]. Like *V. popenoei*, Fennel's 3-way hybrid had a relatively large content of hexanal (19.4%) and 2-hexenal, (E)- (79%) volatiles. Interestingly, the pure muscadine cultivar Cowart appeared distinct and well separated from the rest of the genotypes in the first PC. 'Cowart' is a fresh market muscadine cultivar with a strong fruity aroma that was released in 1968. It has a relatively high amount of ester and alcohol, as well as low aldehydes class volatiles compared to other genotypes (Figures 3 and 5). 'Lane' was separated on the second PC, and this separation was correlated with its high level of monoterpene-class volatile compounds. The primary monoterpene compounds with a positive correlation with PC2 were 3-carene, D-limonene, beta-pinene, Citronellol, Geranic acid, and Citral. Cultivars Lane, Paulk, and Ruby Crisp share the same pedigree ('Supreme' × 'Tara'). Additionally, breeding selection Ga. 1-6-14 has one parent in common ('Scarlet' × 'Tara'), and breeding selection AM-195 was derived from a cross between selections AM-19 (Supreme × Tara) and AM-1 (open pollination of Tara). These genotypes were clustered close to each other in a PCA plot that indicated that their genetic similarity may also affect their volatile composition. Ga. 18-5 is a complex hybrid that is 43.75% *Euvitis* (Figure S1), and it has a prominent aroma that differs from most muscadines. Ga. 18-5 has significantly higher concentrations of esters (Table S3), such as ethyl acetate (fruity); butanoic acid and ethyl ester (fruity, juicy, and pineapple); 2-butenoic acid, ethyl ester, and (Z)- (fruity, mango-like); 2-hexenoic acid and ethyl ester (pleasant, rum-like, fruity, green, and sweet with a juicy undertone); and heptanoic acid and ethyl ester (fruity and grape-like). The genotypes 'Oh My!', 'Tarheel', 'Golden Isles', 'Noble', 'Carlos', and 'Magnolia' were concentrated mostly in the third quadrant of the PCA plot along with Fennel's 3-way hybrid and DVIT 2970. These genotypes have a proportionately higher concentration of aldehyde, as well as low alcohol, ester, and terpene-class aroma volatiles. 'Tarheel' and 'Noble' are closely related to each other as 'Noble' is a direct descendent of 'Tarheel'. 'Noble', 'Carlos', and 'Magnolia' muscadines are popular wine cultivars and comprise a large proportion of the muscadine wine industry; in addition, 'Golden Isles' was released as a wine cultivar but was never planted on a wide scale [39]. 'Oh My!' is a seedless cultivar developed from a complex hybridization of muscadine with *V. vinifera* [26]. All these cultivars were high in the aldehyde class and low in the ester- and alcohol-class volatile compound concentrations (Table S3).

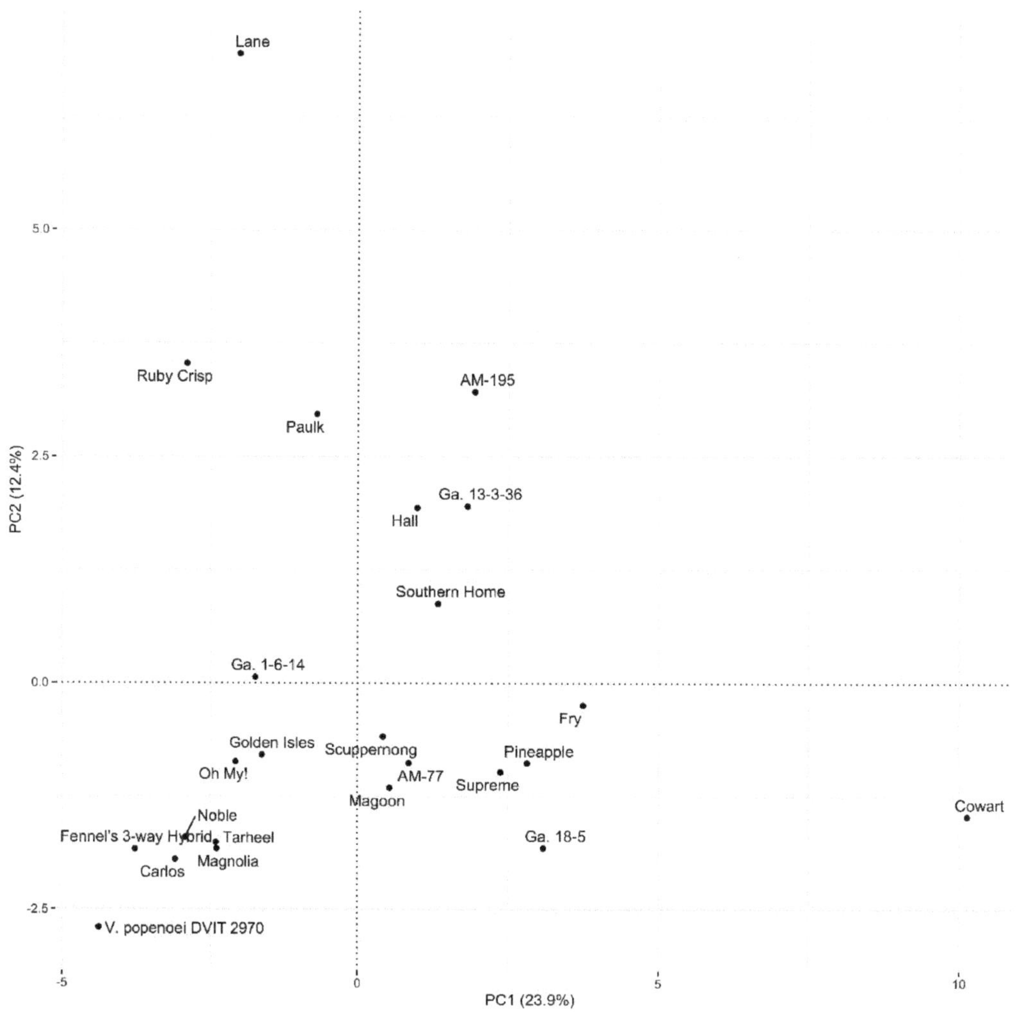

Figure 3. Principal component analysis (PCA) score plot of the aroma volatiles of the twenty-four muscadine genotypes.

To further reduce the dimensionality of the volatile composition data and to better understand how genotypes differ for various chemical classes of aroma volatiles, the concentration of individual volatiles belonging to the same chemical class were grouped together and used to perform the ANOVA and principal component analyses. The muscadine genotypes were significantly different for each chemical class of volatile compounds (Table S6). The PCA results showed a similar genotype distribution pattern to that obtained from the PCA using individual volatile compounds (Figure S2). The first and second PC explained 34.2% and 17.6% of the total variation, respectively (Figure S3). This result is very similar to the PCA results reported by Deng et al. [16], with their PC1 and PC2 explaining 36.7% and 18.9% of the total variation (although they had only five commercial muscadine cultivars included in their analysis). The PCA loading plot showed aldehyde, furan, ester, alcohol, aromatic hydrocarbon, and sesquiterpene as the important classes through which to separate the genotypes in PC1 and the monoterpene class for separation in PC2 (Figure S4).

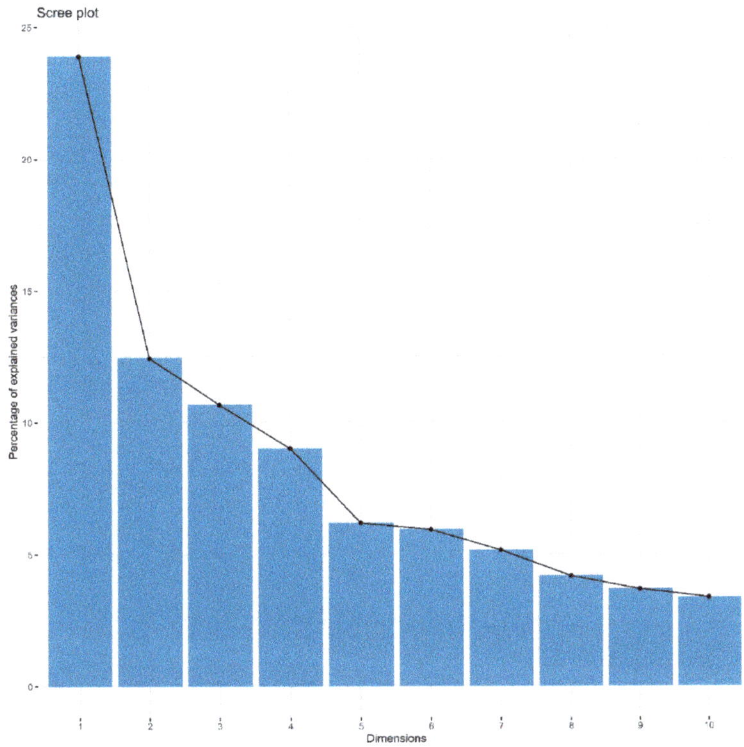

Figure 4. Scree plot showing the percentage of variation in the composition of volatile compounds between genotypes, which was explained by the first ten principal components obtained from PCA.

Aldehydes were the most abundant volatile class in this germplasm, representing about 45.8% of the total volatile abundance (Table S3). Aldehydes have been found to be the predominant volatiles produced in ripe grape berries in studies that investigated a wide variety of grape species [34–36]. Among the aldehydes, the C-6 volatiles hexanal and 2-hexenal, (E)- were the predominant volatiles in this muscadine germplasm, averaging about 45.2% of the volatiles across the samples (Table S3). These compounds produced green aromas and were found across all of the muscadine samples. In an evaluation of berry aroma volatiles in a range of Chinese wild grape species, Rahman et al. [34] identified 45 aroma volatiles, and the C-6 volatiles were the predominant volatiles in most of the grape species studied. Hexenal and 2-hexenal, (E)- are also abundant in *V. vinifera* table grape breeding lines [40].

Esters were the second most abundant volatile class and the most diverse class within the muscadine germplasm studied here, whereby 31 different esters were identified that represented, on average, 44.5% of the total volatile abundance (Table S3). Among the predominant esters, there were the following: ethyl acetate (sweet, grape, and cherry); 2-butenoic acid, ethyl ester, and (Z)- (pungent and sour caramellike); and acetic acid and butyl ester (apple, banana, and glue). These represented about 30.1% of the total volatiles. Comparably, Rahman et al. [34] and Liu et al. [36] found esters to be most abundant in *V. labrusca*, where they represented 15.8% to 24.3% of all volatiles; meanwhile, in other grape species, esters were detected in relatively small amounts or not detected at all. The actual abundance of esters measured in *V. labrusca* ranged from 121.4 ng/g [36] to 592 ng/g [35]. The abundance of esters in muscadine germplasm measured here ranged from 13.3 ng/g to 6162.4 ng/g (Table S3). The relatively high abundance and diversity of the esters in muscadine germplasm with their characteristic fruity and sweet aromas are likely why muscadines are well-known for their fruity aroma [13].

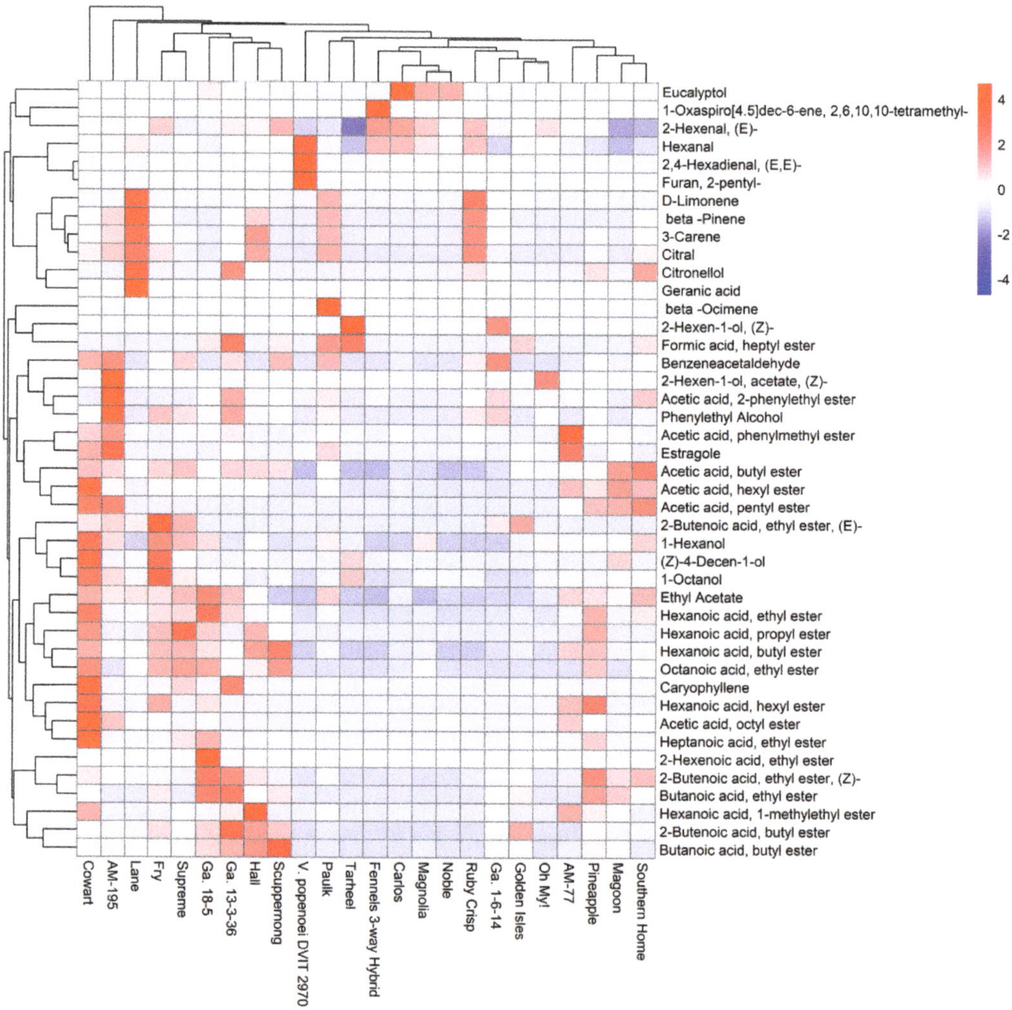

Figure 5. Heatmap generated by the hierarchical clustering of aroma volatile compounds in the 24 muscadine genotypes. Genotypes are presented in columns and aroma volatiles are in rows. The color in each cell represents the relative concentration of the volatile compounds.

Monoterpenes were produced in relatively low abundance in these muscadine accessions, and the predominant monoterpene was C-carene (nutmeg) (Table S3). Amongst the other grape species, monoterpenes are predominantly found in *V. vinifera* accessions. Monoterpenes in ripe *V. vinifera* berries represent 17.4% of all volatiles [36] in terms of the absolute amounts produced. Moreover, 171.5 ng/g is very similar to the average amounts produced in muscadine germplasm (172.4 ng/g), where monoterpenes are only 3.2% of the volatiles produced. This difference in relative abundance highlights the large amount of aldehyde and ester volatiles produced in muscadine in comparison to *V. vinifera*.

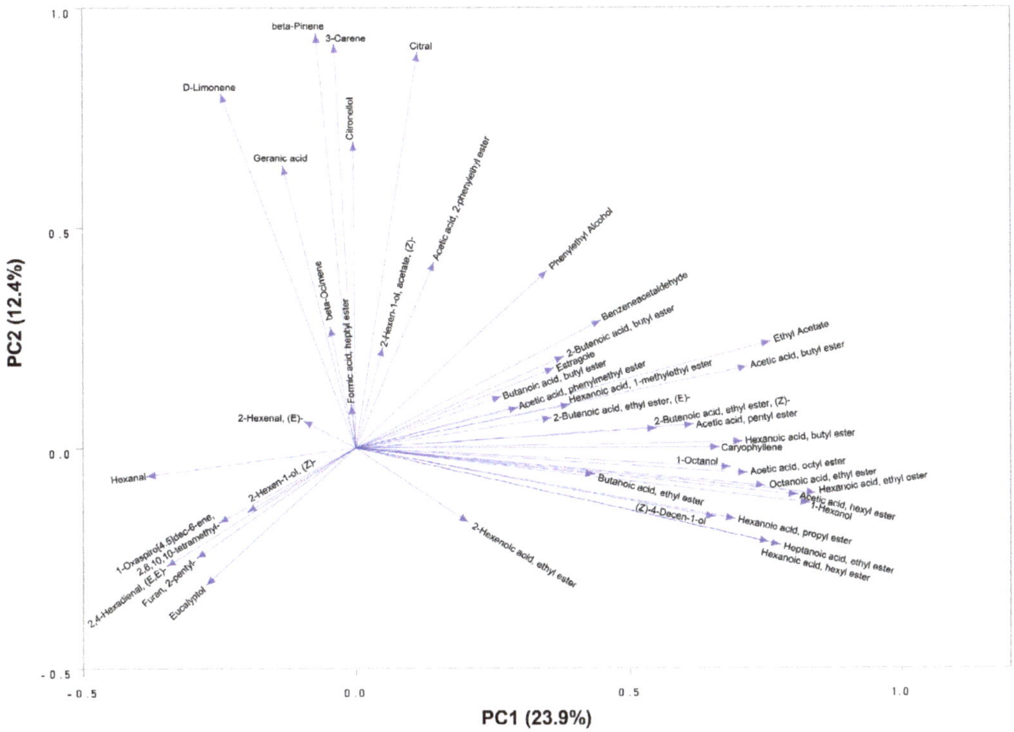

Figure 6. PCA loading biplot showing the contribution of each volatile compounds to the first (PC1) and second (PC2) principal components. Length of arrow represents the amount of contribution made by the volatile compound.

3.3. *Differences in the Aroma Volatile Composition between Black and Bronze Muscadines*

Berry color in muscadines is genetically controlled with bronze-colored berries resulting from a recessive glutathione S-transferase4 (VrGST4) mutation that leads to a lack of anthocyanin pigmentation in the native selection 'Scuppernong' [41,42]. Most muscadine cultivars can be grouped as black or bronze due to their berry color, though there are some cultivars that have red or pink berries. Fresh fruit vineyards typically plant both black and bronze varieties in approximately equal proportions as consumers vary in their preference for color type. We grouped 17 of the pure *V. rotundifolia* muscadine cultivars used in this study into black and bronze categories (Table 1), as well as performed two sample t-tests for all the volatile compounds detected. As expected, the color measurement showed that black and bronze muscadines are significantly different in their lightness scale (L*(C)) (Table S7). We identified six volatile compounds that differed between the black and bronze cultivars (Bonferroni adjusted p-value < 0.05) (Figure 7). Black muscadine cultivars had a significantly higher concentration of ethyl acetate (fruity aroma) and estragole (odor description: sweet, phenolic, anise, harsh, spice, green, herbal, and minty), while two aldehyde-class (2-hexenal, (E)-; 2,4-hexadienal, (E, E)-), and two ester-class (butanoic acid, ethyl ester; 2-butenoic acid, butyl ester) volatiles were significantly more abundant in bronze muscadines. Deng et al. [16] performed a partial least-squares discriminate analysis (PLS-DA) that considered black and bronze muscadine color classes as categorical response variables, as well as found a better separation among the five muscadine grape cultivars for their fruit volatile composition. They identified several volatile compounds as potential metabolic markers to distinguish between black and bronze muscadine samples. Ethyl acetate and 2-hexenal, (E)- were identified in both studies, suggesting the existence of inherent differences between black and bronze muscadines for aromatic volatile composition, as well as providing a basis for the further exploration of the genetic and biosynthetic pathways involved in the volatile composition of black and bronze muscadine cultivars.

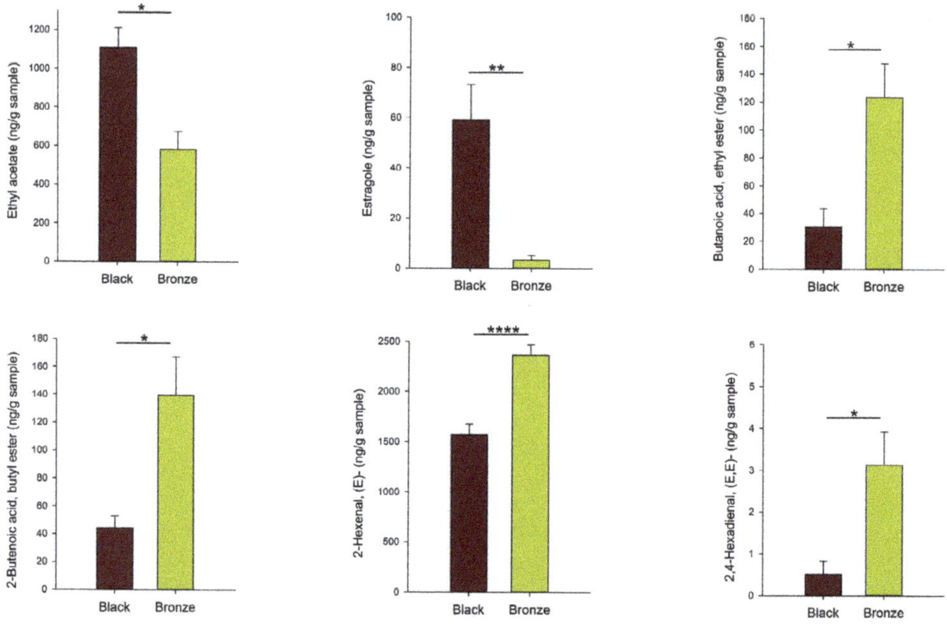

Figure 7. The significantly different (adjusted p-value < 0.05) aroma volatiles between the black (n = 9) and bronze (n = 8) muscadines obtained from the two sample t-tests. * $p \leq 0.05$, ** $p \leq 0.01$, **** $p \leq 0.0001$.

4. Conclusions

Significant variation was detected in the muscadine germplasm for the composition of aroma-related volatiles. Aldehyde, ester, alcohol, and primary monoterpene aroma volatiles were crucial in distinguishing the muscadine genotypes. Aldehydes and furan-class volatiles were more abundant in early muscadine selections, wild genotypes, and juice cultivars, while fresh-market cultivars were mainly characterized by their relatively large quantity of esters, alcohols, and primary monoterpenes. A ripening study revealed a distinctive pattern of aroma volatile composition in the muscadine berries of the cultivar Carlos during fruit ripening. Similar to other studies on fruits and vegetables, we successfully implemented the SPME method of aroma volatile extraction in this study. However, this method might not be ideal for capturing important volatiles, like furaneol, in muscadine grapes. In the future, researchers can explore and incorporate different extraction methods to ensure the successful extraction and detection of furaneol. The results from this study provide an important basis for the utilization of muscadine germplasm in muscadine breeding programs for the selection of parents and the development of cultivars with desired flavor profiles. Similarly, consumer preference studies can be carried out and combined with GC/MS analysis to identify the specific volatiles that contribute to the desired or unwanted flavor in muscadines. Additionally, the pattern of aroma-volatile composition in ripening berries offers valuable information for the development and implementation of metabolic markers to ensure the ripening quality of muscadines.

Supplementary Materials: The following supporting information can be downloaded at: https://www.mdpi.com/article/10.3390/horticulturae9091054/s1, Supplementary information are provided in two separate files: supplementary file.xlsx and supplementary figures.docx. Figure S1: Pedigree of the muscadine selection Ga. 18-5 showing its development via complex hybridization steps. Figure S2: PCA plot of the twenty-four muscadine genotypes obtained using the total amount of volatile compounds in each chemical class as variables. Figure S3: Scree plot showing percentage of variation in the composition of volatile compound class between the genotypes explained by the first seven principal components obtained from PCA. Figure S4: PCA loading biplot showing the contribution of each volatile compounds class to the first (PC1) and second (PC2) principal components. Length of arrow represents the amount of contribution made by the volatile compound class. Table S1: Color, firmness, titratable acidity (TA), and total soluble solid (TSS) contents in the 'Carlos' berries at various stages of maturity. Table S2: Volatile concentration (ng/g of sample) at various stages of berry ripening in the 'Carlos' muscadine cultivar. Table S3: Volatile composition (ng/g of sample) of the muscadine genotypes. Table S4: PC scores of the muscadine genotypes with each volatile compound as a separate variable. Table S5: PCA loadings showing the contribution of each volatile compound to the different principal components. Table S6: Abundance of volatile compounds (ng/g of sample) of the various chemical classes in muscadine genotypes. Table S7: Color, firmness, titratable acidity (TA), and total soluble solid (TSS) contents of muscadine genotypes.

Author Contributions: Conceptualization, P.J.C.; Methodology, P.J.C., A.D., G.B. and O.G.; Software, G.B., O.G. and R.N.C.; Validation, G.B. and A.D.; Formal Analysis, G.B.; Investigation, G.B., O.G., R.N.C. and C.E.E.M.; Resources, P.J.C. and A.D.; Data Curation, G.B.; Writing—Original Draft Preparation, G.B., P.J.C. and O.G.; Writing—Review and Editing, G.B., P.J.C., O.G., A.D. and M.L.W.; Visualization, G.B.; Supervision, P.J.C. and A.D.; Project Administration, P.J.C.; Funding Acquisition, P.J.C., M.L.W., A.D. and G.B. All authors have read and agreed to the published version of the manuscript.

Funding: This research was supported by the Southern Region Small Fruit Consortium (Project ID: 2022 R-10) and by the funding received from the John Ingle Innovation in Plant Breeding award by Gaurab Bhattarai from the University of Georgia, Research Foundation, Inc.

Data Availability Statement: The data presented in this study are available in supplementary file.xlsx and supplementary figures.docx.

Conflicts of Interest: The authors declare no conflict of interest.

References

1. Hedrick, U.P.; Booth, N.O.; Taylor, O.M.; Wellington, R.; Dorsey, M.J. *The Grapes of New York*; 15th Annual Report; State of New York, Department of Agriculture: New York, NY, USA, 1908; Volume 3, Part II.
2. Bailey, L.H. *The Standard Cyclopedia of Horticulture*; Macmillan: New York, NY, USA, 1908.
3. Moore, M.O. Classification and systematics of eastern North American *Vitis L.* (*Vitaceae*) north of Mexico. *SIDA Contrib. Bot.* **1991**, *14*, 339–367.
4. Radford, A.E.; Ahles, H.E.; Bell, C.R. *Manual of the Vascular Flora of the Carolinas*; University of North Carolina Press: Chapel Hill, NC, USA, 2010.
5. Wen, J. Vitaceae. In *The Families and Genera of Vascular Plants*; Kubitz, K., Ed.; Springer: Berlin/Heidelberg, Germany, 2007; pp. 467–479.
6. Hoffmann, M.; Conner, P.; Brannen, P.; Burrack, H.; Mitchem, W.; Cline, B.; Perkins-Veazie, P.; Poling, B. Muscadine Grape Production Guide for the Southeast. North Carolina State Ext. AG-94. 29 January 2021. Available online: https://smallfruits.org/files/2020/07/muscadine-grape-production-guide-southeast.pdf (accessed on 5 October 2022).
7. Hickey, C.C.; Smith, E.D.; Cao, S.; Conner, P. Muscadine (*Vitis rotundifolia* Michx., syn. *Muscandinia* rotundifolia (Michx.) small): The resilient, native grape of the southeastern US. *Agriculture* **2019**, *9*, 131. [CrossRef]
8. Wang, S.Y.; Lin, H.S. Antioxidant activity in fruits and leaves of blackberry, raspberry, and strawberry varies with cultivar and developmental stage. *J. Agric. Food Chem.* **2000**, *48*, 140–146. [CrossRef] [PubMed]
9. Pastrana-Bonilla, E.; Akoh, C.C.; Sellappan, S.; Krewer, G. Phenolic content and antioxidant capacity of muscadine grapes. *J. Agric. Food Chem.* **2003**, *51*, 5497–5503. [CrossRef] [PubMed]
10. Bartoshuk, L.M.; Klee, H.J. Better fruits and vegetables through sensory analysis. *Curr. Biol.* **2013**, *23*, R374–R378. [CrossRef]
11. Colantonio, V.; Ferrão, L.F.V.; Tieman, D.M.; Bliznyuk, N.; Sims, C.; Klee, H.J.; Munoz, P.; Resende, M.F., Jr. Metabolomic selection for enhanced fruit flavor. *Proc. Natl. Acad. Sci. USA* **2022**, *119*, e2115865119. [CrossRef]
12. Rodríguez, A.; Alquézar, B.; Peña, L. Fruit aromas in mature fleshy fruits as signals of readiness for predation and seed dispersal. *New Phytol.* **2013**, *197*, 36–48. [CrossRef]
13. Baek, H.H.; Cadwallader, K.R.; Marroquin, E.; Silva, J.L. Identification of predominant aroma compounds in muscadine grape juice. *J. Food Sci.* **1997**, *62*, 249–252. [CrossRef]

14. Lee, B.; Lin, P.C.; Cha, H.S.; Luo, J.; Chen, F. Characterization of volatile compounds in Cowart muscadine grape (*Vitis rotundifolia*) during ripening stages using GC-MS combined with principal component analysis. *Food Sci. Biotechnol.* **2016**, *25*, 1319–1326. [CrossRef]
15. Lamikanra, O. Aroma constituents of muscadine wines1. *J. Food Qual.* **1987**, *10*, 57–66. [CrossRef]
16. Deng, H.; He, R.; Long, M.; Li, Y.; Zheng, Y.; Lin, L.; Dong, L.; Zhang, X.; Liao, M.; Lv, X.; et al. Comparison of the Fruit Volatile Profiles of Five Muscadine Grape Cultivars (*Vitis rotundifolia* Michx.) using HS-SPME-GC/MS Combined with Multivariate Statistical Analysis. *Front. Plant Sci.* **2021**, *12*, 2302. [CrossRef] [PubMed]
17. Gu, X.; Zhang, X.; Wang, K.; Lv, X.; Li, R.; Ma, W. GC–MS Untargeted Analysis of Volatile Compounds in Four Red Grape Varieties (*Vitis vinifera* L. cv) at Different Maturity Stages near Harvest. *Foods* **2022**, *11*, 2804. [CrossRef] [PubMed]
18. Yang, Y.N.; Zheng, F.P.; Yu, A.N.; Sun, B.G. Changes of the free and bound volatile compounds in *Rubus corchorifolius* L. f. fruit during ripening. *Food Chem.* **2019**, *287*, 232–240. [CrossRef] [PubMed]
19. Zhu, X.; Li, Q.; Li, J.; Luo, J.; Chen, W.; Li, X. Comparative study of volatile compounds in the fruit of two banana cultivars at different ripening stages. *Molecules* **2018**, *23*, 2456. [CrossRef] [PubMed]
20. Corpas Iguarán, E.; Taborda Ocampo, G.; Tapasco Alzate, O. Identification of volatile compound markers during the ripening and senescence of lulo (*Solanum quitoense* Lam.). *J. Food Sci. Technol.* **2018**, *55*, 437–442. [CrossRef] [PubMed]
21. Zidi, K.; Kati, D.E.; Bachir-Bey, M.; Genva, M.; Fauconnier, M.L. Comparative study of fig volatile compounds using headspace solid-phase microextraction-gas chromatography/mass spectrometry: Effects of cultivars and ripening stages. *Front. Plant Sci.* **2021**, *12*, 1312. [CrossRef]
22. Senesi, E.; Di Cesare, L.F.; Prinzivalli, C.; Scalzo, R.L. Influence of ripening stage on volatiles composition, physicochemical indexes and sensory evaluation in two varieties of muskmelon (*Cucumis melo* L. var reticulatus Naud). *J. Sci. Food Agric.* **2005**, *85*, 1241–1251. [CrossRef]
23. Poling, E.B.; Mainland, C.M.; Bland, W.T.; Cline, B.; Sorensen, K.A. *Muscadine Grape Production Guide for North Carolina*; AG-94; NC State Extension: Raleigh, NC, USA, 2003; Available online: https://content.ces.ncsu.edu/muscadine-grape-production-guide (accessed on 2 March 2023).
24. Conner, P.J.; Worthington, M.L. Muscadine Grape Breeding. In *Plant Breeding Reviews*; Goldman, I., Ed.; John Wiley & Sons: Hoboken, NJ, USA, 2022. [CrossRef]
25. Mortensen, J.A.; Harris, J.W.; Hopkins, D.L.; Andersen, P.C. 'Southern Home': An interspecific hybrid grape with ornamental value. *HortScience* **1994**, *29*, 1371–1372. [CrossRef]
26. Bloodworth, P.J. Grapevine Plant Named 'JB06-43-6-22'. US Plant Patent 31,010, 11 February 2019.
27. Lanier, M.R.; Morris, J.R. Evaluation of Density Separation for Defining Fruit Maturities and Maturation Rates of Once-over Harvested Muscadine Grapes. *J. Am. Soc. Hortic. Sci.* **1979**, *104*, 249–252. [CrossRef]
28. R Core Team, R. A Language and Environment for Statistical Computing. Available online: https://www.r-project.org/ (accessed on 11 February 2022).
29. Kolde, R. Pheatmap: Pretty Heatmaps. R Package Version 1.0.12. 2019. Available online: https://CRAN.R-project.org/package=pheatmap (accessed on 11 February 2022).
30. Chen, Y.; Sidisky, L.M. Quantification of 4-hydroxy-2, 5-dimethyl-3-furanone in fruit samples using solid phase microextraction coupled with gas chromatography–mass spectrometry. *J. Chromatogr. A* **2011**, *1218*, 6817–6822. [CrossRef]
31. Dutra, M.D.C.P.; de Souza, J.F.; Viana, A.C.; de Oliveira, D.; Pereira, G.E.; dos Santos Lima, M. Rapid determination of the aromatic compounds methyl-anthranilate, 2′-aminoacetophenone and furaneol by GC-MS: Method validation and characterization of grape derivatives. *Food Res. Int.* **2018**, *107*, 613–618. [CrossRef] [PubMed]
32. Kambiranda, D.; Basha, S.M.; Singh, R.K.; He, H.; Calvin, K.; Mercer, R. In depth proteome analysis of ripening muscadine grape berry cv. Carlos reveals proteins associated with flavor and aroma compounds. *J. Proteome Res.* **2016**, *15*, 2910–2923. [CrossRef] [PubMed]
33. Yang, C.; Wang, Y.; Wu, B.; Fang, S.; Li, S. Volatile evolution of three table grapes with different flavour during and after maturation. *Food Chem.* **2011**, *128*, 823–830. [CrossRef]
34. Rahman, F.U.; Nawaz, M.A.; Liu, R.; Sun, L.; Jiang, J.; Fan, X.; Liu, C.; Zhang, Y. Evaluation of volatile aroma compounds from Chinese wild grape berries by headspace-SPME with GC-MS. *Food Sci. Technol.* **2022**, *42*, e54320. [CrossRef]
35. Yang, C.; Wang, Y.; Liang, Z.; Fan, P.; Wu, B.; Yang, L.; Wang, Y.; Li, S. Volatiles of grape berries evaluated at the germplasm level by headspace-SPME with GC-MS. *Food Chem.* **2009**, *114*, 1106–1114. [CrossRef]
36. Liu, X.; Fan, P.; Jiang, J.; Gao, Y.; Liu, C.; Li, S.; Liang, Z. Evolution of volatile compounds composition during grape berry development at the germplasm level. *Sci. Hort.* **2022**, *293*, 110669. [CrossRef]
37. Fennell, J.L. Two new North American species of *Vitis*. *J. Wash. Acad. Sci.* **1940**, *30*, 15–19. Available online: https://www.jstor.org/stable/24530082 (accessed on 11 February 2022).
38. Cao, S.; Stringer, S.; Gunawan, G.; McGregor, C.; Conner, P.J. Genetic diversity and pedigree analysis of muscadine grape using SSR markers. *J. Am. Soc. Hortic. Sci.* **2020**, *145*, 143–151. [CrossRef]
39. Basiouny, F.M.; Himelrick, D.G. *Muscadine Grapes*; ASHS Press: Alexandria, VA, USA, 2001.
40. Maoz, I.; Kaplunov, T.; Raban, E.; Dynkin, I.; Degani, O.; Lewinsohn, E.; Lichter, A. Insights into the chemosensory basis of flavor in table grapes. *J. Sci. Food Agric.* **2020**, *100*, 1405–1417. [CrossRef]

41. Ismail, A.; Gajjar, P.; Park, M.; Mahboob, A.; Tsolova, V.; Subramanian, J.; Darwish, A.G.; El-Sharkawy, I. A recessive mutation in muscadine grapes causes berry color-loss without influencing anthocyanin pathway. *Commun. Biol.* **2022**, *5*, 1012. [CrossRef]
42. Varanasi, A.; Worthington, M.; Nelson, L.; Brown, A.; Chizk, T.M.; Threlfall, R.; Howard, L.; Conner, P.; Figueroa-Balderas, R.; Massonnet, M.; et al. Glutathione S-transferase: A candidate gene for berry color in muscadine grapes (*Vitis rotundifolia*). *G3 Genes | Genomes | Genet.* **2022**, *12*, jkac060. [CrossRef] [PubMed]

horticulturae

Article

Characterization of New Grapevine Varieties Cross-Bred from Monastrell, Authorized for Winemaking in the Warm Region of Murcia (South-Eastern Spain)

Leonor Ruiz-García [1,*], José Ignacio Fernández-Fernández [2,*], Celia Martínez-Mora [1], Juan Daniel Moreno-Olivares [2], María José Giménez-Bañón [2], Diego José Fernández-López [1], Juan Antonio Bleda-Sánchez [2] and Rocío Gil-Muñoz [2,*]

1 Molecular Genetic Improvement Team, Instituto Murciano de Investigación y Desarollo Agrario y Medioambiental (IMIDA), C/Mayor s/n, La Alberca, 30150 Murcia, Spain; celia.martinez@carm.es (C.M.-M.); diegojose.fl@gmail.com (D.J.F.-L.)

2 Oenology and Viticulture Team, Instituto Murciano de Investigación y Desarollo Agrario y Medioambiental (IMIDA), C/Mayor s/n, La Alberca, 30150 Murcia, Spain; srjuanda@hotmail.es (J.D.M.-O.); gimenezba@hotmail.com (M.J.G.-B.); juanantonio.bleda@carm.es (J.A.B.-S.)

* Correspondence: leonor.ruiz@carm.es (L.R.-G.); josei.fernandez@carm.es (J.I.F.-F.); mariar.gil2@carm.es (R.G.-M.)

Abstract: Crossbreeding programs allow the selection of new genotypes with better agronomic and oenological properties for the production of quality wine, and allow the development of a more sustainable form of viticulture. This paper describes the white genotype 'Calblanque', and the red genotypes 'Calnegre', 'Gebas' and 'Myrtia', the first wine grape varieties registered by the Instituto Murciano de Investigación y Desarollo Agrario y Medioambiental (IMIDA) as commercial varieties after confirming the winemaking quality of their grapes in a semi-arid climate with high temperatures. These new varieties have recently been authorized for winemaking in the Region of Murcia. 'Calblanque', 'Calnegre' and 'Gebas' were obtained from crosses between 'Monastrell' and 'Cabernet Sauvignon', and 'Myrtia' from crosses between 'Monastrell' and 'Syrah'. The red genotypes were selected for their phenolic quality—which was very superior to that of the parentals—and for their different harvest dates that allow a staggered harvest and their cultivation in different areas. 'Calblanque' was selected for its good balance of acidity and aromatic profile. The attributes of these new varieties could allow their better adaptation to the effects of climate change on grape and wine quality in warm areas.

Keywords: breeding; quality; distinctness; uniformity; stability; sustainability; wine grape

Citation: Ruiz-García, L.; Fernández-Fernández, J.I.; Martínez-Mora, C.; Moreno-Olivares, J.D.; Giménez-Bañón, M.J.; Fernández-López, D.J.; Bleda-Sánchez, J.A.; Gil-Muñoz, R. Characterization of New Grapevine Varieties Cross-Bred from Monastrell, Authorized for Winemaking in the Warm Region of Murcia (South-Eastern Spain). *Horticulturae* **2023**, *9*, 760. https://doi.org/10.3390/horticulturae9070760

Academic Editors: Antonio Amores-Arrocha and Ana Jiménez-Cantizano

Received: 31 May 2023
Revised: 20 June 2023
Accepted: 29 June 2023
Published: 1 July 2023

1. Introduction

Wine quality is associated with physicochemical parameters of the grape such as the accumulation of minerals, sugar, amino acids and organic acids, and the synthesis of flavor and aroma compounds [1,2]. In wine, the acidity is essential for its conservation and good evolution over time, as well as for its organoleptic properties, so a reduction in total acidity can lead to unbalanced and flat wines [3–5]. Particularly in red wines, there is a relationship between wine quality attributes such as aroma, color and body, and the high phenolic content of the berry [6,7]. Nevertheless, high temperatures have been correlated with a reduction in acidity, and with a greater and faster synthesis of sugars and anthocyanins, although at temperatures above 35 °C, anthocyanins stop accumulating and may even be degraded depending on the variety [8–10].

One of the long-term strategies for adapting wine production to hot climates is the selection of suitable plant material (variety/clone and rootstock) from the existing vine biodiversity [11–17]. Another alternative is the selection of crossbreeding better adapted to the specific conditions of the viticulture zone [14,18,19] while still showing good agronomic properties, grape quality and enological characteristics [20,21]. The need to develop a sustainable viticulture model has led to different grapevine improvement programs designed in order to achieve this goal [14,22].

The Instituto Murciano de Investigación y Desarollo Agrario y Medioambiental (IMIDA) in Murcia (Spain) has been running a program to develop grapevine varieties with better phenolic quality for semi-arid wine-producing areas since 1997, which was initiated by Adrián Martínez-Cutillas who was responsible for the Viticulture and Enology Department at IMIDA. The program is based on new genotypes obtained from crosses between 'Monastrell' and other varieties such as 'Cabernet Sauvignon', 'Syrah', 'Tempranillo', 'Verdejo' or 'Barbera' [23,24]. 'Monastrell' is cultivated in Spain (particularly in the southeast), in France (where it is known as 'Mourvedre'), California (where it is known as 'Mataró'), Chile and in Australia (https://www.oiv.int (accessed on 25 April 2022)). In Spain it is the main variety grown in semi-arid Mediterranean climate zones such as the Jumilla, Bullas and Yecla Denominations of Origin (occupying 81% of the cultivated area). The first outcomes of these classical type of crossbreeding were the 'Calblanque', 'Calnegre', 'Gebas' and 'Myrtia' varieties. These four varieties were added to the list of commercial varieties, both at national and community level, on 25 March 2022 (https://www.boe.es (accessed on 25 March 2022)), via the Spanish Plant Variety Office (OEVV), entrusted by the CPVO to carry out DUS (distinctness, uniformity and stability) tests of vine varieties (*Vitis vinifera* L.) and vine rootstocks. The ampelographic characteristics of the new varieties are available on the website of the Ministry of Agriculture, Fisheries and Food (https://www.mapa.gob.es/ (accessed on 28 March 2022)), in the National and Community Catalogs of the Spanish Office of Vegetable Varieties. These characteristics were monitored for four years (2018–2021) using the CPVO technical protocol (CPVO-TP/050/2) based on the UPOV guidelines and descriptors (TG/50/9) (UPOV 2008; https://www.upov.int/edocs/tgdocs/en/tg050.pdf (accessed on 1 March 2017)). Recently (on 1 March 2023), these varieties were added to the list of varieties that could be grown in the wine-growing area of Murcia for winemaking (https://www.boe.es (accessed on 1 March 2023)). Therefore, the time that has taken from the starting of the breeding program (1997) to the acceptance of the use of these new varieties for the production of wine in the Region of Murcia (2022) has been 26 years. In addition, certified material of these varieties is available for exploitation and multiplication by nurseries, via the Vine Health Certification service of the IMIDA entrusted by the Spanish Ministry of Agriculture, Fisheries and Food to carry out the corresponding tests to evaluate the health status of the vine.

For all of the above, the main objective of this breeding program was to develop varieties with better winemaking quality of their grapes that may be suitable for cultivation in warm Mediterranean climate conditions. We hypothesized that some offspring from the crossings of these parental lines could inherit better agronomic and oenological properties for the production of quality wine in our climate conditions. The attributes of these new varieties could allow their better adaptation to the effects of climate change in semi-arid areas and the development of sustainable viticulture.

2. Materials and Methods

2.1. Study Site and Origin of the Genotypes

Supplementary Tables S1 and S2 show the average distribution of some meteorological parameters recorded in the experimental farm 'Hacienda Nueva' (38°06′40.7″ N; 1°40′50.3″ W; altitude 433 m) where the crossbreeding was carried out.

The plant material include four new genotypes: three selected from crosses between 'Monastrell' (M) and 'Cabernet Sauvignon' (C)—'Calblanque' (MC180), 'Calnegre' (MC80) and 'Gebas' (MC98)—and one between 'Monastrell' (M) and 'Syrah' (S)—'Myrtia' (MS10). All genotypes were unequivocally identified by the analysis of eight simple sequence repeat (SSR) markers via PCR [25] as shown in Table 1, confirming the parental varieties: 'Monastrell' and 'Cabernet Sauvignon' for 'Calblanque', 'Calnegre' and 'Gebas', and 'Monastrell' and 'Syrah' for 'Myrtia'.

Table 1. Genetic profile of the parental varieties and the four new varieties analysed with eight microsatellite loci.

Variety	VMC1A12		VMC8G6		VVMD27		VVMD5		VMC1E11		VMC5E9		VVMD28		VVIV67	
Monastrell	119	137	139	173	177	187	223	238	188	194	214	228	243	256	357	364
Cabernet S.	121	150	161	165	173	187	229	238	192	196	195	218	233	235	364	372
Syrah	137	150	169	173	187	189	223	229	196	206	218	222	217	227	361	381
Calblanque	121	137	165	173	187	187	223	229	188	192	195	228	233	243	357	372
Calnegre	121	137	165	173	187	187	223	238	194	196	195	214	233	256	357	364
Gebas	119	121	139	165	173	177	223	238	188	196	214	218	235	243	357	372
Myrtia	137	150	139	169	187	189	223	229	188	206	214	218	217	256	357	381

Alleles expressed in base pairs (bp).

2.2. Experiment Set Up

In a first phase, evaluation of a total of 1591 offspring from the crossings began when the vines were three years old (one plant per genotype from the germination of a seed), at the same locations as the crossing. This phase of selection was based on the quality of the grape and on the adequate agronomic behavior of the plant. In a second phase, twenty-five plants per genotype preselected in the first phase were grafted onto 110-Ritcher rootstocks for a more comprehensive study in which the quality of the wine was also included. 'Calnegre' was grafted in 2003, before Calblanque' (2007), 'Gebas' (2012) and 'Myrtia' (2012). The cultivation techniques—training system, fertilizer use, phytosanitary treatments and soil maintenance—were the same throughout the experimental plot.

The selection criteria of the genotypes for their registration as commercial varieties were based on different dates of ripening, in order to allow a staggered harvest, and in the quality of the grape and wine, using 'Monastrell' as the reference cultivar in the area. Concerning the grape quality, in the case of red grapes were selected genotypes with pH values ≤ 3.8, content in anthocyanins > 2000 mg kg^{-1} berry and total phenols > 2700 mg kg^{-1} berry. In the case of white grapes genotypes were selected with pH values ≤ 3.5 and content of malic acid > 2.0 g/L. About the wine quality, the parameters used were the total polyphenol index (TPI) and color intensity (CI), looking for crosses with more than 80 TPI and more than 40 CI.

In 2017, ten scions per selected genotype were grafted onto 110-Ritcher rootstocks. In 2018, the grafted genotypes were sent to the OEVV qualified technical testing center for conducting DUS examinations for four years (2018–2021). Previously, it was verified by serological methods (DAS-ELISA test) that these plants were free of viruses [26], such as three grapevine leafroll-associated viruses (GLRaV-1, 2, 3), grapevine fleck virus (GFkV), grapevine fanleaf virus (GFLV) and arabis mosaic virus (ArMV). The serological test was carried out by the Vine Health Certification service of the IMIDA. In March 2022 the varieties were added to the list of commercial varieties, and certified as virus-free material thanks to the collaboration of the Spanish Ministry of Agriculture, Fisheries and Food with the IMIDA Vine Health Certification service. Finally, in March 2023 they were added to the list of varieties that could be grown in the wine-growing area Murcia for winemaking (BOE of 1 March 2023).

2.3. Sampling and Measurements in Grapes

The plant material was characterized in triplicate (four plants per replica) over 5 years (2017–2021) by the phenological, agronomic and quality level of grapes and wines. The dates for the different phenological stages—budbreak, flowering, veraison and harvest—for each genotype were recorded [27]. The date of budbreak was considered when vines reached BBCH stage 09 (green shoot tips clearly visible); the date of flowering when vines reached BBCH stage 65 (50% of flowerhoods fallen); the veraison date when vines reached BBCH stage 85 (softening of berries); and the date of harvest when vines reached BBCH stage 89 (physiological maturity). Physiological maturity was deemed to begin when the grape reached its maximum size and its highest concentration of sugars. At this point, the berry begins to decrease in size due to water loss and some dehydrated berries appear in the cluster, the organoleptic maturity of the skin is good, and the seeds are mature (brown color).

For each genotype, total yield (kg/vine) and the weight of 100 randomly selected berries were assessed at harvest time.

The grape quality was assessed at the IMIDA experimental winery. For each replicate, 350 berries were randomly selected from the different areas of the bunches. From this representative sample, 30 berries were taken for the extraction and analysis in triplicate of the total phenolic content (TPC) (mg/kg berry), and of the total anthocyanins (TA) (mg/kg berry) [28]. The rest of the berry sample (320 berries) was crushed, without breaking the seed, and centrifuged. The °Brix value (OIV-MA-AS2-02), total acidity (OIV-MA-AS313-01), must pH (OIV-MA-AS313-15), tartaric acid content [29] and malic acid content (OIV-MA-AS313-11) were analyzed in the must obtained by centrifugation [24].

2.4. Winemaking

Grapes were transported to the winery located in Jumilla (Murcia, Spain), where wines were elaborated in accordance with a traditional vinification protocol in 100 L steel tanks. For red wines, grapes were destemmed, crushed and sulphited (50 mg SO_2/kg). Commercial yeast (Zymaflore FX10 *Saccharomyces cerevisiae*) (Laffort, Bordeaux, France) was used in a dosage of 20 g/100 kg. During alcoholic fermentation (conducted with a temperature adjustment of 25 °C) a daily punching of the tank was made. At the end of alcoholic fermentation, two rackings were carried out, and then pomace was pressed at 1.5 bars in a 75 L tank membrane press. For white wines, the sulphite was added in destemming, crushing, pressing and settling tank. Defanging was static using cold and pectolytic enzymes and then acidity correction was made. There was no strict control of the fermentation temperature, as varietal aromas were sought. Once the alcoholic fermentation was finished, it was racked, sulphited and kept cold. Samples were analyzed in triplicate at the end of alcoholic fermentation.

2.5. Measurements in Wines

The wine characteristics were assessed using different physicochemical parameters following the methodology described by the OIV: alcohol content (OIV-MA-AS312-01), total acidity (OIV-MA-AS313-01), pH (OIV-MA-AS313-15), relative density 20/20 (OIV-MA-AS2-01), total dry extract (OIV-MA-AS2-03B), color intensity and taint (OIV-MA-AS2-07B).

Regarding spectrophotometric parameters, color was measured using the CIELab space, using illuminant D65 and 10° standard observer conditions. The parameters measured were: L* (lightness), a* (from green to red), b* (from blue to yellow) and C* (chroma or saturation) (OIV-MA-AS2-11). These parameters were measured with the spectrophotometer Shimadzu UV-180 (Shimadzu Corporation, Kyoto, Japan). Total polyphenol index (TPI) was analyzed measuring the absorbance at 280 nm [30]. Total anthocyanins (TA) by the method proposed by Ho et al. (2001) [31]. All these parameters were analyzed using the autoanalyzer Miura One (TDI, Barcelona, Spain).

Organoleptic evaluation was carried out using the OIV score sheet for still wines defined in annex 3.1 of Resolution OIV/Concours 332A/2009 (https://www.oiv.int/public/medias/1852/oiv-concours-332a-2009-es-signe.pdf (accessed on 1 March 2017)), and the tasting panel was formed by staff of the Oenological Station previously trained.

2.6. Statistical Analysis

The collected data were subjected to analysis of variance (ANOVA), using StatGraphics Centurion XVI v.16.1.18 software (StatGraphics Technologies, Inc., The Plains, VA, USA). Means were compared according to the LSD test ($p < 0.05$).

3. Results and Discussion

3.1. Ampelographic Characteristics

Some of the ampelographic characteristic are presented in Figure 1 and Supplementary Table S3, which indicates its UPOV descriptors and the note assigned in parentheses.

Figure 1. The young shoot, upper and lower side of the mature leaf, and bunch of 'Monastrell', 'Cabernet Sauvignon', 'Syrah', 'Calblanque', 'Calnegre', 'Gebas' and 'Myrtia'. Images obtained from the website of the Spanish Ministry of Agriculture, Fisheries and Food (https://www.mapa.gob.es/ (accessed on 28 March 2022)).

The tip of the young shoot is fully open in all new varieties, and the flowers have fully developed stamens and fully developed gynoecium, except 'Myrtia', which has reflexed stamens and fully developed gynoecium. The mature leaf is circular with seven lobes in 'Calnegre', pentagonal with five lobes in 'Calblanque' and 'Myrtia,' and pentagonal with three lobes in 'Gebas'. The proportion of the main veins on the upper side of the blade with anthocyanin coloration is absent or very low in 'Calblanque', 'Gebas' and 'Myrtia', and medium in 'Calnegre'. The postrate hairs between the main veins on the lower side of the blade have a medium density in 'Calblanque' and 'Myrtia', and a sparse density in 'Calnegre' and 'Gebas'. The bunch is a medium size and has lax to medium density in 'Calblanque' and 'Gebas', a medium size and lax density in 'Myrtia', and a small size and medium density in 'Calnegre'. The berries have a small size, globose shape, blue black color of skin and a particular, different flavor of muscat, foxy or herbaceous in 'Calnegre', 'Gebas' and 'Myrtia', and a small to medium size, globose shape, yellow green color of skin and no particular flavor in 'Calblanque'. The main color of the woody shoot is yellowy brown in 'Calblanque' and 'Calnegre', orange and brown in 'Gebas', and dark brown in 'Myrtia' (Supplementary Table S3).

The main differences according to the UPOV descriptors between the parentals 'Monastrell' and 'Cabernet Sauvignon' and the new variety 'Calblanque' are in the number of lobes and teeth shape of the mature leaf, bunch size and density, and the color of the skin. In the case of the new varieties 'Calnegre' and 'Gebas', the main differences are in the postrate hairs between the main veins on the lower side of the blade, bunch density and the particular flavor of the berries. Finally, the main differences between the parentals 'Monastrell' and 'Syrah' and the new variety 'Myrtia' are in the sexual organs of the flower, the length and teeth shape of the mature leaf, bunch density, and the particular flavor of the berries (Supplementary Table S3).

The complete ampelographic information is available on the website of the Spanish Ministry of Agriculture, Fisheries and Food (https://www.mapa.gob.es/ (accessed on 28 March 2022)), in the National and Community Catalogs of the Spanish Office of Vegetable Varieties (https://www.mapa.gob.es/app/regVar/BusRegVar.aspx?id=es (accessed on 28 March 2022)).

3.2. Phenological, Agronomic and Qualitative Characteristics

In our experimental conditions, 'Syrah' was the earlier parent for all the phenology-related traits, while 'Monastrell' was the latest parent (Table 2), harvesting 'Syrah' 35 days before 'Monastrell'.

Table 2. Mean data (2017–2021) for the phenological stage dates of parental varieties and the new varieties.

Variety	Budbreak	Flowering	Veraison	Harvest	Harvest Days before Monastrell
Monastrell	22 April cd	02 June	10 August c	27 September c	0 a
Cabernet S.	16 April bc	27 May	05 August bc	09 September b	18 c
Syrah	08 April a	23 May	22 July a	23 August a	35 e
Calblanque	13 April ab	24 May	08 August c	25 August a	33 d
Calnegre	21 April cd	30 May	12 August c	11 September b	16 b
Gebas	24 April d	31 May	09 August c	09 September b	18 c
Myrtia	10 April ab	24 May	29 July ab	23 August a	35 e

Different letters in the same column indicate significant differences among genotypes at the 5% level, according to the LSD's multiple range test.

The length of the growing season (from budbreak to harvest) ranged from the 137 days of 'Syrah' to the 158 days of 'Monastrell'. Taking into account the mean harvest date, none of the new varieties were harvested later than 'Monastrell'. 'Calblanque' (white new variety) was harvested 33 days before 'Monastrell' and had a length of growing season of 134 days.

With respect to the new red varieties, 'Myrtia' was the earliest for all the phenology-related traits, harvesting 35 days before 'Monastrell' (similar to 'Syrah'), and presenting the shortest length of the growing season (135 days). 'Calnegre' and 'Gebas' were harvested 16 and 18 days before Monastrell, respectively, and had a length of growing season of 143 and 138 days, respectively.

The new varieties obtained have different optimal maturation dates, which will allow a staggered harvest in the winery. On the other hand, the variability found in the phenology of these new varieties will allow their adaptation to different growing areas, depending on climatic conditions.

Regarding productivity parameters of the new varieties (Table 3), the white variety 'Calblanque' was the most productive with values of yield similar to 'Monastrell' and 'Cabernet Sauvignon' (its parentals). 'Calnegre' (red variety) was the least productive with values lower than its parentals. The higher productivity of 'Calblanque' and the lower productivity of 'Calnegre' coincided with a higher and lower weight of its berries, respectively. Nevertheless, previous studies of the 'Calnegre' variety, comparing its behavior under controlled deficit irrigation and under rainfed conditions, showed that 'Calnegre' (MC80) is one of the varieties, among those studied, whose production is least reduced under rainfed conditions [21].

Since the climatic and growing conditions are the same for all varieties, the variation in berry weight could be due to differences in cell number and/or cell volume, which are determined by cell division and cell expansion, respectively [32]. This hypothesis could be verified with new experiments.

Quality must parameters were also analyzed during the five years of characterization of these new varieties (Table 3). Anthocyanins are a type of polyphenol from the flavonoid group that is the red pigment found in grape skins and sometimes in the flesh. Nevertheless, the amount and composition of anthocyanins present in them varies greatly depending on the species, variety, maturity, vintage, region of cultivation and many other factors [33]. Our parentals obtained anthocyanin values ranging between 1000 and 1800 mg/kg berries; however, in our new red varieties, we obtained anthocyanin values ranging between almost 3000 and a little more than 3500 mg/kg berries, thus tripling the value obtained in the 'Monastrell' variety or doubling the values of 'Cabernet Sauvignon' or 'Syrah'. The highest anthocyanin content was obtained by 'Myrtia' with an average value of 3533 mg/kg berries, the other two varieties 'Gebas' and 'Calnegre' showed very similar average values of around 3000 mg/kg berries. As can be observed, our new varieties greatly exceeded the values obtained by their parentals and this fact is called transgressive segregation; it means that we are going to find a large number of crossbreeds in which the anthocyanin concentration is not within the range of concentration of their parental phenotypes, which is frequent in intraspecific crosses and in domesticated populations [34].

The same situation could be observed when the TPC (total phenolic compounds) were analyzed. Among parentals, the highest concentrations were observed in 'Syrah' grapes (2114 mg/kg berries) and the lower quantities in 'Monastrell' grapes (1554 mg/kg berries). 'Cabernet Sauvignon' showed intermediate values between both varieties (1905 mg/kg berries). However, the new red varieties again showed values much higher than those obtained by their parents, highlighting among them the 'Calnegre' variety, which was the one that obtained the highest amount of total polyphenols, followed by 'Myrtia' and finally 'Gebas'.

As can be observed, no large statistical differences were found between parentals and new red varieties with respect to °Brix (Table 3). 'Calblanque' showed the lowest °Brix value probably because it is a white variety and this type of variety is usually harvested with less sugar quantity. One of the strategies to alleviate the effect of high temperatures on the increase in sugar content and, therefore, on the increase in alcoholic strength, is the use of late-ripening varieties that avoid plants suffering high temperatures during the ripening period. Nevertheless, our results show that in our climatic conditions, varieties that are harvested even 35 days before 'Monastrell', such as 'Myrtia', reach their optimum maturity with the same sugar content as 'Monastrell' (Table 3).

Table 3. Mean data (2017–2021) of production and grape quality variables of parental varieties at harvest.

Variety	kg per Vine	kg per ha	Weight of 100 Berries	Anthocyanins (mg kg^{-1} Berry)	TPC (mg kg^{-1} Berry)	° Brix	pH	TA (g L^{-1} Tartaric Acid)	Tartaric Acid (g L^{-1})	Malic Acid (g L^{-1})
Monastrell	3.25 ± 0.67 ab	8648 ± 1400 abc	152.2 ± 11.9 c	1061 ± 86 a	1554 ± 139 a	24.1 ± 0.6 bc	3.95 ± 0.30 b	2.88 ± 0.35 a	4.20 ± 0.19 a	1.33 ± 0.11 a
Cabernet S.	3.62 ± 0.67 b	9658 ± 1797 c	107.3 ± 7.7 a	1287 ± 127 a	1905 ± 81 ab	24.3 ± 0.4 bc	3.94 ± 0.20 b	3.25 ± 0.22 a	4.91 ± 0.17 b	1.9 ± 0.13 b
Syrah	3.53 ± 0.50 b	9427 ± 1338 c	125.5 ± 7.8 ab	1791 ± 148 b	2114 ± 172 b	24.7 ± 0.4 c	3.94 ± 0.19 b	3.29 ± 0.48 a	4.51 ± 0.12 ab	2.35 ± 0.08 c
Calblanque	3.37 ± 0.60 ab	9329 ± 1516 bc	135.8 ± 10.4 bc			20.2 ± 0.7 a	3.54 ± 0.04 a	4.81 ± 0.26 b	4.88 ± 0.26 b	2.91 ± 0.10 d
Calnegre	2.05 ± 0.16 a	5401 ± 1506 a	106.3 ± 4.4 a	2925 ± 93 c	3697 ± 69 d	22.9 ± 0.5 b	3.67 ± 0.06 a	3.51 ± 0.18 a	4.81 ± 0.18 b	1.13 ± 0.15 a
Gebas	2.48 ± 0.29 ab	7262 ± 971 abc	121.7 ± 7.2 ab	2934 ± 160 c	3151 ± 213 c	23.7 ± 0.8 bc	3.97 ± 0.09 b	3.10 ± 0.22 a	4.06 ± 0.22 a	2.20 ± 0.06 bc
Myrtia	2.18 ± 0.28 a	5819 ± 739 ab	107.2 ± 5.2 a	3533 ± 241 d	3521 ± 134 cd	24.1 ± 0.4 bc	3.64 ± 0.07 a	4.32 ± 0.35 b	4.90 ± 0.22 b	2.27 ± 0.13 c

Data expressed as mean value ± standard deviation. TPC, total phenolic content; TA, total acidity. Different letters in the same column indicate significant differences among genotypes at the 5% level, according to the LSD's multiple range test.

With respect to other parameters, pH, total acidity and organic acid were also measured at harvest. In recent years, in warm areas such as ours, a pH increase has been observed with respect to the values normally detected some decades ago. With respect to the results found in our parentals and new varieties, we observed that 'Monastrell', 'Cabernet Sauvignon', 'Syrah' and 'Gebas' showed the highest pH values close to 4, in contrast to 'Calblanque', 'Calnegre' and 'Myrtia' that showed the lowest pH values close to 3.6 (Table 3). Regarding total acidity, the highest mean value was found in 'Calblanque', the new white variety, with a much higher value than the parent varieties. The red varieties also showed acidity total values higher than 'Monastrell', the reference variety of the area, standing out among them 'Myrtia' with a value of 4.32 despite being harvested in August when in our area we reached temperatures close to 40 °C. These results are in agreement with those obtained by other authors who previously reported that organic acid concentration and the relative proportions of malate and tartrate varied according to the genotype at the ripe stage [21,35–37].

During ripening, tartaric acid concentration decreases by dilution due to fruit enlargement, while malic acid concentration decreases through both dilution and respiration [38–40]. Our results showed how 'Monastrell' together with 'Gebas' obtained the lowest values of tartaric acid, followed by 'Syrah'. The rest of the studied varieties showed similar content for this organic acid. Finally, the highest values of malic acid were obtained in the new white variety (2.91) followed by 'Syrah' and 'Myrtia', then 'Gebas', 'Cabernet Sauvignon', 'Monastrell' and finally by 'Calnegre'. It is remarkable that the malic acid content was one of the criteria used for the selection of white varieties in our genetic breeding program, searching genotypes with values greater than 2.0 g/L.

3.3. Wine Characteristics

Wine quality is determined by several factors such as the type (or blend) of grape varieties, the terroir, the viticultural practices, the winemaking techniques, and the aging conditions [41–43]. The variety of grapes is a key factor in determining the wine flavor, especially during the production of premium wines. Different physicochemical parameters were analyzed at the end of alcoholic fermentation (Table 4): alcohol content, total acidity, pH, density and total dry extract.

Table 4. Mean physical–chemical data (2017–2021) of the wines at the end of alcoholic fermentation.

Variety	Alcohol (V/V)	TA (g L^{-1} Tartaric)	pH	Relative Density 20/20	Total Dry Extract
Monastrell	13.89 ± 0.41 bc	7.37 ± 0.22 bc	3.41 ± 0.03 a	0.9922 ± 0.0004 a	27.30 ± 0.13 b
Cabernet S.	13.79 ± 0.46 bc	7.12 ± 0.53 abc	3.47 ± 0.03 ab	0.9928 ± 0.0006 a	27.52 ± 2.17 b
Syrah	14.35 ± 0.46 c	6.06 ± 0.46 a	3.59 ± 0.06 bc	0.9923 ± 0.0003 a	28.13 ± 0.71 b
Calblanque	12.09 ± 0.41 a	6.40 ± 0.40 ab	3.37 ± 0.04 a	0.9919 ± 0.0006 a	20.37 ± 0.86 a
Calnegre	12.94 ± 0.41 ab	7.39 ± 0.31 bc	3.42 ± 0.03 a	0.9957 ± 0.0002 b	32.61 ± 0.90 c
Gebas	13.56 ± 0.41 bc	7.25 ± 0.48 bc	3.64 ± 0.04 c	0.9951 ± 0.0004 b	32.21 ± 1.05 c
Myrtia	13.40 ± 0.41 bc	7.89 ± 0.32 c	3.41 ± 0.04 a	0.9948 ± 0.0004 b	31.80 ± 0.88 c

Data expressed as mean value ± standard deviation. TA, total acidity. Different letters in the same column indicate significant differences among genotypes at the 5% level, according to the LSD's multiple range test.

The alcohol content of wine is a consequence of the relative sugar content in grapes and varies depending on the variety of wine, as well as the winemaker [1,2]. As can be observed in Table 4, 'Calblanque' wine showed the lowest alcohol percentage as expected since it comes from a white variety whose wines usually have a lower alcohol content than those from red varieties. With respect to the rest of the wines, it can be observed how 'Calnegre' wine obtained the lowest alcohol percentage and 'Syrah' wine the highest. The rest of the wine varieties obtained intermediate values of alcohol content. It is remarkable that the wines of the new varieties obtained values of alcohol content lower than their parental wines. This could mean that our varieties could be an opportunity to obtain wines with a lower alcohol content in areas as warm as ours by allowing a coupling of phenolic and technological maturity and, at the same time, we could offer to the consumers, wines that are more adapted to their actual tastes.

All the wines studied showed values of total acidity ranging between 6 and 8 g/L of tartaric acid at the end of alcoholic fermentation (Table 4). The lowest value was found in 'Syrah' wines followed by 'Calblanque' wines and the highest value was found in 'Myrtia' wines, despite the fact that at the beginning of the winemaking process, all wines are adjusted to an acidity of 5.5 g/L with tartaric acid.

The pH of the wine is strictly connected with its microbiological and physicochemical stability [44] and it may contribute to the natural selection of microorganisms during winemaking [45,46]. Even the color of red wines may be strongly conditioned by the pH because this variable affects the equilibrium between the different forms of anthocyanins [47,48]. The pH level of a wine ranges from 3 to 4 [49]. The analyzed wines showed values ranging from 3.41 to 3.64, with 'Calblanque', 'Calnegre', 'Myrtia' and 'Monastrell' wines being those that reached the lowest values of pH; however, the highest value was found in 'Gebas' wine with a value of 3.64. In spite of the differences obtained in the different wines, values of pH around 3.6 are very adequate in warm areas such as ours.

Another parameter to take into account when we analyze the quality parameters of wines is relative density. The results showed how wines from parentals and the 'Calblanque' variety obtained the lowest relative density values; however, the rest of the wines from the new red varieties obtained the highest values, being statistically different with respect to the first.

Finally, the dry extract values correspond to all the non-volatile substances contained in it. The results in Table 4 showed data between 20.37 for 'Calblanque' wines and 32.61 for 'Calnegre' wines. As can be observed, the wines of the new red varieties obtained the highest values similar to relative density results.

3.4. Wine Spectrophotometric Characteristics

Spectrophotometric characteristics were evaluated in wines from the new varieties and their parentals at the end of alcoholic and malolactic fermentation. The results corresponding to color intensity, taint, anthocyanins, IPTs and different CIELab parameters are shown in Table 5.

Regarding color intensity (CI), differences among wines were very large. The CI values were higher at the end of alcoholic fermentation compared to those obtained at the end of malolactic fermentation. Due to the considerable chemical changes in the malolactic fermentation mainly driven by the increase in pH and the SO_2 addition, the color parameters were affected in the red wine. CI ranged between 60.28 ('Myrtia') and 14.68 ('Monastrell') at the end of alcoholic fermentation and between 40.02 ('Myrtia') and 11.69 ('Monastrell') at the end of malolactic fermentation. Wines from the parentals showed great differences with respect to the wines of the new varieties. With regards parental wines, 'Syrah' obtained the highest values followed by 'Cabernet Sauvignon' and 'Monastrell'. Regarding new varieties of wines, the wines with the highest CI were those from the 'Myrtia' variety, followed by 'Calnegre' and 'Gebas'. Different authors showed values of CI ranging between 3.65 and 25.7 at the end of malolactic fermentation in a study carried out in 'Monastrell' wines from different wineries from the same geographic area and, within each winery, from wines elaborated based on different market prices [50]. Values ranged from 13.1 to 21.3 in 'Monastrell' wines and 12.2–38.2 in 'Cabernet Sauvignon' wines at the end of alcoholic fermentation in a study carried out over two seasons [51]. As can be noticed with the results shown in Table 5, our new varieties are capable of producing wines with an extraordinarily high color despite being grown in areas with high temperatures and semi-arid conditions.

Table 5. Mean composition (2017–2021) of red wines at the end of alcoholic fermentation (AF) and malolactic fermentation (MF).

Parameters		Monastrell	Cabernet S.	Syrah	Calnegre	Gebas	Myrtia
Color intensity	AF	14.86 ± 0.52 a	19.17 ± 0.65 a	25.87 ± 2.56 b	46.66 ± 1.58 d	40.72 ± 2.31 c	60.28 ± 1.61 e
	MF	11.69 ± 1.70 a	16.00 ± 0.45 ab	18.66 ± 1.09 b	32.50 ± 2.46 d	25.99 ± 1.45 c	40.02 ± 1.97 e
Taint	AF	0.44 ± 0.02 a	0.44 ± 0.02 a	0.39 ± 0.01 a	0.38 ± 0.02 a	0.42 ± 0.03 a	0.39 ± 0.01 a
	MF	0.57 ± 0.03 b	0.56 ± 0.01 b	0.55 ± 0.02 b	0.48 ± 0.01 a	0.54 ± 0.01 b	0.48 ± 0.01 a
Anthocyanins	AF	571.00 ± 36.88 a	698.00 ± 52.84 a	1084.00 ± 58.57 b	1598.00 ± 54.40 c	1526.00 ± 90.92 c	1936.00 ± 252.68 d
	MF	330.47 ± 75.00 a	432.00 ± 54.00 a	692.00 ± 65.83 b	979.00 ± 110.38 c	972.00 ± 91.18 c	1262.00 ± 109.82 d
T.P.C.	AF	43.35 ± 1.86 a	45.10 ± 1.72 a	60.73 ± 3.60 b	94.41 ± 6.95 c	91.84 ± 7.99 c	100.78 ± 4.87 c
	MF	36.70 ± 3.99 a	41.84 ± 1.00 ab	53.83 ± 2.82 b	85.94 ± 8.07 c	82.73 ± 5.80 c	86.05 ± 3.70 c
L*	AF	13.77 ± 0.43 d	8.25 ± 0.81 c	3.91 ± 0.58 b	1.76 ± 0.16 a	1.35 ± 0.56 a	0.84 ± 0.20 a
	MF	14.18 ± 3.07 c	6.86 ± 0.67 b	5.03 ± 0.44 ab	2.57 ± 0.14 a	2.65 ± 0.34 a	1.46 ± 0.42 a
a*	AF	46.10 ± 0.51 e	38.43 ± 1.44 d	26.01 ± 2.55 c	12.74 ± 1.11 b	9.72 ± 4.00 ab	6.14 ± 1.47 a
	MF	44.89 ± 2.48 d	35.89 ± 1.49 c	31.30 ± 1.55 c	18.55 ± 1.03 b	18.91 ± 2.32 b	10.65 ± 3.08 a
b*	AF	23.69 ± 0.73 d	14.22 ± 1.40 c	6.74 ± 1.01 b	3.03 ± 0.27 a	2.33 ± 0.97 a	1.45 ± 0.35 a
	MF	19.15 ± 1.67 d	11.80 ± 1.14 c	8.66 ± 0.76 b	4.43 ± 0.25 a	4.58 ± 0.60 a	2.48 ± 0.69 a
C* (ab)	AF	51.84 ± 0.83 d	41.01 ± 1.82 c	26.88 ± 2.73 b	13.10 ± 1.14 a	10.00 ± 4.11 a	6.31 ± 1.51 a
	MF	48.87 ± 2.64 d	37.83 ± 1.77 c	32.48 ± 1.67 c	19.07 ± 1.08 b	19.46 ± 2.39 b	10.94 ± 3.16 a

Data expressed as mean value ± standard deviation. T.P.C., total phenolic content. Different letters in the same file indicate significant differences among genotypes at the 5% level, according to the LSD's multiple range test.

With respect to taint values at the end of alcoholic fermentation, they were similar for all wines, including those from parentals and new varieties, and indicated no oxidations in any of them. Nevertheless, at the end of malolactic fermentation, 'Calnegre' and 'Myrtia' showed the lowest values in comparison with the rest of wines, which showed upper values but were similar among them (Table 5).

Other phenolic parameters measured in wines were total anthocyanin and total phenolic compounds also measured previously in grapes at harvest. As can be observed in Table 5, the concentrations of anthocyanins shown in the wines of the new varieties were much higher than those shown by their parentals. Specifically, the 'Myrtia' wines, which were those with the highest anthocyanin content, doubled the content obtained by 'Syrah' wines and quadrupled that obtained by 'Monastrell' wines, both at the end of alcoholic and malolactic fermentation. The 'Calnegre' and 'Gebas' wines showed values of around 1500 mg/L of total anthocyanins, which were also higher than those obtained by the wines of their parentals ('Monastrell' and 'Cabernet Sauvignon'), although slightly less than in 'Myrtia' wines. It is known that the typical concentrations of free anthocyanins in full-bodied young red wines are around 500 mg/L, but can in some cases be higher than 2000 mg/L [52–54] as shown in the wines of the new varieties. Similar results were found in other works by different authors; Gil-Muñoz et al. (2018) [55], in a study carried out during two consecutive seasons, showed values of total anthocyanin in wines elaborated with varieties cross-bred from 'Monastrell' that ranged from 799.5 to 2206.4 mg/L during 2015 and from 1636.3 to 2210.2 mg/L in 2016; with the values reached being very different in function according to the analyzed season. In addition, Gil-Muñoz et al. (2021) [56], in an experiment carried out during three consecutive seasons in 'Monastrell' and 10 crossbreeds, showed how most of them had a higher anthocyanin concentration in wines with respect to 'Monastrell' wines, although differences were found between years.

Regarding IPTs, as was the case with anthocyanins, the final values were higher for the alcoholic fermentation than for malolactic fermentation, and they were also higher in wines from the new varieties compared to those from the parents. 'Syrah', 'Cabernet Sauvignon' and 'Monastrell' wines showed values of 60.73, 45.10 and 43.35 at the end of alcoholic fermentation, and 53.83, 41.84 and 36.70 at the end of malolactic fermentation, respectively. As can be observed, 'Monastrell' wines always obtained the lowest content of IPTs. With respect to the new varieties of wines, IPTs values ranged between 91.84 and 100.78 at the end of alcoholic fermentation and between 82.73 and 86.05 at the end of malolactic fermentation, with 'Myrtia' wines being those that obtained the highest values and 'Gebas' the lowest. 'Calnegre' showed intermediate values. Again, the results found in the wines of the new red varieties were quite a bit higher than those found in the wines of the parentals.

Finally, concerning the color measured at the end of alcoholic fermentation using the CIELab space (Table 5), the highest value of L* was reached by 'Monastrell' wines indicating they were the clearest wines, followed by 'Cabernet Sauvignon' wines and finally by 'Syrah' wines. New variety wines obtained lower L* values than parental wines, indicating that these wines were darker than parental wines; nevertheless, the lowest L* value was shown in 'Myrtia' wines, followed by 'Gebas' wines and finally by 'Calnegre' wines. With respect to final malolactic fermentation, the L* values increased slightly except in the case of 'Cabernet Sauvignon' wines, but the trend was similar to that found at the end of alcoholic fermentation. Regarding a* and b* parameters, which indicate red and yellow colors, respectively, different results were shown in the different wines analyzed. In general, values were lower at the end of alcoholic fermentation than malolactic fermentation, with the exception of 'Monastrell' and 'Cabernet Sauvignon' wines where the opposite happened. Paladines-Quezada et al. (2019) [51] showed L* values around 12 in 'Monastrell' wines and ranging between 4 and 11 in 'Cabernet Sauvignon' wines at the end of alcoholic fermentation. With respect to a*, the highest value was found in 'Monastrell' wines and the lowest in 'Myrtia' wines in the two moments analyzed at the end of the alcoholic and malolactic fermentation. With respect to b*, the highest value was found in

'Monastrell' wines, but the lowest results were found in the three new varieties of wines due to no statistical differences being found among them. The last parameter analyzed was C* (chroma), which is a parameter that indicates the contribution of a* (redness) and b* (yellowness), so values of C* close to or higher than 50 correspond to vivid colors. As expected, the results were similar to those obtained for parameters a* and b*, with higher values at the end of malolactic fermentation except for 'Monastrell' and 'Cabernet Sauvignon' wines. We were also able to observe the highest values in 'Monastrell' wines, and the lowest values for 'Myrtia' wines, although similar values were obtained at the end of alcoholic fermentation in the new varieties since no statistically significant differences were shown.

3.5. Wine Sensorial Analysis

The quality and phenolic characteristics of our wines from the new varieties were promising, but we wanted to check if they also had that quality sensorial characteristic, so we carried out a descriptive sensory analysis. The technique of descriptive analysis (DA) provides a quantitative analytical characterization of appearance, aroma, taste and mouthfeel as described in detail elsewhere [57,58].

As can be observed in Figure 2, 'Calblanque' wine was compared to 'Verdejo' wine due to both being white varieties, and in addition, 'Verdejo' is considered a high quality wine among white wines in Spain. Among the descriptive attributes taken into account, color, mouth and aroma characteristics were evaluated. Although the difference was small between both wines, the tasters were able to distinguish them sensorially, giving them different scores. With regard color, 'Calblanque' wines showed a higher value in comparison to 'Verdejo' wines. Regarding nose attributes, intensity was higher in 'Calblanque' wines than in 'Verdejo' wines; however, the quality was superior in the latter in comparison with 'Calblanque'. Intensity, quality and persistence in mouth were higher in 'Calblanque' than 'Verdejo' wines. Finally, harmony, a parameter that alludes to the global perception of the wine, also scored higher in 'Calblanque' wines. Moreno-Olivares et al. (2020) [59], in a study carried out with different white crossbreeds from 'Monastrell', showed results that highlighted how the crosses MT103, MC69 and MC180 ('Calblanque') showed significant differences from and better quality than the 'Verdejo' wine.

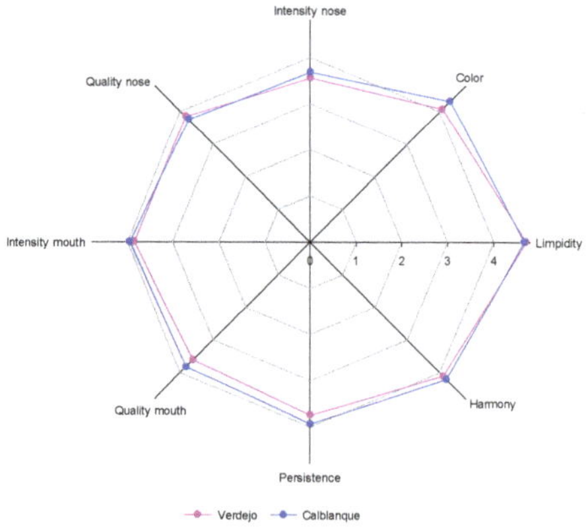

Figure 2. Sensorial analysis of 'Verdejo' and 'Calblanque' wines at the end of malolactic fermentation.

Figure 3 shows the sensorial analysis of the parentals and the new red varieties wines. As can be observed, more differences were found between the new varieties and parental wines. Regarding color, the highest scores were shown in 'Myrtia', 'Calnegre' and 'Gebas' wines. Limpidity was also highest in 'Myrtia' wines, although similar scores were shown in 'Monastrell', 'Syrah', 'Calnegre' and 'Gebas' wines, with 'Cabernet Sauvignon' being the variety with the lower score. With respect to the nose attributes, similar and the highest intensities were found in 'Syrah', 'Myrtia' and 'Calnegre' wines, intermediate values were shown in the 'Monastrell' and 'Gebas' wines, and finally, again, 'Cabernet Sauvignon' wines showed the lowest score. The highest quality nose was found in 'Syrah' and 'Calnegre' wines followed by 'Myrtia' and 'Gebas' and finally by 'Monastrell' and 'Cabernet Sauvignon' wines. It is known that the aromatic profile of many wines depends on the varietal compounds of the grapes that have been employed in their production. As well as lactic acid, the main substrate of malolactic fermentation, during this fermentation, there are a large number of metabolic end products, produced by specific bacterial species/strains that are responsible for modifying the aroma and flavor perception of wine [60]. With respect to mouth characteristics, the highest intensity, quality and persistence was shown by the new varieties and 'Syrah' wines, with the lowest score shown by 'Monastrell' and 'Cabernet Sauvignon' wines. Finally, harmony was superior in 'Syrah' wines, followed by the rest of the new varieties, and then by 'Monastrell' and 'Cabernet Sauvignon' wines.

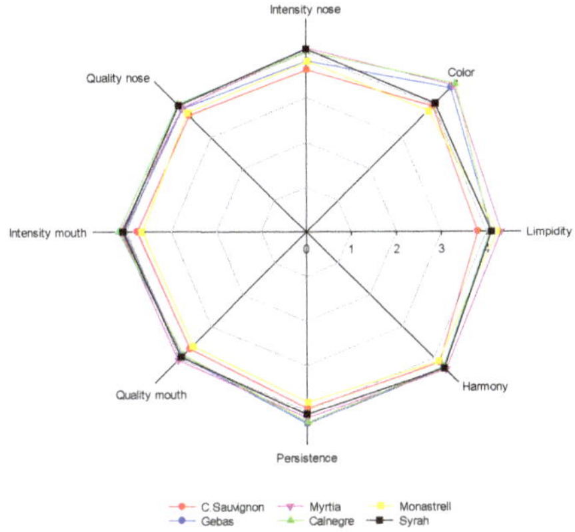

Figure 3. Sensorial analysis of 'Myrtia', 'Gebas', 'Calnegre', 'Syrah', 'Cabernet Sauvignon' and 'Monastrell' wines at the end of malolactic fermentation.

4. Conclusions

Crossbreeding programs generate great genetic variation and allow the selection of new genotypes, as described in this work, that are better adapted to the specific conditions of the viticulture zone. The attributes of the white variety 'Calblanque', and the red varieties 'Calnegre', 'Gebas' and 'Myrtia', registered by the IMIDA as commercial varieties and authorized for winemaking in the Region of Murcia, could allow their better adaptation to the effects of high temperatures on grape and wine quality in semi-arid areas. The red genotypes were selected for their phenolic quality—which was very superior to that of the parentals—and the white variety 'Calblanque' was selected for its good balance of acidity and aromatic profile. From a sensorial point of view, the new varieties of wines also showed high scores in comparison with their parentals.

Therefore, the new varieties described in this work represent a support to the wine sector of the area, which will have an innovative and competitive material of high quality, while maintaining the Mediterranean profile of the wines made with these varieties.

5. Patents

The new varieties are protected at the European level. The ownership and all rights over them belong to IMIDA. The breeders are: Adrián Martínez-Cutillas, José Ignacio Fernández-Fernández, Leonor Ruiz-García, Celia Martínez-Mora, Juan Antonio Bleda-Sánchez and Rocío Gil-Muñoz.

Supplementary Materials: The following supporting information can be downloaded at: https://www.mdpi.com/article/10.3390/horticulturae9070760/s1, Table S1: Cumulative and average values of meteorological parameters recorded in the experimental farm from the starting to the end of the breeding program (1997-2021); Table S2: Cumulative and average values of meteorological parameters recorded in the experimental farm during the evaluation period (2017-2021); Table S3: Ampelographic characteristics of the parental and new varieties indicated with the UPOV descriptors (2008) and the CPVO technical protocol (CPVO-TP/050/2) for distinctness, uniformity and stability tests.

Author Contributions: Conceptualization, J.I.F.-F. and L.R.-G.; methodology, C.M.-M. and D.J.F.-L.; software, J.A.B.-S.; validation, J.I.F.-F., J.A.B.-S. and J.D.M.-O.; formal analysis, M.J.G.-B., C.M.-M. and J.D.M.-O.; resources, L.R.-G. and R.G.-M.; data curation, C.M.-M. and J.I.F.-F.; writing—original draft preparation, L.R.-G. and R.G.-M.; writing—review and editing, L.R.-G. and R.G.-M.; visualization, J.I.F.-F. and J.A.B.-S.; supervision, L.R.-G. and R.G.-M.; project administration, L.R.-G. and R.G.-M.; funding acquisition, L.R.-G. and R.G.-M. All authors have read and agreed to the published version of the manuscript.

Funding: This work was financed by the European Regional Development Fund (80%), with the collaboration of the Region of Murcia (20%), via projects PO07-037, FEDER1420-04, and FEDER1420-29. PCR products were separated by capillary electrophoresis in an ABI Prism 3730 sequencer at the Unidad de Genómica-Universidad Complutense de Madrid/Parque Científico de Madrid, Spain.

Institutional Review Board Statement: Not applicable.

Informed Consent Statement: Not applicable.

Data Availability Statement: All data are presented in the study.

Acknowledgments: The authors thank Ana Fuentes-Denia, José Cayetano Gómez-Martínez, Ana Cebrian and Juan Corredor for their technical assistance; Adrián Yepes-Hita, and José Antonio Martínez-Jiménez for plant management in the field; Carlos V. Padilla, Eliseo Salmerón and Isidro Hita for crop health control.

Conflicts of Interest: The authors declare no conflict of interest.

References

1. Poni, S.; Gatti, P.; Dai, Z.W. Grapevine quality: A multiple choice issue. *Sci. Hortic.* **2018**, *234*, 445–462. [CrossRef]
2. Jackson, R.S. *Wine Science Principles and Applications*, 3rd ed.; Elsevier: Amsterdam, The Netherlands; Academic Press: San Diego, CA, USA, 2008.
3. Droulia, F.; Charalampopoulos, I. A Review on the observed climate change in Europe and its impacts on viticulture. *Atmosphere* **2022**, *13*, 837. [CrossRef]
4. Keller, M. Managing grapevines to optimise fruit development in a challenging environment: A climate change primer for viticulturists. *Aust. J. Grape Wine Res.* **2010**, *16*, 56–69. [CrossRef]
5. Junquera, P.; Lissarrague, J.R.; Jiménez, L.; Linares, R.; Baeza, P. Long-term effects of different irrigation strategies on yield components, vine vigour, and grape composition in cv. Cabernet-Sauvignon (*Vitis vinifera* L.). *Irrig. Sci.* **2012**, *30*, 351–361. [CrossRef]
6. Kennedy, J.A.; Saucier, C.; Glories, Y. Grape and wine phenolics: History and perspective. *Am. J. Enol. Vitic.* **2006**, *57*, 239–248. [CrossRef]
7. Ivanova, V.; Vojnoski, B.; Stefova, M. Effect of winemaking treatment and wine aging on phenolic content in Vranec wines. *J. Food Sci. Technol.* **2011**, *49*, 161–172. [CrossRef] [PubMed]
8. Guan, L.; Dai, Z.; Wu, B.H.; Wu, J.; Merlin, I.; Hilbert, G.; Renaud, C.; Gomès, E.; Edwards, E.; Li, S.H.; et al. Anthocyanin biosynthesis is differentially regulated by light in the skin and flesh of white-fleshed and teinturier grape berries. *Planta* **2016**, *243*, 23–41. [CrossRef]
9. He, F.; Mu, L.; Yan, G.L.; Liang, N.N.; Pan, Q.H.; Wang, J.; Reeves, M.J.; Duan, C.Q. Biosynthesis of anthocyanins and their regulation in colored grapes. *Molecules* **2010**, *15*, 9057–9091. [CrossRef]
10. de Rosas, I.; Deis, L.; Baldo, Y.; Cavagnaro, J.B.; Cavagnaro, P.F. High Temperature Alters Anthocyanin Concentration and Composition in Grape Berries of Malbec, Merlot, and Pinot Noir in a Cultivar-Dependent Manner. *Plants* **2022**, *11*, 926. [CrossRef]
11. van Leeuwen, C.; Destrac-Irvine, A. Modified grape composition under climate change conditions requires adaptations in the vineyard. *OENO One* **2017**, *51*, 147. [CrossRef]
12. Fraga, H.; Malheiro, A.C.; Moutinho-Pereira, J.; Santos, J.A. An overview of climate change impacts on European viticulture. *Food Energy Secur.* **2012**, *1*, 94–110. [CrossRef]
13. Pavlousek, P. Evaluation of drought tolerance of new grapevine rootstock hybrids. *J. Environ. Biol.* **2011**, *32*, 543–549. [PubMed]

14. Reynolds, A.G. *Grapevine Breeding Programs for the Wine Industry: Traditional and Molecular Techniques*, 1st ed.; Reynolds, A., Ed.; Woodhead Publishing: Oxford, UK, 2015; ISBN 978-1-78242-075-0.
15. Berdeja, M.; Nicolas, P.; Kappel, C.; Dai, Z.W.; Hilbert, G.; Peccoux, A.; Lafontaine, M.; Ollat, N.; Gomès, E.; Delrot, S. Water limitation and rootstock genotype interact to alter grape berry metabolism through transcriptome reprogramming. *Hortic. Res.* **2015**, *2*, 15012. [CrossRef] [PubMed]
16. Romero, P.; Botía, P.; Navarro, J.M. Selecting rootstocks to improve vine performance and vineyard sustainability in deficit irrigated Monastrell grapevines under semiarid conditions. *Agric. Water Manag.* **2018**, *209*, 73–93. [CrossRef]
17. Zombardo, A.; Mica, E.; Puccioni, S.; Perria, R.; Valentini, P.; Mattii, G.B.; Cattivelli, L.; Storchi, P. Berry quality of grapevine under water stress as affected by rootstock–scion interactions through gene expression regulation. *Agronomy* **2020**, *10*, 680. [CrossRef]
18. Santos, J.A.; Costa, R.; Fraga, H. New insights into thermal growing conditions of Portuguese grapevine varieties under changing climates. *Theor. Appl. Climatol.* **2018**, *135*, 1215–1226. [CrossRef]
19. Nicolle, P.; Williams, K.A.; Angers, P.; Pedneault, K. Changes in the flavan-3-ol and polysaccharide content during the fermentation of *Vitis vinifera* Cabernet-Sauvignon and cold-hardy *Vitis* varieties Frontenac and Frontenac blanc. *OENO One* **2021**, *55*, 337–347. [CrossRef]
20. Santillán, D.; Garrote, L.; Iglesias, A.; Sotes, V. Climate change risks and adaptation: New indicators for Mediterranean viticulture. *Mitig. Adapt. Strat. Glob. Chang.* **2020**, *25*, 881–899. [CrossRef]
21. Fernández-López, D.J.; Fernández-Fernández, J.I.; Martínez-Mora, C.; Bleda-Sánchez, J.A.; Ruiz-García, L. Productiveness and berry quality of new wine grape genotypes grown under drought conditions in a semi-arid wine-producing Mediterranean Region. *Plants* **2022**, *11*, 1363. [CrossRef]
22. Charlotte Brault, C.; Segura, V.; This, P.; Le Cunff, L.; Flutre, T.; François, P.; Pons, T.; Péros, J.-P.; Doligez, A. Across-population genomic prediction in grapevine opens up promising prospects for breeding. *Hortic. Res.* **2022**, *9*, uhac041. [CrossRef]
23. Ruiz-García, L.; Gil-Muñoz, R.; Martínez-Mora, C.; Bleda, J.A.; Fuentes-Denia, A.; Martínez-Jiménez, J.A.; Martínez-Cutillas, A.; Fernández-Fernández, J.I. Nuevas variedades de vid obtenidas en la Región de Murcia. *Actas Hortic.* **2018**, *80*, 226–229.
24. Fernández-Fernández, J.I.; Gil-Muñoz, R.; Bleda-Sánchez, J.A.; Corredor-Cano, J.; Moreno-Olivares, J.D.; Cebrián-Pérez, A.; Martínez-Balsas, D.; Gómez-Martínez, J.C.; Palencia-Sigüenza, M.S.; Carcelén-Cutillas, J.C.; et al. Nuevas variedades procedentes de Monastrell adaptadas a clima cálido. Cosechas 2016–2019. *Rev. Enólogos* **2020**, *126*, 78–88.
25. Bayo-Canha, A.; Fernández-Fernández, J.I.; Martínez-Cutillas, A.; Ruiz-García, L. Phenotypic segregation and relationships of agronomic traits in Monastrell × Syrah wine grape progeny. *Euphytica* **2012**, *186*, 393–407. [CrossRef]
26. Clark, M.F.; Adams, A.N. Characteristics of the microplate method of enzyme linked immunosorbent assay for the detection of plant viruses. *J. Gen. Virol.* **1977**, *34*, 475–483. [CrossRef] [PubMed]
27. Lorenz, D.H.; Eichhorn, K.W.; Bleiholder, H.; Klose, R.; Meier, U.; Weber, E. Phenological growth stages of the grapevine (*Vitis vinifera* L. ssp. *vinifera)-Codes and descriptions according to the extended BBCH scale. Aust. J. Grape Wine Res.* **1995**, *1*, 100–110.
28. Rustioni, L.; Maghradze, D.; Popescu, C.F.; Cola, G.; Abashidze, E.; Aroutiounian, R.; Brazão, J.; Coletti, S.; Cornea, V.; De-jeu, L.; et al. First results of the European grapevine collections collaborative network: Validation of a standard enocarpological phenotyping method. *Vitis-J. Grapevine Res.* **2014**, *53*, 219–226.
29. García-Barceló, J. *Técnicas Analíticas Para Vinos*, 1st ed.; GAB: Barcelona, Spain, 1990; ISBN 84-404-7827-5.
30. Ribéreau-Gayon, P.; Pontallier, P.; Glories, Y. Some interpretations of colour changes in young red wines during their conservation. *J. Sci. Food Agric.* **1983**, *34*, 505–516. [CrossRef]
31. Ho, P.; Da Silva, M.; Hogg, T.A. Changes in colour and phenolic composition during the early stages of maturation of port in wood, stainless steel and glass. *J. Sci. Food Agric.* **2001**, *81*, 1269–1280. [CrossRef]
32. Fernández, L.; Pradal, M.; Berud, F.; Romieu, C.; Torregrosa, L. Berry size variability in *Vitis vinifera* L. *Vitis* **2006**, *45*, 53–55.
33. Río-Segade, S.; Vázquez, E.S.; Losada, E.D. Influence of ripeness grade on accumulation and extractability of grape skin anthocyanins in different cultivars. *J. Food Compos. Anal.* **2008**, *21*, 599–607. [CrossRef]
34. Hernández-Jiménez, A.; Gil-Muñoz, R.; Ruiz-García, Y.; López-Roca, J.M.; Martinez-Cutillas, A.; Gómez-Plaza, E. Evaluating the polyphenol profile in three segregating grape (*Vitis vinifera* L.) populations. *J. Anal. Methods Chem.* **2013**, *2013*, 572896. [CrossRef] [PubMed]
35. Kliewer, W.K.; Howarth, L.; Omori, M. Concentrations of tartaric acid and malic acids and their salts in *Vitis vinifera* grapes. *Am. J. Enol. Vitic.* **1967**, *18*, 42–54. [CrossRef]
36. Liu, H.-F.; Wu, B.-H.; Fan, P.-G.; Li, S.-H.; Li, L.-S. Sugar and acid concentrations in 98 grape cultivars analyzedby principal component analysis. *J. Sci. Food Agric.* **2006**, *86*, 1526–1536. [CrossRef]
37. Shiraishi, M. Comparison in changes in sugars, organic acids and amino acids during berry ripening of sucrose- and hexose-accumulating grape cultivars. Potential of table crape varieties (*Vitis vinifera* L.). *Eur. J. Hortic. Sci.* **2000**, *73*, 124–130.
38. Lakso, A.N.; Kliewer, W.M. The influence of temperature on malicacid metabolism in grape berries. II. Temperature responses of net dark CO_2 fixation and malic acid pools. *Am. J. Enol. Vitic.* **1978**, *29*, 145–149. [CrossRef]
39. Dai, Z.W.; Ollat, N.; Gomès, E.; Decroocq, S.; Tandonnet, J.P.; Bordenave, L.; Pieri, P.; Hilbert, G.; Kappel, C.; van Leeuwen, C.; et al. Ecophysiological, genetic, and molecular causes of variation ingrape berry weight and composition: A review. *Am. J. Enol. Vitic.* **2011**, *62*, 413–425. [CrossRef]

40. Famiani, F.; Farinelli, D.; Palliotti, A.; Moscatello, S.; Battistelli, A.; Walker, R.P. Is stored malate the quantitatively most important substrate utilised by respiration and ethanolic fermentation in grape berry pericarp during ripening? *Plant Physiol. Biochem.* **2014**, *76*, 52–57. [CrossRef]
41. Anastasiadi, M.; Zira, A.; Magiatis, P.; Haroutounian, S.A.; Skaltsounis, A.L.; Mikros, E. 1H NMR-Based metabonomics for the classification of Greek wines according to variety, region, and vintage. Comparison with HPLC data. *J. Agric. Food Chem.* **2009**, *57*, 11067–11074. [CrossRef]
42. Kupsa, J.; Pavloušek, P.; Kumšta, M.; Lampíř, L. Phenolic profiles of Riesling wines originating from different terroirs of the Czech Republic. *Mitt. Klosterneubg.* **2017**, *67*, 182–193.
43. Parpinello, G.P.; Ricci, A.; Arapitsas, P.; Curioni, A.; Moio, L.; Rio-Segade, S.; Ugliano, M.; Versari, A. Multivariate characterisation of Italian monovarietal red wines using MIR spectroscopy. *OENO One* **2019**, *4*, 741–751. [CrossRef]
44. Usseglio-Tomasset, L. *Chimica Enologica*, 2nd ed.; AEB Edizioni: Brescia, Italy, 1985.
45. Dupuy, P. Les facteurs du développement de l'acescence dans le vin. *Ann. Technol. Agric.* **1957**, *6*, 391–407.
46. Wibowo, D.; Eschenbruch, R.; Davis, C.R.; Fleet, G.H.; Lee, T.H. Occurrence and growth of lactic acid bacteria in wine: A review. *Am. J. Enol. Vitic.* **1985**, *36*, 302–313. [CrossRef]
47. Berg, H.W. Stabilisation des anthocyanes. Comportement de la couleur dans les vins rouges. *Ann. Technol. Agricol.* **1963**, *12*, 247–259.
48. Brouillard, R.; Delaporte, B. Chemistry of anthocyanin pigments. 2. Kinetic and thermodynamic study of proton transfer, hydration, and tautomeric reactions of malvidin-3-glucoside. *J. Am. Chem. Soc.* **1977**, *99*, 8461–8468. [CrossRef]
49. Forino, M.; Picariello, L.; Rinaldi, A.; Moio, L.; Gambuti, A. How must pH affects the level of red wine phenols. *LWT-Food Sci. Technol.* **2020**, *129*, 109546. [CrossRef]
50. Gómez-Plaza, E.; Olmos, O.; Bautista-Ortín, A.B. Tannin profile of different Monastrell wines and its relation to projected market prices. *Food Chem.* **2016**, *1*, 506–512. [CrossRef] [PubMed]
51. Paladines-Quezada, D.F.; Moreno-Olivares, J.D.; Fernández-Fernández, J.I.; Bautista-Ortín, A.B.; Gil-Muñoz, R. Influence of methyl jasmonate and benzothiadiazole on the composition of grape skin cell walls and wines. *Food Chem.* **2019**, *277*, 691–697. [CrossRef] [PubMed]
52. Burns, J.; Mullen, W.; Landrault, N.; Teissedre, P.L.; Lean, M.E.J.; Crozier, A. Variations in the profile and content of anthocyanins in wines made from Cabernet Sauvignon and hybrid grapes. *J. Agric. Food Chem.* **2002**, *50*, 4096–4102. [CrossRef] [PubMed]
53. Mulinacci, N.; Santamaria, A.R.; Giaccherini, C.; Innocenti, M.; Valletta, A.; Ciolfi, G.; Pasqua, G. Anthocyanins and flavan-3-ols from grapes and wines of *Vitis vinifera* cv. Cesanese d'Affile. *Nat. Prod. Res.* **2008**, *22*, 1033–1039. [CrossRef]
54. Nikfardjam, M.S.P.; Márk, L.; Avar, P.; Figler, M.; Ohmacht, R. Polyphenols, anthocyanins, and trans-resveratrol in red wines from the Hungarian Villány region. *Food Chem.* **2006**, *98*, 453–462. [CrossRef]
55. Gil-Muñoz, R.G.; Moreno-Olivares, J.D.; Paladines-Quezada, D.F.; Cebrián-Pérez, A.; Fernández-Fernández, J.I. High anthocyanin level of grape hybrids from Monastrell and their wines. *Int. J. Hort. Agric.* **2018**, *3*, 1–8.
56. Gil-Muñoz, R.; Moreno-Olivares, J.D.; Paladines-Quezada, D.F.; Bleda-Sánchez, J.A.; Cebrían-Pérez, A.; Giménez-Bañón, M.J.; Fernández-Fernández, J.I. Characterization of anthocyanins from intraspecific crosses of Monastrell with other premium varieties. *Front. Nutr.* **2021**, *8*, 664515. [CrossRef] [PubMed]
57. Einstein, M.A. Descriptive techniques and their hybridization. In *Sensory Science Theory and Applications in Foods*; Lawless, H.T., Klein, B.P., Eds.; Dekker: New York, NY, USA, 1991.
58. Stone, H.; Sidel, J.; Oliver, S.; Woolsey, A.; Singleton, C. Sensory evaluation by quantitative descriptive analysis. *Descr. Sens. Anal. Pract.* **1974**, *28*, 24–34.
59. Moreno-Olivares, J.D.; Giménez-Bañón, M.J.; Paladines-Quezada, D.F.; Gómez-Martínez, J.C.; Cebrián-Pérez, A.; Fernández-Fernández, J.I.; Bleda-Sánchez, J.A.; Gil-Muñoz, R. Aromatic characterization of new white wine varieties made from Monastrell grapes grown in South-Eastern Spain. *Molecules* **2020**, *25*, 3917. [CrossRef]
60. Gil-Sánchez, I.; Bartolomé-Suáldea, B.; Moreno-Arribas, M.V. *Malolactic Fermentation. Red Wine Technology*; Academic Press: London, UK, 2019.

horticulturae

Article

Genotypic and Sanitary Characterization of Minority Grapevine Varieties Prospected in Andalusia, Spain

Inmaculada Rodríguez-Torres [1,†], Ana Martín Carrillo [2,†], María del Pilar Ramírez [3], Francisco Jesus Gómez Gálvez [4], Leonardo Velasco Arjona [5], Carlos Padilla [6] and Enrico Cretazzo [2,*]

1 Agroindustry and Food Quality Area, IFAPA Rancho de la Merced Center, Ctra. Cañada de la Loba PK 3,1, 11471 Jerez de la Frontera, Spain; icrodrigueztorres@gmail.com
2 Biotechnology and Plant Breeding Area, IFAPA Rancho de la Merced Center, Ctra. Cañada de la Loba PK 3,1, 11471 Jerez de la Frontera, Spain; ana.martin.carrillo@juntadeandalucia.es
3 Agroalimentary Technologies and Engineering, IFAPA Cabra Center, Antigua Ctra. Cabra-Doña Mencía, Km. 2,5, 14940 Cabra, Spain; mariap.ramirez.perez@juntadeandalucia.es
4 Biotechnology and Plant Breeding Area, IFAPA Alameda del Obispo Center, Avda. Menéndez Pidal s/n, 14080 Cordoba, Spain; franciscoj.gomez.galvez@juntadeandalucia.es
5 Sustainable Plant Protection Area, IFAPA Málaga Center, Cortijo de la Cruz s/n, 29140 Malaga, Spain; leonardo.velasco@juntadeandalucia.es
6 Plant Protection Department, IMIDA, Calle Mayor s/n, 30150 Murcia, Spain; carlosv.padilla@carm.es
* Correspondence: enrico.cretazzo@juntadeandalucia.es
† These authors contributed equally to this work.

Abstract: Andalusia is a Spanish region that is home to numerous minority varieties due to its diversity and territorial extension, offering the local viticulture the possibility of diversifying its wine production. The genotypic characterization of 98 specimens from six areas with a winemaking tradition in Andalusia was carried out between the years 2020 and 2022, by means of thirteen microsatellite markers, including the nine recommended by the OIV. A total of 33 different genotypes were obtained, 20 of which corresponded to profiles of already described varieties (11 of them are of minority cultivars in Andalusia: 'Rojal Tinto', 'Beba', 'Zurieles', 'Rome', 'Hebén', 'Mollar Cano', 'Listán Prieto', 'Listán del Condado', 'Jarrosuelto', 'Negra Dorada', and 'Mantúo de Pilas'), while the other 12 profiles did not match with previously identified varieties. These profiles were registered in the database of the IFAPA "Rancho de la Merced" Germplasm Bank. The eco-geographical groups of the new identified genotypes were determined through an analysis of genetic diversity. The presence of grapevine fanleaf virus, grapevine fleck virus, and grapevine leafroll-associated viruses was also determined due to the requirement of healthy clones of the new varieties for their potential interest to be authorized for cultivation in Spain.

Keywords: genetic resources; molecular markers; local cultivar; certification

Citation: Rodríguez-Torres, I.; Martín Carrillo, A.; Ramírez, M.d.P.; Gómez Gálvez, F.J.; Velasco Arjona, L.; Padilla, C.; Cretazzo, E. Genotypic and Sanitary Characterization of Minority Grapevine Varieties Prospected in Andalusia, Spain. *Horticulturae* **2023**, *9*, 759. https://doi.org/10.3390/horticulturae9070759

Academic Editors: Antonio Amores-Arrocha, Ana Jiménez-Cantizano and Paolo Sabbatini

Received: 19 May 2023
Revised: 9 June 2023
Accepted: 22 June 2023
Published: 1 July 2023

1. Introduction

Andalusia has a surface area of about 87,000 km^2, which represents 17.3% of the Spanish territory. This wide extension entails a great diversity of landscapes and ecosystems. Grapevine (*Vitis vinifera* L.) is one of the most rustic crops and one of the best adapted to the multiple edaphoclimatic conditions of Andalusia. Vineyards can be found in high areas, such as "La Alpujarra" (Granada), located at more than 1000 m; in areas close to the sea, such as the hills of "Marco de Jerez" (Cádiz); and even in regions with high slopes, such as "La Axarquía" (Málaga). There are seven Protected Denominations of Origin (PDO) for still wines in Andalusia: "Jerez-Xérès-Sherry", "Manzanilla de Sanlúcar de Barrameda", "Granada", "Condado de Huelva", "Montilla-Moriles", "Málaga", and "Sierras de Málaga". There are also 16 Protected Geographical Indications (PGIs) and a PDO of flavored wines. Moreover, the region has enormous wine-growing potential due to its plant genetic resources. Numerous bibliographical references show the great heterogeneity of vine varieties that were cultivated before the phylloxera arrived in this region [1,2]. Although this plague ended up destroying more than a million hectares with the consequent loss of many local varieties, it is estimated that about 300 out of the 600 varieties that were originally cultivated before the invasion in 1878 were conserved.

Currently, 56 grape varieties are authorized for cultivation in Andalusia in accordance with the List of Authorized Varieties for Andalusia (Real Decreto 111/2022, of 9 February, which modifies Real Decreto 1338/2018, of 29 October, which regulates the potential of viticultural production). Despite this number, only four white grapes account for 77% of the total cultivated area: 'Palomino Fino', 'Pedro Ximénez', 'Moscatel de Alejandría', and 'Zalema' (unpublished, according to "Consejería de Agricultura, Pesca, Agua y Desarrollo Rural, Junta de Andalucía, 2022). Furthermore, the sum of the 29 least cultivated varieties only represents 2% of the total surface. It can be understood from these data that many varieties are barely cultivated and, therefore, are not well known by viticulturists [3]. In addition, not all the minority cultivated varieties are included in the previously mentioned list, and since their cultivation is not authorized in Spain, their conservation is endangered. For all these reasons, it is crucial to investigate them in situ and to identify, recover, and evaluate the possibility of including some of these varieties in the Andalusian wine heritage. However, as has already been mentioned, in many cases, these varieties are not authorized for cultivation in Spain. Therefore, if there is genuine interest in growing some of these varieties, it would be necessary to carry out the official authorization process. The first step of this process consists of forwarding plants grafted onto '110R' rootstock to the National Reference Collection of Grape Varieties (maintained at IMIDA, Murcia, and funded by MAPA, Madrid). A regulated ampelographic assessment is then carried out when the vines are formed, in which 44 characters (including leaves, shoots, flowers, and fruits) according to CPVO and UPOV will be recorded for at least two years. In addition, to qualify the candidate plant materials, the MAPA requires the absence of grapevine fanleaf virus (GFLV), grapevine leafroll-associated viruses 1 and 3 (GLRaV-1 and -3) and grapevine fleck virus (GFkV) as tested by serology. These are the viruses contemplated by Orden APA/2474/2006, of 27 July, which modifies certain annexes of the Technical Regulation for the Control and Certification of Vine Nursery Plants approved by Real Decreto 208/2003, of 21 February. Therefore, to include a variety in the Register of Commercial Grapevine Varieties in Spain (RVCV) and then obtain authorization for its cultivation in a given Autonomous Community, it is required to identify or proceed through sanitation propagation (i.e., in vitro cultivation) with a clone that is certifiable from the phytosanitary point of view.

It is well known that a big concern in most viticultural areas around the world is the scarceness of varietal diversification [4]. Moreover, frequently, only a very few clones are used for a given variety, resulting in a substantial reduction in genetic variability [5]. In this study, we show a methodology for approaching the initial steps necessary to increase the pool of cultivated grape varieties and clones in a given area. Multiple goals could be pursued: (a) to add a commercial variety to the list of authorized ones in a given region; (b) to recognize new synonymies of already authorized varieties and have available new local clones that are sanitary guaranteed; (c) to recover old varieties, possibly described in the past, that are very well adapted in a given ecosystem; (d) to discover new autochthonous varieties that could provide the basis for climate change adaptation in a given area [6]. Here, we present the results of surveys of minority varieties carried out on 98 specimens in six wine-producing areas spanning Andalusia (Figure 1).

Figure 1. Map of the prospected areas.

2. Materials and Methods

2.1. Areas and Specimens Prospected

Expert staff from local entities such as Regional Agricultural Offices, Rural Development Groups, Regulatory Councils of Denomination of Origin, and autonomous viticulturists and oenologists have intensively collaborated with IFAPA, pointing out the old vineyards as the object of the study and selecting the vines included in the prospection because of their interesting productive and qualitative features. Next, more details are provided about the areas studied:

a. "Altiplano de Granada", a region in the northeast of the Granada province that includes the Huéscar and Baza municipalities, is characterized by its high average altitude (between 700 and 1200 m). Grapevine cultivation occupies an area of about 350 ha and it is widely distributed in more than 1000 small vineyards with an average surface of 0.3 ha, thus showing great fragmentation and being owned by many local winemakers for domestic consumption. This area is protected by the PDO "Granada" and the PGI "Altiplano de Sierra Nevada". Given that an exhaustive survey had been previously carried out in this area, only two plants have been studied in this work [7].

b. "Alpujarra de Granada", located in the southeastern part of the Granada province. It is characterized by its rugged relief, occupying most of the southern face of the Sierra Nevada. The 25 municipalities that make it up are located at an altitude between 1200 and 1500 m. Consisting of 2000 ha of vineyards with slate and stony soils and a slightly limestone subsoil. This region is protected by the PDO "Granada" and the PGI "Cumbres del Guadalfeo". Nine specimens were surveyed in this area from two plots whose vineyards are more than 60 years old, and their origin could be dated back to the pre-phylloxera period.

c. PDO "Montilla-Moriles", in the central area of Andalusia belonging to the Córdoba province, has the Guadalquivir River to the north, the Subbética Mountains to the south, the Genil River to the east, and the Guadajoz River to the west. This zone includes 17 municipalities. This PDO has nearly 6000 ha of vineyards that produce fortified wines (Fino, Amontillado, Oloroso, and Palo Cortado) and sweet wines with the 'Pedro Ximénez' variety. There has been a commitment to the parallel production of other types of wine in the recent years, possibly accompanied by a greater varietal diversification. In this area, thirty-two specimens were surveyed from eight plots located in the municipalities of Montilla, Moriles, and Cabra, with the aim of clarifying the situation of a supposed different variety locally called "Montepila", given the strong interest of the winemaking sector that requires its inclusion in the registry of authorized varieties in Andalusia.

d. "Valle de los Pedroches", a region of 3162 km^2 also located in the Córdoba province, represents the northernmost territory of Andalusia. This area is characterized by its high diversity of landscapes (it goes from the holm oak dehesa to the peneplain until reaching Sierra Morena) and soils (with a predominance of slate in agricultural areas) and an average altitude of between 500 and 700 m. Valle de los Pedroches is protected by the PGI "Córdoba". The vineyards of this area reached their maximum splendor in the 17th and 18th centuries, when they exceeded 2000 ha. It has been estimated that about 1800 ha were destined for the cultivation of vines shortly before the arrival of phylloxera at the end of 1800. Grapevine cultivation started disappearing shortly after this plague, although isolated or semi-isolated vine plants remained in the area, mainly for the production of self-consumption wines. In recent years, there has been a growing interest in these family orchards in order to preserve the biodiversity and traditional landscapes of the area. This indicates that there has been no influence from modern viticulture in this area, making it particularly attractive in terms of prospecting for possibly pre-phylloxera varieties since the introduction and spread of major national and/or international varieties have been avoided. Thirty-six vines were surveyed in this area, most of them isolated specimens located near the dividing boundaries between the plots.

e. "Pago Burujena", a territory located within El Marco de Jerez (PDO "Jerez-Xérès-Sherry" and "Manzanilla-Sanlúcar de Barrameda"), is the most emblematic and important area for the production of fortified and sweet Andalusian wines, having the same typologies mentioned for the PDO "Montilla-Moriles" but using instead the traditional 'Palomino Fino' variety. A greater varietal diversification has been recently pursued in the Jerez DO, and one of the objectives of producers is to rescue pre-phylloxera varieties for the production of both fortified and still wines. In the Jerez viticulture and its surroundings area, "Pagos" (rural places) represent cultivation areas geographically delimited by orographic elements and characterized by uniformity in terms of soil, microclimate, variety, and even the human qualities of the viticulturists. In this context, the definition of Pago Burujena dates back to the 16th century, under the Duchy of Medina Sidonia, and consists of an area of about 22 ha characterized by high quality limestone soils and the presence of different traditional varieties of the area. [8]. Seven specimens from a plot located in the Pago were surveyed.

f. "Moguer", whose municipality is protected by the PDO "Condado de Huelva", is a territory with a high vocation for growing vines, characterized by vineyards at an average altitude of 25 m, loamy or sandy soils with a certain content of lime (which gives them a slightly basic pH), and a Mediterranean climate with Atlantic influences. Around 550 ha of vineyards are currently cultivated, with a high degree of fragmentation of the plots (average surfaces between 2 and 3 ha), many of them older than 40 years. For these reasons, and because of its long winemaking tradition, it is an ideal area to survey for minority varieties linked to possible wine diversification. In the province of Huelva, this production is mostly based on the cultivation of 'Zalema'. Twelve vines over 50 years old belonging to this municipality and collected from a single plot were surveyed.

2.2. Plant Materials

For the genotypic study, shoots were collected from each of the 98 grapevine plants and stored at 4 °C during the winter of 2021. The shoots were placed in buckets with water to force their sprouting later in the spring of 2022. Once sprouted, samples of about 50 mg were taken from the apical meristem of each of them for DNA extraction.

For virus tests, five adult leaves were collected from different shoots of each plant in the late spring of 2022, and quickly stored at 4 °C. From each of the 5 leaves, 1 cm of petiole and a leaf surface of about 2 cm^2 attached to the same petiole were kept apart, stored in the cold and used for the determination of virus infections.

2.3. Genotypic Analysis

The DNeasy 96 Plant Kit (Qiagen, Hilden, Germany) extraction kit was used to purify total DNA. DNA concentration was recorded with the NanoDrop ND-1000 spectrophotometer (NanoDrop Technologies, Inc., Wilmington, NC, USA).

Microsatellite markers (SSR) were chosen for varietal identification since they are one of the most widespread tools for identifying grapevine genetic resources [9]. Thirteen SSR were analyzed, nine of them (VVS2 [10], VVMD5, VVMD7 [11], VVMD25, VVMD27 [12], VVMD28, VVMD32, VrZag62, and VrZag79) are those recommended by the European project GrapeGen06 and the OIV (International Office of Vine and Wine), and the other four (ISV3, ISV4, VVS2, and VMCNG4b9), used by CREA-UTV (Turi, Italy [13]), have been proven to be quite polymorphic in previous studies. For the GrapeGen06 SSR set, data were coded to compare microsatellite profiles as reported by Maul et al. [14] in order to include them in the European Vitis database (http://www.eu-vitis.de/index.php (accessed on 1 May 2023)). Four multiplex PCRs were carried out depending on the annealing temperatures of the different primers, which were purchased from Biomers (Ulm, Germany), using the 6-FAM, Hex, and Atto 550 fluorophores. Reactive volumes and PCR conditions were set according to previous studies [13,15]. The amplicons were separated on the ABI3130 sequencer. Subsequently, allelic sizes were determined using the Gene Mapper program. The International Variety Catalog (VIVC) was chosen as a reference to compare the profiles obtained; other databases were used for this as well. The varieties 'Garnacha', 'Tempranillo', 'Merlot' and 'Syrah' were used to harmonize the size of the alleles and to be able to compare the genotypes. Results were integrated into the IFAPA Rancho de la Merced database, which has around 1500 genotypes.

2.4. Cluster Analysis

A set of 258 genotypes with a clear ancestry inferred by Cretazzo et al. [15] in accordance with the eco-geographic origin of the cultivars [16], was used as a template composed of six groups representing different putative geographical origins. The twelve genotypes not previously described in this work were integrated into this dataset in order to investigate their ancestry. An unweighted neighbor-joining (NJ) tree was constructed based on the Simple Matching dissimilarity index (SM) between the unique genetic profiles using Darwin software package v6.0 [17], according to the default setting recommended by the software, with the only modification being to increase bootstrap replicates up to one thousand to obtain more accuracy.

2.5. Pedigree Analysis

The software CERVUS v.3.0.7 (Field Genetics, London, UK) [18] was used to identify first-order kinship relationships, mother-father-offspring trios, among the unknown grapevine cultivars found in this study and a set of 529 cultivars of *Vitis vinifera* from the Germplasm Bank of the IFAPA Rancho de la Merced whose 13 SSR profiles had been fully characterized [15]. The profiles of the 12 unknown varieties were included as candidates as well, with 541 cultivars, the set of possible parents used. This analysis relies on allele frequencies and is based on the difference in the log-likelihood ratio (LOD) between related and unrelated relationships to assign parentage, combined with a simulation of parentage analysis to determine the confidence of assignments. The parameters considered for the simulation were the following: number of offspring = 100,000; number of candidate parents = 100; proportion of candidate parents sampled = 0.3; prop. loci typed = 0.8, and prop. Loci mistyped = 0.01. Three criterions were considered to establish strict parentage relationships: (i) 10 minimum type loci; (ii) a confidence level of the LOD score higher than 95% (strict) or 80% (relaxed); and (iii) a maximum number of tolerated trio loci mismatches equal to two. The parameters considered for the simulation and criteria were adopted according to the default settings recommended by the software as well as the bibliography [19–21].

2.6. DAS-ELISA Test

Serological tests for the determination of GFLV, GLRaV-1, GLRaV-2, GLRaV-3, and GFkV were performed using the Bioreba (Reinach, Switzerland) and Agritest (Valenzano, Italy) specific DAS-ELISA tests. Although GLRaV-2 analysis is not currently required in certification schemes in Spain, its determination is recommended by the ICVG [22] and its importance in Spain is well known [23]. For virus determination, plant tissue was homogenized in sterile bags with phosphate buffered saline at pH = 7.2–7.4, including 0.2% *w*/*v* diethyldithiocarbamic acid (DIECA) and 2% *w*/*v* polyvinylpyrrolidone average mol wt 10,000 (PVP). Extracts were then analyzed according to Sánchez-Vizcaino et al. [24]. Homogenates from three healthy vines were included as negative controls in all the DAS-ELISA plates. Samples were considered positive when OD405 readings were at least two times the average of controls [25].

2.7. Quantitative RT-PCR

Total RNA was obtained using 100 mg of leaf material that was crushed in liquid nitrogen and homogenized in the lysis buffer from the SpectrumPlant™ Total RNA kit (Sigma-Aldrich Co., San Luis, AZ, USA), and extracted following the manufacturer's recommendations. For each biological replicate, genomic DNA was removed by the On-Column Dnase I Digestion Set (Sigma-Aldrich Co., San Luis, AZ, USA) during the extraction protocol. Total RNA yield and purity were determined using the NanoDrop 2000 spectrophotometer (Thermo Fisher Scientific, Waltham, MA, USA) for A260/A280 ratio verification, expected to range from 1.85 to 2.05. The primer/probe mixes for the one-step TaqMan® RT-PCR protocols were as follows: 20 µL each of the 100 pmol/µL forward and reverse primers and 4 µL of the 100 pmol/µL TaqMan® probe were added to 196 µL of water to bring the final volume to 240 µL. Single-tube TaqMan® RT-PCR reactions (12 µL) were set up in 96-well reaction plates using a TaqMan® core reagent kit (Thermo Fisher Scientific, Waltham, MA, USA) as follows: 6.1 µL of one-step RT-PCR Master Mix, 0.6 µL of primer/probe mix (400 nM primers and 80 nM probe), 0.3 µL of MuLV/RNA inhibitor, and 3 µL of total RNA template in a 12 µL reaction. Reactions were carried out in a Biorad I-cycler (Biorad, Hercules, USA) in a one-step reaction as recommended by Thermo Fisher Scientific (RT-PCR Master Mix procedure). Reverse transcription and amplification conditions were as follows: 45 °C for 35 min, 95 °C for 10 min, followed by 40 cycles of 95 °C for 15 s and 60 °C for 1 min. The data were analyzed quantitatively by measuring the threshold cycles (CT) in the Microsoft Excel program and graphically by an amplification plot. For the GLRaVs, GFkV, and GFLV, the primers and probes used were previously described by Osman et al. [26,27] and Cepin et al. [28], respectively.

3. Results

According to the genetic analyses, 71 of the 98 vines analyzed in this work corresponded to varieties already described in the literature, discerning 21 different genotypes (Table 1). The remaining 27 individuals showed genotypes that had not been previously described; in particular, 12 unknown varieties were identified (Table 2).

Most of the unidentified genotypes fit into the eco-geographical group of varieties of the Mediterranean Iberian Peninsula (Figure 2), according to the clustering deduced by Cretazzo et al. [15]. Remarkably, one of the eight unidentified varieties found in "Valle de los Pedroches" fit into the group of varieties that represents *Prole orientalis*, whose geographic origin extends from the Middle East to East Asia [16,29]. On the other hand, one of the two unidentified varieties found in "La Alpujarra de Granada" fitted into the group of Northern Italian and Southern French varieties.

Table 1. Known varieties found in this study.

Zone	Input Name	Confirmed Variety	Number of Individuals
Altiplano de Granada	Rosada Hornico	Rojal Tinta	1
La Alpujarra de Granada	Rome	Rome	1
	Mollar Cano	Rojal Tinta	1
	Tinta	Rojal Tinta	1
	Tinta Cortijo La Paz	Jacquez	1
	Llaqui	Jacquez	1
	Ricardera	Mantúo de Pilas	1
	Desconocida blanca	Airén	1
PDO Montilla-Moriles	Peñalista	Negra Rayada	1
	Montepila	Zalema	27
	Montepila	Cayetana Blanca	3
	Montepila	Pedro Ximénez	1
Valle de los Pedroches	Risquez	Ahmeur bou Ahmeur	2
	Risquez	Cayetana Blanca	1
	Merino	Cayetana Blanca	2
	Vieja Primera	Cayetana Blanca	2
	Blanca Lagareyes	Alarije	1
	Villaharta Llanos Suelo	Alarije	1
	Tío Kiko Camino	Negra Dorada	1
	Hebén	Hebén	1
	Jarrosuelto	Jarrosuelto	1
	Schiava Grossa	Schiava Grossa	1
	Entreárboles	Zurieles	1
Pago Burujena	Mantúo Castellano	Listán del Condado	3
	Mantúo de Pilas	Alarije	2
	Barcelonés	Alarije	2
Moguer	Mollar Cano	Mollar Cano	1
	Beba	Beba	1
	Listán Prieta	Beba	1
	Moguer	Airén	1
	Moguer	Listán Prieto	2
	Mantúo de Sanlúcar	Listán del Condado	3
	Mesa Plaza Tinta	Alphonse Lavallée	1

Table 2. Unidentified varieties found in this study.

Zone	Input Name	Confirmed Variety	Number of Individuals
Altiplano de Granada	Blanca Hornico	Unidentified 09	1
La Alpujarra de Granada	Tinta Piedras Blancas	Unidentified 10	1
	Plateá	Unidentified 11	1
Valle de los Pedroches	Tinta Amparo	Unidentified 01	10
	Huerta de los Leones	Unidentified 02	2
	Arises	Unidentified 03	4
	Falda de la Sierra	Unidentified 04	1
	Arroyo Lorito	Unidentified 05	1
	Autóctona Miguel	Unidentified 06	1
	Risquez	Unidentified 07	1
	Lagarreyes	Unidentified 12	2
Moguer	Jaén Negro	Unidentified 08	2

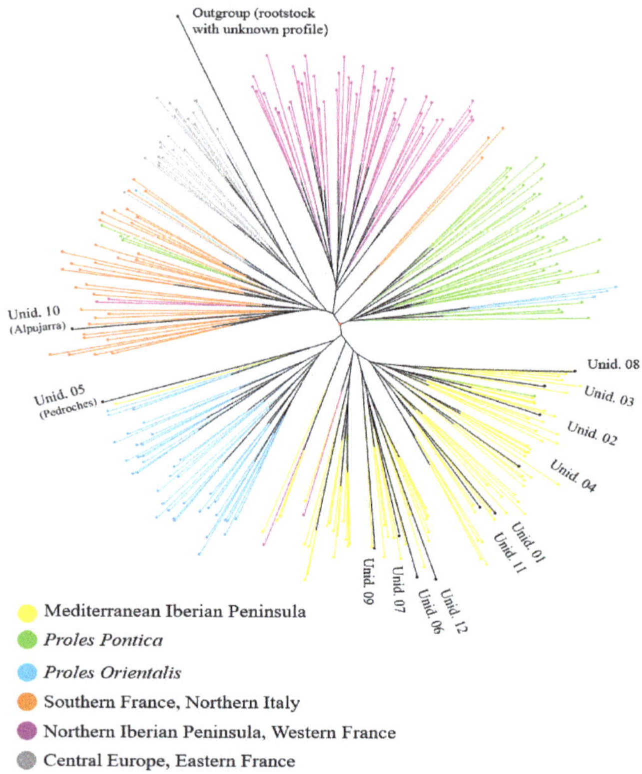

Figure 2. Phylogenetic tree obtained from the 13 SSR profiles of 271 varieties. The 12 unidentified genotypes, indicated with the abbreviation "unid.", detected in this study have been integrated into the set of varieties distributed by eco-geographic groups, according to Cretazzo et al. [15].

A third of the unidentified varieties ('Unidentified 01', 'Unidentified 04', 'Unidentified 06' and 'Unidentified 08') were assigned to a parent-pair with a trio LOD score above the strict critical threshold (14.04). Three unidentified cultivars ('Unidentified 02', 'Unidentified 03' and 'Unidentified 09') were assigned to a trio with a LOD between the critic and the relaxed threshold (10.19). The remaining unidentified cultivars were assigned to trios under the relaxed critical LOD ('Unidentified 07', 'Unidentified 10' and 'Unidentified 11') or showed no trios inferred from the parent-pair analysis ('Unidentified 05' and 'Unidentified 12'). For the trios with a LOD score above the strict critical threshold, the pedigree analysis showed that 'Garrido Fino' and 'Unidentified 07' were inferred as putative parents of 'Unidentified 01'. Furthermore, 'Montúa' and 'Jarrosuelto' would be the parents of 'Unidentified 06' For 'Unidentified 08' the best trio was formed with 'Roal' and 'Gabriela', showing the highest LOD score. Lastly, 'Mencía' and 'Tortosí' would be the parents of 'Unidentified 04'. For the trios with a LOD score between the critic and the relaxed threshold, we only considered the parent with zero pair loci mismatching. This analysis showed 'Montúa' as one of the parents of 'Unidentified 07', 'Gabriela' as one of the parent candidates of 'Unidentified 03', and 'Marfal' as one of the parents of 'Unidentified 11', which would be one of the parents of 'Unidentified 02' as well. Summary statistics obtained from the allele frequency analysis conducted in CERVUS 3.0 are provided in Appendix A, Table A1.

The virus incidence was very different among the surveyed areas (Appendix A, Table A2). Nevertheless, it was always possible to identify virus-free specimens in each area that could be useful as starting material in the process of registering the unregistered varieties in Spain. Additionally, it is also possible to recover potential certified local clones for the already registered varieties in case they are authorized in Andalusia under their main name or under some synonymy in the future.

4. Discussion

Most of the confirmed varieties identified in this study clustered in the Mediterranean-Iberian Peninsula group (Table 1), according to the origin proposed by VIVC and a previous phylogenetic study [15]. Furthermore, varieties 'Rojal Tinta', 'Listán del Condado', 'Alarije', and 'Cayetana Blanca' were identified in more than one of the areas prospected (Table 1). Likewise, ten out of the twelve unidentified varieties fit into the Mediterranean-Iberian Peninsula group (Table 2). Within the "Valle de los Pedroches", it is worth noting that the unidentified genotypes 06 and 07 share a common putative parent ('Montúa'), as well as 289 genotypes 01 (Amparo Tinta) and 07 showed a first-grade relationship (Figure 3). Likewise, we propose another direct relationship between specimens from two different areas (as in the case of Unidentified 11 from "La Alpujarra de Granada" and Unidentified 02 from "Valle de los Pedroches"), as well as a common putative parent ('Gabriela') that is shared by Unidentified 03 (Arises, "Valle de los Pedroches") and Unidentified 08 (Jaén Negro, "Moguer"). All that suggests a long history of natural hybridization, breeding, selection, human-mediated movements of seeds and cuttings, and other factors inside the Andalusian territory.

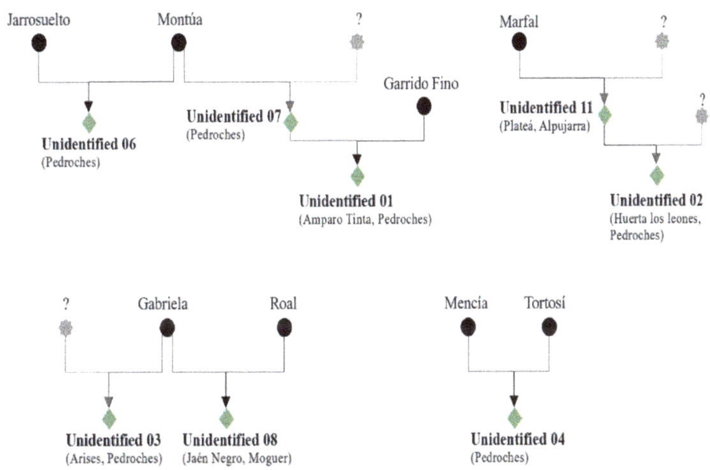

Figure 3. First-order kinship relationships were obtained among the unidentified varieties found in this study. Trios with a LOD score above the strict critical threshold and a minimum pair loci mismatching and parents with zero pair loci mismatching are represented. The character "?" means undetermined parent.

Next, we highlight the most relevant aspects of the results in each area studied:

a. "Altiplano de Granada". One specimen was shown to correspond to 'Rojal Tinta', a suitable variety for producing rosé wines that is not currently authorized in Andalusia but authorized in Castilla-La Mancha, an adjoining Community. Interestingly, two other prospected vines, each in both locations of La Alpujarra, corresponded to 'Rojal Tinta', suggesting the intentional cultivation of this variety in Eastern Andalusia. The microsatellite profile of 'Rojal Tinta' matched the profile described for the 'Rojal' variety in Castilla-La Mancha by Fernández-Gonzalez et al. [30]. The other specimen, white grape, showed an unidentified genotypic profile; thus, additional studies would be necessary to further characterize this variety. It is remarkable the low rate or absence of virus infections in the "Altiplano de Granada" area [7]. Possible explanations rely on the one hand, on the absence of massive introductions of propagation material from other origins due to the marginal viticulture in this area and, on the other hand, on the high altitude that prevents the presence of insect vectors of grapevine leafroll-associated viruses, mainly *Planococcus ficus* and *P. citri*. The results of the ELISA and RT-RT-qPCR tests of the two varieties surveyed in this study have confirmed the absence of virus infections. The presence of healthy phytosanitary material enables the possibility of certifying a local 'Rojal Tinta' clone in Andalusia.

b. "La Alpujarra de Granada". The presence of 'Rojal Tinta' in this area has already been highlighted. Regarding the other genotypes present in this area, there are two local varieties ('Rome' and 'Mantúo de Pilas') for which there is a strong interest among the winemakers to include them in the Register of Commercial Varieties in Spain, followed by authorization in Andalusia. 'Rome' has been already described in other areas of Andalusia, such as "La Axarquía" in Málaga province [31]. In addition, we identified one specimen of 'Airén', the most cultivated white variety in Spain; two other specimens corresponded to 'Jacquez', a direct producer hybrid with red berries (skin and pulp, therefore it is a dyera) and characterized by a high acidity in wines, a highly appreciated and sought-after quality in the indigenous red varieties of Andalusia, which usually lacks it; and two not previously identified genotypes, corresponding to a white grape vine and a red grape one. The red grape vine was shown to be virus-free. "La Alpujarra" is also characterized by vineyards at high altitude. However, unlike Altiplano, viticulture in La Alpujarra has historically been more intensive, and the exchange of plant materials has been frequent. This might explain why four of the vines showed simple or multiple virus infections (Appendix A, Table A2).

c. PDO "Montilla-Moriles" (Córdoba). A total of 31of the 32 vines prospected from eight different locations are named "Montepila" by local viticulturists. The literature regarding the term 'Montepila' is unclear, with several conjectures about the origin of this name (unpublished). Microsatellite analysis allowed us to clarify that it corresponded to 'Zalema', the main white variety grown in the Huelva province, usually known in the Córdoba province as "Torrontés". Moreover, 27 of the prospected specimens showed the "Zalema" genotype, three additional vines matched to 'Cayetana Blanca' (a common variety in several regions of Spain, known in the Córdoba province as "Baladí-Verdejo") and another vine matched to 'Pedro Ximénez'. Therefore, 'Montepila' is not an unidentified minority variety, although it is necessary to further determine by ampelography that it is not a somatic mutant. In a preliminary study by studying 11 basic grouping characters, we found no differences compared with 'Zalema'. 'Zalema' is authorized in Andalusia; however, the synonymy 'Montepila' is not yet considered; consequently, its authorization could be undertaken through a request to the MAPA, supported by its corresponding technical report. The last specimen studied, displaying black berries, corresponded to 'Negra Rayada', a Spanish variety not authorized in the RVCV. This variety has also been found in the Andalusian province of Almería [32]. In one of the eight locations, two vines showing the 'Zalema' genotype resulted virus-free, and, therefore, they could be plausibly proposed as certified clones once the synonymy 'Montepila' is authorized.

d. "Valle de los Pedroches" (Córdoba). Eight previously unidentified genotypes were identified in this area. Two of these genotypes were detected 10 times (in eight different locations) and four times (in four different locations), respectively. Although viticulture has practically disappeared from the Valle de los Pedroches, our survey suggests that there was some intentional cultivation of these two varieties, presumably in the pre-phylloxera period. They consist of red and white varieties that are currently experimentally vinified at IFAPA. According to the classification established by Muñoz-Organero et al. [29], 'Tinta Amparo' and 'Arises' would be novel autochthonous minority varieties. Two additional novel genotypes were found twice;

in both cases, they were collected in the same location, similarly to the other four unidentified genotypes that were found only once. Among the other varieties found in the area, there were two Spanish commercial varieties ('Alarije' and 'Cayetana Blanca'), a table variety ('Ahmeur bou Ahmeur'), a foreign variety ('Schiava Grossa'), and four known minority varieties ('Hebén', 'Jarrosuelto', 'Zurieles', and 'Negra Dorada'). 'Hebén' is a wine grape cultivar described since the 16th century [33]. It was traditionally grown in Andalusia [2], but currently it is residual in the provinces of Córdoba, Granada, Badajoz, Guadalajara, Toledo, and Cádiz [34,35]. There has been no interference through the introduction of foreign propagation material to the area in more than a century since the cultivation of these vineyards was almost abandoned. On the other hand, isolation and altitude might explain the absence of virus infections in the area. Hence, we identified healthy candidates for each variety, suitable as starting material to eventually proceed with the registration process and possible authorization in Andalusia, a strictly necessary condition for its permitted cultivation, even in plots for self-consumption.

e. "Pago Burujena" (Cádiz). There has been a growing interest in recent years in the recovery of varieties grown in the past and suitable for the production of fortified and still wines by the regulatory council of the PDOs "Jerez-Xérès-Sherry" and "Manzanilla-Sanlúcar de Barrameda" [1,36]. In Pago Burujena, one of the most traditional areas in the Jerez DO, clones of 'Mantúo Castellano' and 'Mantúo de Pilas' were apparently maintained in a vineyard for which three and two vines known by these denominations were respectively surveyed. In addition, two vines known as 'Barcelonés' were also collected. 'Mantúo Castellano' was shown to be a synonym of 'Listán del Condado', a variety commonly cultivated in the province of Huelva, while 'Mantúo de Pilas' and 'Barcelonés' was shown to correspond to 'Alarije'. For the former, it can be assumed that it was a naming error, while for the latter, it was a synonymy. In the case of 'Mantúo Castellano', it would be interesting to request the recognition of synonymy since this is the name by which it has traditionally been known in Jerez. In addition, one of the three plants was virus-free, so a certified clone could be derived from it.

f. "Moguer" (Huelva). In this municipality, we identified an specimen of the most cultivated variety in Spain, 'Airén'; two specimens of 'Beba', recently authorized in Andalusia, of significant interest for its cultivation in the Jerez area; a specimen of 'Mollar Cano'; three specimens of 'Listán del Condado'; a specimen of the table variety 'Alphonse Lavallée'; two genetically identical specimens of an unidentified variety locally called "Jaén Negro" (it would be a homonym, since 'Jaén Negro' also refers to a synonym of the 'Jaén Tinto' variety); and two specimens of 'Listán Prieto'. The last one is possibly the most significant result. In his book from 1513, Agricultura General Herrera [33] described the 'Uvas Prietas' variety cultivated in the center of the Iberian Peninsula, which could possibly correspond to 'Listán Prieto'. This variety was also explicitly mentioned in Andalusia in the year 1807 [1]. As a result of the America and the Canary Islands conquests in the 15th century, colonists from the Iberian Peninsula (Galician, Castillan, Andalusian, Extremaduran or even Portuguese) settled down in these lands, bringing grapevines together with their customs and crops [37]. Around 1550, a Jesuit introduced the 'Listán Prieto' variety in Peru [38], and cultivars were introduced in Mexico between the years 1520 and 1540 [39]. In California, it is known as Mission' [40] as an allusion to the vines carried by the friars in their evangelizing work. In the 19th century, phylloxera obliterated almost half of the peninsular vineyards [39], but this insect did not reach the Canary Islands. The use of this variety is widespread in the Canary Islands and in America [41,42]. There is no record of its cultivation on the Iberian Peninsula since the last century [40]. Therefore, the location of several specimens of this variety, some over 50 years old, in the township of Moguer has great historical relevance. 'Listán Prieto' usually produces wines with a high color intensity and good acidity, characteristics highly desired in Andalusian red wines, so its detection in these surveys can support the intention of authorizing it in Andalusia. We identified three virus-free vines in this zone, one belonging to 'Beba', another to 'Listán Prieto' and another to the unidentified "Jaén Negro". They could represent the starting point for obtaining locally certified clones of these varieties.

In this work, we have considered a vine as a certifiable clone only when both ELISA and RT-qPCR analyses tested negative for all viruses considered in the official certification schemes in addition to GLRaV-2. In the case of the unidentified varieties for which a registration process might be requested by viticulturists and winemakers, these same virus-free vines could be forwarded to the National Reference Collection of Grape Varieties for their evaluation. As mentioned above, MAPA requires these plants to be virus-free only through serological assays. In view of these results, we recommended requiring both ELISA and RT-qPCR (and additionally including GLRaV-2) in the analyses, as RT-qPCR is more sensitive than ELISA [43]. We have frequently observed along this study that some samples resulting in negative results in the serological assays tested positive in RT-qPCR (see Appendix A, Table A2, sections La Alpujarra de Granada, PDO Montilla-Moriles, Pago Burujena, and Moguer). Nonetheless, the intrinsic genetic variability of viruses prevents, in some cases, the efficient use of amplification techniques, which could result in false-negatives [44,45]. We provide here two examples for which a negative RT-qPCR corresponded to a positive ELISA (Mantúo de Sanlucar/Listán del Condado for GLRaV-3 and Marenas/Zalema for GLRaV-1) (see Appendix A, Table A2). Therefore, we suggest testing the vines before sending the material for registration in a first stage by serology, which is less expensive and more accessible, and in case of negative results, performing an RT-qPCR test. In conclusion, our results show that it is necessary to pursue research on minority grapevine varieties through prospecting, recovery, and conservation, as well as study their agronomic and oenological characteristics, since they represent a viticultural heritage at both regional and national levels. Finally, this work can be representative of similar studies carried out in other regions, mainly European, but not limited to those where viticulturists and winemakers are interested in recovering and diversifying their local varieties and wines.

Author Contributions: I.R.-T. drafted part of the manuscript and carried out the SSR analysis. A.M.C. drafted part of the manuscript and carried out the pedigree analysis and SSR analysis. M.d.P.R. took part in sample prospecting. F.J.G.G. carried out the pedigree analysis. L.V.A. took part in sampling and prospecting and carried out ELISA tests. C.P. took part in sample prospecting and carried out both ELISA and PCR tests. E.C. designed the study and supervised the manuscript redaction. All authors have read and agreed to the published version of the manuscript.

Funding: This research was financed by the "Conservation of Phytogenetic Resources of Vine. 443 PR.CRF.CRF202200.005" project, co-financed with FEADER funds, and the "Research and Technological Innovation in Viticulture. PP.AVA.AVA2019.016" project, co-financed with FEDER funds.

Acknowledgments: We appreciate the support of the Altiplano Rural Development Group of Granada, the Doñana Regional Agrarian Office, the Regulatory Council of the PDO Jerez-Xérès-Sherry and Manzanilla-Sanlúcar de Barrameda, and the collaboration of César Ortega García, Francisco Javier Domínguez, José Cabral Fernández and Francisco Antonio Torres Rosa. We also give thanks to Isidro Hita, Eliseo Salmerón and Miguel Alcaraz for their assistance in vineyard prospecting.

Conflicts of Interest: The authors declare no conflict of interest.

Abbreviations

The following abbreviations are used in this manuscript:

OIV	International Organisation of Vine and Wine
ELISA	Enzyme-Linked ImmunoSorbent Assay
PDO	Protected Designation of Origin
PGI	Protected Geographical Indications

IMIDA	Instituto Murciano de Investigación y Desarrollo Agrario y Medioambiental
MAPA	Ministerio de Agricultura, Pesca y Alimentación
CPVO	Community Plant Variety Office
UPOV	International Union for the Protection of New Varieties of Plants
GFLV	Grapevine Fanleaf Virus
GLRaV	Grapevine Leafroll associated Virus
GFkV	Grapevine Fleck Virus
DNA	Deoxyribonucleic Acid
SSR	Short Sequence Repeat
CREA-UTV	Consiglio per la ricerca in agricoltura e l'analisi dell'economia agraria
PCR	Polymerase Chain Reaction
VIVC	Vitis International Variety Catalog
IFAPA	Instituto Andaluz de Investigación y Formación Agraria, Pesquera y Alimentaria
RVCV	Registro de Variedades Comerciales de Vid
ICVG	International Council for the Study of Virus and Virus-Like Diseases of the Grapevine

Appendix A

Table A1. Summary statistics obtained from the allele frequency analysis conducted in CERVUS 3.0.

Locus	k	N	HObs	HExp	PIC	NE-1P	NE-2P	NE-PP	NE-I	NE-SI	HW	F (Null)
md7	14	541	0.797	0.803	0.779	0.551	0.375	0.189	0.063	0.364	NS	0.0046
md32	11	540	0.863	0.820	0.797	0.527	0.353	0.173	0.055	0.354	NS	−0.0285
zag62	10	541	0.824	0.807	0.782	0.548	0.371	0.187	0.061	0.362	*	−0.0138
zag79	13	539	0.818	0.854	0.838	0.448	0.287	0.118	0.037	0.332	NS	0.0208
eva2	19	540	0.854	0.865	0.851	0.420	0.264	0.100	0.031	0.326	NS	0.0060
isv2	23	541	0.884	0.851	0.833	0.464	0.300	0.132	0.040	0.335	NS	−0.0212
vvs2	15	541	0.850	0.842	0.824	0.477	0.311	0.137	0.043	0.340	**	−0.0042
md5	11	541	0.854	0.859	0.842	0.449	0.286	0.122	0.036	0.330	NS	0.0022
md27	9	541	0.848	0.829	0.804	0.520	0.347	0.173	0.053	0.349	NS	−0.0122
md25	14	540	0.785	0.767	0.728	0.632	0.454	0.273	0.093	0.390	NS	−0.0118
md28	19	540	0.852	0.870	0.856	0.417	0.262	0.103	0.031	0.323	NS	0.0107
isv4	11	539	0.844	0.814	0.788	0.544	0.368	0.189	0.060	0.358	*	−0.0210
isv3	10	540	0.830	0.673	0.613	0.750	0.590	0.418	0.167	0.456	***	−0.1167

Significance of Hardy-Weinberg equilibrium test (HW): *** = $p < 0.001$; ** = $p < 0.01$; * = $p < 0.05$; NS = $p > 0.05$

Table A2. Viruses detected in each studied area.

Zone	Input Name	Variety	ELISA	PCR
Altiplano de Granada	Rosada Hornico	Rojal Tinta	-	-
	Blanca Hornico	Unidentified 09	-	-
La Alpujarra de Granada	Rome	Rome	-	-
	Mollar Cano	Rojal Tinta	-	GFkV
	Tinta	Rojal Tinta	-	-
	Tinta Cortijo La Paz	Jacquez	-	-
	Llaqui	Jacquez	-	-
	Ricardera	Mantúo de Pilas	-	GfkV, GLRaV-3
	Desconocida Blanca	Airén	GFLV, GLRaV-2	GFLV, GfkV, GLRaV-2
	Tinta Piedras Blancas	Unidentified 10	-	-
	Plateá	Unidentified 11	GFLV	GFLV, GFkV

Table A2. *Cont.*

Zone	Input Name	Variety	ELISA	PCR
PDO Montilla-Moriles	Peñalista	Negra Rayada	m.d.	m.d.
	Cp1 Marenas fila 3 cp5	Zalema	GFkV	GFkV
	Cp2 Marenas fila 4 cp18	Zalema	GFkV, GLRaV-3	GFkV, GLRaV-3
	Marenas fila 7 cp3 cp44	Zalema	GFkV, GLRaV-3	GFkV, GLRaV-2, GLRaV-3
	Cp 4Marenas fila 8 cp16	Zalema	GFkV, GLRaV-1	GFkV
	Cp5 Marenas fila 9 cp20	Zalema	GFkV	GFkV
	Cp9 La Plata fila 7	Zalema	m.d.	m.d.
	La Plata fila 13 p7	Pedro Ximénez	-	GFkV, GLRaV-2
	Cp6 La Plata fila 17	Zalema	GFkV	GFkV, GLRaV-5
	Cp3 La Plata fila 19	Zalema	-	GLRaV-3, GLRaV-5
	Los Rosales 1	Zalema	GFkV	GFkV
	Los Rosales 2	Zalema	-	GFkV, GLRaV-2
	Los Rosales 3	Zalema	-	GFkV, GLRaV-3
	Los Rosales 4	Zalema	GFkV	GFkV, GLRaV-3
	Colección Montepila 1	Zalema	-	GFkV
	Colección Montepila 2	Zalema	-	GFkV
	Montepila 1 recinto 16	Zalema	GFkV	GFkV
	Montepila 2 (recinto 16)	Zalema	-	GFkV
	Montepila 3 recinto 16	Zalema	GFkV, GLRaV-3	GFkV, GLRaV-2, GLRaV-3
	Montepila 4 recinto 16	Zalema	GLRaV-3	GFkV, GLRaV-3
	Montepila los Naranjos 1	Zalema	m.d.	m.d.
	Montepila (2 linde) los Naranjos	Zalema	m.d.	m.d.
	Montepila los Naranjos 3	Zalema	m.d.	m.d.
	Montepila 4	Zalema	m.d.	m.d.
	Montepila los Naranjos 5	Cayetana Blanca	m.d.	m.d.
	La Primilla 1	Zalema	-	-
	La Primilla 2	Cayetana Blanca	m.d.	m.d.
	La Primilla 3	Zalema	-	-
	La Primilla 4	Cayetana Blanca	m.d.	m.d.
	Cuesta Blanca 1	Zalema	m.d.	m.d.
	Cuesta Blanca 2	Zalema	m.d.	m.d.
	Cuesta Blanca 3	Zalema	m.d.	m.d.
Valle de los Pedroches	Risquez 1	Ahmeur bou Ahmeur	-	-
	Risquez 2	Ahmeur bou Ahmeur	-	-
	Risquez 3	Unidentified 07	-	-
	Risquez 4	Cayetana Blanca	-	-
	Vieja Primera	Cayetana Blanca	-	-
	Vieja Primera	Cayetana Blanca	-	-
	Merino 1	Cayetana Blanca	-	-
	Merino 2	Cayetana Blanca	-	-
	Lindero C1	Unidentified 01	-	-
	La Torre Amparo Tinta	Unidentified 01	-	-
	Camino falda de la sierra	Unidentified 01	-	-
	Recio 1° Amparo Tinta	Unidentified 01	-	-
	Recio 2° Amparo TInta	Unidentified 01	-	-
	Recio 3° Amparo TInta	Unidentified 01	-	-
	Recio 4° Amparo TInta	Unidentified 01	-	-
	Recio 5° Amparo TInta	Unidentified 01	-	-
	Tinta Amparo malla Rafael	Unidentified 01	-	-

Table A2. *Cont.*

Zone	Input Name	Variety	ELISA	PCR
	Tinta Amparo huerta Rafael	Unidentified 01	-	-
	Huerta Los Leones C5	Unidentified 02	-	-
	Huerta Los Leones C2	Unidentified 02	-	-
	Isleta (Arises)	Unidentified 03	-	-
	Garrido (Arises)	Unidentified 03	-	-
	Arises cuadra Rafael	Unidentified 03	-	-
	Arises huerta Rafael	Unidentified 03	-	-
	Camino falda de la sierra (Portillo)	Unidentified 04	-	-
	Arroyo Lorito	Unidentified 05	-	-
	Autóctona Miguel	Unidentified 06	-	-
	Lagareyes	Unidentified 12	-	-
	Lagareyes	Unidentified 12	-	-
	Blanca Lagareyes	Alarije	-	-
	Villaharta llanos suelo	Alarije	-	-
	Tío Kiko Camino	Negra Dorada	-	-
	Hebén	Hebén	-	-
	Jarrosuelto	Jarrosuelto	-	-
	Schiava Grossa	Schiava Grossa	-	-
	Entreárboles	Zurieles	-	-
Pago Burujena	Mantúo Castellano	Listán del Condado	GFkV, GLRaV-3	GFkV, GLRaV-2, GLRaV-3
	Mantúo Castellano	Listán del Condado	GFkV, GFLV, GLRaV-3	GFkV, GFLV, GLRaV-2, GLRaV-3
	Mantúo Castellano	Listán del Condado	-	-
	Mantúo de Pilas	Alarije	GFkV	GFkV
	Mantúo de Pilas	Alarije	GFkV	GFkV
	Barcelonés	Alarije	GFkV	GLRaV-2, GLRaV-3, GFkV
	Barcelonés	Alarije	GFkV	GLRaV-2, GLRaV-3, GFkV
Moguer	Mollar Cano	Mollar Cano	GLRaV-3	GLRaV-3
	Listán Prieto	Beba	-	GFkV, GLRaV-3
	Beba	Beba	-	-
	Moguer	Airén	-	-
	Moguer	Listán Prieto	-	-
	Moguer	Listán Prieto	-	GLRaV-3
	Mantúo de Sanlúcar	Listán del Condado	-	GLRaV-3
	Mantúo de Sanlúcar	Listán del Condado	GLRaV-3	-
	Mantúo de Sanlúcar	Listán del Condado	-	GLRaV-3
	Mesa Plaza Tinta	Alphonse Lavallée	-	GLRaV-3
	Jaén Negro	Unidentified 08	GFkV. GFLV	GFkV. GFLV
	Jaén Negro	Unidentified 08	-	-

Virus-free vines are indicated in red.

References

1. Clemente y Rubio, S.R. *Ensayo Sobre las Variedades de la vid Común que Vegetan en Andalucía: Con un índice Etimológico y Tres Listas de Plantas en que se Ca-Racterizan Varias Especies Nuevas*; Imprenta de Villalpando: Madrid, Spain, 1807; 325p.
2. García de los Salmones, N. Memoria general de las sesiones del Congreso y ponencias presentadas. In Proceedings of the Congreso Nacional de Viticultura, Pamplona, Spain, 11–22 July 1912.
3. Rodríguez-Torres, I.; Márquez, M.I.; Ramírez, M.P.; Cretazzo, E. Caracterización molecular e identificación de variedades minoritarias de vid de cultivo tradicional en Andalucía. In Proceedings of the Congreso Nacional de Ciencias Hortícolas, Córdoba, Spain, 17–21 October 2021.
4. OIV. *Distribution of the World's Grapevine Varieties*; OIV: Paris, France, 2017; p. 54. Available online: https://www.oiv.int/public/me4d66ias/5distribution-of-the-worlds-grapevine-varieties.pdf (accessed on 1 May 2023).

5. Cretazzo, E.; Tomás, M.; Padilla, C.; Roselló, J.; Medrano, H.; Padilla, V.; Cifre, J. Incidence of virus infection in old vineyards of local grapevine varieties from Majorca: Implications for clonal selection strategies. *Span. J. Agric. Res.* **2010**, *8*, 409–418. [CrossRef]
6. Espinosa Roldán, F.E.; Muñoz Organero, G.; Uscola Fernández, M.; de Santa María, F.C.S.; De Toda, F.M. Minority grapevine varieties as climate change adaptation strategy: Exploring heat tolerance plasticity. In Proceedings of the 43rd World Congress of Vine and Wine, Ensenada, Mexico, 31 October–4 November 2022.
7. Egea Bartual, C.; León Gutiérrez, J.M.; Lasheras Ocaña, J.; Ramírez Pérez, P. Estudio de las Variedades Locales y Foráneas Cultivadas en la zona del Altiplano de Granada; SERVIFAPA: 2016; 36p. Available online: https://ifapa.junta-andalucia.es/agriculturaypesca/ifapa/ser4v7i4fapaservifapa/19f319c3-6a05-4014-85ee-f5eecb197429 (accessed on 1 May 2023).
8. Cabral Fernández, J. *La viña y el Vino en la Edad Moderna a Través de la Casa Ducal de Medina Sidonia*; Santa Teresa Industrias Gráficas SA: Trebujena, Spain, 2021; 192p.
9. Avramidou, E.V.; Masaoutis, I.; Pitsoli, T.D.; Kapazoglou, A.; Pikraki, M.; Trantas, E.A.; Markantonakis, M.; Doulis, A.G. Analysis of Wine-Producing Vitis vinifera L. Biotypes, Autochthonous to Crete (Greece), Employing Ampelographic and Microsatellite Markers. *Life* **2023**, *13*, 220. [CrossRef] [PubMed]
10. Thomas, M.R.; Scott, N.S. Microsatellite repeats in grapevine reveal DNA polymorphisms when analysed as sequence-tagged sites (STSs). *Theor. Appl. Genet.* **1993**, *86*, 985–990. [CrossRef] [PubMed]
11. Bowers, J.E.; Dangl, G.S.; Vignani, R.; Meredith, C.P. Isolation and characterization of new polymorphic simple sequence repeat loci in grape (*Vitis vinifera* L.). *Genome* **1996**, *39*, 628–633. [CrossRef] [PubMed]
12. Bowers, J.E.; Meredith, C.P. The parentage of a classic wine grape, Cabernet Sauvignon. *Nat. Genet.* **1997**, *16*, 84–87. [CrossRef] [PubMed]
13. Bergamini, C.; Caputo, A.R.; Gasparro, M.; Perniola, R.; Cardone, M.F.; Antonacci, D. Evidences for an alternative genealogy of 'Sangiovese'. *Mol. Biotechnol.* **2013**, *53*, 278–288. [CrossRef]
14. Maul, E.; Sudharma, K.N.; Kecke, S.; Marx, G.; Müller, C.; Audeguin, L.; Boselli, M.; Boursiquot, J.M.; Bucchetti, B.; Cabello, F.; et al. The European Vitis Database (www.eu-vitis.de): A technical innovation through an online uploading and interactive modification system. *Vitis* **2012**, *51*, 79–85.
15. Cretazzo, E.; Moreno Sanz, P.; Lorenzi, S.; Benítez, M.L.; Velasco, L.; Emanuelli, F. Genetic characterization by SSR markers of a comprehensive wine grape collection conserved at Rancho de la Merced (Andalusia, Spain). *Plants* **2022**, *11*, 1088. [CrossRef]
16. Negrul, A.M. Evolution of cultivated forms of grapes. *CR Acad. Sci. URSS* **1938**, *18*, 585–588.
17. Perrier, X.; Jacquemoud-Collet, J.P. DARwin Software. 2006. Available online: http://darwin.cirad.fr/ (accessed on 1 May 2023).
18. Kalinowski, S.T.; Taper, M.L.; Marshall, T.C. Revising how the computer program CERVUS accommodates genotyping error increases success in paternity assignment. *Mol. Ecol.* **2007**, *16*, 1099–1106. [CrossRef]
19. Lassois, L.; Denancé, C.; Ravon, E.; Guyader, A.; Guisnel, R.; Hibrand-Saint-Oyant, L.; Poncet, C.; Lasserre-Zuber, P.; Feugey, L.; Durel, C. Genetic Diversity, Population Structure, Parentage Analysis, and Construction of Core Collections in the French Apple Germplasm Based on SSR Markers. *Plant Mol. Biol. Report.* **2016**, *34*, 827–844. [CrossRef]
20. Khadari, B.; Bakkali, A.E.; Essalouh, L.; Tollon, C.; Pinatel, C.; Besnard, G. Cultivated Olive Diversification at Local and Regional Scales: Evidence from the Genetic Characterization of French Genetic Resources. *Front. Plant Sci.* **2019**, *10*, 1593. [CrossRef]
21. Weng, Z.; Yang, Y.; Wang, X.; Wu, L.; Hua, S.; Zhang, H.; Meng, Z. Parentage analysis in giant grouper (Epinephelus lanceolatus) using microsatellite and SNP markers from genotyping-by-sequencing data. *Genes* **2021**, *12*, 1042. [CrossRef]
22. Martelli, G.P. Grapevine virology highlights 2004–2005. In Proceedings of the 15th ICVG Conference, Stellenbosch, South Africa, 3–7 April 2006.
23. Padilla, V.; Hita, I.; García de la Rosa, S.B.; Padilla, C.V.; Salmerón, E.; Cretazzo, E. Virosis de la vid.Situación por comunidades autónomas. *Vitic. Enol. Prof.* **2007**, *113*, 6–12.
24. Sánchez-Vizcaino, J.M.; Cambra, M. *Técnicas Inmunoenzimáaticas en Patología Animal y Vegetal*; Monografía INIA 29; INIA: Madrid, Spain, 1981; 57p.
25. Boscia, D.; Digiaro, M.; Fresno, J.; Greif, C.; Grenan, S.; Kassemeyer, H.H.; Prota, V.A.; De Sequueira, O.A. ELISA for the detection and identification of grapevine viruses. In *Sanitary Selection of the Grapevine*; Walter, B., Ed.; INRA Editions: Colmar, France, 1997; Volume 86, pp. 129–155.
26. Osman, F.; Leutenegger, C.; Golino, D.; Rowhani, A. Real-time RT-PCR (TaqMan®) assays for the detection of Grapevine Leafroll associated viruses 1-5 and 9. *J. Virol. Methods* **2007**, *141*, 22–29. [CrossRef]
27. Osman, F.; Leutenegger, C.; Golino, D.; Rowhani, A. Comparison of low-density arrays, RT-PCR and real-time TaqMan RT-PCR in detection of grapevine viruses. *J. Virol. Methods* **2008**, *149*, 292–299. [CrossRef]
28. Cepin, U.; Gutiérrez-Aguirre, I.; Balažic, L.; Pompe-Novak, M.; Gruden, K.; Ravnikar, M. A one-step reverse transcription time PCR assay for the detection and quantitation of Grapevine fanleaf virus. *J. Virol. Methods* **2010**, *170*, 47–56. [CrossRef]
29. Muñoz Organero, G.; De Andrés, M.T.; Vargas, A.; Aller, M.; Serrano, M.J.; Cretazzo, E.; Pérez, J.A.; Puertas, M.B.; Gogorcena, Y.; Giménez, R.; et al. La enorme diversidad varietal de vid en España, en proceso de descubrimiento. *ACE Rev. Enol.* **2017**, *170*, ISSN-e 1697-412.
30. Fernández-González, M.; Mena, A.; Izquierdo, P.; Martínez, J. Genetic characterization of grapevine (*Vitis vinifera* L.) cultivars from Castilla La Mancha (Spain) using microsatellite markers. *Vitis* **2007**, *46*, 126–130.
31. Jiménez-Cantizano, A.; Muñoz-Martín, A.; Amores-Arrocha, A.; Sancho-Galán, P.; Palacios, V. Identification of Red Grapevine Cultivars (*Vitis vinifera* L.) Preserved in Ancient Vineyards in Axarquia (Andalusia, Spain). *Plants* **2020**, *9*, 1572. [CrossRef]

32. Tello, J.; Galán, Á.; Rodríguez-Torres, I.; Martínez Zapater, J.M.; Rubio, A.; Ibáñez, J. Prospección, identificación genética y conservación de parras antiguas de Almería. In Proceedings of the IV Jornada del Grupo de Viticultura, Sociedad Española de Ciencias Hortícolas, Pamplona, Spain, 26–28 October 2022.
33. De Herrera, A.G. *Agricultura General*; Corrected and Edited by the Real Sociedad Económica Matritense; Imprenta Real: Madrid, Spain, 1818–1819; 2026p.
34. Cabello, F.; Ortiz, J.M.; Muñoz, G.; Rodríguez-Torres, I.; Benito, A.; Rubio, C.; García, S.; Sáiz, R.; De Andrés, M.T. *Variedades de vid en España*; Comunidad de Madrid y Editorial Agrícola Española: Madrid, Spain, 2019; 519p.
35. Zinelabidine, L.H.; Cunha, J.; Eiras-Dias, J.E.; Cabello, F.; Martínez-Zapater, J.M.; Ibáñez, J. Pedigree analysis of the Spanish grapevine cultivar 'Hebén'. *Vitis* **2015**, *54*, 81–86.
36. Fernández de Bobadilla, G. *Viníferas Jerezanas y de Andalucía Occidental*; Instituto Nacional de Investigaciones Agronómicas: Madrid, Spain, 1956; 271p.
37. Rodríguez-Torres, I. Patrimonio Vitícola de Canarias. In Proceedings of the VII Semana Científica Telesforo Bravo, Instituto de Estudios Hispánicos de Canarias, Puerto de la Cruz, Spain, 21–25 November 2011.
38. Zerolo, J.; Cabello, F.; Espino, A.; Borrego, J.; Ibañez, J.; Rodriguez-Torres, I.; Muñoz-Organero, G.; Rubio, C.; Hernández, M. *Variedades de Vid de Cultivo Tradicional en Canarias*; Instituto Canario de Calidad Agroalimentaria, Gobierno de Canarias: Santa Cruz de Tenerife, Spain, 2006; 224p.
39. Hidalgo, L. *Tratado de Viticultura General*; Mundiprensa: Madrid, Spain, 2002; 1235p.
40. Milla, A.; Cabezas, J.A.; Cabello, F.; Lacombe, T.; Martínez-Zapater, J.M.; Hinrichsen, P.; Cervera, M.T. Determining the Spanish origin of representative ancient American grapevine varieties. *Am. J. Enol. Vitic.* **2007**, *58*, 242–251.
41. Rodríguez-Torres, I. *Variedades de vid Cultivadas en Canarias: Descriptores Morfológicos: Caracterización Morfológica, Molecular, Agronómica y Enológica*; Instituto Canario de Investigaciones Agrarias: Santa Cruz de Tenerife, Spain, 2017; 200p.
42. Lacoste, P. La variedad de uva País (Listán Prieto) en el Cono Sur de América: Trayectoria histórica. *Idesia* **2021**, *39*, 75–84. [CrossRef]
43. Osman, F.; Rowhani, A. Real-time RT-PCR (TaqMan®) assays for the detection of viruses associated with rugose wood complex of grapevine. *J. Virol. Methods* **2008**, *154*, 69–75. [CrossRef]
44. Angelini, E.; Bertazzon, N.; Borgo, M. Diversity among Grapevine leafroll-associated virus 2 isolates detected by heteroduplex mobility assay. *J. Phytopathol.* **2004**, *152*, 416–422. [CrossRef]
45. Bertazzon, N.; Bazzo, I.; Repetto, O.; Borgo, M.; Angelini, E. Advances in molecular detection of GLRaV-1. In Proceedings of the 16th Meeting of the ICVG, Dijon, France, 31 August–4 September 2009.

horticulturae

MDPI

Article

Identification of *Vitis vinifera* L. Local Cultivars Recovered in Andalusia (Spain) by Using Microsatellite Markers

Ana Jiménez-Cantizano [1,*], Anna Puig-Pujol [2] and Rosa Arroyo-García [3]

1 Department of Chemical Engineering and Food Technology, Vegetal Production Area, University of Cadiz, Agrifood Campus of International Excellence (CeiA3), IVAGRO, P.O. Box 40, 11510 Puerto Real, Spain
2 IRTA-INCAVI Institute of Agrifood Research and Technology-Catalan Institute of Vine and Wine, Department of Enological Research, Plaça Àgora 2, 08720 Vilafranca del Penedès, Spain
3 CSIC-INIA (CBGP) Centro de Biotecnología y Genómica de Plantas, UPM-INIA, Parque Científico y Tecnológico de la UPM Campus de Montegancedo, 28223 Madrid, Spain
* Correspondence: ana.jimenezcantizano@uca.es; Tel.: +34-9-5601-6470

Abstract: In Andalusia (Spain), there are different wine regions that have a great recognized tradition. In these regions, the cultivation of the vine is ancient and there are still vineyards planted with local varieties of *Vitis vinifera* L. that have not yet been identified. The aim of this research study was to identify 49 accessions of grapevine collected in the districts of four provinces in Andalusia (Spain). All samples were genotyped with 20 microsatellite markers in order to ascertain the identity and analyze the genetic diversity of the collected material. In total, 30 different genotypes were obtained, 22 of them which were identified with named, known varieties by comparison to the Spanish or European microsatellite databases, and eight which are referred to as new genotypes. All loci were polymorphic, and a total of 159 alleles were detected, ranging from 4 to 12 alleles per locus, with an average allele number of 7.95. The overall observed heterozygosity was 0.763 and was slightly higher than expected (0.715), while the gene diversity per locus varied between 0.167 (VVIN73) and 0.967 (VVMD5). A dendrogram representing the genetic similarities among cultivars was depicted using the UPGMA method to investigate their relationships. The eight new genotypes identified in this research work could represent ancient local varieties in danger of extinction. These new cultivars may be used to determine original wines.

Keywords: grapevine; genetic characterization; synonymies

Citation: Jiménez-Cantizano, A.; Puig-Pujol, A.; Arroyo-García, R. Identification of *Vitis vinifera* L. Local Cultivars Recovered in Andalusia (Spain) by Using Microsatellite Markers. *Horticulturae* **2023**, *9*, 316. https://doi.org/10.3390/horticulturae9030316

Academic Editor: Paolo Sabbatini

Received: 1 February 2023
Revised: 22 February 2023
Accepted: 23 February 2023
Published: 28 February 2023

1. Introduction

The Andalusia (Spain) region, in the south of the Iberian Peninsula, is one of the most ancient and important wine regions in Spain [1]. Archaeological, paleobotanical, and historical sources confirm that grapevines were spread and cultivated for a long time in this area. The presence of the species *Vitis vinifera* L. has been verified by pollen analysis performed in different Phoenician sites of Andalusia located in the provinces of Cádiz, Málaga and Almería [2]. In addition, numerous archaeological remains have been found at these sites, which may be associated with the existence of a wine industry [3–5]. The first evidence of planting techniques characteristic of protohistoric viticulture in the west has been documented in an archaeological site located in Huelva (Andalusia) dating back to the 1st millennium BC [6].

There are many citations that reference the diversity of grapevine (*Vitis vinifera* L.) varieties grown in Andalusia. Roxas Clemente [7], in his paper *Essay of common grapevine varieties that are growing in Andalusia*, includes 119 varieties grouped in two sections and 15 tribes. In 1831, James Busby, considered the "father of Australian viticulture" introduced 678 varieties in Australia [8]. These varieties originated in France and Spain. According to Morilla Critz [8], at least half of these varieties were from Andalusia.

Nevertheless, the genetic diversity of the Andalusia grapevine (*Vitis vinifera* L.) has been declining due to the phylloxera (*Daktulosphaira vitifoliae*) attack of the late 19th century [9], when severe regulations were approved, and the grapevine varieties authorized for wine production were restricted and the vineyard was restructured, frequently stimulated by subsidies. In Spain, previous to this vineyard restructuring, which begin in the 1970s, all vines were grafted in the field with mass-selected *Vitis vinifera* material from older vineyards, which often included different varieties [10]. With the aim of preserving grapevine phytogenetic resources, numerous studies on the surveying, localization, characterization, and maintaining of cultivars in germplasm banks are being carried out worldwide [11–20]. In Andalusia, a germplasm bank was established in 1940, and it was replanted between 1984 and 1987, and the number of accessions substantially increased [21]. Actually, this collection preserves 1417 accessions according to the *Vitis* International Variety Catalogue (VIVC, www.vivc.de accessed on 26 December 2022) [22].

The recovery of autochthonous or local varieties allows a genetic, ecological and agronomic enrichment capable of dealing with various diseases, improving the adaptation to edaphoclimatic conditions [23] or facilitating the adaptation in the face of future market changes [24]. For this reason, the accurate identification of local cultivars and their conservation could prevent their disappearance and preserve them for future needs. Traditionally, the identification of grape varieties has been based on the morphological features of vegetative and reproductive structures [25], but phenotypic traits are not sufficiently reliable for the classification of closely related varieties due to genotype–environment interactions [26]. Therefore, molecular characterization is the favoured technique for varietal identification. At the present time, there are different molecular markers available to carry out a molecular identification of a grape variety. However, microsatellites or Simple Sequence Repeats (SSRs) markers are the most used for this purpose [27,28]. In this sense, microsatellite markers have been widely used to identify and genotype grapevine cultivars collected in old vineyards of the Iberian Peninsula [29–31]. In addition, SSRs have been used for studies of genetic diversity and genetic relationships [32].

The main objective of this research work is focused on the molecular identification of a total of 49 vine accessions collected in old Andalusian vineyards. The genotyping of these accessions could help to detect new local cultivars growing in Andalusia aiming to provide a solid basis to develop a regional germplasm collection to protect local biodiversity.

2. Materials and Methods

2.1. Plant Material

After prospecting more than 200 vineyards throughout the provinces of Almería, Cádiz, and Huelva y Málaga of the Andalusia region (Spain), those plants that were not visually identified as common varieties cultivated in Andalusia were sampled and placed in the germplasm bank at the Rancho de la Merced. This grapevine collection is located in Jerez de la Frontera (Cádiz, Spain) (36°41′10″ N; 6°08′10″ W; alt. 20 m). The list of the 49 accessions used in this study are shown in the Supplemental Table S1. Each accession was identified with a code of three letters and a number. The initials correspond to the name of the municipality where it was collected.

Two internationally known cultivars ('Cabernet Sauvignon' and 'Syrah') were also included to compare the genetic profiles obtained with the different published databases.

2.2. Molecular Analysis

DNA was extracted from young leaves collected from each accession and stored at −80 °C, using a DNeasy Plant Mini Kit (Qiagen, Hilden, Germany). A genotypic characterization was performed for 20 nuclear microsatellite loci located in the 19 linkage groups of grapevine genome VMC1b11, VMC4F3-1 (*Vitis* Microsatellite Consortium); VVMD5, VVMD7, VVMD21, VVMD24, VVMD25, VVMD27, VVMD28, VVMD32 [33,34]; VVS2 [35]; VVIB01, VVIH54, VVIN16, VVIN73, VVIP31, VVIP60, VVIQ52, VVIV37, and VVIV67 [36]. Two multiplex PCR tests were set up to amplify the 20 microsatellite loci in a 20 μL reaction mix according to Vargas et al. [37]. PCR reactions were carried out in the 44 Applied Biosystems 9700 thermocycler.

Amplified products were separated by capillary electrophoresis using an automated sequencer (ABI Prism 3130, Applied Biosystems, Foster City, CA, USA). Fluorescently labelled fragments were detected and sized using GeneMapper v. 3.7 software (Applied Biosystems), and fragment lengths were determined with the help of internal size standards (GeneScan-500 LIZTM, Applied Biosystems, Foster City, CA, USA).

The identification of redundant genotypes was determined by comparing microsatellite genotypes with data contained in the Spanish microsatellite grapevine databases Rancho de la Merced [38–40] and the *Vitis* Germplasm Bank (BGV) at the Finca El Encín (IMIDRA, Alcalá de Henares, Spain) [41–43] and other European databases [22,44]. Genotype comparisons were carried out using the Microsatellite toolkit v. 9.0 software package [45].

2.3. Data Analysis

2.3.1. Genetic Diversity Analyses

For the calculation of the number of alleles (Na), expected (He) and observed (Ho) heterozygosity, frequency of null alleles (r) and probability of identity (PI), the GENALEX software [46] was used. The polymorphism information content (PIC) of each microsatellite loci was determined using an online tool [47].

2.3.2. Genetic Relationships among Cultivars

Genetic distances between grapevine genotypes were calculated as [-ln (proportion shared alleles)] using Microsat [48]. The obtained data was used for the construction of a dendrogram using the programs EXE from the PHYLIP package software [49] and MEGA version 7 [50].

3. Results and Discussion

3.1. Microsatellite Analysis and Genetic Diversity

The molecular analysis performed with the 49 studied accessions resulted in 30 non-redundant genotypes (Table 1). These genotypes were used for the calculation of genetic parameters (Table 2) in order avoid overestimation. A total of 159 alleles, ranging from 12 in VVMD7 and four in VVIN73, were detected, with an average of eight alleles per locus, similar to the mean Na attained by Fernández-González et al. [51]. The most frequent allele was VVIN73-264, which showed a frequency up to 90%, and 27 alleles were unique.

The expected heterozygosity (He, gene diversity) ranged from 0.185 at locus VVIN73 to 0.866 at locus VVIP31, with a mean value of 0.715. The observed heterozygosity (Ho) varied between 0.167 at locus VVIN73 and 0.967 at locus VVMD5. For 16 loci, Ho was higher than He, and the probability of null alleles was always negative, except for VMC4F31, VVMD21, VVMD25, VVMD28, VVIN73 and VVIP60. Samples in which only one single allele per locus was detected were considered as homozygous genotypes instead of heterozygous with a null allele. The VVIN73 and VVIP31 markers displayed the minimum (0.1769) and maximum (0.8522) *PIC* values, respectively. The 20 microsatellite loci showed a mean *PIC* value of 0.67241.

The 20 microsatellite loci used reflected a high discrimination power and a low probability that two randomly chosen individuals had identical genotypes using the 20 loci (PI. 1.74×10^{-19}). This indicates the probability that two of the 30 varieties analyzed randomly were chosen to share the same genotype using the set of these 20 microsatellite loci.

Table 1. Thirty genotypes obtained for the 49 analyzed accessions at 20 microsatellite loci. Allele sizes are given in base pairs.

Cultivars [a]	Accession Code	VVIB01		VMC1b11		VMC4F31		VVMD5		VVMD7		VVMD21		VVMD24		VVMD25		VVMD27		VVMD28		VVMD32		VVIH54		VVIN16		VVIN73		VVIP31		VVIP60		VVIQ52		VVS2		VVIV37		VVIV67	
AFUS ALI	Man-4	291	295	184	184	168	190	224	228	236	246	255	255	209	209	246	252	186	186	260	260	256	270	166	178	151	151	264	264	174	184	318	318	83	85	131	133	151	163	358	362
AHMEUR BOU AHMEUR	Lau-5, Lau-6	291	295	184	184	190	190	228	236	236	246	249	255	209	209	252	264	184	194	250	256	250	254	166	168	151	153	264	264	184	192	322	322	83	89	133	146	161	161	358	366
ALICANTE HENRI BOUSCHET	Lau-8	291	295	182	188	174	206	224	236	236	240	243	249	209	211	240	240	182	194	246	262	248	270	164	168	151	159	256	264	176	184	322	322	83	89	131	144	161	171	358	364
ATTIKA SEEDLESS	Pla-2	291	295	166	184	184	206	236	238	250	252	249	249	213	217	246	252	186	194	246	246	254	270	164	166	151	157	256	264	186	188	320	326	85	85	133	151	163	181	372	379
BEBA	Chu-2	291	295	184	188	188	188	234	238	240	246	249	255	209	211	252	252	182	190	246	260	254	270	164	166	151	153	256	264	190	192	318	322	83	85	133	142	161	163	366	372
BOBAL	Ron-1	291	295	184	188	174	184	226	232	236	240	243	243	209	211	240	264	182	190	236	262	248	270	166	168	151	153	264	264	176	186	326	326	83	85	144	146	159	163	358	362
CARIGNAN NOIR	Lau-17	291	295	174	184	180	188	224	226	236	236	249	253	209	215	240	252	182	186	250	260	248	250	166	166	151	153	264	264	176	176	318	326	83	85	142	144	163	171	362	375
CAYETANA BLANCA	Alb-1, Chu-3, Lau-12, Lau-13, Lau-10, Man-3	291	307	168	188	188	204	232	234	240	246	249	255	209	211	240	252	182	182	236	250	250	254	166	168	151	153	264	264	176	180	322	322	85	89	135	144	163	177	366	375
COJONATA	Ron-2	291	291	184	188	174	188	228	234	236	240	243	255	209	209	240	264	182	194	250	250	250	270	166	168	153	153	264	264	176	184	322	322	83	85	144	146	159	163	358	366
CORNICHÓN BLANC	Ins-2, Pla-1, Rot-2	291	291	166	184	168	204	234	244	244	246	249	265	209	213	238	246	180	182	246	260	256	260	160	166	149	153	264	264	182	196	318	318	89	95	144	149	173	181	348	358
DONA MARÍA	Ins-4	291	291	166	184	190	206	224	226	236	248	255	255	209	213	246	252	186	194	236	270	256	262	166	178	149	151	264	264	174	188	318	322	83	85	133	149	163	175	362	375
IMPERIAL NAPOLEÓN	Ins-3	291	295	166	188	188	190	232	236	246	248	249	255	209	211	238	252	184	194	246	250	254	270	166	166	153	153	264	264	184	186	322	322	85	89	131	133	153	161	362	366
JACQUEZ	Alb-2	289	291	178	184	184	184	226	240	236	238	237	249	209	217	254	256	180	190	232	238	250	250	166	166	149	151	262	270	176	182	316	318	85	85	137	142	159	171	334	364
JAÉN TINTO	Com-2, Lau-2, Lau-7	291	291	184	188	174	188	232	238	236	240	249	249	209	209	240	240	182	190	246	250	254	256	166	168	151	153	264	264	180	192	318	322	85	89	131	144	167	177	362	366
LISTÁN PRIETO	Ron-4	291	291	184	184	168	174	226	238	236	246	243	249	209	209	238	240	186	190	236	246	254	256	166	168	151	151	256	264	176	192	318	322	85	89	131	133	163	167	362	364
MANTEUDO	Chu-1, Man-1, Rot-1	291	307	184	184	168	176	220	224	236	246	243	265	209	211	252	252	182	182	246	246	270	270	166	168	151	151	264	264	176	176	322	322	85	89	142	142	163	167	366	375
MOLINERA	Ins-1	291	291	186	188	168	188	232	236	240	246	249	249	211	211	240	252	182	194	236	260	250	270	166	168	151	153	264	264	186	190	318	322	85	89	135	144	161	163	372	375
PEDRO XIMENES	Chu-5	291	307	168	188	168	174	234	238	236	236	243	249	209	213	240	246	182	186	260	266	248	270	166	166	149	153	264	264	176	188	322	322	89	89	131	144	163	177	364	366
ROAL	Chu-4, Man-2	291	307	168	184	204	206	234	236	246	250	255	255	209	209	240	254	182	184	246	246	250	254	168	168	151	153	264	264	176	180	318	322	85	89	135	142	163	167	358	366
ROME TINTO	Com-4	291	295	184	188	188	190	236	238	236	236	243	249	209	209	240	252	182	194	238	260	254	270	166	166	153	153	264	264	176	190	318	326	83	89	135	144	163	171	366	375
TINTO VELASCO	Lau-1, Lau-14, Lau-15, Lau-18, Lau-19	291	291	172	184	188	206	228	236	232	250	255	255	209	215	238	238	180	186	250	262	250	250	164	166	151	151	264	264	184	190	322	322	89	89	131	131	159	159	358	375
XARELLO	Lau-9	291	295	184	184	168	180	234	238	236	240	249	253	209	209	238	252	182	190	238	260	248	254	166	168	151	153	264	264	190	196	322	326	85	89	131	142	161	163	360	372
Unknown 1 *	Can-1	291	293	174	188	190	190	226	228	240	260	249	255	209	209	240	254	184	194	238	262	250	254	144	166	151	153	264	264	180	180	318	326	85	89	131	144	153	153	358	368
Unknown 2 *	Com-3	291	291	166	188	190	206	228	236	238	246	249	255	209	213	246	254	182	194	246	260	270	270	166	166	151	153	264	264	176	190	318	322	83	83	131	135	153	163	375	375
Unknown 3 *	Lau-3, Lau-4, Lau-16	291	307	168	188	174	176	226	234	236	246	243	255	209	217	240	240	182	186	238	250	254	270	166	166	151	151	264	264	180	192	318	322	83	89	131	131	153	171	364	366
Unknown 4 *	Lau-11	289	295	172	184	180	188	224	226	230	254	255	255	209	213	238	252	186	190	250	262	250	254	150	164	151	153	264	264	174	190	316	316	85	89	131	131	159	161	364	375
Unknown 5 *	Man-5	291	291	166	188	204	204	226	234	244	246	249	265	209	221	244	248	180	186	260	260	248	270	166	176	151	153	264	264	184	192	320	328	85	87	142	151	153	161	358	358
Unknown 6 *	Ron-3	291	307	184	188	168	174	232	234	236	240	249	255	209	209	240	240	182	182	236	250	238	270	166	168	151	153	264	264	186	188	318	326	85	89	131	144	161	177	364	372
Unknown 7 *	Ron-5	291	307	184	188	174	206	234	234	236	246	255	255	209	209	240	254	182	186	246	250	238	254	168	168	151	153	264	264	188	188	316	318	85	89	131	142	167	177	358	364
Unknown 8 *	Ron-6	291	307	184	184	174	188	236	238	232	240	249	249	209	211	252	252	182	186	260	262	250	270	166	166	151	153	264	264	176	186	318	322	87	89	131	142	153	163	366	375
CABERNET SAUVIGNON **		291	291	184	184	174	178	228	238	236	236	249	257	209	217	238	246	176	190	236	238	238	238	166	182	153	153	264	268	188	188	306	314	83	89	137	151	163	163	364	372
SYRAH **		291	295	166	188	174	206	224	228	236	236	247	265	209	215	240	240	190	192	220	230	238	270	164	166	151	153	264	264	182	190	318	318	89	89	131	131	163	165	362	382

[a] Prime names are according to *VIVC*, except those cultivars marked with an asterisk (*). Reference cultivars marked with two asterisks (**).

Table 2. Characterization of 20 microsatellite markers in the 30 genotypes.

Locus	*Na*	*He*	*Ho*	*r*	*PIC*	*PI*
VVIB01	5	0.546	0.700	−0.100	0.5024	0.250
VMC1b11	9	0.709	0.800	−0.053	0.6703	0.123
VMC4F31	9	0.863	0.833	0.016	0.8481	0.034
VVMD5	10	0.861	0.967	−0.057	0.8453	0.035
VVMD7	12	0.793	0.900	−0.060	0.5866	0.069
VVMD21	6	0.691	0.633	0.034	0.6368	0.150
VVMD24	6	0.541	0.600	−0.038	0.5097	0.242
VVMD25	9	0.786	0.733	0.029	0.7852	0.075
VVMD27	6	0.768	0.867	−0.056	0.7676	0.086
VVMD28	10	0.831	0.800	0.017	0.8089	0.050
VVMD32	8	0.790	0.867	−0.043	0.7593	0.075
VVIH54	8	0.607	0.667	−0.037	0.5600	0.201
VVIN16	5	0.585	0.733	−0.094	0.5009	0.256
VVIN73	4	0.185	0.167	0.015	0.1769	0.672
VVIP31	10	0.866	0.867	−0.001	0.8522	0.031
VVIP60	6	0.688	0.600	0.052	0.6369	0.149
VVIQ52	5	0.677	0.833	−0.093	0.6143	0.167
VVS2	9	0.817	0.867	−0.028	0.7938	0.057
VVIV37	11	0.846	0.900	−0.030	0.8132	0.040
VVIV67	11	0.847	0.933	−0.047	0.7798	0.042
TOTAL	159					1.74×10^{-19}
MEAN	7.95	0.715	0.763	−0.029	0.67241	0.140

Number of alleles (*Na*), expected heterozygosity (*He*), observed heterozygosity (*Ho*), Frequency of null alleles (*r*), Polymorphism information content (*PIC*) and Probability of identity (*PI*).

The values obtained from the statistical characterization of the 20 microsatellite loci used in this research study (Table 2) are similar to those obtained in other studies on the genetic characterization of local grapevine cultivars using microsatellite markers [51–53]. Nevertheless, the percentage of new accessions recovered (16.3%) is higher than that obtained by Balda et al. [10] for 45 accessions recovered in Rioja (Spain) (4.4%), Fort et al. for 223 accessions in recovered in Lanzarote (Canary Islands, Spain) (3.6%) [20], and Augusto et al. for 310 accessions recovered in northeast Portugal [32]. This suggests that the grapevine richness of the Andalusian region has not been prospected with the same degree of intensity.

3.2. *Cultivar Analysis*

Most of the analyzed accessions were identified with known grapevine cultivars. The varietal names were assigned based on the comparison with Spanish [38–43] and European [22,44] microsatellite databases and using the genetic profile of reference varieties for adapting the allele sizes. Allele sizes of genotypes obtained for the twenty SSRs loci analyzed are shown in Tables 1 and 2, and the prime names of the identified cultivars according to *VIVC* [22], indicating the code of sampled accession for each cultivar (Table 3). Thirty-nine accessions corresponded to 22 known varieties and the ten accessions remaining (Can-1, Comp-3, Lau-3, Lau-4, Lau-16, Lau-11, Man-5, Ron-3, Ron-5 and Ron-6) to the eight unidentified cultivars. These cultivars showed genotypes that did not match any of the published cultivars in the Spanish and European microsatellite databases consulted in this research. Half of the identified accessions are of Spanish origin according to the *VIVC* database [22] (Table 3), and the country of origin of the rest was France (four accessions), Portugal (three accessions), the United States (one accession), Italy (one accession), Greece (one accession), Algeria (one accession) and Lebanon (one accession). The accessions coded as Lau-3, Lau-4 and Lau-16 showed the same genotype, and they were collected in the same location (Laujar de Andarax, Almería, Spain).

The identified accessions include table and wine grapevine varieties. The table grape varieties, identified by 'Molinera' (Ins-1), 'Imperial Napoleon' (Ins-3) and 'Attika seedless' (Pla-2), have been collected in different regions of the province of Almería (Spain). In this province the cultivation of table grapes was predominant until the 1960s [54]. Furthermore, one hybrid interspecific ('Jacquez') was identified (Table 2). This hybrid was used for the reconstitution of European vineyards [55]. It is currently prohibited from use in Europe.

Table 3. Grapevine material studied with SSR identification, utilization and country of origin of the variety are according to VIVC [21].

Accession Code	Cultivars [a]	Utilization [b]	Country of Origin of the Variety [c]
Alb-1, Chu-3, Lau-12, Lau-13, Lau-10, Man-3	CAYETANA BLANCA	W-T	Spain
Alb-2	JACQUEZ	W	United States
Can-1	Unknown 1 *		
Chu-1, Man-1, Rot-1	MANTEUDO	W	Portugal
Chu-2	BEBA	W-T	Spain
Chu-4, Man-2	ROAL	W-T	Portugal
Chu-5	PEDRO XIMENES	W	Spain
Com-2, Lau-2, Lau-7	JAEN NEGRO	W-T	Spain
Com-3	Unknown 2 *		
Com-4	ROME TINTO		
Ins-1	MOLINERA	W-T	Spain
Ins-2, Pla-1, Rot-2	CORNICHON BLANC	W-T	Italy
Ins-3	IMPERIAL NAPOLEON	T	Spain
Ins-4	DONA MARIA	W-T	Portugal
Lau-1, Lau-14, Lau-15, Lau-18, Lau-19	TINTO VELASCO	W-T	Spain
Lau-3, Lau-4, Lau-16	Unknown 3 *		
Lau-5, Lau-6	AHMEUR BOU AHMEUR	W-T	Algeria
Lau-8	ALICANTE HENRI BOUSCHET	W	France
Lau-9	XARELLO	W	Spain
Lau-11	Unknown 4 *		
Lau-17	CARIGNAN NOIR	W	France
Man-4	AFUS ALI	W-T	Lebanon
Man-5	Unknown 5 *		
Pla-2	ATTIKA SEEDLESS	T	Greece
Ron-1	BOBAL	W	Spain
Ron-2	COJONATA	W	Spain
Ron-3	Unknown 6 *		
Ron-4	LISTAN PRIETO	W-T	Spain
Ron-5	Unknown 7 *		
Ron-6	Unknown 8 *		
	CABERNET SAUVIGNON **	W	France
	SYRAH **	W	France

[a] Prime names are according to VIVC, except those cultivars marked with an asterisk (*). Reference cultivars marked with two asterisks (**). [b] Utilization is according to VIVC: W (Wine grape) and T (Table grape). [c] Country of origin of the variety are according to VIVC.

One genotype (Com-4) was identified as 'Rome Tinto', after comparing it with the microsatellite database from Rancho de la Merced [38,39,56]. This variety was only conserved in the Rancho de la Merced Germplasm bank according to the VIVC database and presents a different genotype to the 'Rome' cultivar published by Ibáñez et al. [41].

Four of the varieties identified, 'Beba', 'Jaén negro', 'Pedro Ximenez' and 'Rome Tinto', were already mentioned by Rojas Clemente [6] as being present in the Andalusian region. This shows the antiquity of the cultivation of these varieties in this region. 'Jaén negro' and 'Rome Tinto' are two red grapevine cultivars that have already been identified in old vineyards in the province of Málaga [57].

Currently, most of the cultivars identified in this work have disappeared from the Andalusia vineyards, and the unique cultivar that is growing in the commercial vineyards is 'Pedro Ximenez'. All of this vegetal material recovered from old vineyards could be interesting for the wine industry in Andalusia or regions with similar agroclimatic conditions. However, many of these identified cultivars are not included in the official register of Spanish grapevine varieties for the community of Andalusia, which would make their cultivation difficult. Recently, 'Beba' has been included as an authorized variety in the regulation of wines of the Protected Designation of Origin "Jerez-Xérès-Sherry" (Spain) [58], as there is some interest in increasing the diversity of wines [59].

In addition, of the identified varieties, the eight new genotypes should be studied and evaluated in order to make their oenological potential and adaptation climate change known among the wine sector. Furthermore, these cultivars could be important genetic resources for future breeding programs.

3.3. Genetic Relationships among Cultivars

Based on the results of the analysis of the microsatellites, the distance matrix was used to carry out a grouping using UPGMA. To characterize the genetic structure of different genotypes obtained and two references varieties ('Cabernet Sauvignon' and 'Syrah'), a dendrogram based on the proportion of shared alleles was constructed. Figure 1 shows the resulting dendrogram of the 30 non-redundant genotypes found in this study.

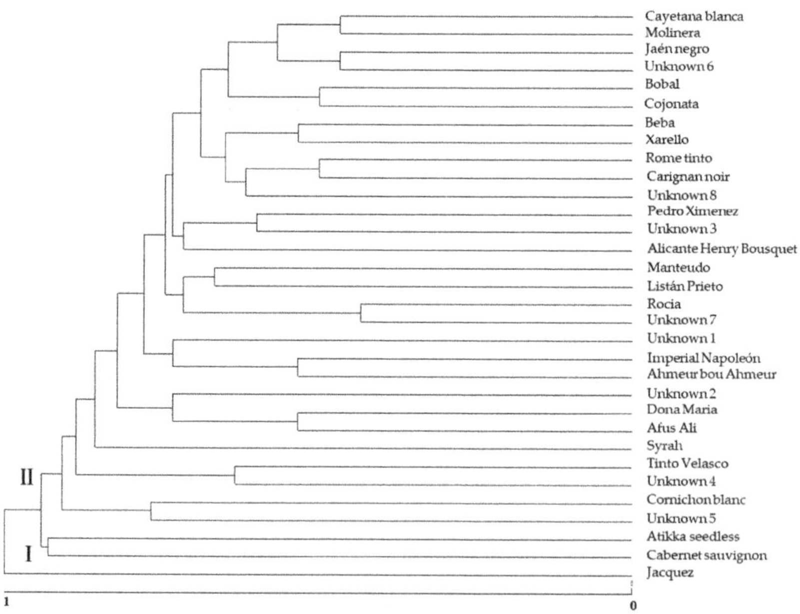

Figure 1. Genetic relationships among the thirty genotypes obtained and two reference cultivars.

SSR analysis allowed for the evaluation of the genetic relationships among European cultivars and unknown accessions recollected in different regions of Andalusia. The dendrogram in Figure 1 shows the existence of two defined groups. Group I includes only one cultivar identified with 'Jacquez', which is a hybrid interspecific of the cross between *Vitis aestivalis* × *Vitis vinifera* [22]. All of the rest of the identified and unknown cultivars are included in group II and are cultivars of the *Vitis vinifera* species. The formation of these two groups may be related to the pedigree of the cultivars. In group II, there is no clear separation of different subgroups in relation to regions of origin as found in other published research papers on Sicilian varieties [60]. Varieties with different countries of origin are grouped in this cluster II (Table 3).

Two varieties, 'Cabernet Sauvignon' and 'Atikka seedless' are markedly distant from the rest of the cultivars in Group II, probably because of a different origin and use. 'Atikka seedless' is considered a seedless variety of Greek origin according to VIVC [22].

Phylogenetic distances of the subgroup where variety "Unknown 6" is included indicates that it could be a wine and table grape, since it is grouped with other grapes that are used as wine and table grapes, such as 'Cayetana Blanca', 'Molinera' and 'Jaén Tinto' (Table 3). The same behavior could be said for the variety "Unknown 7" and the 'Roal', "Unknown 4" and 'Tinto Velasco' or "Unknown 5" and 'Cornichon Blanc'.

4. Conclusions

Forty-nine accessions collected in Andalusia have been described by molecular methods. A total of 83.7% of these accessions analyzed have been identified by comparison to Spanish or European microsatellite databases with known cultivars. However, eight genotypes have not yet been identified and could represent old local cultivars in danger of extinction. All of these genotypes have been preserved within the Rancho de la Merced germplasm bank (Andalusia, Spain).

This study indicates an important biodiversity within the old vineyards from the Andalusia region that provides interesting information for the wine industry and that points out the wide genetic diversity of grapevines which are still unexploited. Our efforts should lead to the protection and study of local grape natural richness.

Supplementary Materials: The following supporting information can be downloaded at: https://www.mdpi.com/article/10.3390/horticulturae9030316/s1, Table S1: List of the 49 accessions collected and their place of origin in the Andalusia region (Spain).

Author Contributions: Conceptualization, A.J.-C., A.P.-P. and R.A.-G.; methodology, A.J.-C., A.P.-P. and R.A.-G.; software, A.J.-C. and R.A.-G.; validation, A.J.-C. and R.A.-G.; formal analysis, A.J.-C.; investigation, A.J.-C., A.P.-P. and R.A.-G.; resources, A.J.-C.; data curation, A.J.-C.; writing—original draft preparation, A.J.-C., A.P.-P. and R.A.-G.; writing—review and editing, A.J.-C., A.P.-P. and R.A.-G.; visualization, A.J.-C., A.P.-P. and R.A.-G.; supervision, A.J.-C., A.P.-P. and R.A.-G. All authors have read and agreed to the published version of the manuscript.

Funding: This research was funded by Ministerio de Ciencia y Tecnología (INIA-Spain), grants number RF2004-00014-00-00, VIN00-036-C6-5X, RF2006-00011-00-00 and RF2007-00017-00-00.

Acknowledgments: The authors would like to thank the IFAPA for facilitating the material conserved in the Rancho de la Merced germplasm bank for the study.

Conflicts of Interest: The authors declare no conflict of interest.

References

1. Pozo-Bayón, M.Á.; Moreno-Arribas, M.V. Sherry wines. In *Advances in Food and Nutrition Research*; Jackson, R.S., Ed.; Academic Press: Cambridge, MA, USA, 2011; Volume 63, pp. 17–40. [CrossRef]
2. Prados Martínez, F. La producción vinícola en el mundo fenicio-púnico. Apuntes sobre cultivo de la vid y consumo del vino a través de las fuentes arqueológicas y literarias. *Gerión* **2011**, *29*, 9–35. [CrossRef]
3. González Rodríguez, R. Excavaciones de Urgencia en el Cerro Naranja (Jerez de la Frontera, Cádiz). In *Anuario Arqueológico de Andalucía, III*; Junta de Andalucía: Sevilla, Spain, 1985; pp. 90–96.

4. Torres, J.R. Las Ánforas Fenicio-Púnicas del Mediterráneo Central y Occidental. Ph.D. Thesis, University of Barcelona, Barcelona, Spain, 1991.
5. Aubet Semmler, M.E.; Buxó Capdevila, R. Los recursos y la economía colonial. En Aubet Semmler, M.E.; et al. ii Cerro del Villar I. El Asentamiento fenicio en la desembocadura del río Guadalhorce y su interacción con el hinterland. *Arqueol. Monogr.* **1999**, *5*, 334–339.
6. Vera, J.C.; Echevarría, A. Sistemas agrícolas del I milenio a.C. en el yacimiento de La Orden-Seminario de Huelva: Viticultura protohistórica a partir del análisis arqueológico de las huellas de cultivo. In Proceedings of the Patrimonio Cultura de la Vid y el Vino, Madrid, Spain, 2013; pp. 95–106.
7. De Rojas Clemente y Rubio, S. *Ensayo Sobre las Variedades de vid Común que Vegetan en Andalucía*; Imprenta Villalpando: Madrid, Spain, 1807; p. 324.
8. Morilla Critz, J. La viticulture de Andalucía en 1831 vista por James Busby, padre de la viticultura Australiana. *Estud. Reg.* **1997**, *49*, 261–298.
9. Fernández de Bobadilla, G. *Viníferas Jerezanas y de Andalucía Occidental*; Instituto Nacional de Investigaciones Agronómicas: Madrid, Spain, 1956.
10. Balda, P.; Ibáñez, J.; Sancha, J.C.; Martínez de Toda, F. Characterization and identification of minority red grape varieties recovered in Rioja, Spain. *Am. J. Enol. Vitic.* **2014**, *65*, 148–152. [CrossRef]
11. Fatahi, R.; Ebadi, A.; Bassil, N.; Mehlenbacher, S.A.; Zamani, Z. Characterization of Iranian grapevine cultivars using microsatellite markers. *Vitis* **2003**, *42*, 185–192.
12. Martínez, L.E.; Cavagnaro, P.F.; Masuelli, R.W.; Zuñiga., M. SSR-based assessment of genetic diversity in South American *Vitis vinifera* varieties. *Plant Sci.* **2006**, *170*, 1036–1044. [CrossRef]
13. De Mattia, F.; Imazio, S.; Grassi, F.; Lovicu, G.; Tardaguila, J.; Failla, O.; Maitt, C.; Scienza, A.; Labra, M. Genetic characterization of Sardinia grapevine cultivars by SSR markers analysis. *J. Int. Sci. Vigne et du Vin.* **2007**, *41*, 175–184. [CrossRef]
14. Lacombe, T.; Boursiquot, J.M.; Laucou, V.; Dechesne, F.; Varès, D.; This, P. Relationships and genetic diversity within the accessions related to Malvasia held in the Domaine de Vassal grape germplasm repository. *Am. J. Enol. Vitic.* **2007**, *58*, 124–131. [CrossRef]
15. Dzhambazova, T.; Tsvetkov, I.; Atanassov, I.; Rusanos, K.; Martínez Zapater, J.M.; Atanassov, A.; Hvarleva, T. Genetic diversity in native Bulgarian grapevine germplasm (*Vitis vinifera* L.) based on nuclear and chloroplast microsatellite polymorphisms. *Vitis* **2009**, *48*, 115–121.
16. De Olivera, G.L.; De Souza, A.P.; De Oliver, F.A.; Zucchi, M.I.; De Souza, L.M.; Moura, M.F. Genetic structure and molecular diversity of Brazilian grapevine germplasm: Management and use in breeding programs. *PLoS ONE* **2020**, *15*, e0240665. [CrossRef]
17. Upadhayay, A.; Aher, L.B.; Shinde, M.P.; Mundankar, K.Y.; Datre, A.; Karibasappa, G.S. Microsatellite analysis to rationalize grape germplasm in India and development of a molecular database. *Plant Genet. Resour.* **2013**, *11*, 225–233. [CrossRef]
18. Popescu, C.F.; Maul, E.; Dejeu, L.C.; Dinu, D.; Gheorge, R.N.; Laucou, V.; Lacombe, T.; Migliaro, D.; Crespan, M. Identification and characterization of Romanian grapevine genetic resources. *Vitis* **2017**, *56*, 173–180.
19. Marsal, G.; Méndez, J.J.; Mateo, J.M.; Ferrer, S.; Canals, J.M.; Zamora, F.; Fort, F. Molecular characterization of *Vitis vinifera* L. local cultivars from volcanic areas (Canary Islands and Madeira) using SSR markers. *OENO One* **2019**, *53*, 667–680. [CrossRef]
20. Fort, F.; Marsal, G.; Mateo-Sanz, J.M.; Pena, V.; Canals, J.M.; Zamora, F. Molecular characterisation of the current cultivars of *Vitis vinifera* L. in Lanzarote (Canary Islands, Spain) reveals nine individuals which correspond to eight new varieties and two new sports. *OENO One* **2022**, *56*, 281–295. [CrossRef]
21. García de Luján, A.; Lara Benítez, M. *La colección de vides del Rancho de la Merced*; Junta de Andalucía, Consejería de Agricultura y Pesca: Sevilla, Spain, 1997; p. 157.
22. Vitis International Variety Catalogue. Available online: www.vivc.de (accessed on 26 December 2022).
23. Pastor, I.; Rovira, J.M. La Investigación arqueológica en los procesos de recuperación de variedades de uva autóctonas tradicionales. In Proceedings of the Patrimonio Cultura de la Vid y el Vino, Madrid, Spain, 2013; pp. 71–79.
24. Sancho-Galán, P.; Amores-Arrocha, A.; Palacios, V.; Jiménez-Cantizano, A. Identification and characterization of white grape varieties autochthonous of a warm climate region (Andalusia, Spain). *Agronomy* **2020**, *10*, 205. [CrossRef]
25. Boursiquot, J.M.; This, P. Les nouvelles techniques utilisées en ampélographie: Informatique et marquage. *J. Int. Sci. Vigne Vin* **1996**, *610*, 13–23.
26. Emanuelli, F.; Lorenzi, S.; Grzeskowiak, L.; Catalano, V.; Stefanini, M.; Troggio, M.; Myles, S.; Martínez-Zapater, J.M.; Zyprian, E.; Moreira, F.M.; et al. Genetic diversity and population structure assessed by SSR and SNP markers in large germplasm collection of grape. *BMC Plant Biol.* **2013**, *13*, 39. [CrossRef]
27. Sefc, K.M.; Lefort, F.; Grando, M.S.; Scott, K.D.; Steinkellner, H.; Thomas, M.R. Microsatellite markers for grapevine: A state of the art. In *Molecular Biology and Biotechnology of Grapevine*; Roubelakis-Angelakis, K.A., Ed.; Kluwer Academic: Dordrecht, The Netherlands, 2001; pp. 433–463.
28. This, P.; Jung, A.; Boccacci, P.; Borrego, J.; Botta, R.; Costantini, L.; Crespan, M.; Dangl, G.S.; Eisenheld, C.; Ferreira-Monteiro, F.; et al. Development of a standard set of microsatellite reference alleles for identification of grape cultivars. *Theor. Appl. Genet.* **2004**, *109*, 1448–1458. [CrossRef]
29. González-Andrés, F.; Martín, J.P.; Yuste, J.; Rubio, J.A.; Arranz, C.; Ortiz, J.M. Identification and molecular biodiversity of autochthonous grapevine cultivars in the "Comarca del Bierzo", León, Spain. *Vitis* **2007**, *46*, 71–76.

30. Jiménez-Cantizano, A.; Lara, M.; Ocete, M.E.; Ocete, R. Short communication: Characterization of the relic Almuñécar grapevine cultivar. *Span. J. Agric. Res.* **2012**, *10*, 454–460. [CrossRef]
31. Mena, A.; Martínez, M.; Fernández-González, M. Recovery, identification and relationships by microsatellite analysis of ancient grapevine cultivars from Castilla-La Mancha: The largest wine growing region in the world. *Genet. Resour. Crop Evol.* **2014**, *61*, 625–637. [CrossRef]
32. Augusto, D.; Ibáñez, J.; Pinto-Sintra, A.L.; Falco, V.; Leal, F.; Martínez-Zapater, J.M.; Oliveira, A.A.; Castro, I. Grapevine Diversity and Genetic Relationships in Northeast Portugal Old Vineyards. *Plants* **2021**, *10*, 2755. [CrossRef] [PubMed]
33. Bowers, J.E.; Dangl, G.S.; Vignani, R.; Meredith, C.P. Isolation and characterization of new polymorphic simple sequence repeat loci in grape (*Vitis vinifera* L.). *Genome* **1996**, *39*, 628–633. [CrossRef] [PubMed]
34. Bowers, J.E.; Dangl, G.S.; Meredith, C.P. Development and characterization of additional microsatellite markers for grape. *Am. J. Enol. Vitic.* **1999**, *50*, 243–246. [CrossRef]
35. Thomas, M.R.; Scott, N.S. Microsatellite repeats in grapevine reveal DNA polymorphisms when analyzed as sequence-tagged sites (STSs). *Theor. Appl. Genet.* **1993**, *86*, 985–990. [CrossRef] [PubMed]
36. Merdinoglu, D.; Butterlin, G.; Bevilacqua, L.; Chiquet, V.; Adam-Blondon, A.F.; Decroocq, S. Development and characterization of a large set of microsatellite markers in grapevine (*Vitis vinifera* L.) suitable for multiplex PCR. *Mol. Breed.* **2005**, *15*, 349–366. [CrossRef]
37. Vargas, A.M.; Velez, M.D.; De Andrés, M.T.; Laucou, V.; Lacombe, T.; Boursiquot, J.M.; Borrego, J.; Ibáñez, J. Corinto bianco: A seedless mutant of Pedro Ximenes. *Am. J. Enol. Vitic.* **2007**, *58*, 540–543. [CrossRef]
38. Jiménez-Cantizano, A. Caracterización Molecular del Banco de Germoplasma de vid del Rancho de la Merced. Ph.D. Thesis, University of Cádiz, Cádiz, Spain, 2014. Available online: http://hdl.handle.net/10498/17919 (accessed on 12 November 2022).
39. Jiménez-Cantizano, A.; García De Luján, A.; Arroyo-García, R. Molecular characterization of table grape varieties preserved in the Rancho de la Merced Grapevine Germplasm Bank (Spain). *Vitis* **2018**, *57*, 93–101.
40. Cretazzo, E.; Moreno Sanz, P.; Lorenzi, S.; Benítez, M.L.; Velasco, L.; Emanuelli, F. Genetic Characterization by SSR Markers of a Comprehensive Wine Grape Collection Conserved at Rancho de la Merced (Andalusia, Spain). *Plants* **2022**, *11*, 1088. [CrossRef] [PubMed]
41. Ibáñez, J.; Vargas, A.M.; Palancar, M.; Borrego, J.; De Andrés, M.T. Genetic Relationships among Table-Grape Varieties. *Am. J. Enol. Vitic.* **2009**, *60*, 35–42. [CrossRef]
42. Vargas, A.M.; De Andrés, M.T.; Borrego, J.; Ibáñez, J. Pedigrees of fifty tables-grape cultivars. *Am. J. Enol. Vitic.* **2009**, *60*, 525–531. [CrossRef]
43. De Andrés, M.T.; Benito, A.; Pérez-Rivera, G.; Ocete, R.; López, M.A.; Gaforio, L.; Muñoz, G.; Cabello, F.; Martínez-Zapater, J.M.; Arroyo-García, R. Genetic diversity of wild grapevine populations in Spain and their genetic relationships with cultivated grapevines. *Mol. Ecol.* **2012**, *21*, 800–816. [CrossRef]
44. Lacombe, T.; Boursiquot, J.M.; Laucou, V.; Di Vecchi-Staraz, M.; Péros, J.P.; This, P. Large-scale parentage analysis in an extended set of grapevine cultivars (*Vitis vinifera* L.). *Theo. Appl. Genet.* **2013**, *126*, 401–414. [CrossRef]
45. Park, S.D.E. Trypanotolerance in West African Cattle and the Population Genetic Effects of Selection. Ph. D. Thesis, University of Dublin, Dublin, Ireland, 2001.
46. Peakall, R.; Smouse, P. GENALEX 6: Genetic analysis in Excel. Population genetic software for teaching and research. *Mol. Ecol. Res.* **2006**, *6*, 288–295. [CrossRef]
47. Bi 'nkowski, J.; Miks, S. Gene-Calc. 2018. Available online: www.gene-calc.pl. (accessed on 20 February 2023).
48. Minch, E.; Ruiz-Linares, A.; Goldstein, D.; Feldman, M.; Cavalli-Sforza, M. *Microsat 1.5d: A Computer Program for Calculating Various Statistics on Microsatellite Allele Date*; Washington State University: Pullman, WA, USA, 1997.
49. Felsenstein, J. Phylogeny inference package. *Cladistics* **1989**, *5*, 164–166.
50. Kumar, S.; Stecher, G.; Tamura, K. MEGA 7: Molecular evolutionary genetics analysis version 7.0 for bigger datasets. *Mol. Biol. Evol.* **2015**, *33*, 1870–1874. [CrossRef]
51. Fenández-González, M.; Mena, A.; Izquierdo, P.; Martínez, J. Genetic characterization of grapevine (*Vitis vinífera* L.) cultivars from Castilla La Mancha (Spain) using microsatellite markers. *Vitis* **2007**, *46*, 126–130.
52. Žulj Mihaljević, M.; Šimon, S.; Pejić, I.; Carka, F.; Sevo, R.; Kojić, A.; Gaši, F.; Tomić, L.; Jovanović Cvetković, T.; Maletić, E.; et al. Molecular characterization of old local grapevine varieties from South East European countries. *Vitis* **2013**, *52*, 69–76.
53. Basheer-Salimia, R.; Lorenzi, S.; Batarseh, F.; Moreno-Sanz, P.; Emanuelli, F.; Grando, M.S. Molecular identification and genetic relationships of Palestinian grapevine cultivars. *Mol. Biotechnol.* **2014**, *56*, 546–556. [CrossRef]
54. Gómez-Espín, J.M.; Gil Meseguer, E. La exportación española de uva de mesa en la segunda mitad del siglo XX. *Papeles Geografía* **1987**, *13*, 87–104.
55. Santiago, J.L.; González, I.; Gago, P.; Alonso-Villaverde, V.; Boso, S.; Martínez, M.C. Identification of and relationships among a number of teinturier grapevines that expanded across Europe in the early 20th century. *Aust. J. Grape and Wine Res.* **2008**, *14*, 223–229. [CrossRef]
56. Jiménez-Cantizano, A.; Amores-Arrocha, A.; Gutiérrez-Escobar, R.; Palacios, V. Identification and relationship of the autochthonous 'Romé' and 'Rome Tinto' grapevine cultivars. *Span. J. Agric. Res.* **2018**, *16*, e07SC02. [CrossRef]

57. Jiménez-Cantizano, A.; Muñoz-Martín, A.; Amores-Arrocha, A.; Sancho-Galán, P.; Palacios, V. Identification of Red Grapevine Cultivars (*Vitis vinifera* L.) Preserved in Ancient Vineyards in Axarquía (Andalusia, Spain). *Plants* **2020**, *9*, 1572. [CrossRef] [PubMed]
58. Orden de 4 de Octubre de 2022, Boletín Oficial de la Junta de Andalucía (BOJA). Available online: https://www.juntadeandalucia.es/eboja/2022/195/s54.html (accessed on 3 January 2023).
59. Sancho-Galán, P.; Amores-Arrocha, A.; Palacios, V.; Jiménez-Cantizano, A. Genetical, Morphological and Physicochemical Characterization of the Autochthonous Cultivar 'Uva Rey' (*Vitis vinifera* L.). *Agronomy* **2019**, *9*, 563. [CrossRef]
60. Carimi, F.; Mercati, F.; Abbate, L.; Sunseri, F. Microsatellite analyses for evaluation of genetic diversity among Sicilian grapevine cultivars. *Genet. Resour. Crop Evol.* **2010**, *57*, 703–719. [CrossRef]

horticulturae

MDPI

Article

Genetic Diversity and Population Structure Analysis of Anatolian Kara Grapevine (*Vitis vinifera* L.) Germplasm Using Simple Sequence Repeats

Nur Arslan [1,†], Funda Yılmaz Baydu [2,†], Nahid Hazrati [1], Canan Yüksel Özmen [1], Onur Ergönül [3], Tamer Uysal [3], Ahmet Semih Yaşasın [3], Cengiz Özer [3], Yılmaz Boz [4], Yusuf Serhat Kuleyin [1] and Ali Ergül [1,*]

[1] Biotechnology Institute, Ankara University, 06135 Ankara, Turkey; nuryildirim84@gmail.com (N.A.); nahid.hazrati@yahoo.com (N.H.); cananyuksel555@gmail.com (C.Y.Ö.); serhatkuleyin@gmail.com (Y.S.K.)

[2] Department of Horticulture, Agriculture Faculty, Ankara University, 06110 Ankara, Turkey; fundayilmaz_@hotmail.com

[3] Ministry of Agriculture and Forestry, General Directorate of Agricultural Researches and Policies, Viticulture Research Institute, 59200 Tekirdağ, Turkey; onur.ergonul@tarim.gov.tr (O.E.); tameru48@gmail.com (T.U.); ahmetsemih.yasasin@tarimorman.gov.tr (A.S.Y.); ozercengiz59@gmail.com (C.Ö.)

[4] Ministry of Agriculture and Forestry, General Directorate of Agricultural Researches and Policies, Atatürk Horticultural Central Research Institute, 77100 Yalova, Turkey; yilmaz_boz@hotmail.com

* Correspondence: ergul@ankara.edu.tr

† These authors contributed equally to this work.

Abstract: Grape (*Vitis vinifera* L.) is among the most important commercial fruit species grown worldwide in terms of its economic value. Anatolia (Turkey), located in the favorable climate zone for viticulture, has a rich grape genetic potential due to its location at the intersection of the grapevine gene centers. In Turkish Kara grape germplasm, there are problems in terms of accuracy during the production phase due to the inadequacies in ampelographic definitions, and also very little information is available on genetic analysis of Kara grape germplasm. This study carried out genetic analysis of 49 Kara grape cultivars from six regions (sub-populations) of Turkey and 3 reference cultivars using 22 microsatellite loci (SSR), and ampelographic analysis were also performed concerning 39 OIV descriptors. In the SSR analysis, the average number of alleles per locus was 8.91, ranging from 4 to 13; four synonymous and five homonymous cases were also identified. In the population structure analysis, the genetic differentiation (*Fst*) values among six populations were moderate. In the BAPS analysis, all populations except Central Anatolia were found to be highly admixed with each other, and in the FCA analysis, the East Anatolia population was completely separated. In the multilocus lineages (MLLs) analysis, a total of three accessions were matched to different accessions as clone assignment. In this study, SSR-based genetic characterization of the Turkish Kara grape germplasm was revealed for the first time, and it is thought that the obtained data will help other grape genetic characterization studies and contribute to viticulture research in other areas such as breeding, protection and variety registration.

Keywords: *Vitis vinifera* L.; Kara grapevine; SSR; population structure; Anatolia; Turkey

Citation: Arslan, N.; Yılmaz Baydu, F.; Hazrati, N.; Yüksel Özmen, C.; Ergönül, O.; Uysal, T.; Yaşasın, A.S.; Özer, C.; Boz, Y.; Kuleyin, Y.S.; et al. Genetic Diversity and Population Structure Analysis of Anatolian Kara Grapevine (*Vitis vinifera* L.) Germplasm Using Simple Sequence Repeats. *Horticulturae* **2023**, *9*, 743. https://doi.org/10.3390/horticulturae9070743

Academic Editors: Antonio Amores-Arrocha and Ana Jiménez-Cantizano

Received: 20 May 2023
Revised: 15 June 2023
Accepted: 20 June 2023
Published: 24 June 2023

1. Introduction

The grapevine (*Vitis vinifera* L.), apart from being one of the most extensively cultivated fruit trees in the world, is also a fascinating subject for history and evolutionary studies [1,2]. The plant is an extremely important resource, not only in terms of its fruit, but also because of the presence of secondary metabolites contained in its cellular structure. Resveratrol is one of these secondary metabolites which acts like an antioxidant, protecting the body against high risks [3–5].

Anatolia, located in Turkey, is rich in wild grape (*Vitis vinifera* ssp. sylvestris) varieties [6,7], and diversity among germplasm of these varieties has gradually resulted in the creation of a potent cultivated variety (*Vitis vinifera* ssp. sativa). The coastlines of the Eastern Black Sea and Eastern Anatolian regions, except for the highlands, are involved in economic viticulture. According to data from the FAO 2020, 77.1 million tons of grape production has been conducted on an area of 6.9 million hectares worldwide. Turkey ranks fifth in terms of area with 400,000 hectares (5.85%) of cultivated land and sixth in terms of grape production with 4.2 million tons (5.32%) in the world.

As an important exporting crop in Turkey, more than 1200 cultivars of this species are gathered in Turkey's National Collection Repository (TNCR) by Tekirdağ Viticulture Research Institute-Tekirdağ for the identification and protection of grape genetic potential. In this collection, numerous ampelographic investigations have been performed, and as a result of ampelographic researches, cultivars with different or similar morphological features but same names have been encountered. The most important of them, especially those consumed as fresh edible grapes, include 10 to 50 prevalent cultivars like White, Black, Amasya, Dimrit, Parmak and Razakı. Due to the insufficiency of ampelographic discrimination/identification of Kara (or Siyah, called Black in English) grape cultivars, its nomenclature has been inaccurate and its synonyms/homonyms are undetermined. This has complicated the correct identification of the number of grape cultivars. In addition to its nutritional value, black grapes have accumulations of effective antioxidant substances, especially in the skin and seeds, which are very important for health [8].

Development of genetic markers is considered a big step forward, because they are not influenced by environmental conditions, nor by the type of the sample or the developmental stage, and thus provide distinctive information [9,10]. DNA markers are widely used in germplasm characterization and variety creation, and also in clone identification through parent analysis [11,12]. A common criterion of a suitable or applicable marker is the degree of polymorphism that can highlight differences between cultivars and clones [13]. Due to the high polymorphism, repeatability and predominant quality of SSR microsatellites, they have most often been selected for genetic analysis of Vitis cultivars [14,15].

SSR microsatellite markers, or Simple Sequence Repeats (SSR), are genomic repetitive regions in the category of abundant Short Tandem Repeats (STRs) scattered throughout the genome [16,17]. Indicatively, Velasco et al. [18] estimated the number of SSRs in the genome of highly heterozygous individuals of *Vitis vinifera* to be about 89,000.

Recently, it was concluded that the utilization of 20 SSR markers tends to be sufficient to distinguish existing cultivars or to solve synonymy and homonymy issues [19]. In this work, the identical, synonymous and homonymous status of 49 Kara grape cultivars derived from 30 provinces of six different eco-geographic sub-populations were distinguished using 22 SSR loci and 39 ampelographic characterization descriptors (OIV: International Organization of Vine). Furthermore, correlation between cultivars and eco-geographical sub-population (region) distribution was illustrated by genetic relations, different population structure approaches and clonal analysis.

2. Materials and Methods

2.1. Plant Material

Overall, 52 cultivars, including 49 Kara grape cultivars (Table S1) and 3 reference cultivars (CS: Carbernet Sauvignon, M: Merlot, PN: Pinot noir), were analyzed in this study (Figure 1). These grape cultivars were obtained from the National Grapevine Germplasm Vineyard at the Institute of Viticulture in Tekirdağ, Turkey.

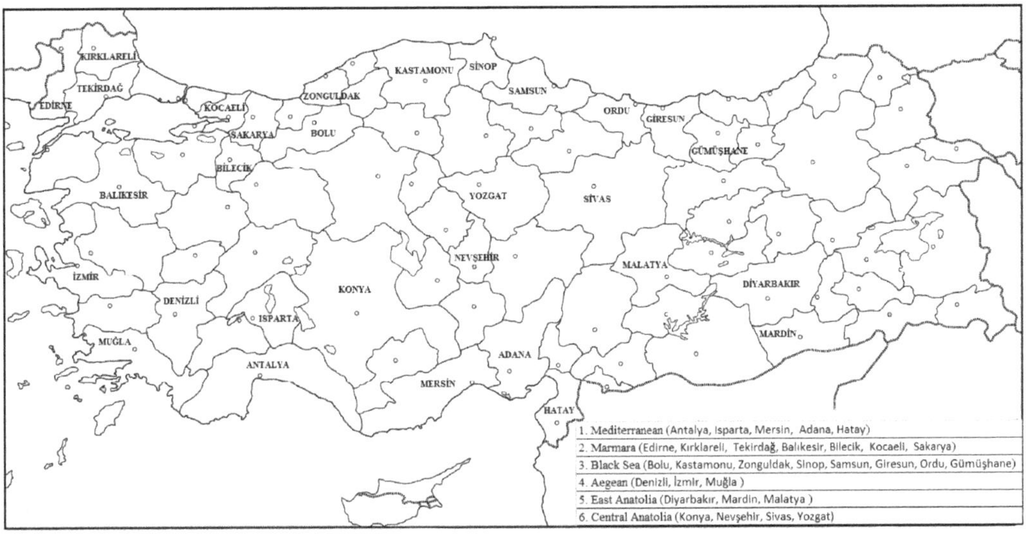

Figure 1. Provinces from which Kara grape cultivars were collected (each sub-population and the provinces belonging to each sub-population are given in the below right part of the figure).

2.2. OIV Data Analysis

Ampelographic characters of these grape cultivars were determined according to the Descriptors of Grape norms from the IBPGR (International Board for Plant Genetic Resources) [20]. A total of thirty nine major OIV descriptors were used for a set of 49 Kara grape cultivars. Information related to OIV descriptors (characteristic description) and OIV data codes is given in Table S2. The morphometric data were constructed using standard OIV codes. However, the standard OIV coding was converted into a data format that could be analyzed. The OIV 016 character was removed from the data file as it contained no variation (all cultivars were coded as 1). PAST (Paleontological Statistics Version 3.22) was used for the data analysis [21].

2.3. DNA Isolation

DNA was extracted from leaf tissue as described by Lefort et al. [22]. Determination of DNA concentration and quality was performed according to Akçay et al. [23] and Ergül et al. [24] using an ND-1000 spectrophotometer (NanoDrop Technologies, Wilmington, DE, USA) and agarose (1%, w/v) gel electrophoresis. Isolated DNA was stored at −20 °C until PCR reactions were performed.

2.4. SSR Analysis and Capillary Electrophoresis

22 SSR loci, named VVS1, VVS2 [25], VVMD5, VVMD7, VVMD21, VVMD24, VVMD27, VVMD28, VVMD31 [26], vrZAG21, vrZAG47, vrZAG64, vrZAG112, vrZAG62, vrZAG79, vrZAG83 [27], VMC2H4, VMC2C3 [28] and VVIH54, VVIB01 [29], VVMD25, VVMD32 [26], were used in this study. Nine loci, called the "core SSR marker set" (VVS2, VVMD5, VVMD7, VVMD27, ZAG62, ZAG79, VVMD25, VVMD32), directly allow comparisons of allele sizes from different grape cultivars analyzed in different studies [30,31]. Information on the 22 SSR loci used (SSR locus name, primer sequence and references) is given in Table S3.

Allele size detections, PCR amplifications and capillary electrophoresis conditions were conducted according to Akçay et al. [23] and Yılmaz et al. [32]. In the PCR reactions, fluorescent-labeled D4 (blue), D3 (green), and D2 (black) forward primers allocated to each SSR locus were applied. A negative control (distilled water instead of DNA) was used for checking for the presence of possible contamination in the PCR reactions. Samples for

DNA amplification were subjected to PCR for 3 min at 94 °C, 1 min at 94 °C, or 1 min at 48–66 °C, depending on the degree of primer binding (annealing temperature), then for 2 min at 72 °C, and kept at 72 °C for 10 min during the last cycle. This was carried out for 35 cycles. Amplification (with 100 bp DNA ladder, Invitrogen™, Waltham, MA, USA), and control of SSR loci products once their PCR steps were completed, were performed using electrophoresis on a 2% agarose gel.

After diluting the PCR products with SLS (Sample Loading Solution) solution in suitable proportions (20 μL), Genomelab DNA Standard Kit-400 (0.4 μL) was added to the mixture, and then electrophoresis was applied in the CEQ 8800XL capillary DNA analysis system (Beckman Coulter, Fullerton, CA, USA). Determination of the peak sizes (bp) was carried out using the fragment analysis software of the system. "Cabernet Sauvignon (CS)", "Merlot (M)" and "Pinot noir (PN)" were included as reference cultivars. PCR and SSR analyses were performed at least twice to indicate reproducibility of the results. The heterozygosity or homozygosity of PCR fragments was visualized considering the types and colors of each SSR locus peak after capillary electrophoresis.

2.5. Genetic Analysis

2.5.1. SSR Analysis

In this study, allele number (*n*), allele frequency, expected (*He*) and observed heterozygosity (*Ho*) and detection probability (*PI*, Probability of Identity) values were determined as genetic parameters with the IDENTITY 1.0 program [33]. This program was used to detect the same genotypes. The proportion of shared alleles was determined by the 'ps (option 1-(ps))' method described by Bowcock et al. [34], and the similarity ratio index was calculated using the Microsat program [35]. Genetic similarities among a total of 52 studied cultivars were determined with the NTSYS program (version 2.02g, Exeter Software, Setauket, NY, USA) using the UPGMA (Unweighted Pair-Group Method using Arithmetic means) method [36].

Additionally, to confirm international synonymous and homonymous cases, the genetic profiles of 49 cultivars were determined using the nine core SSR marker set (VVS2, VVMD5, VVMD7, VVMD27, ZAG62, ZAG79, VVMD25, VVMD32) and compared with the European Vitis database (www.eu-vitis.de (accessed on 15 May 2021)) and Vitis International Variety Catalogue VIVC (www.vivc.de (accessed on 10 June 2021)).

2.5.2. Population Genetic Analysis

Population genetic parameters of Kara grape sub-populations (population information of cultivars can be seen in Table S1) were estimated using the Arlequin software Ver. 5.3 program according to the method of Excoffier and Lischer [37]. The genetic relationship dendrogram of the cultivars was created on the basis of the genetic distance of Nei, using the NTSYS-pc (Numerical Taxonomy and Multiware Analysis System) analysis program [38]. FCA was performed to find the presence of any population structure among eight grape populations using Genetix 4.05 [39]. Gene flows (*Nm*) among populations were estimated using Genetix 4.05, and genotypes were analyzed using the STRUCTURE 2.3 program. For each run, 100,000 replicates of "burn-in" followed by an additional 100,000 MCMC (Markov Chain Monte Carlo) replicates of data collection were conducted. A linkage model based on known distances among microsatellite loci and a model of correlated allele frequencies were used, and the data were analyzed with the STRUCTURE 2.3 program. Bayesian Population genetic analysis was also applied using the BAPS software to visualize population structure and admixture (http://www.helsinki.fi/bsg/software/BAPS (accessed on 21 March 2020)) [40].

We also analyzed the population structure with the STRUCTURE 2.3 program to estimate the possible RPPs. The grape populations were analyzed according to Pereira-Lorenzo et al. [41], and were based on the population structure analyzed with the STRUCTURE 2.3 program [42] using the same computing parameters except that the K level was calculated for K = 2–9 unknown RPPs with 25 replicates. We also used STRUCTURE-HARVESTER [43] to estimate the best K value supported by the current data [44]. Similar to Ergül et al. [24], the number of accessions strongly assigned to each RPP was determined based on the *qI* probabilities (probability of membership) higher than 80%.

2.5.3. Clonal Analysis

For clonal differentiation, the GenAlEx v6.5 program [45] was used to identify Multilocus Genotypes (MLGs) in the populations. Effective alleles (*Ne*), number of different alleles (*Na*), observed heterozygosity (*Ho*), Nei's [46] unbiased expected heterozygosity (*uHe*) and private alleles summary (*PAS*) values for each population were determined by using the same program.

In addition, a histogram of pairwise distances created by the software GenoType v1.2 [47] was applied to determine whether somatic mutations were present. Simpson's diversity and possible number of clones (representing clone number) based on multilocus lineage (MLLs) calculations were conducted by using the GenoDive v1.1 program [47]. Furthermore, genotypic diversity (*div*), effective number of genotypes (accessions) (*eff*), evenness (*eve*) and Shannon–Wiener (*shw*) diversity index values were calculated by using the GenoDive v1.1 program.

In analysis of the MLGs, various mutational threshold or T values (T represents the maximum distance allowed to identify a clone among the individuals with the same "multilocus genotype (accession)" value) were tested (e.g., from threshold = 0 to threshold = 10) to minimize the mutational problems and potential scoring errors. The groupings of MLGs within MLLs were evaluated, and accessions with the similar values of the mutational thresholds were thought to represent the clones.

3. Results

3.1. OIV Data Analysis

The detailed ampelographic definition presented in this work highlights clear morphological differentiation between studied Kara grape cultivars characterized using 39 OIV descriptors (Table S2).

The dendrogram (Figure 2A) of the studied OIV descriptors ultimately consists of two sub-groups, in which the character OIV 223 (Berry: shape) alone constitutes a single-member group. However, the larger sub-group with 37 members consists of two sub-groups, one of which is the OIV 004 (Young Shoot: density of prostrate hairs on tip) character alone. In other words, cultivars are more diverse in terms of OIV 223 and OIV 004 traits, and there was no significant correlation between these traits and the other examined traits. On the other hand, the OIV 66-4 (Mature leaf: length of vein N5) and OIV 225 (Berry: color of skin) properties showed correlation with each other and defined the shorter length of vein N5 in Kara grape cultivars. Additionally, OIV 51-2 (Young leaf: color of the upper side—leaf 4–6) was in correlation with the OIV 230 (Berry: color of flesh) character and determined that cultivars with green and yellow leaves contain colorless flesh. Besides the correlation of OIV 83-1 (Mature leaf: shape of base of upper leaf sinuses) and OIV 83-2 (Mature leaf: shape of base of lower leaf sinuses) with OIV 007 (Young leaf: density of prostrate hairs between veins), and other descriptors were identified (Figure 2A).

Using all accessions, two major clusters were identified by the cluster analysis: group 1 (46 accessions) and group 2 (3 accessions) (Figure 2B). The most dissimilar cultivars grouped together showing the longer branches in the tree. The minor cluster consisted of accessions: Miri Kara (Dendrogram no: 17) and Acı Kara (Dendrogram no: 5), which were linked to the Bulgar Karası (Dendrogram no: 15), while Bulgar Karası (Dendrogram no: 15) had the least similarity with the remaining 48 cultivars, although the larger group was divided to two sub-groups in dendrogram. The most similar accessions were Kokulu Kara (Dendrogram no: 12) and Beyaz Saplı Kara (Dendrogram no: 13), both of which are from the Marmara region. Finally, it should be noted that the results of the genetic and phenotypic similarity graphs were not completely consistent.

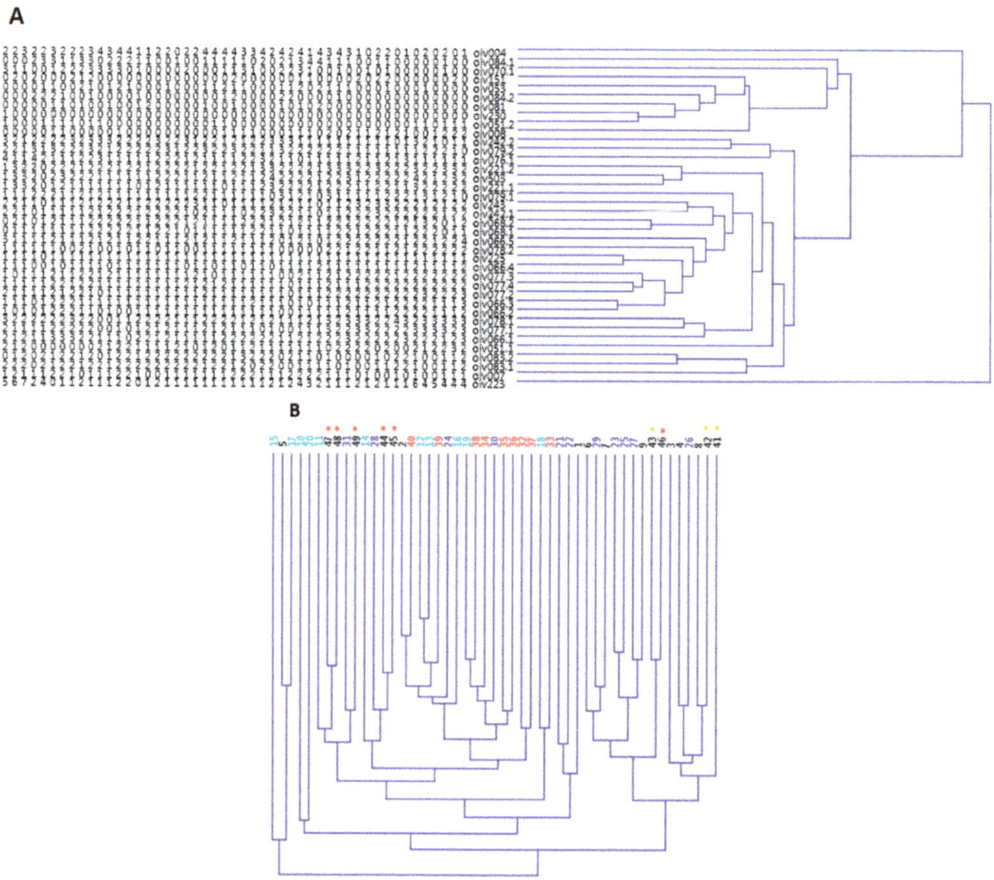

Figure 2. Cluster analysis dendrogram of OIV characters. (**A**) The cluster analysis dendrogram shows the relationship between Kara grape individuals (Table S2 can be used for the dendrogram sample numbers (dendrogram no) information. Green numbers: Marmara; Blue numbers: Black sea; Red numbers: Aegean; Black numbers: Mediterranean; Black numbers with yellow star: East Anatolia; Black numbers with red stars: Central Anatolia) (**B**).

The scatter plot showing the individuals with the sub-group population names is presented in Figure 3. It shows the scattering of the different grape samples and individuals belonging to different geographic sub-groups, analyzed in the bi-dimensional space determined by PCA 1 and PCA 2. In Figure 3, we do not see any groupings among Kara sub-populations based on the OIV data, but it can be seen that the Mediterranean accessions (black circles) accumulate on one side of the graph while the accessions of the Marmara region (blue squares) are gathered on the other side of the graph. However, among these, the Black Sea accessions (white diamonds) are well distributed among different sub-groups of accessions.

The biplot with loadings of each OIV characters is presented in Figure 4, showing which characters contributed to the separation of the individuals. It can be easily seen that OIV223 and OIV 077.1 (Mature leaf: length of teeth N2) have the highest loadings on the positive side, while OIV 004 and OIV 084.1 (Mature leaf: density of prostrate hairs—lower side) have the highest loadings on the negative side. Additionally, the first axis was defined by the density of prostrate hairs on the tip in young shoots (OIV 004) and length of teeth N2 in mature leaves (OIV 077-1), respectively.

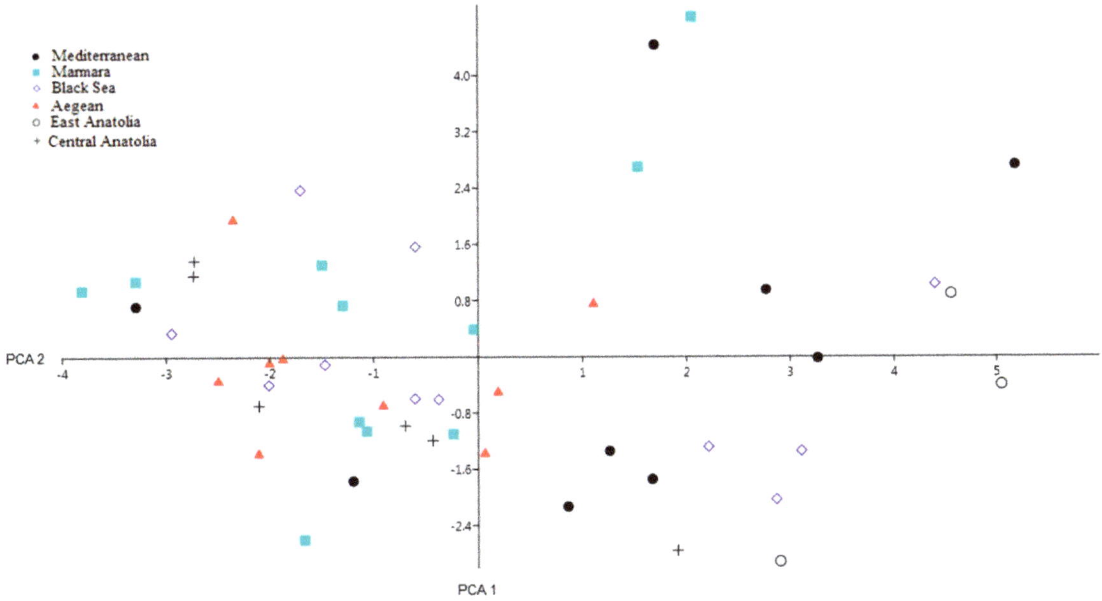

Figure 3. PCA scatter plot of Kara sub-populations. The PCA plot was determined for the first two principal components (PCA 1 and PCA 2). Each sample is indicated as a colored shape, according to the population information.

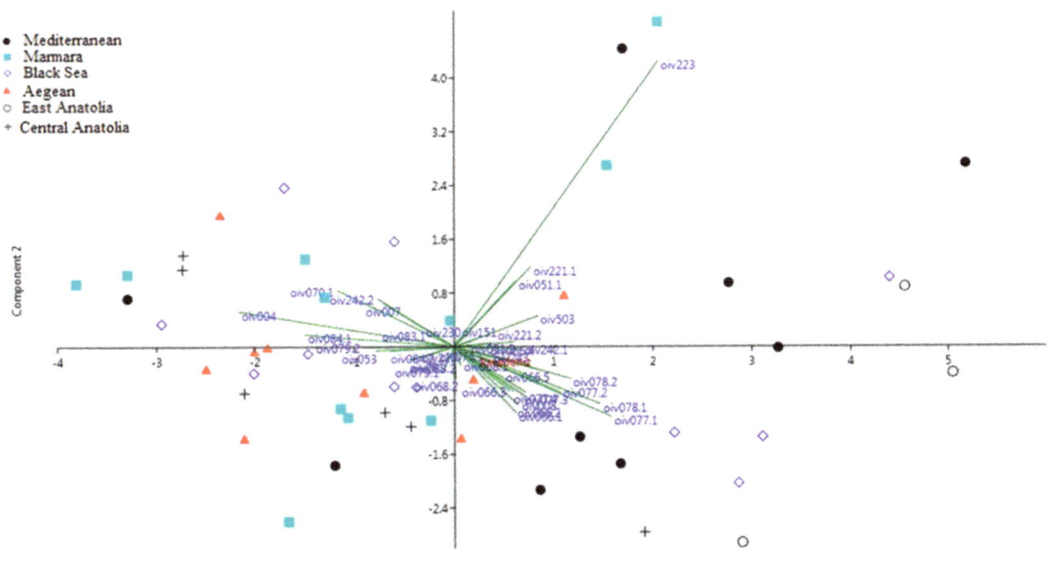

Figure 4. Biplot showing both scattering and the loadings of all OIV characters used in the analysis. Each sample is indicated as a colored shape according to the sub-population information.

3.2. SSR Analysis

A total of 196 alleles were approved in the 22 mentioned SSR loci, and the mean allele number (*n*) was found to be 8.91. The most and least informative loci were determined to be VMC2H4, with 13 alleles, and VVIB01, with 4 alleles, respectively (Table 1).

Table 1. SSR loci, allele number (*n*), expected heterozygosity (*He*), observed heterozygosity (*Ho*), probability of identity (*PI*).

No	SSR Loci	*n*	*He*	*Ho*	*PI*
1	VVS1	7	0.551	0.500	0.315
2	VVS2	12	0.843	0.885	0.075
3	VVMD5	9	0.8036	0.769	0.111
4	VVMD7	7	0.775	0.808	0.162
5	VVMD21	7	0.743	0.692	0.190
6	VVMD24	9	0.682	0.750	0.198
7	VVMD27	9	0.823	0.769	0.099
8	VVMD28	12	0.812	0.769	0.110
9	VVMD31	9	0.714	0.692	0.196
10	VrZAG21	8	0.709	0.750	0.185
11	VrZAG47	11	0.831	0.731	0.090
12	VrZAG62	10	0.778	0.904	0.140
13	VrZAG64	8	0.802	0.846	0.122
14	VrZAG79	9	0.815	0.846	0.100
15	VrZAG83	6	0.660	0.673	0.277
16	VrZAG112	9	0.621	0.615	0.219
17	VMC2H4	13	0.871	0.962	0.056
18	VMC2C3	7	0.696	0.712	0.240
19	VVIH54	11	0.745	0.500	0.132
20	VVIB01	4	0.526	0.558	0.557
21	VVMD25	8	0.789	0.846	0.136
22	VVMD32	11	0.820	0.577	0.094
	Total	**196**	**16.411**	**16.154**	**3.803**
	Average	**8.91**	**0.746**	**0.734**	**0.173**

Observed (*Ho*) and expected (*He*) heterozygosity rates were 0.734 and 0.746, respectively. The highest *Ho* was determined in VMC2H4, with a rate of 0.962, and the lowest was determined in VVS1 and VVIH54, with a rate of 0.500. *He* rates for VMC2H4 and VVIB01 were 0.871 and 0.526, respectively. These results prove the high polymorphism characteristics of these loci in grapes. VMC2H4 (0.056) and VVS2 (0.075) loci showed the lowest *PI* rates and were identified as being the most informative among the studied loci (Table 1).

3.2.1. Genetic Relations among the Kara Grape Cultivars

In this study, four synonymous (identical genotypes called by the different names) and five homonymous (different genotypes called by the same name) cases were determined. However, no identical (same names and same SSR profiles) cases were found in the study. The similarity cases detected in 49 Kara grape cultivars are shown in Table 2.

Table 2. Identical, synonymous and homonymous cultivars identified based on SSR analysis.

No	Identical	Synonymous	Homonymous
		(Cultivar Name/List no/Population (Region)-Location (Province))	
1	-	Patlak Kara/45/Central Anatolia-Sivas Siyah Üzüm/47/Central Anatolia-Yozgat	Kara Üzüm/1/Mediterranean-Adana-Kara Üzüm/15/Black Sea-Gümüşhane-Kara Üzüm/16/Mediterranean-Hatay-Kara Üzüm/23/Aegean-İzmir-Kara Üzüm/25/Black Sea-Kastamonu-Kara Üzüm/26/Marmara-Kırklareli-Kara Üzüm/49/Black Sea-Zonguldak
2	-	Deli Kara/3/Marmara-Balıkesir Yerli Kara/41/Marmara-Sakarya	Siyah Üzüm/12/East Anatolia-Diyarbakır-Siyah Üzüm/17/Mediterranean-Hatay-Siyah Üzüm/29/Central Anatolia-Konya-Siyah Üzüm/31/East Anatolia-Malatya-Siyah Üzüm/32/East Anatolia-Mardin-Siyah Üzüm/42/Black Sea-Samsun-Siyah Üzüm/43/Black Sea-Sinop-Siyah Üzüm/44/Black Sea-Sinop/-Siyah Üzüm/47/Central Anatolia-Yozgat
3	-	Eski Kara/11/Aegean-Denizli Yerli Kara/34/Aegean-Muğla	Ekşi Kara/10/Aegean-Denizli-Ekşi Kara/48/Central Anatolia-Yozgat
4	-	Siyah Üzüm/12/East Anatolia-Diyarbakır Kara Üzüm/26/Marmara-Kırklareli	Yerli Kara/34/Aegean-Muğla-Yerli Kara/41/Marmara-Sakarya
5	-	-	Katı Kara/36/Aegean-Muğla-Katı Kara/38/Black Sea-Ordu-Katı Kara/39/Black Sea-Ordu

In comparison with the European Vitis database (www.eu-vitis.de (accessed on 15 May 2021)) and Vitis International Variety Catalogue VIVC (www.vivc.de (accessed on 10 June 2021)), it was found that one cultivar (Siyah Üzüm, List no: 17) was synonymous with the "Muscat Hamburg" cultivar originating in the United Kingdom (Variety number VIVC: 8226, Accession number: DEU098-1980-274).

3.2.2. Genetic Structure Analysis among Kara Sub-Populations

The expected (H*exp*) and observed (H*obs*) heterozygosity values considering all sub-populations are presented in Table 3. Among the Kara grape sub-populations, the highest mean number of alleles was observed in the Marmara sub-population (6.72), while the lowest one was observed in the East Anatolia sub-population (3.31). However, H*exp* and H*obs* values were found to be approximately close to each other in all Kara sub-populations (~0.700) except East Anatolia, Central Anatolia and reference sub-populations (Table 3).

The Kara grape cultivars and reference (PN-CS-M) population were divided into two groups in the factorial correspondence analysis (FCA) (Figure 5A). In the FCA analysis of seven sub-populations, it was seen that the reference cultivars and the East Anatolia cultivars emerged as separate groups, while the remaining cultivars formed a cluster (Figure 5B).

Table 3. The expected and observed heterozygosity values for Kara grape sub-populations.

Sub-Population	Number of Cultivars (n)		Heterozygosity		Polymorphic Locus		Mean of Alleles/Locus
			Hexp	*Hobs*	*p* (0.95)	*p* (0.99)	
Mediterranean	9		0.700	0.732	1.0000	1.0000	5.27
		Std. error	0.086	0.170			
Marmara	11		0.752	0.743	1.0000	1.0000	6.72
		Std. error	0.084	0.155			
Black Sea	11		0.700	0.714	1.0000	1.0000	5.95
		Std. error	0.148	0.178			
Aegean	9		0.651	0.717	1.0000	1.0000	4.54
		Std. error	0.118	0.226			
East Anatolia	3		0.596	0.727	1.0000	1.0000	3.31
		Std. error	0.170	0.284			
Central Anatolia	6		0.613	0.765	1.0000	1.0000	3.72
		Std. error	0.125	0.255			
References	3		0.611	0.772	1.0000	1.0000	3.50
		Std. error	0.151	0.238			

Std. error: Standard error.

In the Bayesian population structure (BAPS) analysis, similarly to the FCA, the Kara grape population and reference population were clearly separated from each other (Figure 6A). In the BAPS analysis based on individuals (as six sub-populations and one reference population) (Figure 6B), it was observed that the Kara sub-populations were admixed to some extent (Figure 6B). In the BAPS analysis based on individuals and populations, all sub-populations except Central Anatolia were found to be highly admixed with each other (Figure 6C).

Genetic distance values are given in Table 4. The reference population had high genetic distance values relative to the Kara sub-populations. Excluding the reference population, the highest genetic distance values among the Kara sub-populations were observed between the East Anatolia and Aegean populations (0.478), and between the Central Anatolia and East Anatolia (0.470) populations. The lowest genetic distance values were recorded between the Black Sea and Marmara populations (0.147), and between the Black Sea and Mediterranean (0.149) populations.

Table 4. Genetic distance between sub-populations.

Sub-Populations	Mediterranean	Marmara	Black Sea	Aegean	East Anatolia	Central Anatolia	Ref.
Mediterranean	-						
Marmara	0.149	-					
Black Sea	0.170	0.147	-				
Aegean	0.248	0.180	0.195	-			
East Anatolia	0.300	0.311	0.352	0.478	-		
Central Anatolia	0.229	0.232	0.157	0.231	0.470	-	
Ref.	0.632	0.519	0.699	0.641	0.815	0.757	-

Ref: references.

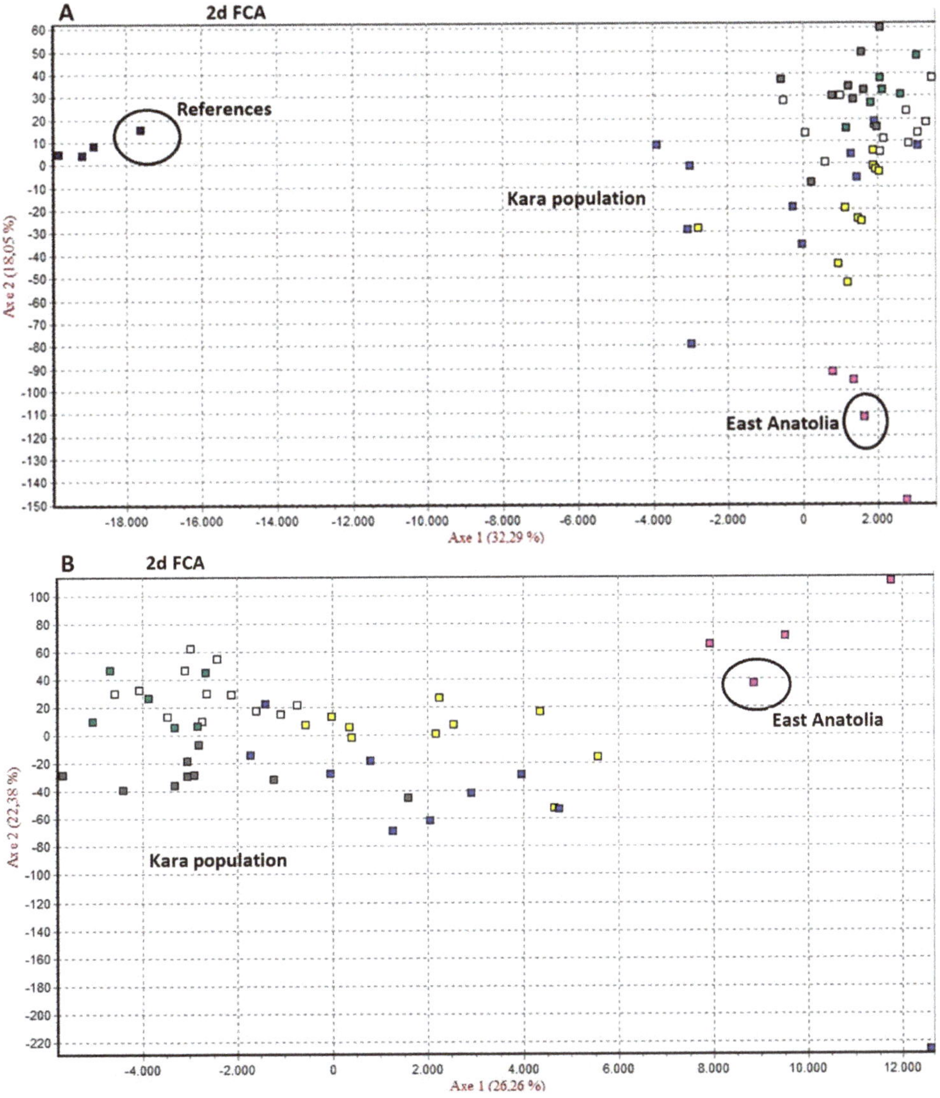

Figure 5. Factorial correspondence analysis (FCA). Six Kara sub-populations with reference population based on individuals (1. Mediterranean—yellow, 2. Marmara—blue, 3. Black Sea—white, 4. Aegean—gray, 5. East Anatolia—pink, 6. Central Anatolia—dark blue) (**A**). Six Kara sub-populations without reference population. The points in the circle show the average values of the reference population and East Anatolia population (**B**).

The genetic differentiation (*Fst*) values and gene flow (*Nm*) values are given in Table 5. *Fst* values among sub-populations were moderate, and the East Anatolia sub-population showed especially significant genetic differentiation with the Aegean (0.087, $p < 0.001$) and Central Anatolia (0.084, $p < 0.05$) sub-populations. Table 5 also indicates that the reference population showed significant genetic differentiation with the Central Anatolia (0.147, $p < 0.05$) and Aegean (0.122, $p < 0.05$) sub-populations. The *Fst* values explain the high *Nm* among the sub-populations. The gene flow (*Nm*) between Kara sub-populations and the reference population is extremely limited, and the highest *Nm* value (4.08) was observed between the reference and Marmara populations. Among Kara sub-populations, the highest *Nm* value (84.93) was observed between the Marmara and Black Sea sub-populations, and the lowest *Nm* value (2.28) was observed between the Central Anatolia and East Anatolia sub-populations (Table 5).

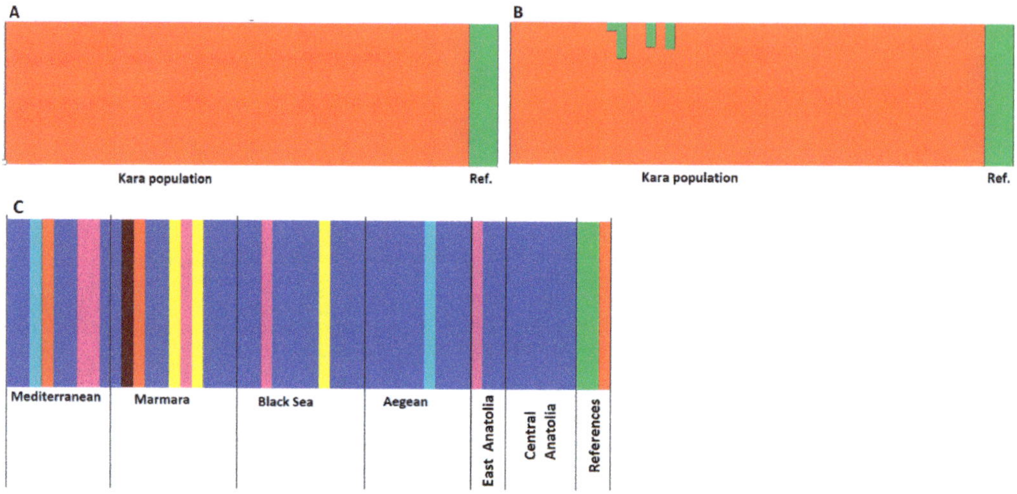

Figure 6. Bayesian analysis of population structure (BAPS) of Kara grape cultivars (Kara population) and reference (Ref.) populations (**A**). Bayesian analysis of population structure based on individuals of Kara grape cultivars (Kara population) and reference (Ref.) cultivars (**B**). Bayesian analysis of population structure based on individuals and sub-populations (sub-populations are separated by black vertical lines) (**C**).

Table 5. Pairwise *Fst* values and gene flow (*Nm*) among the Kara sub-populations.

Sub-Population (*Fst/Nm*)	Mediterranean	Marmara	Black Sea	Aegean	East Anatolia	Central Anatolia	Ref.
Mediterranean	-						
Marmara	0.00027/İnf.	-					
Black Sea	0.01284/21.03	0.00500/84.93	-				
Aegean	0.04291 ***/5.47	0.02003/12.82	0.03029 */7.96	-			
East Anatolia	0.01655/15.66	0.01518/25.11	0.03902 */6.50	0.08739 ***/2.48	-		
Central Anatolia	0.03113/6.97	0.03068/7.80	0.00898/22.23	0.04072 */5.08	0.08424 */2.28	-	
Ref.	0.09815 ***/2.29	0.06173 */4.08	0.11655 ***/1.91	0.12209 */1.71	0.12159/1.68	0.14787 */1.28	-

* $p < 0.05$, *** $p < 0.001$, Inf: infinite, Ref.: References.

The structural analysis of the whole dataset of Kara grape cultivar sub-populations reached a maximum K value at K = 2, as estimated by STRUCTURE-HARVESTER, which corresponds to two main groups (RPP1 and RPP2) of genotypes (Figure S1). Reconstructed panmictic populations were realized based on the Bayesian model-based clustering methods. We also calculated the number of genotypes strongly assigned to each of the two

RPPs based on the qI (probability of membership) probabilities greater than 80%. The RPP distribution of the cultivars (individuals) is given in Figure 7.

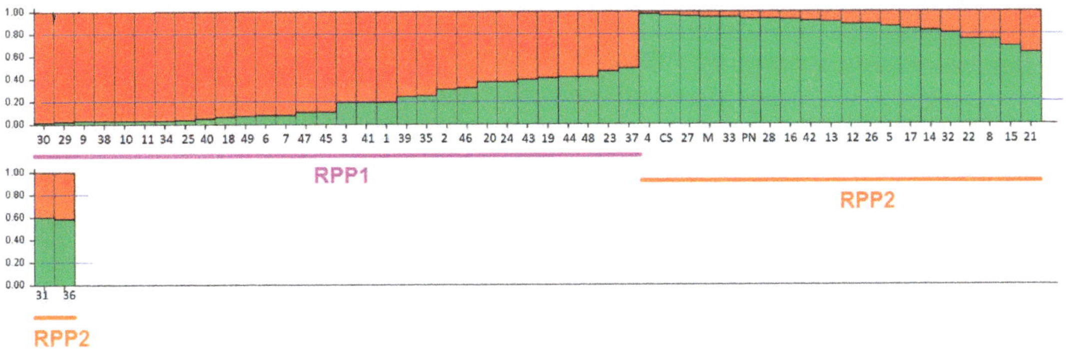

Figure 7. Individual representation of RPPs and the percent probabilities of each individual in RPP groups (Table S1 can be used for the sample numbers (List no.) information).

Overall, results showed that 34 (65%) genotypes were assigned to each RPP with higher than 80% probability, and 35% either did not belong to the representative RPPs or showed a low probability of membership. Table 6 summarizes the results of membership probabilities and the representative populations that formed the RPPs. All members of the reference and East Anatolia sub-populations were grouped in RPP2, and all members of the Central Anatolian sub-population were grouped in RPP1. All other sub-population genotypes were seemingly admixed and were grouped in both RPP1 and RPP2.

Table 6. Classification of Kara grape cultivars by STRUCTURE using 22 SSR loci in K = 2 reconstructed panmictic populations.

RPP	Number of Genotypes	$qI > 0.8$	$qI < 0.8$	The Most Representative Populations and the Number of Individuals
RPP1	30	18 (60%)	12 (40%)	Mediterranean (5), Marmara (5), Black Sea (7), Aegean (7), Central Anatolia (6)
RPP2	22	16 (73%)	6 (27%)	Mediterranean (4), Marmara (6), Black Sea (4), Aegean (2), East Anatolia (3), References (3)
Overall	52	34 (65%)	18 (35%)	All populations

qI: coefficient coancestry.

3.2.3. Clonal Analysis

In the MLG analysis, except for three reference cultivars, 12 different multi-locus genotypes (MLGs) were determined, while 37 unique genotypes were detected. According to the distribution of the sub-populations, the highest MLG number was determined as 3 in the Marmara sub-population, while the lowest was 1 in East Anatolia. The number of different alleles (*Na*) was highest in the Marmara (*Na*: 6.72) and Black Sea (*Na*: 5.95) sub-populations, in direct correlation with the sub-population number (Table 7).

Table 7. Multi-locus (MLG) number, number of different alleles (*Na*), effective alleles (*Ne*), observed heterozygosity (*Ho*), unbiased expected heterozygosity (*uHe*) and private alleles summary (*PAS*) values found in different grape sub-populations.

Sub-Population (Sub-Population Number)	MLG	*Na*	*Ne*	*Ho*	*uHe*	*PAS* (Locus No: Alleles (bp))
Central Anatolia (6)	2	3.72	2.83	0.76	0.66	VVMD32:146
East Anatolia (3)	1	3.31	2.75	0.72	0.71	VVS1:181, VVMD28:281, VVMD31:201, ZAG83:203
Mediterranean (9)	2	5.27	3.62	0.73	0.74	VVS2:141, VVMD7:230, VVMD24:221, ZAG62:190, ZAG112:229, VVMD5:271
Marmara (11)	3	6.72	4.44	0.74	0.78	VVS1:193, VVMD5:231, VVMD7:258, VVMD24:201, VVMD28:219, ZAG21:195, ZAG47:183, ZAG62:210, ZAG83:183, ZAG112:247, VMC2h4:196, VVIh54:142 and154, VVIb01:300, VVMD25:237 and247, VVMD32:258
Black Sea (11)	2	5.95	3.99	0.71	0.73	VVMD5:219,VVMD28:245, ZAG21:211, ZAG47:169, ZAG62:208, ZAG64: 144, VMC2h4:208 and210; VMC2c3:187, VVMD32:262
Aegean (9)	2	4.54	3.14	0.71	0.68	VVMD21:250, VVIb01:308
Total	12	24.97	20.77	4.37	4.3	-

Effective alleles (*Ne*) ranged from 2.75 (East Anatolia) to 4.44 (Marmara) among sub-populations, and *Ho* values were found to be greater than *uHe* values in the same sub-population comparisons (except for Mediterranean and Marmara sub-populations). The highest private alleles summary (*PAS*) value was found in Marmara, with 17 alleles at 15 different SSR loci, and the lowest *PAS* value was found in Central Anatolia, with 1 allele at 1 SSR locus; no correlation was shown between *PAS* values and the sub-population number (Table 7).

Considering the clonal diversity values, it was seen that all accessions of the East Anatolia, Mediterranean and Black Sea sub-populations were unique genotypes. Genotypic diversity (*div*) values, also known as expected heterozygosity, were found to be similar to each other (approximately 0.9) in all sub-populations, while the evenness (*eve*) value, which shows the distribution profiles of individuals within the sub-population, had the lowest amount in the Aegean (0.920) sub-population (Table 8).

The highest evenness value was observed in East Anatolia (1.00), Mediterranean (1.00) and Black Sea (1.00), which reflects the equal frequency of all genotypes in these sub-populations. The Shannon–Wiener (*shw*) value was found to be higher (1.041) in the Black Sea sub-population compared to other populations, and this value shows that this population has a high diversity (Table 8).

In the multi-locus lineages (MLLs) analysis, it was determined that there were some small differences in the number of different MLLs, especially between T = 0 and T = 3 threshold values, and only one match difference occurred between T = 1 and 2.

Table 8. Number of genotypes (accessions) (*gen*)/clonality, effective number of genotypes (accessions) (*eff*), genotypic diversity (*div*), evenness (*eve*), and Shannon–Wiener (*shw*) values (for threshold = 2) determined in multi-locus lineages (MLLs) analysis.

Sub-Population (Sub-Population Number)	Number of Genotypes (Accessions) (*gen*)/Clonality	Effective Number of Genotypes (Accessions) (*eff*)	Genotypic Diversity (*div*)	Evenness (*eve*)	Shannon–Wiener (*shw*)
Central Anatolia (6)	5/1	4	0.933	0.900	0.677
East Anatolia (3)	3/0	3	1	1	0.477
Mediterranean (9)	9/0	9	1	1	0.954
Marmara (11)	10/1	9	0.981	0.930	0.986
Black Sea (11)	11/0	11	1	1	1.041
Aegean (9)	8/1	7	0.972	0.920	0.887

There was no difference in the number of MLLs between T = 2 and 3. For this reason, results for T = 2 were considered as the threshold value, and a total of three clones were detected. Detailed information (accession name, accession no, etc.) of the clones determined in the T = 2 threshold value are given in Table 9.

Table 9. Cultivar names and other cultivar names assigned to different accessions at the T = 2 threshold value based on MLLs analysis.

Matched Number	Accession Name (List No-Region, Province)	Matches at T = 2 (List No-Region, Province)
1	Kokulu Kara (5-Marmara, Bilecik)	Siyah Üzüm (17-Mediterranean, Hatay)
2	Yerli Kara (34-Aegean, Muğla)	Eski Kara (11-Aegean, Denizli)
3	Patlak Kara (45-Central Anatolia, Sivas)	Kara Üzüm (26-Marmara, Kırklareli)

4. Discussion

4.1. OIV Data Analysis

Similar to the results of previous investigations [48–50], the highest correlations were observed between traits of the same category (i.e., leaf and berry traits). In this sense, the characters OIV 66-2 (Mature leaf: length of vein N2) and OIV 66-3 (Mature leaf: length of vein N3) were the most strongly correlated descriptors, and showed similar variations in different accessions. The other correlation was seen between OIV 221-1 (Berry length) and OIV 221-2 (Berry width) properties and OIV 503 (Single berry weight); this showed that the single berry weight is more strongly related to the length of the berry. Additionally, a high correlation between OIV 242-1 (Length of berry seed) and OIV 243 (Weight of berry seed) descriptors was detected. Previously, Lamine et al. [51], in a multivariate analysis and clustering of Tunisian autochthonous grapes, reported that OIV 225 (color of the berry skin) and OIV 230 (color of the berry flesh) descriptors were the most strongly correlated characters in terms of fruit characteristics, and characters OIV 079-1 and OIV 079-2 were the most strongly correlated descriptors in terms of leaf characteristics. In fact, morphological markers have a higher degree of genomic coverage and most individual phenotypic markers are multigenic [51]. Morphological investigations of grapevine have previously been carried out by Lamine et al. [51] and Khalil et al. [52].

Shoot, leaf and berry descriptors have been generally used as powerful tools for discrimination of grapevine cultivars. The number of shoot descriptors (n = 4) in our study was not enough to separate grapevine cultivars. The number of leaf descriptors used in this work (n = 25) was higher than those reported previously by Sabir et al. [53], who used 12 descriptors, Khalil et al. [52], who used 9 descriptors, Dilli et al. [54], who used 22 descriptors, and Knezović et al. [55], who used 7 descriptors. Furthermore, nine ampelographic descriptors were evaluated for berry characterization in this study, and the highest amount of variation was attained by berry descriptors. The number of descriptors in the studies by Dilli et al. [54], Knezović et al. [55] and Khalil et al. [52] was 12, 15, 4 and 9, respectively.

Finally, in order to determine the table cultivars of grape, the berry shape variable, representing the most independent variance segment, was selected, along with the variables of berry weight, berry seed weight, berry length and berry width. These traits are important both for the consumer and producers in terms of berry size, seed size and berry shape [52]. As a result, considering the above characteristics, the cultivars Kara üzüm (List no: 16), Acı kara (List no: 18), İyi Bağ Karası (List no: 28) and Sofra Karası (List no: 33) were identified as table grapes, due to the high (5.1 g) or very high (6.3 g or more) weight of their berries, their small seeds (35–50 mg), the berry length (21.5–25 mm or more), berry width (17.5–21.0 mm) and cylindrical and elliptic shapes. In addition, the cultivars Pat kara (List no: 20) and Patlak kara (List no: 45), with the above specifications in the size of berry, but with a roundish berry shape, were among the special table cultivars. Kara üzüm (List no:26), from Kırklareli (Marmara sub-population), has a berry with colored flesh, and so could be a suitable candidate for use in the fruit juice industry. All cultivars had blue and black berry skin. In table cultivars, a deep red to dark purple color is preferred to light-colored selections, and a round to oval shape and a strong Muscat flavor are welcomed [56].

Doligez et al. [57] indicated the lack of co-localization between QTLs for berry and seed traits, suggesting that the genetic control of both traits may be partly dissociated. In this sense, direct correlation between seed traits and berry traits, like berry dimensions, were not observed. This is in accordance with the findings of Khalil et al. [52].

4.2. SSR Analysis

In this study, the number of different alleles for 22 SSR was 196, ranging from 4 (VVIB01) to 13 (VMC2H4) per locus. The VMC2H4 locus has been shown among the loci with high allele numbers [28,58], and previous studies also showed that the VVS2 locus [58–60] and VVMD28 locus [58,59,61] were among the most informative loci (Table 1). Furthermore, the VVIB01 locus has been used in genetic diversity studies of grapevine accessions from Southeast [62] and Eastern Turkey [63], and also the genetic identification and characterization of Armenian grapevine cultivars [60], in which it showed the lowest allele number among the loci. In this study, the VVIB01 locus, with four alleles, involved the lowest number of alleles, showing similar results to those of previous studies.

Given that the average number of alleles in each locus is influenced by the number of samples and the number of examined loci, this amount varies from 5 to 20 in different analyses [13,60,64], similar to our study (average number of alleles: 8.91, Table 1). This value was found to be 5.75 by Eyduran et al. [63] in the genetic characterization of autochthonous grapevine cultivars from Eastern Turkey, whereas it was recorded as 20.95 by Riaz et al. [61] in accessions from around the Mediterranean basin and Central Asia.

Ho is the proportion of heterozygous individuals in the analyzed sample; expected heterozygosity (*He*) or genetic diversity shows the percentage of the population that would be heterozygous if an accidental cross occurred between individuals. The highest amount of *Ho* was found in locus VMC2H4 and VrZAG62, implying high genetic diversity in the mentioned SSR loci. Additionally, the average *Ho* value (0.73) obtained in this survey for the Kara grape population was found to be very similar to average values for East Anatolia grapes (0.71, 0.75 and 0.73) [62,63,65], to Armenian grapevine cultivars (0.74) [60] and also to Mediterranean basin and Central Asian cultivars, which had an average of 0.74 [61]. Moreover, the high average value of heterozygosity (Table 3) may be seen by the high number of crosses in the variety set and random pollination, which is consistent with findings from previous studies [66].

4.3. Genetic Relations among the Kara Grape Cultivars

There are over 10,000 cultivars of grapes in the world, and the grouping of these cultivars is mostly confused due to homonymous, synonymous, limited or inaccurate historical information, etc. [67]. In this study, which might contribute to the conservation of national genetic resources, four synonymous and five homonymous cases were found as a result of SSR analysis. The synonymous cultivars were Patlak Kara (List no: 45) and Siyah Üzüm (List no: 47), from Sivas and Yozgat provinces, respectively; Deli Kara (List no: 3) and Yerli Kara (List no: 41) from Balıkesir and Sakarya provinces, respectively; and Eski Kara (List no: 11) and Yerli Kara (List no: 34) from Denizli and Muğla, respectively. These cultivars are from the same region, but some of their morphological characteristics (Table S2) are different. Siyah Üzüm (List no: 12) and Kara Üzüm (List no: 26) from the synonymous cultivars were collected from Diyarbakır (East Anatolia sub-population) and Kırklareli (Marmara sub-population) provinces, respectively, and also gave different morphological results in some traits from separate regions. In this case, it can be concluded that the same cultivar has incorrectly been nominated by a different name. Furthermore, the fact that these cultivars are taken from different places with similar morphological data supports their synonymous cases.

Kara Üzüm (List no: 1), Kara Üzüm (List no: 15), Kara Üzüm (List no: 16), Kara Üzüm (List no: 23), Kara Üzüm (List no: 25) Kara Üzüm (List no: 26) and Kara Üzüm (List no: 49) were found to be homonymous cultivars. All homonymous cultivars were collected from different provinces, and there was no relationship between them (Tables 2 and S1). The fact that homonymous cultivars are not the same could be caused by incorrect nominating. In addition, variations (clones, types, etc.) of each variety could occur over time. Molecular studies have shown that the majority of homonymous cultivars are clonal variations [68], and these findings support our results.

The highest similarity ratio among cultivars was found to be 95.5% between Kokulu Kara (List no: 5) and Siyah Üzüm (List no: 17) and between Siyah Üzüm (List no: 29) and Yerli Siyah (List no: 30) cultivars. Kokulu Kara (List no: 5) and Siyah Üzüm (List no: 17) were different from each other in one allele of the VVMD7 locus, and Siyah Üzüm (List no: 29) and Yerli Siyah (List no:30) were different from each other in one allele of the VVMD28 locus. However, the Kokulu Kara (List no: 5) and Siyah Üzüm (List no: 17) cultivars have relatively different morphological features, and they are quite different in terms of their sampling place. On the other hand, morphological data in the Siyah Üzüm (List no: 29) and Yerli Siyah (List no: 30) cultivars support the genetic data (genetic similarity. These two cultivars were collected from the same location and are morphologically separated from each other only in berry shape (Table S1).

Except for the examples mentioned above, the highest similarity ratio was found to be 75% in Kara Üzüm (List no: 23) and Sıkkara (List no: 24) cultivars, while they differed in 11 loci. Moreover, the regions from which these two cultivars were collected were the same, their maturation dates were close to each other and they were similar in terms of other morphological characteristics. (Table S2).

In this study, a relatively low similarity ratio was determined among the Kara grape cultivars. On the other hand, there was no direct correlation between the genetic relations (genetic similarity dendrogram) of the cultivars and the original cultivation regions. The genotypes belonging to six sub-populations (regions) show a nested distribution in the dendrogram, and the homonymous character in some cultivars is an indicator of a gene flow that occurs naturally or by transport in the regions in ancient times. This is supported by the results obtained in Table 5.

4.4. Genetic Structure Analysis among Kara Sub-Populations

Despite the large number of synonymous and clonal relationships among the accessions, we observed a high level of genetic variation. In the total germplasm collection, the average genetic diversity quantified by the expected heterozygosity (0.752) and the number of alleles in each locus (6.7) was higher for Marmara (Table 3). Likewise, the highest value of observed heterozygosity was revealed for the Central Anatolia and Marmara sub-populations (Table 3).

As shown in Table 5, *Fst* values that are significantly greater than zero indicate a deficiency of differentiation among these populations, probably as a consequence of genetic drift, gene flow, mating systems, selection, and mutations [69]. Previously, Ergül et al. [7] emphasized that limited genetic differentiations observed in Anatolian grape cultivars might be a reflection of the area's long history of grape cultivation and material exchanges between provinces of Turkey. In this survey, the *Nm* value among Kara sub-populations ranged from 2.28 to an infinite amount, indicating a high degree of gene flow and continuous distribution of genes, and also limited differentiation between sub-populations (Table 5). On the other hand, as expected, the highest levels of pairwise genetic differentiation values were observed between Anatolian grapes and the reference populations (PN-CS-M), which is in accordance with Yılmaz et al. [32].

Genetic distances and genetic similarities are two important parameters to measure when assessing the genetic diversity of the population. Results of genetic distance (genetic similarity, %) between populations based on Nei [46] revealed that Kara grape sub-populations of the Aegean and East Anatolia regions were the least genetically similar (52.2%), with a maximum genetic distance (0.478) (Table 4). Among these, the minimum genetic distance (0.147) was found between Black Sea and Marmara regions.

To understand the effect of geographic distance on genetic structure, genetic distances among studied populations were examined and the results showed that the genetic distance was not exactly correlated with the geographic distance, suggesting that the geographic distance is not the principal factor influencing genetic differentiation among the Kara sub-populations of Turkey. For instance, these results showed that the genetic relationship of Mediterranean germplasm with Marmara germplasm was close. These results are consistent with a long history of cultivation and increased selection pressure by humans for Kara cultivars. In addition, the East sub-population was found to have the most genetic distance with other groups.

It is accepted that, the consequences of gene introgression from sympatric populations are strongly dependent on the extent of gene flow. In the Bayesian analysis of population structure, the results showed that genotypes similar to the reference cultivars were found in sub-populations, and also similar genotypes were seen between different regions (Figure 6C). When the same data was analyzed collectively, it is clear that the sub-population with the most different genotypes is Marmara, whereas the Central region was found to be a non-admixing region. In FCA analysis, dispersion in Marmara is also clearly seen. According to the BAPS results, this indicates the introgression of genes from different regions to this region, while they were least similar to those from the Eastern and reference accessions. Nevertheless, the highest genetic distance was observed between Anatolian and reference accessions, which is consistent with the results of Wang et al. [70], who showed that Cabernet Sauvignon and Merlot were clustered together in a discrete cluster separately from the remaining local Anatolian cultivars. A high population differentiation value was observed between the grapes from East Anatolia region and the reference cultivars (0.815). High differentiation values were observed between Anatolian and reference cultivars (PN-CS-M), which could be explained by factors such as geographic isolation, variety evolution, and restricted gene flow.

As mentioned in the BAPS analysis, genotypes similar to reference cultivars (PN-CS-M) were also found in sub-populations, and similar genotypes were seen among different regions. This structure was not influenced by geography with an Eastern–Western gradient and human usage factor, as already identified by Laucou et al. [71] using 10 K genome-wide

SNPs (Single nucleotide polymorphisms). Additionally, the large number of genotypes (40%) remained in a large admixed group. This could be due to several factors, such as: (i) the low genetic differentiation [72]; (ii) a departure from the underlying assumptions of the Bayesian model (random mating population), which are probably not met in this cultivated grape population; (iii) true admixing in some cultivars if they are directly descended from spontaneous or man-made crosses between parents belonging to separate ancestral groups; and finally (iv) a computational difficulty for STRUCTURE to assign individuals to groups in the presence of a large number of informative markers [71].

4.5. Clonal Analysis

In viticulture, it is known that grape cultivars usually consist of separate clones sharing common phenotypic characteristics which are grouped as a cluster of cultivars. Clonal polymorphism within perennial species is mainly accepted to be associated with naturally occurring mutations during plant growth [73]. However, if clones belonging to a cluster have sufficiently different characteristics, they are considered different cultivars. Therefore, these clones, which are morphologically very similar, are very difficult to distinguish by visual observation [74]. On the other hand, SSR markers have frequently been used in clone analysis of different plant species in recent years, as they are highly informative about the level of heterozygosity between and within grapevine cultivars [75–78].

In the analysis of 70 Italian grape genotypes with 13 SSR markers, 39 unique genotypes were determined, while in our study [78], interestingly, 46 unique genotypes were determined among 49 grapevine cultivars belonging to six different regions (sub-populations). On the other hand, in the analysis performed with 11 SSR markers on 164 Cypriot indigenous grape genotypes, a total of 83 multi-locus genotypes (MLGs) were identified [77]. In this study, however, 12 MLGs were determined, which could be attributed to the fact that the genotypic abundance feature associated with the number of MLGs in genotypes is slightly lower, despite the different sub-populations.

Heterozygosity values (*uHe* and *Ho*) determined in MLG analysis provide information about the genetic diversity and relationships between multi-locus genotypes [79]. In our study, unbiased expected heterozygosity (*uHe*) and observed heterozygosity (*Ho*) values were found to be very close to each other and both heterozygosity values were found to be more than 0.66 in all six populations in MLG analysis. This situation reveals that there might be possible genetic variation among multi-locus genotypes in each population [79].

5. Conclusions

Anatolia (Turkey) is among the most important countries in the world in terms of vineyard areas and grape production due to the suitability of climatic and growing conditions. In addition, Turkey, as one of the countries wherein *Vitis vinifera* L. was first cultivated, possesses a rich grapevine gene potential that has emerged over time through natural hybridization.

In this study, ampelographic analysis of 49 Kara grapevine germplasms (from six different regional sub-populations), which have high antioxidant content and also economic importance, was performed with 39 OIV descriptors. Genetic characterization and population structure analysis were also carried out using 22 SSR markers. As a result of OIV analysis, it was determined that especially OIV 223 (Berry: shape) and OIV 004 (Young Shoot: density of prostrate hairs on tip) were the defining features in the Kara grape OIV dendrogram. In the PCA analysis, a clear grouping profile was not determined among the Kara grape sub-populations.

As a result of SSR analysis, the high number of homonyms and the low number of detected clones and synonyms reveals the genetic richness of the Kara grape germplasm. Among the sub-populations, the East Anatolia sub-population especially differed from other sub-populations according to FCA and genetic distance analysis, while the Central Anatolia sub-population showed a homogeneous profile in the BAPS analysis. It is useful to screen the detected clonal genotypes with SNP-level techniques such as GBS (Genotyping-by-Sequencing).

Evaluation of the genetic diversity in grape germplasms is of great importance in terms of improving quality traits of interest and identifying gene sources. In this sense, it is thought that the data from this study will contribute to genetic characterization, genetic protection and different breeding programs in grapes.

Supplementary Materials: The following supporting information can be downloaded at: https://www.mdpi.com/article/10.3390/horticulturae9070743/s1, Table S1: Some ampelographic characteristics and original collection locations of Kara grape accessions used in this study, Table S2: Information related to OIV descriptors (characteristic description) and OIV data codes, Table S3: SSR loci information (SSR locus name, primer sequence (F: forward primer, R: reverse primer) and references) [25–29], Figure S1: Population structure of predefined Kara grape sub-populations based on 22 SSR data (**A**). The graph showing the best K value as 2 (the highest peak) (**B**). Reconstructed panmictic populations (RPPs) groups general distribution (**C**).

Author Contributions: Conceptualization, N.A. and A.E.; methodology F.Y.B. and C.Y.Ö.; data curation, N.H., C.Y.Ö., O.E. and T.U.; writing—original draft preparation, N.A. and F.Y.B.; writing—review and editing, N.H.; provided plant collections, O.E., T.U. and A.S.Y.; supervision, A.E.; formal analysis Y.B.; investigation, Y.S.K. and C.Ö.; visualization C.Ö., Y.S.K. and A.S.Y. All authors have read and agreed to the published version of the manuscript.

Funding: This study was supported by the Scientific and Technical Research Council of Turkey (TUBITAK) (Grant No. 105 G 078). A portion of the current study was conducted as thesis of the first author.

Data Availability Statement: All data are included in the manuscript.

Conflicts of Interest: The authors declare no conflict of interest.

References

1. Ekhvaia, J.; Akhalkatsi, M. Morphological variation and relationships of Georgian populations of *Vitis vinifera* L. subsp. sylvestris (C.C. Gmel.). Hegi. *Flora-Morphol. Distrib. Funct. Ecol. Plants* **2010**, *205*, 608–617. [CrossRef]
2. Naqinezhad, A.; Ramezani, E.; Djamali, M.; Schnitzler, A.; Arnold, C. Wild grapevine (*Vitis vinifera* subsp. sylvestris) in the Hyrcanian relict forests of northern Iran: An overview of current taxonomy, ecology and palaeorecords. *J. For. Res.* **2018**, *29*, 1757–1768. [CrossRef]
3. da Silva Haas, I.C.; Toaldo, I.M.; de Gois, J.S.; Borges, D.L.; de Oliveira Petkowicz, C.L.; Bordignon-Luiz, M.T. Phytochemicals, monosaccharides and elemental composition of the non-pomace constituent of organic and conventional grape juices (*Vitis labrusca* L.): Effect of drying on the bioactive content. *Plant Foods Hum. Nutr.* **2016**, *71*, 422–428. [CrossRef] [PubMed]
4. Glevitzky, I.; Dumitrel, G.A.; Glevitzky, M.; Pasca, B.; Otrisal, P.; Bungau, S.; Cioca, G.; Pantis, C.; Popa, M. Statistical analysis of the relationship between antioxidant activity and the structure of flavonoid compounds. *Rev. Chim.* **2019**, *70*, 3103–3107. [CrossRef]
5. Taymaz, E.R.; Uslu, M.E. Investigation and Determination of the Presence of Resveratrol in the Leaves of 99 Grapevine Varieties from the Western Anatolia Region in Turkey. *JMPB* **2022**, *1*, 1–8.
6. Arroyo-García, R.; Ruiz-García, L.; Bolling, L.; Ocete, R.; López, M.A.; Arnold, C.; Ergul, A.; Söylemezoğlu, G.; Uzun, H.I.; Cabello, F.; et al. Multiple origins of cultivated grapevine (*Vitis vinifera* L. ssp. *sativa*) based on chloroplast DNA polymorphisms. *Mol. Ecol.* **2006**, *15*, 3707–3714. [CrossRef] [PubMed]
7. Ergül, A.; Perez-Rivera, G.; Söylemezoğlu, G.; Kazan, K.; Arroyo-Garcia, R. Genetic diversity in Anatolian wild grapes (*Vitis vinifera* subsp. sylvestris) estimated by SSR markers. *Plant Genet. Resour.* **2011**, *9*, 375–383. [CrossRef]
8. Samah, M.I.; Sahar, S.A.S.; Khaled, A.S.; Hoda, M.H.A. Phenolic compounds and antioxidant activity of white, red, black grape skin and white grape seeds. *Life Sci. J.* **2012**, *9*, 3464–3474.
9. Nadeem, M.A.; Nawaz, M.A.; Shahid, M.Q.; Doğan, Y.; Comertpay, G.; Yıldız, M.; Hatipoğlu, R.; Ahmad, F.; Alsaleh, A.; Labhane, N.; et al. DNA molecular markers in plant breeding: Current status and recent advancements in genomic selection and genome editing. *Biotechnol. Biotechnol. Equip.* **2018**, *32*, 261–285. [CrossRef]
10. Vezzulli, S.; Doligez, A.; Bellin, D. Molecular Mapping of Grapevine Genes. In *The Grape Genome. Compendium of Plant Genomes*; Cantu, D., Walker, M., Eds.; Springer: Cham, Switzerland, 2019; pp. 103–136.
11. Upadhyay, A.; Saboji, M.D.; Reddy, S.; Deokar, K.; Karibasappa, G.S. AFLP and SSR marker analysis of grape rootstocks in Indian grape germplasm. *Sci. Hortic.* **2007**, *112*, 176–183. [CrossRef]
12. Meneghetti, S.; Costacurta, A.; Morreale, G.; Calò, A. Study of intra-varietal genetic variability in grapevine cultivars by PCR-derived molecular markers and correlations with the geographic origins. *Mol. Biotechnol.* **2012**, *50*, 72–85. [CrossRef]

13. Jahnke, G.; Májer, J.; Varga, P.; Szőke, B. Analysis of clones of pinots grown in Hungary by SSR markers. *Sci. Hortic.* **2011**, *129*, 32–37. [CrossRef]
14. Huang, H.; Lu, J.; Ren, Z.; Hunter, W.; Dowd, S.E.; Dang, P. Mining and validating grape (*Vitis* L.) ESTs to develop EST-SSR markers for genotyping and mapping. *Mol. Breed.* **2011**, *28*, 241–254. [CrossRef] [PubMed]
15. Emanuelli, F.; Lorenzi, S.; Grzeskowiak, L.; Catalano, V.; Stefanini, M.; Troggio, M.; Myles, S.; Martinez-Zapater, J.M.; Zyprian, E.; Moreira, F.M.; et al. Genetic diversity and population structure assessed by SSR and SNP markers in a large germplasm collection of grape. *BMC Plant Biol.* **2013**, *13*, 39. [CrossRef]
16. Kantartzi, S.K. (Ed.) *Microsatellites Methods and Protocols, in Methods in Molecular Biology*; Humana Press: Totowa, NJ, USA, 2013; Volume 1006.
17. Tello, J.; Forneck, A. Use of DNA markers for grape phylloxera population and evolutionary genetics: From RAPDs to SSRs and beyond. *Insects* **2019**, *10*, 317. [CrossRef] [PubMed]
18. Velasco, R.; Zharkikh, A.; Troggio, M.; Cartwright, D.A.; Cestaro, A.; Pruss, D.; Pindo, M.; FitzGerald, L.M.; Vezzulli, S.; Reid, J.; et al. A high quality draft consensus sequence of the genome of a heterozygous grapevine variety. *PLoS ONE* **2007**, *2*, e1326. [CrossRef]
19. Stavrakaki, M.; Bouza, D.; Biniari, K. Differentiation of Greek grapevine cultivars (*Vitis vinifera* L.) based on the combination of ampelographic description and microsatellite markers. *Genet. Resour. Crop. Evol.* **2020**, *67*, 21–40. [CrossRef]
20. Anonim. *Descriptor for Grapevine (Vitis spp.)*; International Plant Genetic Resources Institute: Roma, Italy, 1997; p. 58.
21. Hammer, Ø.; Harper, D.A.T.; Ryan, P.D. PAST: Paleontological Statistics Software Package for Education and Data Analysis. *Palaeontol. Electron.* **2001**, *4*, 9.
22. Lefort, F.; Lally, M.; Thompson, D.; Douglas, G.C. Morphological traits microsatellite fingerprinting and genetic relatedness of a stand of elite oaks (*Q. Robur* L.) at Tuallynally, Ireland. *Silvae Genet.* **1998**, *47*, 257–262.
23. Akçay, M.E.; Burak, M.; Kazan, K.; Yüksel, C.; Mutaf, F.; Bakir, M.; Ayanoğlu, H.; Ergül, A. Genetic analysis of Anatolian pear germplasm by simple sequence repeats. *Ann. Appl. Biol.* **2014**, *164*, 441–452. [CrossRef]
24. Ergül, A.; Büyük, B.P.; Hazrati, N.; Yılmaz, F.; Kazan, K.; Arslan, N.; Özmen Yüksel, C.; Aydın Soydam, S.; Bakir, M.; Tan, N.; et al. Genetic characterization and population structure analysis of Anatolian figs (*Ficus carica* L.) by SSR markers. *Folia Hortic.* **2021**, *33*, 49–78. [CrossRef]
25. Thomas, M.R.; Scott, N.S. Microsatellite repeats in grapevine reveal DNA polymorphisms when analyzed as sequence-tagged sites (STSs). *Theor. Appl. Genet.* **1993**, *86*, 985–990. [CrossRef]
26. Bowers, J.E.; Dangl, G.S.; Meredith, C.P. Development and characterization of additional microsatellite DNA markers for grape. *Am. J. Enol. Vitic.* **1999**, *50*, 243–246. [CrossRef]
27. Sefc, K.M.; Regner, F.; Turetshek, E.; Glössl, J.; Steınkellner, H. Identification of microsatellite sequences in *Vitis riparia* and their applicability for genotyping of different Vitis species. *Genome* **1999**, *42*, 367–373. [CrossRef] [PubMed]
28. Goto-Yamamoto, N.; Mouri, H.; Azumi, M.; Edwards, K.J. Development of grape microsatellite analysis including oriental cultivars. *Am. J. Enol. Vitic.* **2006**, *57*, 105–108. [CrossRef]
29. Merdinoglu, D.; Butterlin, G.; Baur, C.; Balthazard, D.J. Comparison of RAPD, AFLP and SSR (microsatellite) markers for genetic diversity analysis in *Vitis vinifera* L. *Acta Hortic.* **2000**, *528*, 193–197. [CrossRef]
30. This, P.; Jung, A.; Boccacci, P.; Borrego, J.; Botta, R.; Constantini, L.; Crespan, M.; Dangl, G.S.; Eisenheld, C.; Ferreria-Montteiro, F.; et al. Development of a standard set of microsatellite reference alleles for identification of grape cultivars. *Theor. Appl. Genet.* **2004**, *109*, 1448–1458. [CrossRef]
31. Maul, E.; Sudharma, K.N.; Kecke, S.; Marx, G.; Müller, C.; Audeguin, L.; Bosellı, M.; Boursiquot, J.M.; Bucchetti, B.; Cabello, F.; et al. The European Vitis Database (www.eu-vitis.de)—A technical innovation through an online uploading and interactive modification system. *Vitis* **2012**, *51*, 79–86.
32. Yılmaz, F.; Shidfar, M.; Hazrati, N.; Kazan, K.; Yüksel Özmen, C.; Uysal, T.; Özer, C.; Yaşasın, A.S.; Söylemezoğlu, G.; Boz, Y.; et al. Genetic analysis of central Anatolian grapevine (*Vitis vinifera* L.) germplasm by simple sequence repeats. *Tree Genet. Genomes* **2020**, *16*, 55. [CrossRef]
33. Wagner, H.W.; Sefc, K.M. *Identity 1.0. Centre for Applied Genetics*; University of Agricultural Science: Vienna, Austria, 1999.
34. Bowcock, A.M.; Ruiz-Linares, A.; Tomfohrde, J.; Minch, E.; Kidd, J.R.; Cavalli-Sforza, L.L. High resolution of human evolutionary trees with polymorphic microsatellites. *Nature* **1994**, *368*, 455–457. [CrossRef]
35. Minch, E.; Ruiz-Linares, A.; Goldstein, D.B.; Feldman, M.; Cavallisforza, L.L. *Microsat (Version 1.4d): A Computer Program for Calculating Various Statistics on Microsatellite Allele Data*; Stanford University: Stanford, CA, USA, 1995.
36. Rohlf, F. *NTSYS-pc: Numerical Taxonomy and Multivariate Analysis System*; Version 2.1; Exeter Software: Setoukat, NY, USA, 2000.
37. Excoffier, L.; Lischer, H.E.L. Arlequin suite ver 3.5: A new series of programs to perform population genetics analyses under Linux and Windows. *Mol. Ecol. Resour.* **2010**, *10*, 564–567. [CrossRef]
38. Rohlf, F. *NTSYS-pc: Numeric Taxonomic Systems*; Version 2.20; Exeter Software, Applied Biostatistics Inc.: Setoukat, NY, USA, 2004.
39. Belkhir, K.; Borsa, P.; Goudet, J.; Bonhomme, F. *GENETIX, Logiciel sous Windows TM pour la Génétique des Populations*; Laboratoire Génome et Populations, CNRS UPR 9060; Université de Montpellier II: Montpellier, France, 1998.
40. Corender, J.; Marttinen, P.; Siren, J.; Tang, J. Enhanced Bayesian Modelling in BAPS software for learning genetic structure of populations. *BMC Bioinform.* **2008**, *9*, 539. [CrossRef]

41. Pereira-Lorenzo, S.; Ramos-Cabrer, A.M.; Ferreira, V.; Díaz-Hernández, M.B.; Carnide, V.; Pinto-Carnide, O.; Rodrigues, R.; Velázquez-Barrera, M.E.; Rios-Mesa, D.; Ascasíbar-Errasti, J.; et al. Genetic diversity and core collection of Malus × domestica in northwestern Spain, Portugal and the Canary Islands by SSRs. *Sci. Hortic.* **2018**, *240*, 49–56. [CrossRef]
42. Pritchard, J.K.; Stephens, M.; Donnelly, P. Inference of population structure using multilocus genotype data. *Genetics* **2000**, *155*, 945–959. [CrossRef] [PubMed]
43. Earl, D.A.; Von Holdt, B.M. Structure harvester: A website and program for visualizing structure output and implementing the Evanno method. *Conserv. Genet. Resour.* **2012**, *4*, 359–361. [CrossRef]
44. Evanno, G.; Regnaut, S.; Gould, J. Detecting the number of clusters of individuals using the software STRUCTURE: A simulation study. *Mol. Ecol.* **2005**, *14*, 2611–2620. [CrossRef]
45. Peakall, R.; Smouse, P.G. 6.5: Genetic analysis in Excel. Population genetic software for teaching and researchdan update. *Bioinformatics* **2012**, *28*, 2537–2539. [CrossRef] [PubMed]
46. Nei, M. Estimation of average heterozygosity and genetic distance from a small number of individuals. *Genetics* **1978**, *89*, 583–590. [CrossRef]
47. Meirmans, P.G.; van Tienderen, P.H. Genotype and Genodive: Two programs for the analysis of genetic diversity of asexual organisms. *Mol. Ecol. Notes* **2004**, *4*, 792–794. [CrossRef]
48. Leão, P.C.D.S.; Cruz, C.D.; Motoike, S.Y. Genetic diversity of table grape based on morphoagronomic traits. *Sci. Agric.* **2011**, *68*, 42–49. [CrossRef]
49. Khadivi-Khub, A.; Salimpour, A.; Rasouli, M. Analysis of grape germplasm from Iran based on fruit characteristics. *Rev. Bras. Bot.* **2014**, *37*, 105–113. [CrossRef]
50. Tello, J.; Torres-Pérez, R.; Grimplet, J.; Carbonell-Bejerano, P.; Martínez-Zapater, J.M.; Ibáñez, J. Polymorphisms and minihaplotypes in the *VvNAC26* gene associate with berry size variation in grapevine. *BMC Plant Biol.* **2015**, *15*, 253. [CrossRef]
51. Lamine, M.; Zemni, H.; Ziadi, S.; Chabaane, A.; Melki, I.; Mejri, S.; Zoghlami, N. Multivariate analysis and clustering reveal high morphological diversity in Tunisian autochthonous grapes (*Vitis vinifera*): Insights into characterization, conservation and commercialization. *Oeno One* **2014**, *48*, 111–122. [CrossRef]
52. Khalil, S.; Tello, J.; Hamed, F.; Forneck, A. A multivariate approach for the ampelographic discrimination of grapevine (*Vitis vinifera*) cultivars: Application to local Syrian genetic resources. *Genet. Resour. Crop. Evol.* **2017**, *64*, 1841–1851. [CrossRef]
53. Sabir, A.; Tangolar, S.; Buyukalaca, S.; Kafkas, S. Ampelographic and molecular diversity among grapevine (*Vitis* spp.) cultivars. *Czech J. Genet.* **2009**, *45*, 160–168. [CrossRef]
54. Dilli, Y.; Ünal, A.; Kesgin, M.; İnan, M.S.; Söylemezoğlu, G. Comparison of ampelographic characteristics of some important grape varieties are grown in the Aegean Region, rootstock and clones. *TURKJANS* **2014**, *1*, 1546–1553.
55. Knezović, Z.; Mandić, A.; Perić, N.; Beljo, J.; Žulj Mihaljević, M. Morphological and genetic characterization of vine grape cultivars of Herzegovina. *CREBSS* **2017**, *3*, 1–9. [CrossRef]
56. Wang, Z.; Zhou, J.; Xu, X.; Perl, A.; Chen, S.; Ma, H. Adoption of table grape cultivars: An attribute preference study on Chinese grape growers. *Sci. Hortic.* **2017**, *216*, 66–75. [CrossRef]
57. Doligez, A.; Bertrand, Y.; Farnos, M.; Grolier, M.; Romieu, C.; Esnault, F.; Dias, S.; Berger, G.; François, P.; Pons, T.; et al. New stable QTLs for berry weight do not colocalize with QTLs for seed traits in cultivated grapevine (*Vitis vinifera* L.). *BMC Plant Biol.* **2013**, *13*, 217. [CrossRef]
58. Vouillamoz, J.F.; McGovern, P.E.; Ergul, A.; Söylemezoğlu, G.; Tevzadze, G.; Meredith, C.P.; Grando, M.S. Genetic characterization and relationships of traditional grape cultivars from Transcaucasia and Anatolia. *Plant Genet. Resour.* **2006**, *4*, 144–158. [CrossRef]
59. Ibañez, J.; Andrés, M.T.; Molino, A.; Borrego, J. Genetic study of key Spanish grapevine varieties using microsatellite analysis. *Am. J. Enol. Vitic.* **2003**, *54*, 22–30. [CrossRef]
60. Nebish, A.; ris Ochssner, I.; Maul, E.; Topfer, R.; Hausmann, L.; Hovhannisyan, A.; Devejyan, H.; Melyan, G.; Aroutiounian, R. Genetic identification and characterization of Armenian grapevine cultivars. In Proceedings of the BIO Web of Conferences, 40th World Congress of Vine and Wine, Sofia, Bulgaria, May 29–June 2 2017; Volume 9.
61. Riaz, S.; De Lorenzis, G.; Velasco, D.; Koehmstedt, A.; Maghradze, D.; Bobokashvili, Z.; Musayev, M.; Zdunic, G.; Laucou, V.; Walker, M.A.; et al. Genetic diversity analysis of cultivated and wild grapevine (*Vitis vinifera* L.) cultivars around the Mediterranean basin and Central Asia. *BMC Plant Biol.* **2018**, *18*, 137. [CrossRef] [PubMed]
62. Karataş, D.D.; Karatas, H.; Laucou, V.; Sarikamiş, G.; Riahi, L.; Bacilieri, R.; This, P. Genetic diversity of wild and cultivated grapevine accessions from southeast Turkey. *Hereditas* **2014**, *151*, 73–80. [CrossRef]
63. Eyduran, P.S.; Ercisli, S.; Akin, M.; Eyduran, E. Genetic characterization of autochthonous grapevine cultivars from Eastern Turkey by simple sequence repeats (SSRs). *Biotechnol. Biotechnol. Equip.* **2015**, *30*, 26–31. [CrossRef]
64. Cipriani, G.; Spadotto, A.; Jurman, I.; Di Gaspero, G.; Crespan, M.; Meneghetti, S.; Frare, E.; Vignani, R.; Cresti, M.; Morgante, M.; et al. The SSR-based molecular profile of 1005 grapevine (*Vitis vinifera* L.) accessions uncovers new synonymy and parentages, and reveals a large admixture. *Theor. Appl. Genet.* **2010**, *12*, 1569–1585. [CrossRef]
65. Boz, Y.; Bakır, M.; Çelikkol, B.P.; Kazan, K.; Yılmaz, F.; Çakır, B.; Aslantaş, Ş.; Söylemezoğlu, G.; Yaşasın, A.S.; Özer, C.; et al. Genetic characterization of grape (*Vitis vinifera* L.) germplasm from Southeast Anatolia by SSR markers. *Vitis* **2011**, *50*, 99–106.
66. Shen, Z.; Zhang, K.; Ma, L.; Duan, J.; Yan, A. Analysis of the genetic relationships and diversity among 11 populations of Xanthoceras sorbifolia using phenotypic and microsatellite marker data. *Electron. J. Biotechnol.* **2017**, *26*, 33–39. [CrossRef]

67. Myles, S.; Boyko, A.R.; Owens, C.L.; Brown, P.J.; Grassi, F.; Aradhya, M.K.; Prins, B.; Reynolds, A.; Chia, J.-M.; Ware, D.; et al. Genetic structure and domestication history of the grape. *Proc. Natl. Acad. Sci. USA* **2011**, *108*, 3457–3458. [CrossRef] [PubMed]
68. Ergül, A.; Kazan, K.; Aras, S.; Çevik, V.; Çelik, H.; Söylemezoğlu, G. AFLP analysis of genetic variation within the two economically important Anatolian grapevine (*Vitis vinifera* L.) varietal groups. *Genome* **2006**, *49*, 467–495. [CrossRef]
69. Hou, L.; Cui, Y.; Li, X.; Chen, W.; Zhang, Z.; Pang, X.; Li, Y. Genetic Evaluation of Natural Populations of the Endangered Conifer Thuja koraiensis Using Microsatellite Markers by Restriction-Associated DNA Sequencing. *Genes* **2018**, *9*, 218. [CrossRef]
70. Wang, L.; Zhang, J.; Liu, L.; Zhang, L.; Wei, L.; Hu, D. Genetic diversity of grape germplasm as revealed by microsatellite (SSR) markers. *Afr. J. Biotechnol.* **2015**, *14*, 990–998.
71. Laucou, V.; Launay, A.; Bacilieri, R.; Lacombe, T.; Adam-Blondon, A.F.; Beârard, A.; Chauveau, A.; de Andrés, M.T.; Hausmann, L.; Ibáñez, J.; et al. Extended diversity analysis of cultivated grapevine *Vitis vinifera* with 10K genome-wide SNPs. *PLoS ONE* **2018**, *13*, e0192540. [CrossRef]
72. Latch, E.K.; Dharmarajan, G.; Glaubitz, J.C.; Rhodes, O.E. Relative performance of Bayesian clustering software for inferring population substructure and individual assignment at low levels of population differentiation. *Conserv. Genet.* **2006**, *7*, 295–302. [CrossRef]
73. Pelsy, F. Molecular and cellular mechanisms of diversity within grapevine varieties. *Heredity* **2010**, *104*, 331–340. [CrossRef] [PubMed]
74. Aradhya, M.K.; Dangl, G.S.; Prins, B.H.; Boursiquot, J.-M.; Walker, M.A.; Meredith, C.P.; Simon, C.J. Genetic structure and differentiation in cultivated grape *Vitis vinifera* L. *Genet. Res.* **2003**, *81*, 179–192. [CrossRef] [PubMed]
75. Barazani, O.; Westberg, E.; Hanin, N.; Dag, A.; Kerem, Z.; Tugendhaft, Y.; Hmidat, M.; Hijawi, T.; Kadereit, J.W. A comparative analysis of genetic variation in rootstocks and scions of old olive trees–A window into the history of olive cultivation practices and past genetic variation. *BMC Plant Biol.* **2014**, *14*, 146. [CrossRef]
76. Xu, N.N.; Jiang, K.; Biswas, S.R.; Tong, X.; Wang, R.; Chen, X.Y. Clone configuration and spatial genetic structure of two Halophila ovalis populations with contrasting internode lengths. *Front. Ecol. Evol.* **2019**, *7*, 170. [CrossRef]
77. Grigoriou, A.; Tsaniklidis, G.; Hagidimitriou, M.; Nikoloudakis, N. The Cypriot indigenous grapevine germplasm is a multi-clonal varietal mixture. *Plants* **2020**, *9*, 1034. [CrossRef] [PubMed]
78. Zombardo, A.; Storchi, P.; Valentini, P.; Ciofini, A.; Migliaro, D.; Crespan, M. Recovery, molecular characterization, and ampelographic assessment of marginal grapevine germplasm from southern umbria (Central Italy). *Plants* **2021**, *10*, 1539. [CrossRef]
79. Harris, A.M.; De Giorgio, M. An unbiased estimator of gene diversity with improved variance for samples containing related and inbred individuals of any ploidy. *G3-Genes Genom Genet.* **2017**, *7*, 671–691. [CrossRef]

horticulturae

MDPI

Article

European Grapevine Cultivars and Rootstocks Show Differential Resistance to *Xylella fastidiosa* Subsp. *fastidiosa*

Sara Martínez [1], Maite Lacuesta [2], Juan Bautista Relloso [1], Ana Aragonés [1], Ana Herrán [1] and Amaya Ortiz-Barredo [1,*]

[1] Department of Plant Production and Protection, NEIKER-Basque Institute for Agricultural Research and Development, Basque Research and Technology Alliance (BRTA), Arkaute Agri-Food Campus, 01192 Arkaute, Spain; sara_mtzc@hotmail.com (S.M.); jbautista@neiker.eus (J.B.R.); aaragones@neiker.eus (A.A.); aherran@neiker.eus (A.H.)

[2] Plant Physiology Department, Pharmacy Faculty, University of the Basque CountryP° de la Universidad 7, 01006 Vitoria-Gasteiz, Spain; maite.lacuesta@ehu.eus

* Correspondence: aortizb@neiker.eus

Abstract: Several *Xylella fastidiosa* subsp. *fastidiosa* (ST1) strains that cause Pierce's disease were isolated from grapevine in Spain. In this study, we applied an approach to assess PD susceptibility among 24 different well-known *Vitis vinifera* subsp. *vinifera* cultivars and five rootstocks belonging to different species of the genus *Vitis*. Both were commonly commercialized, representing about 75% of the cultivated area in Spain. This method incorporated disease severity, disease progression, and water potential from the stem xylem. The trials were carried out under field and greenhouse conditions. The virulence of the *Xff* strain XYL 2055/17 was significantly higher than that of strain XYL 2177/18. However, while this difference in strain virulence did not seem to modify the susceptibility profiles of the cultivars, disease severity could be climate dependent. This work established two significantly different groups of European cultivars of grapevine characterized by high and low susceptibility to *Xff* ST1: cultivars with high susceptibility, including reference cultivars such as Tempranillo and Tempranillo Blanco, and cultivars with high resistance, such as Hondarrabi Zuri and Cabernet Sauvignon. Cultivar susceptibility was independent of the rootstock on which they were grafted. No conclusive data were found regarding the potential of water loss as an early detection test prior to symptom onset. This study provides a framework with which to advance cultivar susceptibility studies under different environmental conditions.

Keywords: Pierce's disease; *Xylella fastidiosa*; grapevine; Scholander pressure chamber; quarantine plant pathogen; water potential

Citation: Martínez, S.; Lacuesta, M.; Relloso, J.B.; Aragonés, A.; Herrán, A.; Ortiz-Barredo, A. European Grapevine Cultivars and Rootstocks Show Differential Resistance to *Xylella fastidiosa* Subsp. *fastidiosa*. *Horticulturae* **2023**, *9*, 1224. https://doi.org/10.3390/horticulturae9111224

Academic Editors: Antonio Amores-Arrocha and Ana Jiménez-Cantizano

Received: 9 October 2023
Revised: 8 November 2023
Accepted: 10 November 2023
Published: 12 November 2023

1. Introduction

Xylella fastidiosa (*Xf*) [1] is a vascular plant pathogen with an extremely large host range [2]. To date, this species has been found to infect 690 plant species classified under 306 genera and 88 families (EFSA, 2023; Sicard, 2018; Trkulja, 2022) [2–4]. Major plant hosts (https://gd.eppo.int/taxon/XYLEFA/hosts (accessed on 8 October 2023)) include two ornamental species (*Nerium oleander* and *Polygala myrtifolia*) and six major crops: citrus, coffee, almond, peach, olive, and grapevine (*Vitis vinifera*). The diverse diseases caused by *Xf* have significant economic impacts and agricultural management consequences [5]. Since 2019, *Xf* has been included as a Priority Pest under European Regulations (Regulation (EU) 2019/1702). One of the characteristics that makes *Xf* a very dangerous pathogen, besides its very high genetic plasticity, is its ability to be transmitted by likely all species of the two main xylem-feeder insect groups: leafhoppers (Cicadellidae subfamily Cicadellinae) and spittlebugs (Cercopoidea families, Aphrophoridae, Cercopidae, and Clastopteridae) [3].

From a taxonomic point of view, *Xf* is a complex species with several subspecies and sequence types (ST). *X. fastidiosa* subsp. *fastidiosa* (*Xff*), which belongs to ST1, causes Pierce's disease (PD) and diverse syndromes in several plant hosts [6,7]. PD was first described in California in 1892, where it remains a significant problem for the grape industry. A devastating PD outbreak in the Temecula Valley in Southern California during the late 1990s was associated with the establishment of a very efficient vector in the area, which prompted an intense investigation of the pathogen and the development of programs to manage PD and the vectors involved [3]. Control of this disease is difficult and, despite significant projected savings of around USD 200 million under the PD Control Program [8], a corresponding increase in efficacy will require the deployment of diverse improved strategies that should include the management of water stress and appropriate plant resistance [6,9].

The main factors causing PD symptoms appear to be related to water stress and the blockage of xylem vessels by the bacterial biofilm, as well as the resulting production of tyloses and gums by the plant, which causes hydraulic dysfunctions, leading to desiccation and plant death within a few years [10,11]. The development of symptoms may not be entirely due to the vessel's occlusion. Instead, the pathogenesis may be more complex and occur starting from the earliest stages of infection, before the colonization of vessels [12,13]. The time period between inoculation and the appearance of symptoms in plants varies depending on the plant species and age and occurs between three and four months, extending beyond a year after the initial infection [13], with longer periods observed in woody plants compared to herbaceous plants [14]. The symptoms typically begin when environmental conditions are generally hot and dry and plants are subjected to water stress and other physiological responses, which vary in different varieties [15,16].

Several techniques are available to assess the water status of plants via physiological indicators [17,18]. Leaf water potential is recognized as one of the most important indexes to evaluate the water status of plants, providing high theoretical value and important information for multiple applications to quantify critical physiological processes, including drought responses [19–21]. Predawn leaf water potential (PLWP) and stem water potential (SWP) were found to be simple and precise indicators for assessing the grapevine water status [22–25]. Drought tolerance is known to be determined in part by the factors related to water transport within the plant, and there is a direct relationship between hydraulic conductivity and the xylem physiological status [26]. The data provided by this technology suggest early infection indicators that may be related to PD resistance factors. Moreover, the water status data obtained via destructive measurement were correlated with the data provided by other nondestructive and portable devices applied in the open field for grapevine [27].

PD resistance has only been identified in several wild grape species endemic to the Americas, including *V. arizonica/candicans* and *Muscadinia rotundifolia*, whereas all assayed European *V. vinifera* genotypes are susceptible to this disease [28–32]. Recent research has compared PD tolerance/resistance within the three major Eurasian pedigree lineages and provides a benchmark for PD susceptibility levels for some of the most widespread table and wine grape cultivars [33]. The complexity of the genomic architecture of resistance to the bacterium and the role of climate in shaping this resistance [32] have also required the use of additional research efforts to evaluate a larger number of *V. vinifera* cultivars from different genetic pools under different environmental conditions. These studies obtained information on the vascular and anatomical characteristics of these grapevine cultivars to develop tools for assessing and predicting the PD susceptibility [10].

After the first outbreak in 2013 in olive trees [34], on the Balearic Islands of Spain, *Xff* was first reported in October 2016 to cause PD and also infect other hosts [35]. Since the first occurrence of PD in Europe, the presence and spread of this bacterium in Mediterranean crops and plant species in the natural and urban landscapes have become a major risk. For this reason, Italy, France, and Spain are now taking specific phytosanitary measures aimed at the eradication or containment of this disease. These measures are based on exhaustive surveys and monitoring PD symptoms to achieve a quick removal of problematic vines.

The recent outbreaks of PD in Europe have driven the pursuit of new and more effective control strategies for both PD and other diseases caused by *Xf*, resulting in the proposal of promising and innovating methods that could facilitate crop sustainability with little to no economic, social, or environmental risks [9]. These methods include techniques conferring systemic resistance in vines among others [36–38]. Nevertheless, to achieve the appropriate management of PD in Europe, it will be necessary to evaluate both the aggressiveness of the different *Xff* isolates and the resistance responses of the numerous clones and cultivars of *Vitis* planted in this continent. Furthermore, given the expected variability in PD epidemiology dependent on agro-climatic factors and the expression of pathogenicity factors [33,39], these studies should be conducted under standardized climatic conditions and correlated with the agro-climatic variability of viticultural areas.

The main objective of this work is to describe the severity of PD using a local inoculum and grapevine cultivars and rootstocks; evaluate disease progression under different climatic conditions; and establish the relationship between symptomatology, xylem pressure changes, and the detection of the bacterium using a standardized molecular methodology. Cultivars and rootstocks were chosen to provide data that could allow the implementation of clonal selection processes. The disease severity, water potential, and time course of PCR detection were studied under four different weather environments, including a climate-controlled greenhouse and open field conditions.

2. Materials and Methods

2.1. Terminology and General Overview of the Trials

Grapevine plants for new plantations are normally marketed as grafted. The same variety can be grafted onto different rootstocks to achieve better adaptation to the soil, phenological development, or resistance to endemic pests. In this study, European grapevine varieties (*Vitis vinifera* subsp. *vinifera*) grafted onto different rootstocks were evaluated. To avoid discrepancies in the terminology, a variety grafted onto the rootstock is referred to as the "*Cultivar*". Plants corresponding to varieties used as rootstock and belonging to different species of the genus *Vitis* were also studied; the term used in this work to designate such species is "*Rootstock*".

The pathogen and handling conditions are worth to be considered as set out in the current European regulations. *Xf* is in the category of "Priority quarantine pathogen" (Regulation (EU) 2016/2031 and Commission Delegated Regulation (EU) 2019/829), which means that we are dealing with one of the 20 most dangerous pathogens for European wine-growing areas as well as for other agricultural areas with major crops. Consequently, its handling, including for scientific purposes, has been extremely restricted even under regional rules (limited to areas where the disease has been widely detected or in accredited facilities (such as high biosafety greenhouses) to avoid any risk of pathogen spread.

The varietal selection and the way the trials were conducted correspond to the following chronology: The first trial in 2019 was carried out in the Mallorca, Balearic Islands, under field conditions. The area for testing had to be restricted and carried out with local *Xf*-isolated strains (XYL 2055/17 and XYL 2177/18). Until then (at the time the trials started), there was no reference to the strains' virulence in different cultivars or rootstocks. A total of 28 *Cultivars* and *Rootstocks* were evaluated. That amount (28) was selected due to their vast degree of establishment (approx. 75%) in the most important wine-growing areas of Spain. Once the results of the first trial were analyzed, and according to the space availability restrictions, a second trial was carried out in the following year (2020). It was conducted during the vegetative period of the grapevine, in the same location and under the same operational conditions. In this year, due to the above-mentioned space availability restrictions, the number of cultivars to be evaluated was reduced to 22 in total vs. 28. Due to the results obtained in the previous year regarding the strain virulence, the susceptibility was assessed just against the most virulent strain (XYL 2055/17). Considering the results obtained in these two trials carried out under field conditions (2019 and 2020), it was decided to carry out a new trial during a new growing season in 2021. In this case, the objective of this third trial was evaluating the influence of climate conditions on disease progression. This is the reason behind the restricted number of plants in this trial (five *Cultivars* and one *Rootstock*). The assay was carried out in parallel under field conditions (the same location in the Balearic Islands) and also in a greenhouse located in the Basque Country. This particular study under the greenhouse condition (in the Basque Country) was located in a free-disease area of Spain and in a facility accredited for handling infected plant material with this harmful pathogen. The 2021 trials were carried out with the only strain that we were authorized to import for this type of experimental trials (IVIA 5770). The cultivars evaluated each year as well as the inoculated strains are detailed in Table 1, included in the results section.

Table 1. Maximum Severity Index value (max. SI) recorded in the plants of each *Cultivar* and *Rootstock* evaluated in all of the trials carried out in this study at 8 and 16 WPI (max. SI-8/max. SI-16). The first line and the columns below it shows the year in which the trials were carried out, the second line shows where the trials were conducted, and the third line corresponds to the *Xff* strain artificially inoculated (XYL2055/17, XYL2177/18, and IVIA5770) in 24 grapevine *Cultivars* and five *Rootstocks*. The *Rootstock* column indicates the rootstock on which each cultivar was grafted.

Max SI 8/Max SI 16	*Rootstock*	2019		2020	2021	
		Field				Greenhouse
		XYL 2177/18	XYL 2055/17	XYL 2055/17	IVIA 5770	IVIA 5770
Airen	R110	3/5	4/5	2/5	-	-
Albarino	R110	2/5	5/5	1/3	-	-
Bobal	R110	4/5	3/5	-	-	-
Cabernet Sauvignon	R110	1/5	2/5	1/5	-	-
Chardonnay	R110	5/5	5/5	1/5	1/5	0/1
Garnacha	R110	4/5	5/5	3/5	1/5	0/2
Garnacha Tintorera	R110	5/5	5/5	3/5	-	-
Garnacha Tintorera	P1103	5/5	5/5	3/5	-	-
Graciano	R110	2/5	4/5	1/5	-	-
Hondarrabi Zuri	SO4	3/5	3/5	1/2	1/5	0/2
Hondarribi Beltza	196-17 Cl	2/5	3/5	0/3	-	-
Macabeo	R110	3/5	3/5	-	-	-
Malvasia	R110	3/5	4/5	-	-	-
Mencia	R110	4/5	4/5	-	-	-
Monastrell	R110	2/5	3/5	-	-	-
Pedro Ximenez	R110	2/5	3/5	2/5	-	-
Pinot Noir	R110	1/5	3/5	1/5	3/5	0/3
Tempranillo	R110	4/5	4/5	1/5	3/5	0/4
Tempranillo	SO4	4/5	5/5	3/5	-	-
Tempranillo	41B-MGt	4/5	5/5	1/5	-	-
Tempranillo Blanco	R110	5/5	5/5	3/5	-	-
Tempranillo RJ43	R110	3/5	5/5	4/5	-	-
Tempranillo RJ78	R110	3/5	5/5	3/5	-	-
Verdejo	R110	3/5	2/5	-	-	-
Rootstock 196-17Cl		3/5	3/5	1/5	-	-
Rootstock 41B-MGt		2/3	2/4	1/4	-	-
Rootstock P1103		1/5	1/5	1/3	-	-
Rootstock R110		1/2	1/2	2/2	0/1	0/1
Rootstock Ru140		1/5	1/5	1/3	-	-

2.2. Plant Material and Facilities

A total of 24 *Cultivars* and five *Rootstocks* [40,41] were selected based on their representativeness in the Spanish Protected Designations of Origin (DOs) or Protected Geographical Indications (PGIs) [28] to evaluate their resistance and disease progression.

During the years 2019, 2020, and 2021, three trials were carried out in an open field (see the climate variables in Figure 1a). One additional trial in a greenhouse was carried out in 2021. These plants under controlled conditions were grown with 70% relative humidity; a day/night temperature of 25 °C ± 2 °C and 18 °C ± 2 °C, respectively; and the absence of additional artificial lighting. All trials were conducted during the leaf development from May (BBCH 11-First leaf unfolded and spread away from shoot) to September (BBCH 19-nine or more leaves unfolded), ensuring a minimum daily light period of 16 h during the first four months after inoculation. After each trial, the plants were destroyed, so each annual test was implemented with new healthy plants.

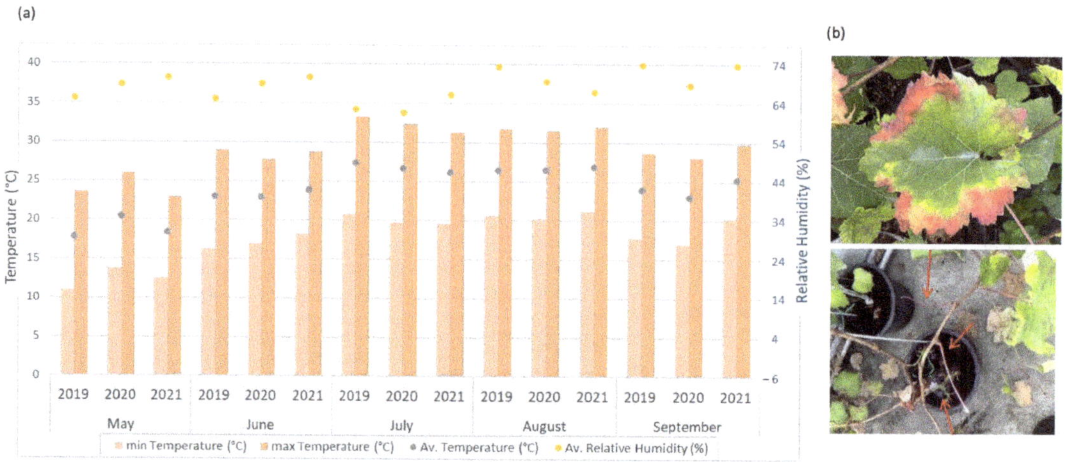

Figure 1. (**a**) Monthly average, maximum, and minimum temperature (°C), and monthly average of relative humidity (%) for the three years and the evaluation period (from June to September) in which the field trials were conducted to evaluate the *Cultivar*'s susceptibility to *Xff*. (**b**) Characteristic symptoms of PD recorded in the field and greenhouse trials. Early symptoms after inoculation: reddish spots surrounded by a yellowish chlorotic halo (top) and defoliation indicated by red arrows as late symptoms (bottom).

Pre-grafted and rooted plants (one-bud cuttings) from the nursery were kept at 4 °C before planting. The plants were potted (6 L pots) into a mixture of peat and siliceous sand (3:1). The peat (pH 6.0 and 90% organic dry matter) was supplemented with calcium carbonate (7 g/L) and NPK14-10-18 fertilizer (1.5 g/L). The plants were selected for the trial if they showed the same phenotypic profile after a period of 20–30 days. Irrigation was applied as necessary to maintain field capacity and no phytosanitary treatment was applied.

The greenhouse was accredited as a temporal confinement greenhouse for *Xff* tasks; NEIKER's facilities are located in this area on the northern peninsula of Spain (42°51′10.2″ N 2°37′29.7″ W). The trials using an open field were conducted in Palma de Mallorca, Spain, at the Balearic Governmental Center for Agricultural Improvement "SEMILLA" (39°35′23.5″ N 2°39′51.1″ E). The plants were randomly distributed in 12-plant rows along an insect-proof net tunnel and exposed to environmental temperatures.

2.3. In Planta Evaluation of PD Susceptibility

Inoculations were performed as follows: *Xff* ST1 strains (IVIA5770, XYL 2055/17 and XYL 2177/18) were isolated from *Vitis vinifera* on the Balearic Islands. These strains were previously characterized for several scientific grapevine tests [7,29,42,43]. The bacteria were cultured on a buffered charcoal yeast extract (BCYE) solid medium at 28 °C for 7–10 days. A bacterial suspension was prepared in phosphate-buffered saline (PBS) and adjusted to $OD_{600} = 0.3$, resulting in a final concentration of approximately 1×10^8 CFU.mL^{-1}. Next, 10 μL of that suspension was used to inoculate the plants following a general procedure consisting of the pin-prick method [44,45]. In total, 9 out of 12 plants per genotype were inoculated, while the remaining three plants were mechanically injured in the same way as those prepared in PBS, but without an inoculum.

The Severity Index (SI) was assessed on a scale of 0 to 5, as previously described by Su et al. in 2013 [46], based on the number of affected and symptomatic leaves (Figure 1b) as follows: asymptomatic leaves (0), 1–2 symptomatic leaves (1), 3–4 symptomatic leaves (2), 5–7 symptomatic leaves (3), 8–10 symptomatic leaves (4), and more than 10 symptomatic leaves (5). The SI data were recorded every 15 days during the period between 8 weeks post-inoculation (WPI) and 16 WPI. Based on these data, it was possible to calculate the disease progression expressed as the absolute area under the disease progression curve (AUDPC) and transformed by the relative AUDPC (rAUDPC) The AUDPC units, as indicators of resistance or susceptibility, are not easily interpretable. In an effort to standardize the AUDPC, same researchers often use rAUDPC. The rAUDPC is calculated by dividing the AUDPC by the "maximum potential AUDPC". The maximum potential AUDPC is simply the AUDPC a cultivar would have if it had had 100% infection all days during the reading period. Considering that the period in which AUDPC was evaluated in our experiments was 56 days (from 8 to 16 WPI) and the maximum SI value is 5, the maximum value of AUDPC would be 280 square units [47].

2.4. Detection of *Xf*

Each leaf was taken between three and five nodes above the inoculation point. Only petioles and veins were used for DNA extraction, amounting to a minimum of 0.5 g fresh tissue. The CTAB protocol for DNA extraction was applied [48]. Real-time PCR (RTi-PCR) for *Xf* detection was carried out using the primers XF-F: 5′-CAC GGC TGG TAA CGG AAG A-3′, XF-R: 5′-GGG TTG CGT GGT GAA ATC AAG-3′, and XF-P: 5′-FAM-TCG CAT CCC GTG GCT CAG TCC-BHQ-1-3′ [48,49], using TaqMan Universal Master Mix (Applied Biosystems®, Thermo Fisher Scientific Inc., Waltham, MA, USA). The final optimized reaction conditions were as follows: RTi-PCR reactions were performed in 20 μL reaction volumes containing 10 μL of 2X TaqMan universal master mix (Applied Biosystems), 300 nM *Xf* sense (XF-F) and antisense (XF-R) primers, 100 nM 6′FAM/BHQ-labeled XF-P probe, ultra-pure bovine serum albumin (BSA) at 300 ng/μL (Invitrogen), and 2 μL of total DNA template. The optimal thermocycling conditions were as follows: 50 °C for 2 min and 94 °C for 10 min, then 40 cycles for 10 s and 62 °C for 40 s, using Quant Studio TM 5 model QS5STD. All samples were amplified in duplicate. Threshold values were applied automatically using the QuantStudio™ Design and Analysis Software (Applied Biosystems). Positive amplification was determined using a crossing threshold (Ct value < 38 cycles). Bacterial re-isolation was accomplished from the same tissue sample RTi-PCR analyzed. The recovered colonies were scraped from the BCYE growth medium and resuspended in a potassium phosphate buffer, prior to being assessed via RTi-PCR.

2.5. Water Potential (ψ) Assessment

Ψ was assessed only in the greenhouse trial using the Scholander pressure chamber (PMS Instrument Company, Albany, NY, USA). The equipment was set up using a pressure of 3 bar in the N_2 gas cylinder and a gas flow rate of 5 bar in the equipment. Measurements were performed systematically between 9:30 and 11:30 am, two hours after sunrise, as recommended by Knipfer et al. (2020) [50]. Leaves were cut from positions 2, 3, or 4 above the inoculation point, depending on phenological development and using a scalpel blade. Immediately after excision, the petiole was placed in the Scholander chamber and the pressure equilibrium technique described by Castander et al. (2020) [51] was used to measure Ψ in MPa. The Ψ data were recorded during the period between 8 WPI and 21 WPI.

2.6. Statistical Analysis

rAUDPC and water potential (Ψ) were analyzed in each trial to assess the effect of cultivars and rootstocks. The rAUDPC data were normalized via arcsine of the square root of the proportion and then subjected to analysis of two-way ANOVA (Cultivar × strain, Cultivar × year, or Rootstock × Strain for each *Cultivar*), followed by Fisher's Least Significant Difference (LSD) test (R software, version 4.2.2) in the case of no significant interactions. Differences at $p < 0.05$ were considered significant. Additional descriptive data in boxplots, comparisons based on estimated marginal means by Tukey's tests for comparing two means, and Pearson correlation coefficients using average trait values were also obtained via the free Jamovi software version 2.3.28.

3. Results

3.1. Pierce's Disease Susceptibility in Vitis Vinifera Cultivars and Rootstocks

During 2019, we evaluated the susceptibility of 24 different grapevine *Cultivars* and five *Rootstocks* to two different strains of *Xff* (XYL 2055/17 and XYL 2177/18) that were previously isolated in the same geographical area where the trial was conducted (Balearic Islands). The rAUDPC data were recorded during the period between 8 WPI and 16 WPI, which corresponded to a maximum AUDPC of 280 u^2 for calculating the rAUDPC. Inoculated plants in which the bacterium was not detected via RT-PCR were not considered for an rAUDPC assessment.

We observed significant differences in the susceptibility of *Cultivars* and *Rootstocks* to each of the two strains of *Xff* without significative interaction (Figure 2). The average rAUDPC for all *Cultivars* combined was significantly higher for strain XLY 2055/17 (0.58) than for strain XYL 2177/18 (0.46), indicating the higher virulence of strain XLY 2055/17 (Figure 2a). In this case, the correlation between the rAUDPC values of both *Xff* strains was high, with a significant coefficient of $r = 0.903$ ($p < 0.05$). The significant difference between *Xff* strains indicates that one of the factors involved in the cultivar resistance was determined using the bacterial genotype. This relation could be established under a linear regression ($R^2 = 0.8156$) expressed as:

$$\text{rAUDPC}_{\text{XYL 2177/18}} = 0.8437\ \text{rAUDPC}_{\text{XYL 2055/17}} - 0.0266.$$

The two *Cultivars* studied were grafted onto three different *Rootstocks*. However, this grafting did not appear to produce any variation in the intrinsic susceptibility of the grafted *Cultivar*. Disease progression, expressed as the rAUDPC values for Tempranillo and Garnacha Tintorera, was independent of the rootstock upon which the plants were grafted (Figure 3). Therefore, hereafter, we refer to each *Cultivar* independently from the *Cultivars* that were grafted.

To establish the resistance levels to *Xff*, a new trial was repeated under the same conditions the following year (2020) via inoculation with only the most virulent strain, XYL 2055/17, according to the results obtained the year prior (Figure 2). We observed significant differences among *Cultivars* in their individual susceptibility to the pathogen and also significant differences between the years (Figure 4). This statistical result indicates that the factors involved in grapevine *Cultivars'* resistance could be dependent on climate (or environmental), but not for all varieties or with the same determination attending to the significant interaction between both factors (Cultivar and year) (Figure 4a). The *Cultivars* Tempranillo Blanco, Tempranillo RJ43, Tempranillo RF78, Tempranillo, Malvasía, Monastrel, Mencía, Garnacha Tintorera, and Macabeo yielded the highest rAUDPC values, with no significant differences among them. Conversely, Pedro Ximenez, Albariño, Verdejo, Bobal, Pinot Noir, Hondarrabi Zuri, H. Beltza, and Cabernet Souvignon presented the highest resistance to *Xff* (Figure 4b).

(a)

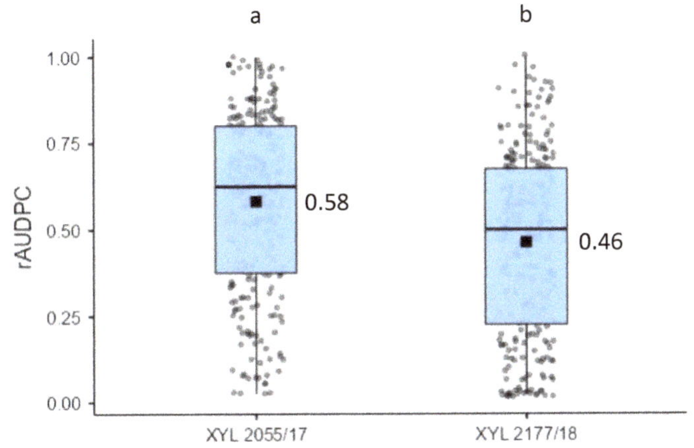

(b)

ANOVA-trAUDPC

	Sum of Squares	df	Mean Square	F	p
Cultivar	30.8	28	1.1	20.92	< 0.001
Strain	3.15	1	3.15	59.84	< 0.001
Cultivar x Strain	2.02	28	0.0722	1.37	0.1
Residuals	23.51	447	0.0526		

Figure 2. Descriptive and statistical parameters for the evaluation of grapevine *Cultivars'* susceptibility to *Xff* strains based on an estimation of the rAUDPC recorded from 8 WPI to 16 WPI for 24 *Cultivars* and five *Rootstocks* artificially inoculated with XYL 2055/17 and XYL 2177/18. (**a**) Each boxplot reports the second and third quartiles, with median (line) and mean values (dot) indicated as squares and points showing all values considered. The *Xff* strains followed by different letters indicate significant differences among the mean rAUDPC values based on Fisher's Least Significant Difference (LSD) test ($p < 0.05$). (**b**) Statistical parameters after two-way ANOVA analysis of rAUDPC previously transformed using arcsine of the square root of the proportion (trAUDPC).

In this study, the five most common *Rootstocks* upon which European varieties are grafted were shown to be less susceptible to *Xff* than most commercial *Cultivars* of *V. vinifera*. This result was essentially confirmed under our conditions because all *Rootstocks* (R110, 41B-MGt, Ru140, P1103, and 196-17 Cl) presented very low susceptibility to the pathogen, with R110 yielding the lowest rAUDPC for *Xff* (Figure 4a).

3.2. Disease Severity and Disease Progression (rAUDPC) under Open Field and Controlled Greenhouse Conditions

To better estimate and establish the degree of susceptibility, it is necessary to evaluate plants during several campaigns or under different climate conditions. Accordingly, we repeated the assays for one additional year. However, due to operational limitations, we decided to perform the inoculations with only five *Cultivars* that showed different levels of susceptibility to *Xff* in previous field trials together with one *Rootstock*. In this case, we used the *Xff* ST1 strain IVIA 5770, the only strain that current regulatory restrictions allow to be manipulated outside of the demarcated areas in Spain.

During 2021, two new trials were carried out in parallel: one in the same location as the two previous trials (Balearic Islands), and the other under controlled conditions in a greenhouse located in a different climatic area of northern Spain. Special care was taken to ensure that the plants for both trials belonged to the same grafted lot and were maintained without fertilizer or water restrictions.

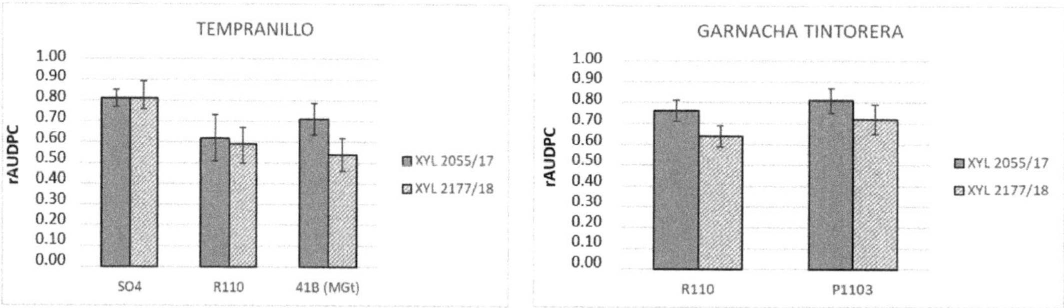

ANOVA-trAUDPC (a)

	Sum of Squares	df	Mean Square	F	p
Rootstock	0.2262	2	0.1131	3.111	0.055
Strain	0.0774	1	0.0774	2.129	0.152
Rootstock x Strain	0.0388	2	0.0194	0.533	0.591
Residuals	1.5634	43	0.0364		

ANOVA-trAUDPC (b)

	Sum of Squares	df	Mean Square	F	p
Rootstock	0.0434	1	0.0434	0.882	0.355
Strain	0.1649	1	0.1649	3.351	0.076
Rootstock x Strain	0.0031	1	0.0031	0.063	0.802
Residuals	1.5748	32	0.0492		

Figure 3. Mean rAUDPC values obtained for Tempranillo (**a**) and Garnacha Tintorera (**b**) *Cultivars* grafted onto different *Rootstocks* (SO4, R110, 41 B, and P1103). The plants were artificially inoculated with two *Xff* strains: XYL 2055/17 (solid grey) and XYL 2177/18 (grey grid). The data were recorded over one year (2019) in open field over 8 weeks, from 8 to 16 WPI. Error bars represent the standard error of the mean. Statistical parameters after two-way ANOVA analysis of rAUDPC previously transformed using arcsine of the square root of the proportion (trAUDPC) shown in the table conclude no significant differences among Rootstocks, Strains, and their interaction (Rootstock*Strain).

The results obtained in this new trial carried out in 2021 using a restricted number of *Cultivars* and *Rootstocks* confirm those obtained in the two previous years. All the *Cultivars* and *Rootstocks* were susceptible to *Xff*, in this case, against the strain *Xff* IVIA 5770, with clear pathogen resistance depending on climatic conditions. The rAUDPC values were significantly higher in four of the five *Cultivars* tested: Tempranillo, Garnacha, Pinot Noir, and Chardonay (Figure 5). The *Cultivars* in which the rAUDPC values were not significatively different between the field and greenhouse were Hondarrabi Zuri and *Rootstock* R110 (Figure 5). Both *Cultivars* belonged to the group with the lowest rAUDPC values recorded in previous field trials, as shown in Figure 4a.

Four months (16 WPI) after inoculation, the SI recorded under field conditions exceeded level 3. However, under greenhouse conditions, this level was not exceeded in any of the *Cultivars* studied (Figure 6). In both the greenhouse and the field, it was possible to observe significant differences in the level of *Cultivar* susceptibility to *Xff* (rAUDPC). However, in the greenhouse, the average of this indicator did not clearly discriminate between the resistance levels of the *Cultivars* studied. Only Tempranillo yielded a significantly different rAUDPC than that recorded in *Rootstock* R110 (Figure 6).

Importantly, after inoculation, the asymptomatic period was shorter in the open field than that in the greenhouse. An asymptomatic to symptomatic transition occurred between 10 and 12 WPI in the greenhouse and after 8 WPI in the open field (Figure 6).

Table 1 shows the maximum SI recorded in the inoculated plants, in both the field and the greenhouse. Two months after inoculation (16 WPI), all *Cultivars* of *Vitis vinifera* tested in the field recorded the maximum value of SI (5). Only two of the five evaluated *Rootstocks* (41 B-MGt and R110) did not obtain the same maximum SI value. In the greenhouse, no *Cultivar* yielded the maximum SI value (5), and only Tempranillo reached level 4 (Table 1).

(a)

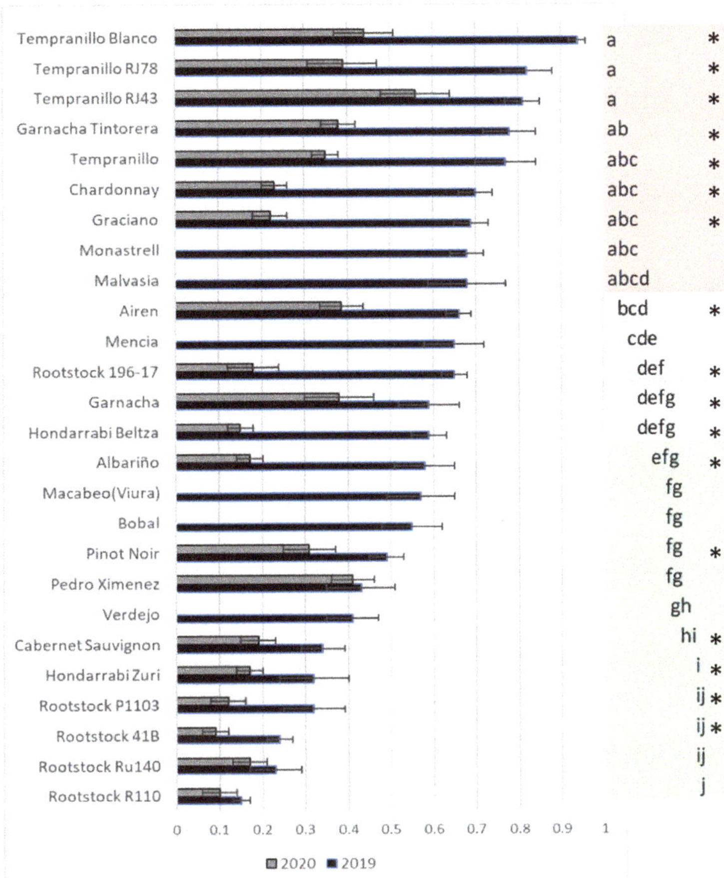

(b) ANOVA-trAUDPC

	Sum of Squares	df	Mean Square	F	p
Cultivar	15.1	19	0.7945	21.21	< 0.001
Year	10.13	1	10.1301	270.47	< 0.001
Cultivar x Year	3.07	19	0.1617	4.32	< 0.001
Residuals	13.6	363	0.0375		

Figure 4. (**a**) Mean rAUDPC values obtained for *Cultivars* and *Rootstocks* artificially inoculated with *Xff* ST1 XYL 2055/17, recorded over two consecutive years (2019 and 2020) for 8 weeks, from 2 months (8 WPI) to 4 months (16 WPI). Error bars represent the standard error of the mean. *Cultivars* followed by the same letter indicate non-significant differences among rAUDPC values based on Fisher's LSD test ($p < 0.05$) for just the 2019 trial. The most susceptible cultivars with the highest rAUDPC values are marked in red, while cultivars with the lowest values are marked in green. The number of inoculated plants per *Cultivar* considered to obtain the mean rADPC varied from seven to nine. (**b**) Statistical parameters after two-way ANOVA analysis of rAUDPC, previously transformed using arcsine of the square root of the proportion (trAUDPC), shown in the table conclude significant differences among *Cultivars*, year of trial conduction, and significative interaction between both factors (Rootstock*Strain). The asterisks indicate significant differences ($p < 0.05$) by Tukey's test between rAUDPC registered in 2019 and 2020 at each *Cultivar*.

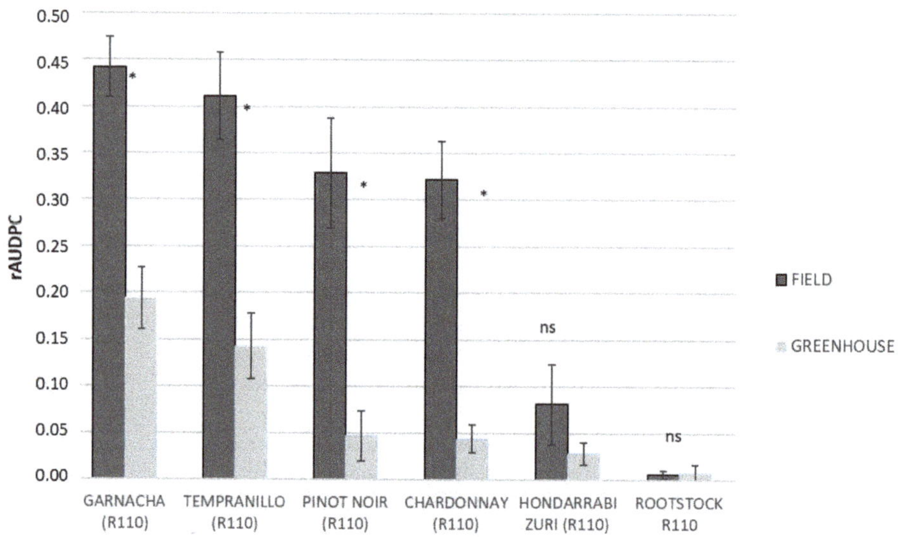

Figure 5. Mean rAUDPC values obtained for five *Cultivars* and one *Rootstock* artificially inoculated with *Xff* IVIA 5770, recorded over one year (2021) in two locations under both an open field (FIELD) and controlled conditions (GREENHOUSE). The rAUDPC was calculated based on the SI progress for 8 weeks, from 8 to 16 WPI. Error bars represent the standard error of the mean. The number of inoculated plants per *Cultivar* considered to obtain the mean rADPC was nine. Asterisks indicate the level of significance: *, $p < 0.05$; ns, not significant) based on Tukey-test between the means values obtained in Field (dark grey) and Greenhouse (light grey) for each *Cultivar*.

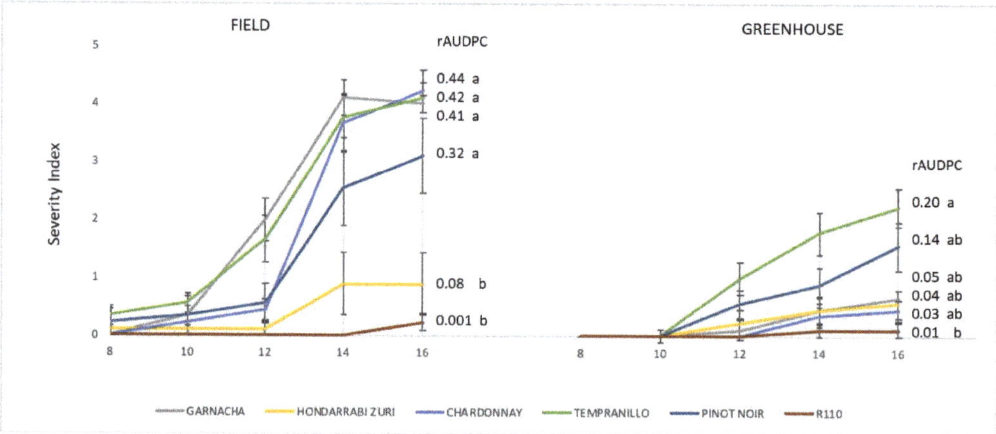

Figure 6. Disease progression in an open field (Field) and greenhouse-controlled conditions (Greenhouse) from 8 to 16 WPI. The plants were artificially inoculated with the *Xff* IVIA 5770 strain. Values are the means of 13 replicates, and error bars represent the standard error of the Severity Index (SI). The rAUDPC derivates from the time progression of the SI as shown in columns. Cultivars followed by the same letter indicate non-significant differences between the rAUDPC values based on Fisher's LSD test ($p < 0.05$).

3.3. Water Potential Progression during Xff Infection and Disease Development

To associate PD symptomatology and bacterial detection with the effects of infection on the water status of the plants, the Ψ differences between treatments (control and inoculated) were also analyzed for 19 weeks, from 2 WPI to 21 WPI, in plants growing under controlled conditions in the greenhouse. Ψ varied as the phenological stage of the plants progressed, but significant differences between inoculated and non-inoculated plants were observed from 12 WPI onwards (Figure 7). However, this result was not clearly established in all tested *Cultivars*. In two of the five cultivars (Garnacha and Pinot Noir), significant differences were not found, even at 21 WPI (Figure 8a). In the *Rootstock* R110 and Hondarrabi Zuri, which both presented the lowest severity index in the greenhouse (Figure 5), it was possible to find significant differences in Ψ between inoculated and healthy plants (Figure 8a). These results suggest no relationship between the severity of the disease and variation in water potential, at least in some *Cultivars*. On the other hand, at 12 WPI, the first period with significant differences in Ψ between inoculated and control plants, 41% of inoculated plants already indicated positive bacterial detection, as shown in Figure 8b.

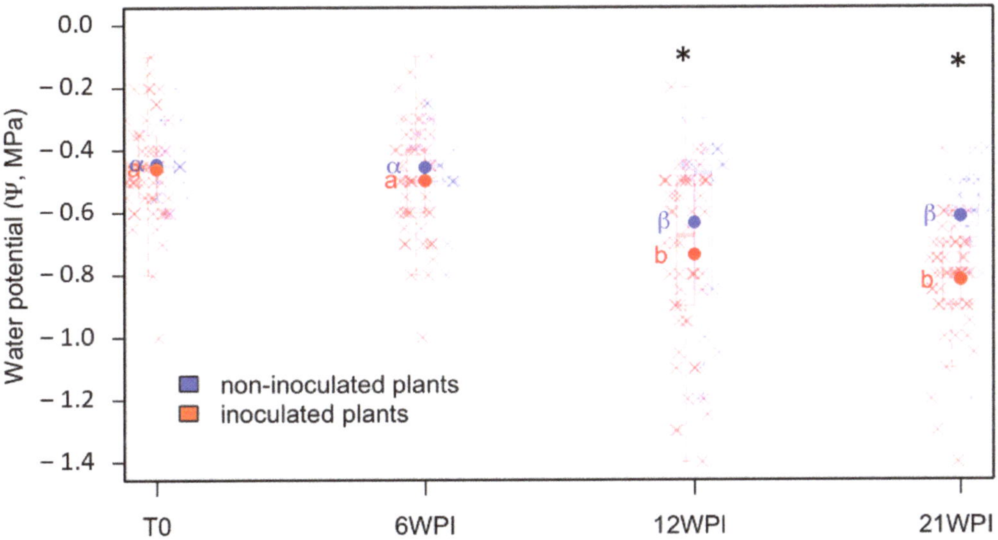

Figure 7. Water potential (Ψ) variation over time under greenhouse-controlled conditions. Latin letters correspond to average Ψ of the inoculated plants, while Greek letters represent the non-inoculated plants. The same letter indicates non-significant differences among mean Ψ values based on Fisher's LSD test ($p < 0.05$). The asterisks indicate significant differences ($p < 0.05$) by Tukey's test between inoculated and non-inoculated plants at each time point: T0 (before inoculation), 6, 12, and 21 WPI. Individual measurements for each plant are shown with "×" signs. The plants were artificially inoculated with *Xff* IVIA 5770.

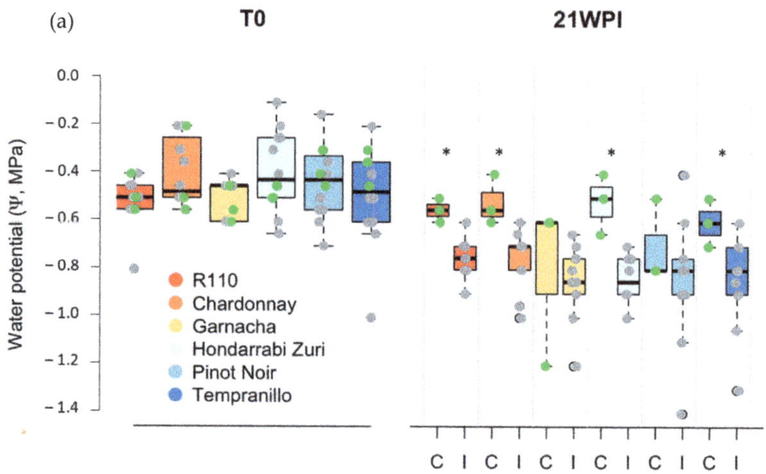

(b)

	(%) RTi-PCR+									
	2 WPI	4 WPI	6 WPI	8 WPI	10 WPI	12 WPI	14 WPI	16 WPI	18 WPI	21 WPI
R110	0	0	0	22	22	22	22	33	89	100
Chardonnay	11	11	44	44	44	44	44	44	78	100
Garnacha	0	11	11	22	22	22	44	56	78	100
Hondarrabi Zuri	0	22	22	22	22	33	33	44	78	100
Pinot Noir	0	44	44	44	44	44	67	78	78	100
Tempranillo	0	44	44	44	44	78	89	100	100	100
Total	2	22	28	33	33	41	50	59	84	100

Figure 8. Progression of the water potential (Ψ) and RT-PCR detection of *Xff* IVIA 5770 artificially inoculated in grapevine plants. The trial was carried out under controlled conditions in a greenhouse. (**a**) Ψ recorded before (T0) and after 21 WPI for artificial inoculation with *Xff* IVIA 5770 in five grapevine *Cultivars* and one *Rootstock* (R110) and 21 WPI under greenhouse-controlled conditions. Each boxplot reports the second and third quartiles, with median values (line) in the square and points showing outliers (green for non-inoculated (C) and grey for inoculated plants (I)). The asterisks indicate significant differences ($p < 0.05$) by Tukey's test between I and C plants at 21 WPI. (**b**) Percentage of artificially inoculated plants in which the bacterium was detected via RTi-PCR from 2 to 21 WPI.

4. Discussion

As previously described, strains of *Xff* ST1, isolated for the first time in the Balearic Islands, were able to cause disease in some of the main grapevine varieties planted in different Spanish regions [29]. The 24 *Cultivars* studied in this work were found to be susceptible. However, it was also possible to clearly establish different levels of resistance among cultivars. The results obtained agree with those previously published, showing different degrees of resistance in certain grapevine cultivars, most likely influenced by their different pedigrees (linked to their shared centers of domestication) and xylem anatomical features [33], or an engineered innate immune defense in grapevines [52].

Notably, the two determining factors of grapevine cultivar resistance to *Xff* in this study were the bacterial genotype (*Xff* strain) and environmental climate. The virulence of the *Xff* strain XYL 2055/17 was significantly higher than that of strain XYL 2177/18.

Plant–pathogen interactions are multifaceted processes mediated by the pathogen- and plant-derived molecules. Thus, the differential virulence between strains is well known. In this work, we detected this condition in two strains of *Xff* isolated independently in the same geographical area over consecutive years, which could corroborate the results of the genetic variability of *Xff* isolated in the Balearic Islands recently presented by Dr. Landa's research team in 2022 [7]. Despite the differences in virulence found between the two strains in this study, resistance was highly correlated between cultivars. Thus, strain virulence does not appear to modify the susceptibility profiles.

In the absence of a greater number of trials under different agroclimatic conditions, this work established, for the first time, two significantly different groups of European cultivars of *Vitis vinifera,* characterized as having high and low susceptibility to *Xff*. The results indicate that Tempranillo and Tempranillo Blanco can be considered as a reference cultivar (variety) (among others) with high susceptibility compared to Hodarrabi Zuri and Cabernet Sauvignon with lower susceptibility.

The results obtained in this work demonstrate that the *Rootstock* commonly used to graft commercial varieties of grapevine are also susceptible to PD caused by *Xff* European isolates. However, the level of susceptibility was significantly lower than that of most commercial *Cultivars*. Several studies showed that rootstocks affect scion responses in many different ways. Rootstocks, for example, can influence scion vigor and phenology and confer differential tolerance to drought and disease in various crops [53,54]. Recent relevant studies [55] proposed that a low root mass may incite resource-limiting conditions to activate carbohydrate metabolic pathways, which reciprocally interact with plant immune system genes to elicit differential levels of cultivar susceptibility in bacterial pathogen–plant interaction. However, the low susceptibility found in the *Rootstocks* studied was not found to confer lower susceptibility among the cultivars. Our results indicate that the susceptibility of one grapevine cultivar to PD could be independent of the rootstock on which the cultivars were grafted.

To ensure higher control over environmental variables, it is common to evaluate cultivars for their resistance to pathogens in growth chambers or greenhouses before the evaluations under field conditions. However, the susceptibility ratings under these two conditions were previously found to show either a significant correlation [56,57] or no correlation [58]. Additionally, in certain cases, greenhouse evaluations tend to overestimate susceptibility [59]. In the end, screening for resistance to pathogens could be more accurate when conducted in open fields or greenhouses, depending on the expression of plant defense-related genes after bacterial inoculation or the presence of climatic conditions more conducive to bacterial growth [60]. Therefore, since inoculation with a quarantine pathogen such as *Xff* is more practical in controlled facilities, we evaluated the correlation between plant susceptibility to *Xff* among five cultivars and one *Rootstock* in a greenhouse and in the field. The results showed that disease progression was significantly slower in the greenhouse than in the open field. Consequently, the rAUPC found in grapevines grown in this facility was no greater than 0.25 units from 2 to 4 months after inoculation. This result is not surprising since in other pathogenicity models of *Xff* in greenhouses, disease severity did not reach the levels found in the open field, as in the case of *Xff* in southern highbush blueberry (*Vaccinium* sp.) [61] and almond [43].

Under the current disease scenario, investigators can handle tasks with *Xf*-infected plant material only under confined, approved conditions to avoid the spread of the disease, denoted as temporal confinement stations for quarantine pathogens (Regulation (EU) 2016/2031). Biosafety greenhouses are the only facilities that can be used for this purpose. Taking into account our results, it would be possible to discriminate between highly and moderately susceptible cultivars with reference to Tempranillo *Cultivars*. Tempranillo grafted onto Rootstock R110 was able to reach a severity index of 4 after 16 WPI, as shown by our results. Together with Tempranillo, Pinot Noir (both grafted onto R110) presented the earliest symptoms. The first symptoms were recorded at 12 WPI. This result confirms our selection of Chardonnay and Pinot Noir varieties as indicator plants in the pathogenicity tests and suggests the inclusion of the Tempranillo and Garnacha varieties in standards for the Diagnostic Protocol of *Xff*.

The differences in varietal susceptibility levels highlighted in this work under open field conditions, together with previous work showing that disease incidence and severity may be related to agricultural management, suggest that PD in Europe could be, in the future, managed as a chronic disease as it is now managed in American winegrowing areas [32]. Several scientific results support this theory, including crop management (Moralejo, 2019 [39]), which found positive results when using innovative synthesized chemical treatments [37,42], the efficacy of foliar-applied biological treatments [62], insect vector deterrence [63], and others [9]. We should also consider the wide variety of biostimulants appearing on the fertilizer and agro-sanitary markets whose contributions to vineyard health in different agroclimatic grapevine areas remain to be studied [64,65].

It is well known that PD in grapevine is related to a decrease in water conductance, water potential, and hydraulic conductance [10,11,66]. Compared to healthy leaves developed under the same conditions, *Xf*-infected plants showed a reduction in available water and nutrients, resulting in wilting and death due to stress. This result was corroborated using the Scholander test in woody plants [12]. In this study, we were able to associate this physiological effect with the onset of symptomatology and *Xff* infection. However, based on our results, it is not easy to consider this parameter as an early detection test for *Xff* infection prior to the symptomatic period. The differences in water potential started to become significantly distinct between healthy and infected plants once the bacteria were detectable via standard RTi-PCR. In addition, the results obtained show differences among cultivars; thus, it was not possible to establish a significant relationship between infection and potential water in all cultivars. This alternative technology would be discarded as an alternative diagnostic measure unless further research is carried out to discriminate its potential use for different cultivars, evaluation times, and agroclimatic conditions. We cannot ignore these new and non-destructive technologies as they create the possibility to survey, and potentially geo-reference, large numbers of plants in real time.

Undoubtably, grapevine agro-management in novel environments coupled with *Cultivar* resistance will play a determining role in PD control in Europe. Indeed, such methods are already being prioritized in other crops [67]. Considering the data presented in this work, several agro-climatic parameters should be tested to enhance breeding programs and establish a discriminative threshold for cultivar selection. At the same time, we should explore new technologies for the early detection of differential physiological responses and phenotyping of foliar disease severity under controlled conditions [68,69].

5. Conclusions

The 24 European *Vitis vinifera* cultivars studied in this work, representing more than 70% of the cultivated area in Spain, are susceptible to the PD caused by current *Xff* ST1 strains detected in Europe. After carrying out trials over three consecutive years, we obtained consistent results indicating two significantly different groups of resistance which may not be influenced by the rootstock upon which the plant is grafted. Importantly, *Cultivar* susceptibility is clearly influenced by the environmental conditions under which the plants are grown. The results of cultivar resistance evaluations may differ between the field and the greenhouse conditions.

Tempranillo could be included as references in breeding programs for *Xff* as indicators of highly susceptible *Cultivars*, since under field and greenhouse conditions, these variety presented the largest SI and rAUDPC values four months after inoculation.

Author Contributions: Conceptualization, A.O.-B. and M.L.; methodology, S.M., A.A. and A.H.; formal analysis, J.B.R.; investigation, S.M.; resources, A.O.-B. and J.B.R.; data curation, S.M.; writing—original draft preparation, S.M. and A.O.-B.; writing—review and editing, A.O.-B. and S.M.; supervision, M.L. and A.O.-B.; funding acquisition, A.O.-B. All authors have read and agreed to the published version of the manuscript.

Funding: This research was partially funded by AEI-INIA Spain, through a NEIKER-Girona University agreement (Laboratory of Plant Pathology, Institute of Food and Agricultural Technology-CIDSAV-XaRTA): Grant number E-RTA2017-00004-C06-03; Basque Government through Grupo de Investigación del Sistema Universitario Vasco: Grant numbers T1682-22 and 00039-IDA2021-45; S.M. was a recipient of a PhD research grant from Department of Economic Development, Sustainability and Environment of the Basque Government (2018–2022): Grant reference SARA MARTINEZ 2018.

Institutional Review Board Statement: Not applicable.

Data Availability Statement: All data are contained in the article.

Acknowledgments: We thank Ander Castander and Diego Llamazares for their assistance with the Scholander pressure chamber methodology.

Conflicts of Interest: The authors declare no conflict of interest. The funders had no role in the design of the study; the analyses or interpretation of the data; the writing of the manuscript; or the decision to publish the results.

References

1. Wells, J.M.; Raju, B.C.; Hung, H.-Y.; Weisburg, W.G.; Mandelco-Paul, L.; Brenner, D.J. *Xylella fastidiosa* gen. nov., sp. nov: Gram-negative, xylem-limited, fastidious plant bacteria related to *Xanthomonas* spp. *Int. J. Syst. Bacteriol.* **1987**, *37*, 136–143. [CrossRef]
2. European Food Safety Authority (EFSA); Gibin, D.; Pasinato, L.; Delbianco, A. Update of the *Xylella* Spp. Host Plant Database—Systematic Literature Search up to 31 December 2022. *EFSA J.* **2023**, *21*, 8061. [CrossRef]
3. Sicard, A.; Zeilinger, A.R.; Vanhove, M.; Schartel, T.E.; Beal, D.J.; Daugherty, M.P.; Almeida, R.P.P. *Xylella fastidiosa*: Insights into an emerging plant pathogen. *Annu. Rev. Phytopathol.* **2018**, *56*, 181–202. [CrossRef]
4. Trkulja, V.; Tomić, A.; Iličić, R.; Nožinić, M.; Milovanović, T.P. *Xylella fastidiosa* in Europe: From the introduction to the current status. *Plant Pathol. J.* **2022**, *38*, 551–571. [CrossRef]
5. Frem, M.; Fucilli, V.; Nigro, F.; El Moujabber, M.; Abou Kubaa, R.; La Notte, P.; Bozzo, F.; Choueiri, E. The potential direct economic impact and private management costs of an invasive alien species: *Xylella fastidiosa* on Lebanese wine grapes. *NeoBiota* **2021**, *70*, 43–67. [CrossRef]
6. Hopkins, D.L.; Purcell, A.H. *Xylella fastidiosa*: Cause of Pierce's Disease of grapevine and other emergent Diseases. *Plant Dis.* **2002**, *86*, 1056–1066. [CrossRef]
7. Velasco-Amo, M.P.; Arias-Giraldo, L.F.; Olivares-García, C.; Denancé, N.; Jacques, M.-A.; Landa, B.B. Use of traC Gene to Type the Incidence and Distribution of pXFAS_Plasmid-Bearing Strains of *Xylella fastidiosa* subsp. *fastidiosa ST1 in Spain. Plants* **2022**, *11*, 1562. [CrossRef] [PubMed]
8. Alston, J.M.; Fuller, K.; Kaplan, J.D.; Tumber, K. The Costs and Benefits of Pierce's Disease Research in the California Winegrape Industry. In Proceedings of the Agricultural and Applied Economics Association (AAEA) Conferences, 2013, Annual Meeting, Washington, DC, USA, 4–6 August 2013; p. 37. [CrossRef]
9. Kyrkou, I.; Pusa, T.; Ellegaard-Jensen, L.; Sagot, M.-F.; Hansen, L.H. Pierce's Disease of grapevines: A review of control strategies and an outline of an epidemiological model. *Front. Microbiol.* **2018**, *9*, 2141. [CrossRef]
10. Chatelet, D.S.; Wistrom, C.M.; Purcell, A.H.; Rost, T.L.; Matthews, M.A. Xylem structure of four grape varieties and alternative hosts to the xylem-limited bacterium *Xylella fastidious*. *Ann. Bot.* **2011**, *108*, 73–85. [CrossRef] [PubMed]
11. Pérez-Donoso, A.G.; Lenhof, J.J.; Pinney, K.; Labavitch, J.M. Vessel Embolism and tyloses in early stages of Pierce's Disease: Embolism and tyloses in Pierce's Disease. *Aust. J. Grape Wine Res.* **2016**, *22*, 81–86. [CrossRef]
12. Raicavoli, J.; Ingel, B.; Blanco-Ulate, B.; Cantu, D.; Roper, C. *Xylella fastidiosa*: An examination of a re-emerging plant pathogen: *Xylella fastidiosa*. *Mol. Plant Pathol.* **2018**, *19*, 786–800. [CrossRef]
13. Carluccio, G.; Greco, D.; Sabella, E.; Vergine, M.; De Bellis, L.; Luvisi, A. Xylem embolism and pathogens: Can the vessel anatomy of woody plants contribute to *X. fastidiosa* resistance? *Pathogens* **2023**, *12*, 825. [CrossRef]
14. Lopes, S.A.; Teixeira, D.C.; Fernandes, N.G.; Ayres, A.J.; Torres, S.C.Z.; Barbosa, J.C.; Li, W.B. An experimental inoculation system to study citrus- *Xylella fastidiosa* interactions. *Plant Dis.* **2005**, *89*, 250–254. [CrossRef] [PubMed]
15. Almeida, R.P.P.; Nunney, L. How Do plant diseases caused by *Xylella fastidiosa* emerge? *Plant Dis.* **2015**, *99*, 1457–1467. [CrossRef] [PubMed]
16. Goodwin, P.H.; DeVay, J.E.; Meredith, C.P. Physiological responses of *Vitis vinifera* Cv. "Chardonnay" to infection by the Pierce's Disease bacterium. *Physiol. Mol. Plant Pathol.* **1988**, *32*, 17–32. [CrossRef]
17. Fernández, J. Plant-Based methods for irrigation scheduling of woody crops. *Horticulturae* **2017**, *3*, 35. [CrossRef]
18. Ihuoma, S.O.; Madramootoo, C.A. Recent Advances in crop water stress detection. *Comput. Electron. Agric.* **2017**, *141*, 267–275. [CrossRef]
19. Virnodkar, S.S.; Pachghare, V.K.; Patil, V.C.; Jha, S.K. Remote sensing and machine learning for crop water stress determination in various crops: A critical review. *Precis. Agric.* **2020**, *21*, 1121–1155. [CrossRef]
20. Rodriguez-Dominguez, C.M.; Forner, A.; Martorell, S.; Choat, B.; Lopez, R.; Peters, J.M.R.; Pfautsch, S.; Mayr, S.; Carins-Murphy, M.R.; McAdam, S.A.M.; et al. Leaf water potential measurements using the pressure chamber: Synthetic testing of assumptions towards best practices for precision and accuracy. *Plant Cell Environ.* **2022**, *45*, 2037–2061. [CrossRef]
21. Gálvez Pavez, R.; Callejas Rodríguez, R.; Reginato Meza, G. Comparación de la cámara de presión tipo scholander modelo pump-up respecto a la cámara de presión tradicional en vides de mesa. *Idesia* **2011**, *29*, 175–179. [CrossRef]
22. Choné, X. Stem water potential is a sensitive indicator of grapevine water status. *Ann. Bot.* **2001**, *87*, 477–483. [CrossRef]

23. Williams, L.E.; Araujo, F.J. Correlations among predawn leaf, midday leaf, and midday stem water potential and their correlations with other measures of soil and plant water status in *Vitis vinifera*. *J. Am. Soc. Hortic. Sci.* **2002**, *127*, 448–454. [CrossRef]
24. Santesteban, L.G.; Miranda, C.; Marín, D.; Sesma, B.; Intrigliolo, D.S.; Mirás-Avalos, J.M.; Escalona, J.M.; Montoro, A.; De Herralde, F.; Baeza, P.; et al. Discrimination ability of leaf and stem water potential at different times of the day through a meta-analysis in grapevine (*Vitis vinifera* L.). *Agric. Water Manag.* **2019**, *221*, 202–210. [CrossRef]
25. Suter, B.; Triolo, R.; Pernet, D.; Dai, Z.; Van Leeuwen, C. Modeling stem water potential by separating the effects of soil water availability and climatic conditions on water status in grapevine (*Vitis vinifera* L.). *Front. Plant Sci.* **2019**, *10*, 1485. [CrossRef]
26. Tyree, M.T.; Zimmermann, M.H. Hydraulic architecture of whole plants and plant performance. In *Xylem Structure and the Ascent of Sap*; Springer Series in Wood Science; Springer: Berlin/Heidelberg, Germany, 2022; pp. 175–214. [CrossRef]
27. Diago, M.P.; Fernández-Novales, J.; Gutiérrez, S.; Marañón, M.; Tardaguila, J. Development and validation of a new methodology to assess the vineyard water status by on-the-go near infrared spectroscopy. *Front. Plant Sci.* **2018**, *9*, 59. [CrossRef]
28. OIV–International Organization of Vine and Wine. Distribution of the world's grapevine varieties. In *Focus OIV 2017*; OIV–International Organization of Vine and Wine: Dijon, France, 2017; p. 53, ISBN 979-10-91799-89-8.
29. Giménez-Romero, A.; Galván, J.; Montesinos, M.; Bauzà, J.; Godefroid, M.; Fereres, A.; Ramasco, J.J.; Matías, M.A.; Moralejo, E. Global predictions for the risk of establishment of Pierce's Disease of grapevines. *Commun. Biol.* **2022**, *5*, 1389. [CrossRef]
30. Fritschi, F.B.; Lin, H.; Walker, M.A. *Xylella fastidiosa* population dynamics in grapevine genotypes differing in susceptibility to Pierce's Disease. *Ann. J. Enol. Vitic.* **2007**, *58*, 326–332. [CrossRef]
31. Riaz, S.; Huerta-Acosta, K.; Tenscher, A.C.; Walker, M.A. Genetic characterization of *Vitis* germplasm collected from the southwestern us and Mexico to expedite Pierce's Disease-resistance breeding. *Theor. Appl. Genet.* **2018**, *131*, 1589–1602. [CrossRef]
32. Morales-Cruz, A.; Aguirre-Liguori, J.; Massonnet, M.; Minio, A.; Zaccheo, M.; Cochetel, N.; Walker, A.; Riaz, S.; Zhou, Y.; Cantu, D.; et al. Multigenic resistance to *Xylella fastidiosa* in wild grapes (*Vitis* spp.) and its implications within a changing climate. *Commun. Biol.* **2023**, *6*, 580. [CrossRef]
33. Deyett, E.; Pouzoulet, J.; Yang, J.-I.; Ashworth, V.E.; Castro, C.; Roper, M.C.; Rolshausen, P.E. Assessment of Pierce's Disease susceptibility in *Vitis vinifera* cultivars with different pedigrees. *Plant Pathol.* **2019**, *68*, 1079–1087. [CrossRef]
34. Schneider, K.; Van Der Werf, W.; Cendoya, M.; Mourits, M.; Navas-Cortés, J.A.; Vicent, A.; Oude Lansink, A. Impact of *Xylella fastidiosa* subspecies *pauca* in European olives. *Proc. Natl. Acad. Sci. USA* **2020**, *117*, 9250–9259. [CrossRef] [PubMed]
35. Latest Developments of Xylella Fastidiosa in the EU Territory. Available online: https://ec.europa.eu/food/plants/plant-healthand-biosecurity/legislation/control-measures/xylella-fastidiosa/latest-developments-xylella-fastidiosa-eu-territory_en (accessed on 6 October 2023).
36. Castro, C.; Massonnet, M.; Her, N.; DiSalvo, B.; Jablonska, B.; Jeske, D.R.; Cantu, D.; Roper, M.C. Priming grapevine with lipopolysaccharide confers systemic resistance to Pierce's Disease and identifies a peroxidase linked to defense priming. *New Phytol.* **2023**, *239*, 687–704. [CrossRef] [PubMed]
37. Badosa, E.; Planas, M.; Feliu, L.; Montesinos, L.; Bonaterra, A.; Montesinos, E. Synthetic peptides against plant pathogenic bacteria. *Microorganisms* **2022**, *10*, 1784. [CrossRef]
38. Avosani, S.; Nieri, R.; Mazzoni, V.; Anfora, G.; Hamouche, Z.; Zippari, C.; Vitale, M.L.; Verrastro, V.; Tarasco, E.; D'Isita, I.; et al. Intruding into a conversation: How behavioral manipulation could support management of *Xylella fastidiosa* and its insect vectors. *J. Pest Sci.* **2023**, *1*, 1–17. [CrossRef]
39. Moralejo, E.; Borràs, D.; Gomila, M.; Montesinos, M.; Adrover, F.; Juan, A.; Nieto, A.; Olmo, D.; Seguí, G.; Landa, B.B. Insights into the epidemiology of Pierce's Disease in vineyards of Mallorca, Spain. *Plant Pathol.* **2019**, *68*, 1458–1471. [CrossRef]
40. Maul Team. Vitis International Variety Catalogue. 2023. Available online: www.vivc.de (accessed on 6 October 2023).
41. Ollat, N.; Bordenave, L.; Tandonnet, J.P.; Boursiquot, J.M.; Marguerit, E. Grapevine rootstocks: Origins and perspectives. In Proceedings of the I International Symposium on Grapevine Roots, Rauscedo, Italy, 16–17 October 2014; pp. 11–22. [CrossRef]
42. Moll, L.; Badosa, E.; Planas, M.; Feliu, L.; Montesinos, E.; Bonaterra, A. Antimicrobial peptides with antibiofilm activity against *Xylella fastidiosa*. *Front. Microbiol.* **2021**, *12*, 753874. [CrossRef]
43. Baró, A.; Montesinos, L.; Badosa, E.; Montesinos, E. Aggressiveness of Spanish Isolates of *Xylella fastidiosa* to almond plants of different cultivars under greenhouse conditions. *Phytopathology* **2021**, *111*, 1994–2001. [CrossRef] [PubMed]
44. Hill, B.L.; Purcell, A.H. Multiplication and Movement of *Xylella fastidiosa* within Grapevine and Four Other Plants. *Phytopathology* **1995**, *85*, 1368–1372. [CrossRef]
45. Almeida, R.P.P.; Pereira, E.F.; Purcell, A.H.; Lopes, J.R.S. Multiplication and movement of a citrus strain of *Xylella fastidiosa* within sweet orange. *Plant Dis.* **2001**, *85*, 382–386. [CrossRef]
46. Su, C.; Chang, C.J.; Chang, C.-M.; Shih, H.-T.; Tzeng, K.-C.; Jan, F.-J.; Kao, C.-W.; Deng, W.-L. Pierce's Disease of grapevines in Taiwan: Isolation, cultivation and pathogenicity of *Xylella fastidiosa*. *J. Phytopathol.* **2013**, *161*, 389–396. [CrossRef]
47. Simko, I.; Piepho, H.-P. The area under the disease progress stairs: Calculation, advantage, and application. *Phytopathology* **2012**, *102*, 381–389. [CrossRef] [PubMed]
48. OEPP/EPPO. PM 7/24 (4) *Xylella fastidiosa*. *EPPO Bull.* **2019**, *49*, 175–227. [CrossRef]
49. Harper, S.J.; Ward, L.I.; Clover, G.R.G. Development of LAMP and Real-Time PCR methods for the rapid detection of *Xylella fastidiosa* for quarantine and field applications. *Phytopathology* **2010**, *100*, 1282–1288. [CrossRef] [PubMed]

50. Knipfer, T.; Bambach, N.; Hernandez, M.I.; Bartlett, M.K.; Sinclair, G.; Duong, F.; Kluepfel, D.A.; McElrone, A.J. Predicting stomatal closure and turgor loss in woody plants using predawn and midday water potential. *Plant Physiol.* **2020**, *184*, 881–894. [CrossRef]
51. Castander-Olarieta, A.; Moncaleán, P.; Pereira, C.; Pěnčík, A.; Petřík, I.; Pavlović, I.; Novák, O.; Strnad, M.; Goicoa, T.; Ugarte, M.D.; et al. Cytokinins are involved in drought tolerance of *Pinus radiata* plants originating from embryonal masses induced at high temperatures. *Tree Physiol.* **2021**, *41*, 912–926. [CrossRef]
52. Dandekar, A.M.; Gouran, H.; Ibáñez, A.M.; Uratsu, S.L.; Agüero, C.B.; McFarland, S.; Borhani, Y.; Feldstein, P.A.; Bruening, G.; Nascimento, R.; et al. An engineered innate immune defense protects grapevines from Pierce's Disease. *Proc. Natl. Acad. Sci. USA* **2012**, *109*, 3721–3725. [CrossRef]
53. Warschefsky, E.J.; Klein, L.L.; Frank, M.H.; Chitwood, D.H.; Londo, J.P.; von Wettberg, E.J.B.; Miller, A.J. Rootstocks: Diversity, domestication, and impacts on shoot phenotypes. *Trends Plant Sci.* **2016**, *21*, 418–437. [CrossRef]
54. Kumar, P.; Rouphael, Y.; Cardarelli, M.; Colla, G. Vegetable grafting as a tool to improve drought resistance and water use efficiency. *Front. Plant Sci.* **2017**, *8*, 1130. [CrossRef]
55. Singh, J.; Fabrizio, J.; Desnoues, E.; Pereira Silva, J.; Busch, W.; Khan, A. Root system traits impact early fire blight susceptibility in apple (*Malus* × *domestica*). *BMC Plant Biol.* **2019**, *19*, 579–593. [CrossRef]
56. Heaton, J.B.; Dullahide, S.R.; Tancred, S.J.; Zeppa, A.G.; McWaters, A.D. Comparison of orchard and glasshouse tests in screening of apple breeding progeny for resistance to *Venturia inaequalis* in Queensland from 1985 to 1994. *Australas. Plant Pathol.* **1995**, *24*, 243–248. [CrossRef]
57. Martinez-Bilbao, A.; Ortiz-Barredo, A.; Montesinos, E.; Murillo, J. *Venturia inaequalis* resistance in local Spanish cider apple germplasm under controlled and field conditions. *Euphytica* **2012**, *188*, 273–283. [CrossRef]
58. MacHardy, W.E. *Apple Scab: Biology, Epidemiology and Management*; APS Press: St. Paul, MN, USA, 1996.
59. Martínez-Bilbao, A.; Ortiz-Barredo, A.; Montesinos, E.; Murillo, J. Evaluation of a cider apple germplasm collection of local cultivars from Spain for resistance to fire blight (*Erwinia amylovora*) using a combination of inoculation assays on leaves and shoots. *HortScience* **2009**, *44*, 1223–1227. [CrossRef]
60. Fred, A.K.; Kiswara, G.; Yi, G.; Kim, K.-M. Screening Rice Cultivars for Resistance to Bacterial Leaf Blight. *J. Microbiol. Biotechnol.* **2016**, *26*, 938–945. [CrossRef]
61. Oliver, J.E.; Cobine, P.A.; De La Fuente, L. *Xylella fastidiosa* isolates from both subsp. *multiplex* and *fastidiosa* cause disease on southern highbush blueberry (*Vaccinium* sp.) under greenhouse conditions. *Phytopathology* **2015**, *105*, 855–862. [CrossRef]
62. Das, M.; Bhowmick, T.S.; Ahern, S.J.; Young, R.; Gonzalez, C.F. Control of Pierce's Disease by phage. *PLoS ONE* **2015**, *10*, e0128902. [CrossRef] [PubMed]
63. Rashed, A.; Kwan, J.; Baraff, B.; Ling, D.; Daugherty, M.P.; Killiny, N.; Almeida, R.P.P. Relative susceptibility of *Vitis vinifera* cultivars to vector-borne *Xylella fastidiosa* through time. *PLoS ONE* **2013**, *8*, e55326. [CrossRef]
64. Irani, H.; ValizadehKaji, B.; Naeini, M.R. Biostimulant-induced drought tolerance in grapevine is associated with physiological and biochemical changes. *Chem. Biol. Technol. Agric.* **2021**, *8*, 5. [CrossRef]
65. Monteiro, E.; Gonçalves, B.; Cortez, I.; Castro, I. The role of biostimulants as alleviators of biotic and abiotic stresses in grapevine: A review. *Plants* **2022**, *11*, 396. [CrossRef]
66. Sarcina, L.; Macchia, E.; Loconsole, G.; D'Attoma, G.; Bollella, P.; Catacchio, M.; Leonetti, F.; Di Franco, C.; Elicio, V.; Scamarcio, G.; et al. Fast and reliable electronic assay of a *Xylella fastidiosa* single bacterium in infected plants sap. *Adv. Sci.* **2022**, *9*, 2203900. [CrossRef]
67. León, L.; De La Rosa, R.; Arriaza, M. Prioritization of olive breeding objectives in Spain: Analysis of a producers and researchers survey. *Span. J. Agric. Res.* **2021**, *19*, e0701. [CrossRef]
68. Surano, A.; Abou Kubaa, R.; Nigro, F.; Altamura, G.; Losciale, P.; Saponari, M.; Saldarelli, P. Susceptible and resistant olive cultivars show differential physiological response to *Xylella fastidiosa* infections. *Front. Plant. Sci.* **2022**, *13*, 968934. [CrossRef] [PubMed]
69. Alves, K.S.; Guimarães, M.; Ascari, J.P.; Queiroz, M.F.; Alfenas, R.F.; Mizubuti, E.S.G.; Del Ponte, E.M. RGB-Based Phenotyping of foliar disease severity under controlled conditions. *Trop. Plant Pathol.* **2022**, *47*, 105–117. [CrossRef]

horticulturae

MDPI

Article

Multiplication, Phenological Period and Growth Vigor of Thirty-One Grapevine Rootstocks and the Role of Parentage in Vigor Heredity

Yonggang Yin *, Bin Han, Minmin Li, Nan Jia, Changjiang Liu, Yan Sun, Yingjie Wang, Qian Gao and Zijuan Guo *

Changli Institute of Fruit Research, Hebei Academy of Agriculture and Forestry Sciences, Qinhuangdao 066600, China
* Correspondence: yinyonggang00@hotmail.com (Y.Y.); grpaesworks@outlook.com (Z.G.)

Abstract: Knowledge about the growth vigor of grapevine rootstocks is required for scion-based rootstock selection and rootstock breeding. We performed this trial aiming to evaluate the multiplication and growth vigor of several rootstocks. Thirty-one rootstock genotypes were compared on their multiplication characteristics, phenological periods, and growth indicators across three consecutive seasons. The results suggested that the cuttings of most rootstocks had callus-forming indices (CFIs) over 0.5 except for '188-08' (0.28). The rooting rate of '420A' was 5%, while that of the rest of the rootstocks was greater than 48%. The internode lengths of the one-year-old vines were positively correlated with those (as well as cane lengths and pruning weights) of the adult vines. These rootstocks were grouped into three clusters based on the growth measurements across three seasons. Eight combinations of genetic backgrounds showed various effects on the growth indicators. The high-vigor cluster includes '1103P', '5BB', '225Ru', etc.; the medium-vigor cluster includes 'Dogridge', '101-14M', 'Fercal', etc.; and the low-vigor cluster includes 'Gloire', '3309C', 'Ganzin1', etc. The *Vitis berlandieri* parentage showed a higher vigor heredity, while the *V. riparia* showed a lower vigor heredity. These findings would contribute to rootstock nursery construction and provide references for vigor-based rootstock selection for grafts and parent selection for rootstock breeding.

Keywords: *Vitis*; callus; internode; pruning weight; phenological periods; grapevine; genetic background; meteorology

Citation: Yin, Y.; Han, B.; Li, M.; Jia, N.; Liu, C.; Sun, Y.; Wang, Y.; Gao, Q.; Guo, Z. Multiplication, Phenological Period and Growth Vigor of Thirty-One Grapevine Rootstocks and the Role of Parentage in Vigor Heredity. *Horticulturae* **2023**, *9*, 241. https://doi.org/10.3390/horticulturae9020241

Academic Editors: Ana Jiménez-Cantizano and Antonio Amores-Arrocha

Received: 13 January 2023
Revised: 6 February 2023
Accepted: 7 February 2023
Published: 10 February 2023

1. Introduction

Rootstocks have acted a vital role in grapevine (*Vitis vinifera* L.) cultivation, ever since using them to protect against phylloxera infections in the European vineyards in the nineteenth century [1]. After that, many grapevine rootstock cultivars have been bred and the advantages of rootstocks such as nematode resistance, drought tolerance, salinity tolerance, and lime tolerance have been recognized [2–5]. Furthermore, rootstocks can alter the yield and berry composition of grapes [6,7] by affecting bud fertility, fruit set, berry weight, and nutrition uptake [8–10].

Rootstocks were not taken seriously due to the absence of phylloxera attacks in the viticultural history of China [11], until scientists realized their value in resisting environmental stresses and improving fruit quality. Furthermore, numerous grapevine rootstocks have been successively introduced [12]. In the meantime, several rootstocks have been bred locally since 1984 by using native wild species like *V. amurensis* [11]. Grapevine grafting has become a popular topic in both academic and commercial areas in recent years. Grafted vine materials are also well-sold for their improved fruit quality or yield. And the grafted plants are easy to survive and tend to grow vigorously after grafting tender shoots on a vigorous rootstock. Some sayings in recent years, like the positive effects of a few rootstocks on the popular fresh grape cultivar, 'Shine Muscat' (*V. labruscana* × *V. vinifera*), have driven the price to surge for cultivars '3309C', 'Kober 5BB', and the unverified '3309M'. This should be attributed to the limited planting area of these rootstocks before their value was noted.

The nursery industry in China mainly consists of individual households, cooperatives, or small-scale companies, and very few of them have rootstock nurseries. The promising market for grafts encourages them to construct their own rootstock nurseries, either to meet their own rootstock demand, screen potential rootstocks, or even breed new rootstocks. However, most rootstocks are just conserved as germplasm resources by research institutes after being introduced, especially the lesser-known cultivars and the locally bred ones [13]. To better utilize these rootstocks, their features need to be investigated.

Terroir is critical to vine performance [14]; thus, regional rootstock screening is necessary. Several rootstocks have been tested for nutrition uptake, cold resistance, virus detection, and impact on scion yield and fruit composition in regions such as the Mediterranean, Eastern Canada, and Algeria [15–18]. Recently in China, the primary core collection of grape genetic resources was constructed based on plant phenotypic traits by the national grapevine repository [13]. Responses to stresses like copper and sodium chloride have been tested on many rootstock cultivars [19,20]. The tolerance to cold or waterlogging has also been tested on the major cultivars [21–23]. However, essential knowledge of the multiplication, phenology, and growth vigor of these rootstocks is rarely reported.

The native species *V. amurensis* has been well utilized in northern China for its good cold-hardiness and perfect flowers, but *V. amurensis* is weak and hard to root [11]. Breeders made intraspecific crosses and interspecific hybridizations with species like *V. riparia* to overcome the weaknesses. The improved cultivars include 'Shanhe' lines ('Shanhe 1', 'Shanhe 2', 'Shanhe 3', and 'Shanhe 4') [24], but their planting characteristics are still required to be investigated. Those lesser-known rootstock cultivars can be good candidates for specific viticultural purposes or be used as promising parents just as *V. amurensis* is used to breed new cultivars. Thus, much detail on their growth and multiplication is required as well. Some cultivars such as '5BB', 'SO4' and '1103P' are recognized as easy to grow, even in cold regions. However, their multiplication information, which is vital for a rootstock nursery, has rarely been assessed publicly.

Urged by these questions, we multiplied 31 grapevine rootstocks and then investigated their phenological periods and growth performances for three consecutive seasons. The main objective was to evaluate the multiplication and growth characteristics of these rootstocks. We also aimed to understand the role of genetic background in transmitting vigor. We conducted this trial in Changli, northern China, a major region that produces fresh grapes, wine, and planting materials. This is a first for most of the tested rootstocks. The results would deepen our understanding of these rootstocks. Furthermore, the results could help nursery growers in selecting rootstock cultivars.

2. Materials and Methods

2.1. Multiplication Evaluation of the Plant Materials

Healthy dormant canes of 31 grapevine rootstocks (Table 1) were acquired in the winter of 2010. Canes in bundles were sprayed with lime sulfur solution (2%) and then stored under one meter of moist soil (30–40% moisture content) until the spring, when they were ready to propagate. Canes were cut into cuttings with two fully developed buds (the schematic diagram of the test is shown in Figure 1). The desirable cuttings were approximately 15 cm long and 0.7 cm thick. The basal end of the section was cut at a sharp angle of about 30°, and the upper end was cut flat, and each end was 2 cm below or above the closest bud (Figure 1). Every ten cuttings were tied up to form one bundle. Four bundles of cuttings with tags were soaked in distilled water (20 ± 2 °C) for 5 h, and then their basal ends were dipped into a rooting solution containing 20% naphthylacetic acid (NAA) and 30% heteroauxin (IAA) (Aibidi Biotechnology, Beijing, China) for 3 min right before placing them on the nursery bed. The nursery bed was made of 10 cm of sand with a controllable electric heating system at the bottom. Gaps between cuttings were filled with sand, and the upper buds were exposed to the air. The bottom temperature was set at

26 ± 2 °C, and the air temperature was 15 ± 2 °C. The moisture content of the sand was kept at 50–60% by spraying water on the bed surface two times a day.

Table 1. Thirty-one grapevine rootstocks used in the trial and their abbreviations and parentage information.

Rootstock	Abbreviation	Parentage
'Millardet et de Grasset 101-14'	'101-14M'	*V. riparia* × *V. rupestris*
'Paulsen 1103'	'1103P'	*V. berlandieri* × *V. rupestris*
'Richter 110'	'110R'	*V. berlandieri* × *V. rupestris*
'Couderc 1202'	'1202C'	*V. vinifera* × *V. rupestris*
'Ruggeri 140'	'140Ru'	*V. berlandieri* × *V. rupestris*
'Couderc 1613'	'1613C'	(*V. riparia* × *V. longii*) × 'Othello'
'Castel 188-08'	'188-08'	*V. monticola* × *V. riparia*
'Ruggeri 225'	'225Ru'	*V. berlandieri* × *V. riparia*
'Couderc 3309'	'3309C'	*V. riparia* × *V. rupestris*
'Millardet et de Grasse 420A'	'420A'	*V. berlandieri* × *V. riparia*
'Téléki 5 A'	'5A'	*V. berlandieri* × *V. riparia*
'Kober–Téléki 5BB'	'5BB'	*V. berlandieri* × *V. riparia*
'Téléki 5C'	'5C'	*V. berlandieri* × *V. riparia*
'Téléki 8B'	'8B'	*V. berlandieri* × *V. riparia*
'Berlandieri Rességuier No. 2'	'BR2'	*V. berlandieri*
'Riparia Barrett 50'	'Barrett50'	*V. riparia*
'Riparia Beaumont'	'Beaumont'	*V. riparia*
'Beta'	'Beta'	*V. riparia* × [(*V. labrusca* × *V. vinifera*) × *V. labrusca*]
'Dog ridge'	'Dogridge'	*V.* × *champinii* (*V. rupestris* × *V. candicans*)
'Fercal INRA Bordeaux'	'Fercal'	(*V. berlandieri* × *V. vinifera*) × (*V. berlandieri* × *V. longii*)
'Ganzin 1' ('AxR 1')	'Ganzin1'	*V. vinifera* × *V. rupestris*
'Riparia Gloire de Montpellier'	'Gloire'	*V. riparia*
'Rupestris du Lot'	'du Lot'	*V. rupestris*
'Saltcreek' ('Ramsey')	'Saltcreek'	*V.* × *champinii* (*V. rupestris* × *V. candicans*)
'Rupestris Scheele'	'Shadi'	*V. rupestris*
'Shanhe 1'	'Shanhe1'	*V. amurensis* × *V. riparia*
'Shanhe 2'	'Shanhe2'	*V. amurensis* × *V. riparia*
'Shanhe 3'	'Shanhe3'	*V. amurensis* × *V. riparia*
'Shanhe 4'	'Shanhe4'	*V. amurensis* × *V. riparia*
'Téléki-Fuhr Selektion Oppenheim No.4'	'SO4'	*V. berlandieri* × *V. riparia*
'Wumao'	'Wumao'	*V. berlandieri* × *V. riparia*

After being nursed for 30 days, a total of four bundles (ten cuttings for each bundle as one replicate) for each cultivar were randomly selected for multiplication evaluation. The cutting with visible green tissue or shoot was defined as "budbreak", and the bud length was measured using a vernier caliper. The cutting with visible root(s) was defined as "rooted", and the root(s) number was recorded. We defined the callus-forming grade (i) as the callus covering percentage on the section surface: 0, 0%; 1, 1–20%; 2, 21–40%; 3, 41–60%; 4, 61–80%; 5, 81–100%. One well-trained assistant rated the cuttings visually. For each replicate, the budbreak rate, rooting rate, and callus-forming index [CFI, modified from the salt damage index [19,25]] were calculated. CFI = $(0 \times N_0 + 1 \times N_1 + 2 \times N_2 + 3 \times N_3 + 4 \times N_4 + 5 \times N_5)/(5 \times 10)$, where N_i represents the number of cuttings with the corresponding callus-forming grade (i, i = 0–5), and one bundle of ten cuttings is one replicate.

Figure 1. Schematic diagram of the present trial. The multiplication of grapevine rootstocks was evaluated after the cuttings were nursed for 30 days in 2011 and then the cuttings were transplanted into the field at Kongzhuang Station. As the "young vines" grew in the field for 90 days, both above-ground and underground growth were determined. Meanwhile, the nursery plants were harvested. In late March 2012, the plants were transplanted to the germplasm repository at Shigezhuang Station. During the years 2016–2018, the phenological periods including bud-break, initial flowering and the end of flowering of the grapevine rootstocks ("adult vines") were recorded; growth indicators including trunk diameter, cane diameter, cane length, pruning weight, etc., were determined after leaves fell.

2.2. Evaluation of Cutting Development

Well-developed cuttings were root-dipped in a mixture of water and bioagent (*Agrobacterium vitis* E26, 1/2 in volume) to prevent crown gall infections before being transplanted in the field on April 17, 2011 (Figure 1). The plants grew at the Kongzhuang experimental station (39°42′29″ N, 119°05′41″ E; altitude 14 m), Changli Institute of Fruit Research. All the plants were managed uniformly, and only one stronger shoot was kept for each plant. The pest and disease control followed the local standards. Vine materials were harvested, and growth parameters were measured after leaves fell on November 30. Ten plants for each cultivar were carefully dug out, and the soil was carefully washed off the roots. Shoot basal diameter and root diameter were measured using a digital vernier caliper. Roots with a diameter of over 2 mm were defined as thick roots and were counted. The thick root proportion was calculated as the ratio of the number of thick roots to the number of total roots in percentage. The lengths of the shoot and root were determined using a tape measure. The lignified proportion of the shoot was defined as cane length/shoot length in percentage. Total root length and average root length are the sum and mean of the lengths, respectively, of the individual roots for each vine. The soil at the site is sandy loam.

2.3. Growth Measurements for the Vines

The vine materials were harvested with two buds retained (Figure 1). The plant materials were preserved in a ditch, which was then covered with moist soil in the shade. In the spring of 2012, they were planted in the grapevine repository at Shigezhuang experimental station (39°45′01″ N, 119°12′44″ E; altitude 20 m), Changli Institute of Fruit Research. Vines were spaced at 0.7 × 4 m and trained on a pergola trellis at a height of 1.8 m. In the planting year, only one vigorous shoot was retained for trunk establishment for each plant. In the first winter, the single cane was pruned at the lower cordon, about one meter high. In the second growing season, the top 5–6 shoots on each plant were allowed to develop until the winter, when each newly formed cane was pruned to two buds. Similarly in the next season, only one shoot was retained on each spur, and around 5–6 new shoots in total were kept. All the vines were generally unearthed several days before the traditional Qingming Festival (4 or 5 April), specifically, 30 March 2016, 1 April 2017 and 1 April 2018. These management changes were repeated in the following years. Investigation and determination were conducted from the fifth to seventh growing seasons (2016–2018). After leaves fell, for each plant, the trunk basal diameter (30 cm above the ground), cane length, cane basal diameter, and shoot length were measured, and the internodes were counted before canes were pruned and weighed. Internode length is defined as cane length/internode number, and lignified proportion is defined as cane length/shoot length. A total of ten plants for each rootstock cultivar were randomly selected and determined. All the vines were managed uniformly and were buried with soil to overcome the cold winter.

2.4. Phenological Periods Investigation for the Vines

During each growing season, the phenological periods for each rootstock cultivar were investigated based on the modified E-L system for grapevine growth stages [26]. The budbreak period is defined as when around 5% of the total buds show green tips (stage E-L 4). Initial flowering is defined as when around 5% of the flower clusters show caps-off (E-L 19). The flowering period ends when the caps fall off (E-L 26). Phenological period recording started on the first day of the year (DOY).

Meteorological data were obtained from the local weather bureau and averaged monthly from April to October (Figure S1 and Table S1). The soil is sandy loam, and its composition was determined (Table S2).

2.5. Statistical Analysis

Data analyses were performed using SPSS 20 (IBM Corp., Armonk, NY, USA). Normality and homogeneity of variance were tested before analysis. A one-way or two-way ANOVA followed by Tukey's post hoc test was employed to compare the means at $p < 0.05$. The nonparametric median test followed by multiple pairwise comparisons was adopted to compare the medians at an adjusted $p < 0.05$ by Bonferroni correction. Principle component analysis (PCA), correlation analysis, hierarchical cluster analysis (HCA), and figure construction were conducted in OriginPro 2018 (OriginLab Corp., Northampton, MA, USA).

3. Results

3.1. Multiplication Characteristics of the Cuttings

The status of cutting development varied largely among these rootstocks after being nursed for 40 days (Table 2). The callus-forming indices (CFIs) ranged from 0.28 ('188-08') to 0.97 (5C). Thirteen rootstocks, or approximately 42% of the total, achieved high CFIs of over 0.9. '188-08' was the only one that attained a CFI less than 0.5, that is, other rootstocks could form calluses in more than half the area of the cutting plane. The average rooting rate of '420A' was 5%, markedly lower than those of other rootstocks, which were greater than 48%. Especially with 'Fercal' and '5A', whose rooting rates reached 100%. The average root number per cutting was found to be positively related to the rooting rate. It ranged from 0.2 to 10.2, with a median of 2.9. Correspondingly, '5A' produced the most roots, while '420A' and 'BR2' produced less than one root on average.

The budbreak rate and bud length lie in ranges of 37.5–100% and 0.42–3.58 cm, respectively. The budbreak rates of two-thirds of the rootstocks reached 80%, especially for 'Beta' which got a rate of 100% and the largest bud length of 3.58 cm. Furthermore, buds of '225Ru' and 'Wumao' germinated less than 50% and were the shortest in length.

3.2. Growth of the One-Year-Old Vine in the Field

Rootstocks performed diversely in both overground and underground growth-related traits after growing in the field for 90 days (Tables 3 and 4). The shoot diameter of '225Ru', which reached 10.1 mm, was significantly larger than that of other rootstocks, which ranged from 5.6 ('Saltcreek') to 8.1 ('Shanhe4') mm. The shoot lengths ranged from 72.7 to 150.3 cm, and 80% of those were longer than one meter (Table 3). Cultivars with longer shoots generally had longer lignified parts, showing a positive correlation between the variables

(Figure S2). Averagely, the lignified shoot length and total shoot length of '110R' were the largest, more than twice those of 'Gloire', '1613C', or 'Shadi'. The variation in lignification rates was relatively small, with a range of 73–100%. Shoots of 'Gloire', 'Beaumont', and '3309C' were less lignified, lower than 75%. The internode lengths ranged from 3.2 to 7.8 cm, being higher on varieties with longer shoots and lower on those with shorter shoots.

Table 2. Multiplication traits of the grapevine rootstock cuttings after being nursed for 40 days.

Rootstock	CFI	Rooting Rate (%)	Root Number	Budbreak Rate (%)	Bud Length (cm)
'101-14M'	0.94 ± 0.06 a–d	62.5 ± 15.0 d–g	1.3 ± 0.1 no	85.0 ± 10.0 b–f	1.05 ± 0.07 k–o
'1103P'	0.54 ± 0.01 m	83.0 ± 17.0 abc	4.2 ± 0.3 de	92.8 ± 4.9 a–d	1.17 ± 0.08 jkl
'110R'	0.72 ± 0.07 i–l	87.5 ± 12.6 ab	3.7 ± 0.4 ef	82.5 ± 9.6 c–g	1.08 ± 0.05 klm
'1202C'	0.92 ± 0.04 a–d	90.0 ± 8.2 ab	5.8 ± 0.4 b	72.5 ± 9.6 ghi	0.95 ± 0.03 mno
'140Ru'	0.87 ± 0.05 b–g	87.5 ± 5.0 ab	3.5 ± 0.3 fg	92.5 ± 9.6 a–d	1.35 ± 0.05 j
'1613C'	0.76 ± 0.04 h–k	83.3 ± 6.1 abc	2.6 ± 0.1 h–k	71.8 ± 5.6 ghi	1.86 ± 0.16 d–g
'188-08'	0.28 ± 0.07 n	61.3 ± 13.7 d–g	1.6 ± 0.2 l–o	97.3 ± 5.5 ab	1.84 ± 0.15 e–h
'225Ru'	0.88 ± 0.05 a–e	90.0 ± 8.2 ab	4.5 ± 0.4 d	37.5 ± 9.6 l	0.45 ± 0.03 r
'3309C'	0.94 ± 0.03 a–d	62.5 ± 9.6 d–g	2.3 ± 0.2 i–l	92.5 ± 5.0 a–d	1.68 ± 0.09 ghi
'420A'	0.78 ± 0.08 f–j	5.0 ± 5.8 h	0.2 ± 0.3 q	67.5 ± 5.0 ij	0.74 ± 0.06 pq
'5A'	0.70 ± 0.04 jkl	100 ± 0 a	10.2 ± 0.7 a	95.0 ± 5.8 abc	1.94 ± 0.14 def
'5BB'	0.92 ± 0.04 a–d	67.5 ± 5.6 c–f	2.9 ± 0.2 ghi	85.3 ± 5.0 b–f	1.24 ± 0.12 jk
'5C'	0.97 ± 0.03 a	67.5 ± 5.0 c–f	2.2 ± 0.1 i–l	90.0 ± 8.2 a–e	1.22 ± 0.11 jk
'8B'	0.93 ± 0.02 a–d	80.0 ± 14.1 bcd	5.7 ± 1.9 b	60.0 ± 0 jk	0.63 ± 0.03 q
'BR2'	0.77 ± 0.11 g–j	48.0 ± 22.2 g	0.5 ± 0.2 pq	85.5 ± 17.1 b–f	0.87 ± 0.09 nop
'Barrett50'	0.86 ± 0.06 c–g	91.8 ± 8.4 ab	5.3 ± 0.7 bc	92.8 ± 5.3 a–d	2.05 ± 0.25 cd
'Beaumont'	0.85 ± 0.05 d–h	55.0 ± 10.0 fg	2.6 ± 0.2 h–k	85.0 ± 5.8 b–f	1.05 ± 0.09 k–o
'Beta'	0.68 ± 0.06 kl	77.5 ± 9.6 bcd	3.7 ± 0.5 ef	100 ± 0 a	3.58 ± 0.26 a
'Dogridge'	0.76 ± 0.02 h–k	75.0 ± 12.9 b–e	1.9 ± 0.1 k–n	90.0 ± 8.2 a–e	1.10 ± 0.03 klm
'Fercal'	0.66 ± 0.03 l	100 ± 0 a	5.9 ± 0.3 b	75.0 ± 5.8 f–i	2.28 ± 0.29 b
'Ganzin1'	0.80 ± 0.08 e–i	77.5 ± 12.6 bcd	3.3 ± 0.4 fgh	95.0 ± 5.8 abc	1.64 ± 0.12 hi
'Gloire'	0.96 ± 0.03 ab	80.0 ± 8.2 bcd	3.7 ± 0.3 ef	92.5 ± 5.0 a–d	1.79 ± 0.09 f–i
'du Lot'	0.87 ± 0.09 a–f	85.0 ± 12.9 abc	4.9 ± 0.4 cd	70.0 ± 8.2 hij	0.87 ± 0.11 nop
'Saltcreek'	0.94 ± 0.05 a–d	85.0 ± 5.8 abc	4.9 ± 0.5 cd	52.5 ± 5.0 k	0.85 ± 0.11 op
'Shadi'	0.93 ± 0.09 a–d	57.5 ± 15.0 efg	1.4 ± 0.1 mno	80.0 ± 14.1 d–h	1.07 ± 0.10 k–n
'Shanhe1'	0.91 ± 0.06 a–d	77.5 ± 9.6 bcd	2.6 ± 0.3 h–k	87.5 ± 5.0 a–e	1.63 ± 0.11 i
'Shanhe2'	0.85 ± 0.07 d–h	75.0 ± 10.0 b–e	1.9 ± 0.2 k–n	95.0 ± 5.8 abc	1.90 ± 0.17 def
'Shanhe3'	0.78 ± 0.08 f–j	90.0 ± 0 ab	2.8 ± 0.3 g–j	92.5 ± 5.0 a–d	2.02 ± 0.18 cde
'Shanhe4'	0.95 ± 0.05 abc	55.0 ± 12.9 fg	1.0 ± 0.1 op	90.0 ± 11.5 a–e	2.16 ± 0.18 bc
'SO4'	0.94 ± 0.05 a–d	62.5 ± 22.2 d–g	2.1 ± 0.1 j–m	77.5 ± 5.0 e–i	0.98 ± 0.08 l–o
'Wumao'	0.90 ± 0.05 a–e	72.5 ± 15.4 b–f	4.9 ± 0.4 cd	40.0 ± 3.3 l	0.42 ± 0.03 r

Note: Data shown are means ± standard error, n = 4. Different lowercase letters within each column represent significant differences at $p < 0.05$ by Tukey's test. CFI, callus-forming index.

Table 3. Growth indicators of the shoot 90 days after planting the grapevine rootstock cuttings in the field (the young vine).

Rootstock	Shoot Basal Diameter (mm)	Cane Length (cm)	Internode Length (cm)	Shoot Length (cm)	Lignified Proportion (%)
'101-14M'	6.9 ± 0.9 b–f	125.8 ± 25.8 abc	6.5 ± 1.4 bcd	126.2 ± 25.5 a–f	99.6 ± 1.1 a
'1103P'	6.4 ± 0.8 c–f	112.6 ± 29.9 b–g	6.6 ± 1.5 bc	112.6 ± 29.9 c–g	100 ± 0 a
'110R'	7.1 ± 1.4 b–e	148.6 ± 18.7 a	7.8 ± 1.0 a	150.3 ± 18.2 a	98.9 ± 2.6 a
'1202C'	7.4 ± 1.3 bcd	124.8 ± 22.8 a–d	5.2 ± 0.7 efg	143.0 ± 27.2 ab	89.0 ± 16.5 abc
'140Ru'	6.5 ± 0.8 c–f	120.3 ± 31.7 a–e	5.4 ± 0.8 efg	126.1 ± 29.0 a–f	95.3 ± 10.0 ab
'1613C'	5.9 ± 1.2 ef	72.7 ± 25.4 jkl	3.7 ± 0.9 ij	72.7 ± 25.4 i	100 ± 0 a
'188-08'	6.7 ± 0.6 c–f	91.8 ± 24.4 e–k	4.8 ± 0.6 fgh	100.2 ± 25.0 fgh	91.6 ± 13.1 abc
'225Ru'	10.1 ± 2.6 a	96.0 ± 23.9 d–j	6.1 ± 1.3 b–e	111.0 ± 31.1 c–g	87.9 ± 11.8 abc
'3309C'	6.9 ± 1.7 b–f	85.1 ± 26.7 g–l	3.8 ± 0.9 ij	113.2 ± 24.3 c–g	74.7 ± 12.8 d
'420A'	6.7 ± 1.2 c–f	135.8 ± 30.6 ab	5.2 ± 0.7 efg	137.3 ± 31.5 abc	99.1 ± 3.0 a
'5A'	7.4 ± 1.2 bcd	119.8 ± 24.3 a–e	6.0 ± 0.8 b–e	125.8 ± 24.7 a–f	95.5 ± 6.7 ab
'5BB'	7.3 ± 1.3 bcd	101.9 ± 30.4 c–i	6.3 ± 1.0 b–e	106.8 ± 33.2 d–g	96.4 ± 8.4 ab
'5C'	6.4 ± 1.1 c–f	80.7 ± 23.2 h–l	4.7 ± 1.2 f–i	100.5 ± 24.6 fgh	81.9 ± 21.4 cd

Table 3. *Cont.*

Rootstock	Shoot Basal Diameter (mm)	Cane Length (cm)	Internode Length (cm)	Shoot Length (cm)	Lignified Proportion (%)
'8B'	6.9 ± 1.1 b–f	97.9 ± 16.8 c–j	5.4 ± 0.8 efg	121.0 ± 15.5 b–f	82.1 ± 17.0 cd
'BR2'	6.5 ± 1.7 c–f	110.3 ± 42.6 b–g	6.0 ± 1.3 b–e	115.5 ± 32.7 b–g	92.5 ± 15.0 abc
'Barrett50'	6.8 ± 1.1 b–f	127.4 ± 21.9 abc	5.4 ± 1.1 d–g	135.5 ± 14.2 a–d	93.6 ± 8.7 abc
'Beaumont'	6.2 ± 1.1 def	78.7 ± 32.4 i–l	3.8 ± 0.7 ij	105.5 ± 28.7 e–h	73.5 ± 15.7 d
'Beta'	7.0 ± 0.6 b–e	134.4 ± 14.9 ab	6.1 ± 0.4 b–e	136.3 ± 15.8 abc	98.7 ± 2.7 a
'Dogridge'	7.8 ± 1.5 bc	116.3 ± 28.4 b–f	4.6 ± 0.8 ghi	116.3 ± 28.4 b–g	100 ± 0 a
'Fercal'	6.1 ± 1.6 def	87.2 ± 30.2 g–k	4.1 ± 1.2 hij	98.1 ± 32.1 f–i	90.7 ± 16.9 abc
'Ganzin1'	6.5 ± 1.1 c–f	107.8 ± 37.5 b–h	5.5 ± 1.5 d–g	116.9 ± 28.7 b–g	89.7 ± 15.8 abc
'Gloire'	7.2 ± 2.0 b–e	57.8 ± 27.0 l	3.7 ± 1.2 ij	78.6 ± 29.2 hi	73.0 ± 16.0 d
'du Lot'	6.3 ± 0.5 def	86.7 ± 21.0 g–k	6.1 ± 1.6 b–e	88.4 ± 22.5 ghi	98.5 ± 4.1 a
'Saltcreek'	5.6 ± 0.5 f	88.8 ± 31.5 f–k	6.2 ± 0.7 b–e	88.8 ± 31.5 ghi	100 ± 0 a
'Shadi'	6.8 ± 0.9 b–f	63.7 ± 31.9 kl	3.2 ± 0.7 j	78.7 ± 28.4 hi	83.9 ± 26.6 bcd
'Shanhe1'	7.0 ± 1.2 b–e	116.2 ± 19.6 b–f	7.0 ± 0.9 ab	122.2 ± 17.4 a–f	95.4 ± 10.3 ab
'Shanhe2'	6.1 ± 1.1 def	132.6 ± 18.1 ab	6.9 ± 0.7 ab	132.6 ± 18.1 a–e	100 ± 0 a
'Shanhe3'	7.2 ± 0.6 b–e	116.7 ± 18.0 b–f	5.7 ± 0.8 c–f	118.1 ± 18.1 b–f	98.9 ± 2.7 a
'Shanhe4'	8.1 ± 1.0 b	121.3 ± 40.0 a–e	5.2 ± 1.1 efg	125.9 ± 39.0 a–f	95.8 ± 4.9 ab
'SO4'	6.5 ± 0.6 c–f	118.3 ± 11.4 b–e	5.5 ± 0.6 d–g	124.6 ± 11.9 a–f	95.2 ± 6.8 ab
'Wumao'	6.2 ± 1.0 def	137.2 ± 13.7 ab	6.9 ± 0.9 ab	142.7 ± 12.6 ab	96.2 ± 5.4 ab

Note: Data shown are means ± standard error, n = 10. Different lowercase letters within each column represent significant differences at $p < 0.05$ by Tukey's test.

Table 4. Growth indicators of the roots 90 days after planting the grapevine rootstock cuttings in the field (the young vine).

Rootstock	Number of Roots per Cutting	Thick Root Number	Thick Roots Proportion (%)	Average Root Length (cm)	Total Root Length (cm)
'101-14M'	8.2 ± 2.66 k	3.4 ± 1.8 c–g	41.9 ± 16.3 bc	12.6 ± 3.8 m	99.3 ± 34.1 k
'1103P'	13.3 ± 3.84 e–j	3.0 ± 1.8 e–h	24.9 ± 19.1 d–j	28.1 ± 6.8 b–h	381.4 ± 156.1 d–h
'110R'	13.9 ± 4.38 d–j	5.3 ± 1.7 ab	39.9 ± 12.5 bcd	12.8 ± 2.1 m	178.5 ± 65.1 jk
'1202C'	10.0 ± 2.49 ijk	3.3 ± 1.6 d–g	33.6 ± 13.3 b–g	23.9 ± 4.5 f–j	241.8 ± 84.6 hij
'140Ru'	14.7 ± 5.38 c–i	3.3 ± 0.8 d–g	26.5 ± 13.2 d–j	35.1 ± 7.3 ab	492.6 ± 142.4 a–d
'1613C'	15.6 ± 7.37 a–h	1.6 ± 1.3 gh	15.6 ± 17.0 ijk	21.8 ± 9.8 h–l	297.3 ± 160.1 f–j
'188-08'	20.3 ± 4.52 ab	2.8 ± 1.4 e–h	14.6 ± 8.9 jk	19.0 ± 2.8 j–m	381.7 ± 87.8 d–h
'225Ru'	19.6 ± 5.13 abc	3.7 ± 1.4 b–f	21.1 ± 11.3 g–k	30.8 ± 7.0 a–f	585.3 ± 148.6 a
'3309C'	17.6 ± 3.44 a–e	1.4 ± 1.3 h	8.2 ± 6.8 k	31.8 ± 7.4 a–e	558.9 ± 181.5 abc
'420A'	13.3 ± 3.13 e–j	4.1 ± 1.4 b–f	30.5 ± 5.6 b–i	24.4 ± 10.1 f–j	334.1 ± 208.4 e–i
'5A'	16.3 ± 4.27 a–g	5.2 ± 1.5 a–d	34.2 ± 15.0 b–g	34.9 ± 10.2 ab	547.8 ± 125.4 abc
'5BB'	20.7 ± 7.92 a	4.0 ± 2.8 b–f	20.8 ± 12.5 g–k	24.8 ± 4.8 e–j	509.6 ± 206.4 a–d
'5C'	18.7 ± 1.63 a–d	4.2 ± 1.2 b–f	22.7 ± 7.9 f–k	26.9 ± 4.2 c–i	500.8 ± 76.7 a–d
'8B'	16.4 ± 3.95 a–f	4.0 ± 1.8 b–f	24.7 ± 12.5 e–j	12.2 ± 2.2 m	198.8 ± 58.0 ijk
'BR2'	8.3 ± 2.63 k	5.3 ± 2.6 abc	61.3 ± 10.6 a	31.6 ± 16.5 a–e	271.8 ± 173.4 g–j
'Barrett50'	17.8 ± 4.49 a–e	4.3 ± 1.6 b–f	25.8 ± 11.8 d–j	25.6 ± 4.2 e–j	444.3 ± 87.6 a–f
'Beaumont'	16.1 ± 4.26 a–g	2.7 ± 1.0 e–h	18.0 ± 7.8 h–k	16.1 ± 3.1 lm	258.7 ± 74.6 g–j
'Beta'	11.5 ± 3.34 f–k	3.6 ± 1.3 b–f	32.3 ± 11.0 b–h	33.8 ± 6.8 abc	386.9 ± 128.8 d–h
'Dogridge'	14.4 ± 4.38 d–j	4.0 ± 1.9 b–f	28.6 ± 13.5 b–j	17.1 ± 4.6 klm	257.9 ± 136.0 g–j
'Fercal'	11.9 ± 6.17 f–k	4.3 ± 2.4 b–f	39.6 ± 23.4 b–e	28.6 ± 4.4 b–h	338.3 ± 168.0 e–i
'Ganzin1'	11.2 ± 5.51 g–k	2.9 ± 1.9 e–h	30.4 ± 17.1 b–i	29.9 ± 10.0 b–g	331.7 ± 185.2 e–j
'Gloire'	15.2 ± 2.39 b–h	3.3 ± 1.7 d–g	22.0 ± 11.3 g–k	20.3 ± 3.0 i–l	306.9 ± 59.0 f–j
'du Lot'	17.9 ± 5.14 a–e	3.6 ± 2.1 b–f	22.0 ± 16.7 g–k	22.6 ± 6.3 h–l	424.3 ± 215.0 b–f
'Saltcreek'	9.6 ± 2.72 jk	2.4 ± 1.0 fgh	27.9 ± 14.8 c–j	28.2 ± 7.0 b–h	262.8 ± 74.1 g–j
'Shadi'	15.1 ± 3.81 c–h	3.7 ± 1.4 b–f	26.1 ± 11.4 d–j	15.8 ± 2.0 lm	239.4 ± 67.5 hij
'Shanhe1'	17.3 ± 4.24 a–e	4.4 ± 2.1 b–e	26.3 ± 11.4 d–j	25.0 ± 6.7 e–j	431.9 ± 153.3 b–f
'Shanhe2'	15.3 ± 4.64 b–h	6.2 ± 2.3 a	43.0 ± 14.5 b	37.6 ± 6.0 a	570.0 ± 192.4 ab
'Shanhe3'	15.7 ± 3.43 a–g	3.6 ± 0.7 b–f	23.6 ± 4.9 f–j	25.9 ± 6.3 d–j	408.7 ± 136.0 c–g
'Shanhe4'	18.4 ± 5.99 a–e	3.3 ± 1.7 d–g	19.7 ± 13.5 g–k	26.8 ± 4.3 d–i	479.9 ± 133.0 a–e
'SO4'	16.1 ± 4.72 a–g	3.9 ± 1.2 b–f	26.1 ± 10.1 d–j	32.7 ± 6.6 a–d	525.6 ± 200.4 a–d
'Wumao'	10.6 ± 3.41 h–k	3.7 ± 1.2 b–f	37.8 ± 16.3 b–f	23.5 ± 4.8 g–k	248.2 ± 85.7 hij

Note: Data shown are means ± standard error, n = 10. Different lowercase letters within each column represent significant differences at $p < 0.05$ by Tukey's test.

The total number of roots per plant averaged 14.9 and varied largely among rootstocks (Table 4). '5BB', '188-08', and '225Ru' grew sufficient roots for approximately 20 per plant, while '101-14M' and 'BR2' generated about eight roots for each plant. Root-rich cultivars tended to have fewer thick roots than root-poor varieties, which resulted in a negative correlation between total root number and thick root proportion (Figure S2). The total root length ranged from 99.3 to 585.3 cm, to which the root number and/or average root length contributed (Figure S2). The total root length of '101-14M' (99.3 cm) was far below the average length of 370.8 cm, while those of '225Ru', 'Shanhe2', '3309C', '5A', etc., were far over 500 cm.

3.3. Phenological Periods of the Rootstocks

Phenological periods are essential for grapevine breeding, and they varied among 31 rootstock genotypes (Figure 2). Budbreak occurred mostly between 98–110 days of the year (DOY). Budbreak of the Shanhe series (Shanhe1–4) occurred the earliest, 11 days before that of '1613C'. The range of the budbreak period was smaller than that of the florescence period which started at 127–145 DOY, indicating a maximum gap of three weeks between the earliest varieties (Shanhe series, 'BR2', and '3309C') and the latest ones ('Saltcreek' and '1613C'). Flowering lasted mostly for 7–8 days, but it lasted for only 4 days on '420A', 'Shadi', and 'Beaumont', and over 10 days on '1202C', 'Shanhe4', '110R', etc. Furthermore, the advanced flowering period in 2017 is evident in Figure 2.

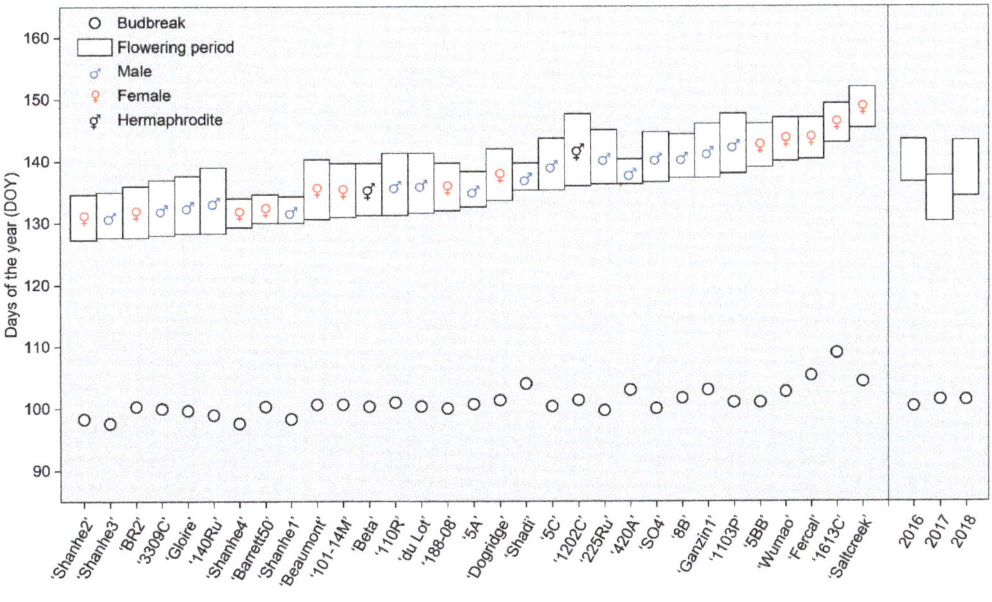

Figure 2. Phenological periods of 31 grapevine rootstocks in North China in 2016–2018. DOY, day of the year. Male, female and hermafrotide indicate the sex of the flower of the rootstocks.

3.4. Growth of the Adult Vine

Growth-related indices varied significantly among rootstocks and seasons (Table 5). Rootstock exerted a larger effect on lignified proportion and internode length, while the season effect was larger on other traits. The interaction effect of rootstock by season was also significant (Table S3).

Table 5. Comparison of growth indicators among 31 grapevine rootstocks in three growing seasons (2016–2018) (the adult vine).

Season	Rootstock	Trunk Diameter (mm)	Cane Diameter (mm)	Total Cane Length (cm)	Total Shoot Length (cm)	Lignified Proportion (%)	Internode Length (cm)	Pruning Weight (g)
2016	'101-14M'	24.98 ± 4.55 a–d	11.42 ± 1.72 ab	1901.8 ± 863.0 a–e	2169.2 ± 940.1 abc	86.3 ± 6.6 b–e	8.62 ± 0.85 g–l	1051.0 ± 427.5 a–f
	'1103P'	21.76 ± 2.17 bcd	8.79 ± 1.05 e–i	2768.3 ± 696.0 a	2836.0 ± 761.7 a	98.4 ± 2.1 a	10.84 ± 0.90 b–f	1033.1 ± 203.8 a–f
	'110R'	21.76 ± 4.50 bcd	9.87 ± 1.20 b–i	1881.1 ± 784.2 a–e	2087.6 ± 861.8 abc	89.3 ± 4.1 a–e	10.18 ± 1.10 b–h	1189.0 ± 471.6 a–d
	'1202C'	21.11 ± 2.98 bcd	9.54 ± 0.72 b–i	1202.0 ± 327.0 b–g	1416.4 ± 332.3 bcd	84.0 ± 8.1 cde	5.49 ± 0.87 m	711.0 ± 172.5 b–i
	'140Ru'	25.69 ± 2.10 abc	10.89 ± 0.63 a–f	2102.3 ± 534.3 abc	2262.9 ± 563.7 ab	92.4 ± 3.4 a–d	10.32 ± 0.47 b–h	1116.0 ± 291.0 a–e
	'1613C'	19.76 ± 0.98 cd	10.29 ± 1.00 a–i	496.0 ± 41.6 g	953.0 ± 75.6 d	51.3 ± 1.5 g	7.57 ± 0.11 j–m	433.3 ± 20.2 ghi
	'188-08'	19.18 ± 2.06 cd	9.09 ± 0.90 c–i	1416.1 ± 361.2 b–g	1511.7 ± 400.7 bcd	93.9 ± 4.1 abc	8.80 ± 0.77 f–l	634.4 ± 170.9 d–i
	'225Ru'	23.62 ± 2.16 bcd	10.41 ± 0.93 a–i	1584.2 ± 397.2 b–f	1828.9 ± 342.5 a–d	86.6 ± 7.3 b–e	10.16 ± 0.96 b–h	1228.0 ± 320.9 abc
	'3309C'	20.86 ± 4.02 bcd	8.53 ± 0.80 ghi	1694.9 ± 369.6 b–f	1953.1 ± 370.9 a–d	86.1 ± 8.4 b–e	7.02 ± 0.69 lm	670.0 ± 126.0 c–i
	'420A'	23.96 ± 6.35 bcd	10.53 ± 0.71 a–g	1659.2 ± 694.8 b–f	1859.4 ± 711.6 a–d	88.9 ± 9.4 a–e	9.81 ± 1.22 c–i	977.0 ± 403.0 a–h
	'5A'	22.14 ± 3.09 bcd	10.74 ± 1.82 a–g	1533.3 ± 730.9 b–f	1644.7 ± 736.2 bcd	90.5 ± 8.3 a–d	10.76 ± 1.48 b–g	1000.0 ± 461.4 a–g
	'5BB'	25.74 ± 4.56 abc	11.25 ± 1.32 abc	1502.6 ± 477.8 b–f	1734.7 ± 511.9 bcd	86.5 ± 4.3 b–e	12.17 ± 1.07 ab	1027.0 ± 303.8 a–f
	'5C'	21.27 ± 3.38 bcd	9.68 ± 0.70 b–i	1724.8 ± 185.3 b–f	1798.2 ± 214.4 a–d	95.8 ± 2.9 ab	10.28 ± 0.81 b–h	961.0 ± 261.5 a–h
	'8B'	21.44 ± 3.72 bcd	9.33 ± 1.01 b–i	1567.7 ± 540.5 b–f	1698.9 ± 553.2 bcd	92.0 ± 4.6 a–d	11.09 ± 0.80 b–e	859.0 ± 280.2 a–i
	'BR2'	21.97 ± 2.03 bcd	9.52 ± 0.61 b–i	2184.5 ± 497.4 ab	2265.2 ± 534.0 ab	96.6 ± 2.0 ab	11.79 ± 0.66 abc	1364.0 ± 279.1 a
	'Barrett50'	21.53 ± 0.59 bcd	10.93 ± 0.75 a–e	1331.4 ± 527.4 b–g	1559.4 ± 681.6 bcd	87.2 ± 4.0 a–e	11.28 ± 0.71 a–e	906.0 ± 398.4 a–i
	'Beaumont'	25.00 ± 5.02 a–d	10.75 ± 2.07 a–g	1774.3 ± 728.0 a–f	2154.2 ± 738.0 abc	78.0 ± 15.2 ef	9.29 ± 1.82 e–k	1238.0 ± 482.0 abc
	'Beta'	25.75 ± 2.62 abc	11.16 ± 1.84 a–d	1552.7 ± 533.3 b–f	1695.2 ± 554.3 bcd	91.5 ± 5.6 a–d	9.71 ± 2.61 c–j	1018.0 ± 424.7 a–f
	'Dogridge'	27.32 ± 2.67 ab	10.39 ± 1.44 a–i	1419.2 ± 284.1 b–g	1715.8 ± 362.1 bcd	83.2 ± 3.6 cde	7.12 ± 0.52 klm	1028.0 ± 186.0 a–f
	'Fercal'	23.31 ± 5.36 bcd	10.53 ± 2.92 a–g	1183.1 ± 709.9 c–g	1359.0 ± 806.6 bcd	88.3 ± 6.3 a–e	9.48 ± 1.63 d–j	514.0 ± 381.1 f–i
	'Ganzin1'	19.20 ± 4.74 cd	9.01 ± 1.22 c–i	1293.7 ± 159.9 b–g	1635.9 ± 192.8 bcd	78.3 ± 3.0 ef	6.85 ± 0.53 lm	553.0 ± 69.8 e–i
	'Gloire'	19.43 ± 4.06 cd	8.21 ± 0.82 hi	982.4 ± 436.2 d–g	1183.6 ± 493.2 cd	81.8 ± 5.8 def	7.84 ± 0.71 i–l	368.0 ± 153.3 i
	'du Lot'	18.23 ± 2.96 d	8.94 ± 1.13 d–i	1570.3 ± 659.8 b–f	1697.1 ± 709.7 bcd	92.3 ± 4.4 a–d	11.03 ± 1.66 b–e	814.0 ± 363.4 a–i
	'Saltcreek'	18.05 ± 3.23 d	10.49 ± 0.73 a–h	811.2 ± 606.0 fg	910.6 ± 686.2 d	89.4 ± 4.8 a–e	8.41 ± 0.87 h–l	404.0 ± 342.5 hi
	'Shadi'	31.46 ± 5.90 a	12.47 ± 1.25 a	1172.0 ± 339.7 c–g	1561.4 ± 354.4 bcd	72.8 ± 8.4 f	7.09 ± 0.74 lm	920.0 ± 202.5 a–i
	'Shanhe1'	22.50 ± 2.67 bcd	8.16 ± 0.46 i	1977.2 ± 703.2 a–d	2099.3 ± 774.5 abc	93.8 ± 4.3 abc	11.61 ± 0.37 a–d	867.0 ± 322.8 a–i
	'Shanhe2'	24.42 ± 4.18 a–d	10.26 ± 0.59 a–i	1620.8 ± 285.6 b–f	1715.4 ± 305.7 bcd	94.6 ± 3.1 abc	9.96 ± 0.96 c–i	779.0 ± 133.0 b–i
	'Shanhe3'	23.90 ± 2.69 bcd	8.62 ± 0.86 f–i	1762.5 ± 376.1 b–f	1881.4 ± 383.2 a–d	92.7 ± 4.8 a–d	11.15 ± 1.37 b–e	779.0 ± 172.3 b–i
	'Shanhe4'	24.47 ± 7.31 a–d	9.88 ± 0.69 b–i	956.9 ± 353.5 efg	1137.1 ± 365.6 cd	83.2 ± 6.5 cde	8.73 ± 2.14 f–l	433.3 ± 155.3 ghi
	'SO4'	22.27 ± 2.90 bcd	10.31 ± 1.04 a–i	1903.7 ± 500.7 a–e	2151.4 ± 567.7 abc	87.9 ± 3.8 a–e	9.97 ± 0.91 b–i	1267.0 ± 319.1 ab
	'Wumao'	23.65 ± 2.48 bcd	11.52 ± 1.15 ab	1722.0 ± 470.3 b–f	1874.4 ± 563.3 a–d	93.0 ± 6.9 a–d	13.39 ± 1.35 a	1091.0 ± 363.4 a–e
2017	'101-14M'	23.00 ± 5.29 ab	10.10 ± 1.91 a–g	610.0 ± 417.5 cd	799.0 ± 493.3 cd	72.2 ± 14.2 e	7.78 ± 1.38 f–k	374.5 ± 274.1 cd
	'1103P'	21.75 ± 3.22 ab	7.81 ± 1.72 ghi	1523.8 ± 730.3 ab	1673.8 ± 785.3 abc	91.9 ± 4.9 abc	10.21 ± 1.39 a–f	775.6 ± 480.4 a–d
	'110R'	24.94 ± 4.48 ab	10.73 ± 1.34 a–f	1079.7 ± 340.7 a–d	1373.9 ± 483.9 a–d	81.5 ± 8.8 b–e	9.88 ± 1.03 a–g	683.0 ± 244.7 a–d
	'1202C'	25.51 ± 3.21 ab	10.85 ± 1.08 a–e	1077.0 ± 312.1 a–d	1283.0 ± 399.6 a–d	85.0 ± 5.5 b–e	7.03 ± 0.34 h–k	655.0 ± 187.0 a–d
	'140Ru'	26.94 ± 6.26 ab	9.02 ± 0.80 c–i	1254.0 ± 397.8 a–d	1517.0 ± 474.7 a–d	79.9 ± 9.0 cde	8.89 ± 1.08 a–j	595.0 ± 183.3 a–d
	'1613C'	25.23 ± 4.50 ab	10.54 ± 0.06 a–f	1115.0 ± 233.3 a–d	1275.0 ± 176.8 a–d	90.5 ± 10.6 abc	7.29 ± 0.74 g–k	620.0 ± 141.4 a–d
	'188-08'	29.18 ± 5.99 a	10.83 ± 1.17 a–e	1338.2 ± 511.5 abc	1486.0 ± 575.0 a–d	91.4 ± 3.2 abc	8.47 ± 0.56 b–k	612.8 ± 244.6 a–d
	'225Ru'	26.11 ± 3.80 ab	9.34 ± 0.90 a–h	1568.0 ± 524.3 ab	1862.2 ± 593.7 ab	85.1 ± 3.7 b–e	10.77 ± 1.11 a–d	992.5 ± 384.4 ab
	'3309C'	20.48 ± 2.80 ab	8.65 ± 0.60 e–h	461.5 ± 203.2 d	616.5 ± 266.1 d	75.7 ± 8.6 de	6.85 ± 1.22 ijk	215.0 ± 92.7 d
	'420A'	27.07 ± 8.83 ab	11.58 ± 1.64 ab	942.1 ± 423.0 a–d	1037.8 ± 463.3 bcd	91.4 ± 5.9 abc	8.63 ± 0.99 b–j	746.5 ± 399.0 a–d
	'5A'	23.43 ± 4.60 ab	10.90 ± 1.65 a–e	943.3 ± 519.1 a–d	1013.3 ± 570.6 bcd	95.0 ± 5.3 ab	10.49 ± 0.59 a–e	590.8 ± 275.8 a–d
	'5BB'	27.12 ± 3.48 ab	11.32 ± 1.05 a–d	1634.2 ± 476.5 a	2020.9 ± 498.2 a	81.3 ± 6.0 b–e	11.52 ± 1.29 a	1096.1 ± 473.2 a
	'5C'	23.48 ± 2.94 ab	10.76 ± 1.24 a–e	1380.0 ± 330.3 abc	1512.0 ± 413.1 a–d	92.6 ± 5.3 abc	10.78 ± 1.33 a–d	802.5 ± 215.7 abc
	'8B'	24.24 ± 4.35 ab	8.75 ± 1.50 e–i	1275.7 ± 516.3 abc	1400.0 ± 558.9 a–d	91.7 ± 4.7 abc	11.15 ± 2.60 ab	685.0 ± 357.6 a–d
	'BR2'	24.23 ± 4.79 ab	9.13 ± 1.45 b–i	837.7 ± 353.1 bcd	983.8 ± 428.3 bcd	83.3 ± 7.7 b–e	9.50 ± 2.05 a–i	485.7 ± 221.3 bdc
	'Barrett50'	-	-	-	-	-	-	-
	'Beaumont'	28.94 ± 6.12 a	10.61 ± 1.50 a–f	1112.4 ± 375.6 a–d	1264.0 ± 390.9 a–d	87.5 ± 6.2 a–d	7.95 ± 0.67 e–k	638.5 ± 336.2 a–d
	'Beta'	30.04 ± 4.37 a	11.74 ± 1.28 a	895.7 ± 595.9 a–d	941.4 ± 620.3 cd	95.1 ± 5.3 ab	9.14 ± 1.03 a–i	750.0 ± 476.6 a–d
	'Dogridge'	25.97 ± 2.84 ab	11.08 ± 1.38 a–e	1029.4 ± 373.6 a–d	1095.4 ± 414.0 bcd	94.5 ± 5.1 ab	8.13 ± 1.48 d–k	833.0 ± 421.3 abc
	'Fercal'	29.33 ± 7.31 a	11.66 ± 1.92 a	912.4 ± 375.1 a–d	1113.7 ± 415.1 bcd	79.3 ± 6.4 cde	7.77 ± 1.52 f–k	735.5 ± 476.3 a–d
	'Ganzin1'	21.10 ± 8.74 ab	8.91 ± 1.26 d–i	943.9 ± 372.4 a–d	1252.9 ± 435.0 a–d	72.0 ± 9.1 e	5.95 ± 0.66 k	450.0 ± 232.8 bdc
	'Gloire'	22.86 ± 6.08 ab	9.69 ± 1.17 a–h	927.2 ± 574.1 a–d	1120.2 ± 666.8 a–d	82.1 ± 6.3 b–e	6.81 ± 1.16 ijk	387.5 ± 236.9 cd
	'du Lot'	21.60 ± 3.68 ab	9.17 ± 0.55 b–i	1516.6 ± 495.5 ab	1648.4 ± 506.7 abc	91.4 ± 5.7 abc	9.12 ± 1.41 a–i	724.6 ± 299.7 a–d
	'Saltcreek'	17.65 ± 3.37 b	11.48 ± 2.08 abc	816.7 ± 279.3 bcd	816.7 ± 279.3 cd	100.0 ± 0.0 a	8.29 ± 0.84 c–k	790.0 ± 236.4 a–d
	'Shadi'	26.19 ± 8.32 ab	10.70 ± 1.28 a–f	1143.0 ± 694.4 a–d	1452.5 ± 741.3 a–d	75.5 ± 6.5 de	7.21 ± 1.28 g–k	862.5 ± 411.2 abc
	'Shanhe1'	24.70 ± 4.98 ab	6.73 ± 0.44 i	832.7 ± 363.2 bcd	1022.1 ± 495.4 bcd	83.0 ± 9.7 b–e	8.87 ± 1.41 a–j	370.6 ± 158.0 cd
	'Shanhe2'	25.23 ± 4.94 ab	10.46 ± 1.21 a–f	703.7 ± 230.1 cd	860.2 ± 323.3 cd	84.7 ± 5.9 b–e	9.54 ± 0.86 a–h	432.0 ± 170.9 bdc
	'Shanhe3'	26.78 ± 4.04 ab	7.55 ± 1.06 hi	1059.0 ± 301.6 a–d	1222.6 ± 339.8 a–d	87.5 ± 11.0 a–d	10.94 ± 3.56 abc	474.5 ± 162.6 bdc
	'Shanhe4'	27.92 ± 8.98 a	8.26 ± 0.73 f–i	783.8 ± 320.2 bcd	944.5 ± 356.3 cd	84.0 ± 7.6 b–e	6.39 ± 1.39 jk	400.0 ± 203.6 cd
	'SO4'	23.37 ± 6.71 ab	9.40 ± 1.28 a–h	959.0 ± 372.7 a–d	1175.0 ± 512.1 a–d	82.4 ± 8.6 b–e	9.89 ± 1.28 a–g	615.5 ± 196.7 a–d
	'Wumao'	24.02 ± 3.49 ab	11.10 ± 1.44 a–e	998.2 ± 184.8 a–d	1086.4 ± 208.9 bcd	91.6 ± 7.7 abc	10.32 ± 1.41 a–f	749.5 ± 181.3 a–d
2018	'101-14M'	28.40 ± 4.90 a–d	10.54 ± 1.16 ab	1035.7 ± 558.2 abc	1258.0 ± 564.9 ab	78.2 ± 16.5 c–i	8.28 ± 1.15 e–h	642.0 ± 413.0 abc
	'1103P'	27.45 ± 6.41 a–d	8.72 ± 1.91 b–h	1101.5 ± 699.2 abc	1237.5 ± 751.3 ab	64.0 ± 5.9 ijk	12.12 ± 2.01 ab	712.5 ± 410.4 abc
	'110R'	24.80 ± 5.13 a–d	9.18 ± 1.30 a–h	1282.6 ± 585.4 ab	1444.2 ± 637.6 a	89.1 ± 5.9 a–g	10.80 ± 0.95 a–e	864.0 ± 394.4 ab
	'1202C'	19.31 ± 3.54 cd	7.38 ± 0.54 h	833.7 ± 288.7 abc	1205.5 ± 403.6 ab	68.9 ± 6.5 hij	8.53 ± 1.48 d–h	490.0 ± 174.6 abc
	'140Ru'	29.33 ± 2.68 abc	9.31 ± 1.06 a–h	1216.5 ± 468.1 abc	1351.5 ± 486.3 ab	89.6 ± 5.2 a–f	10.23 ± 2.21 a–f	580.5 ± 254.2 abc
	'1613C'	29.25 ± 5.11 abc	9.29 ± 0.71 a-h	615.0 ± 49.5 bc	1115.0 ± 162.6 ab	56.5 ± 10.6 jk	6.77 ± 1.59 h	405.0 ± 106.1 bc
	'188-08'	27.66 ± 4.45 a–d	8.34 ± 1.30 c–h	1140.3 ± 441.3 abc	1273.0 ± 518.6 ab	90.1 ± 6.8 a–e	8.56 ± 1.09 d–h	484.0 ± 283.6 abc
	'225Ru'	31.43 ± 2.88 a	9.82 ± 1.17 a–g	1382.0 ± 338.7 ab	1593.0 ± 427.8 a	87.7 ± 4.4 a–g	12.64 ± 0.93 a	979.5 ± 326.2 a
	'3309C'	20.54 ± 3.30 bcd	8.82 ± 0.91 a–h	821.5 ± 284.9 abc	906.0 ± 296.2 b	89.6 ± 5.9 a–f	8.01 ± 1.10 fgh	375.5 ± 129.6 bc
	'420A'	29.80 ± 7.75 ab	8.97 ± 0.98 a–h	1063.0 ± 507.0 abc	1106.0 ± 509.7 ab	95.5 ± 5.7 a	9.25 ± 1.05 c–h	658.5 ± 248.3 abc
	'5A'	26.74 ± 3.74 a–d	9.83 ± 1.89 a–f	1004.0 ± 544.2 abc	1246.8 ± 567.8 ab	76.6 ± 22.0 e–i	10.16 ± 1.92 a–f	750.0 ± 416.1 abc
	'5BB'	27.39 ± 6.47 a–d	10.24 ± 1.16 a–d	1317.0 ± 431.1 ab	1408.0 ± 457.2 ab	93.5 ± 2.3 ab	12.29 ± 0.88 ab	883.0 ± 335.9 ab
	'5C'	27.61 ± 4.86 a–d	8.39 ± 0.93 b–h	1199.2 ± 524.7 abc	1457.3 ± 603.0 a	80.9 ± 6.5 b–h	11.20 ± 3.10 a–d	441.0 ± 318.2 abc
	'8B'	23.68 ± 2.88 a–d	9.12 ± 1.08 a–h	1449.7 ± 467.3 a	1608.8 ± 521.4 a	91.1 ± 4.4 a–d	11.69 ± 0.99 abc	781.7 ± 285.3 abc
	'BR2'	27.85 ± 4.72 a–d	9.21 ± 0.93 a–h	1286.3 ± 492.3 ab	1322.3 ± 503.3 ab	97.6 ± 2.4 a	11.21 ± 1.38 a–d	858.5 ± 345.4 ab
	'Barrett50'	24.27 ± 3.88 a–d	9.36 ± 1.38 a–h	1119.5 ± 514.2 abc	1150.5 ± 533.1 ab	96.9 ± 4.0 a	10.63 ± 1.64 a–f	574.5 ± 323.1 abc
	'Beaumont'	29.43 ± 7.68 abc	9.41 ± 1.42 a–h	974.0 ± 494.4 abc	1141.0 ± 491.0 ab	84.4 ± 11.3 a–g	8.36 ± 0.99 e–h	499.5 ± 318.0 abc
	'Beta'	26.92 ± 3.41 a–d	10.92 ± 1.07 a	1234.1 ± 355.2 abc	1275.1 ± 380.7 ab	97.5 ± 1.8 a	9.30 ± 1.23 c–h	758.5 ± 216.5 abc
	'Dogridge'	26.62 ± 5.68 a–d	9.49 ± 0.92 a–h	1068.5 ± 328.1 abc	1143.5 ± 398.2 ab	96.3 ± 5.1 a	7.34 ± 0.52 gh	716.0 ± 300.5 abc
	'Fercal'	33.69 ± 7.54 a	10.36 ± 1.41 abc	1154.5 ± 558.6 abc	1272.0 ± 689.8 ab	92.2 ± 5.8 abc	9.07 ± 1.29 c–h	668.0 ± 526.4 abc
	'Ganzin1'	19.48 ± 3.47 cd	7.96 ± 0.84 e–h	747.9 ± 237.7 abc	968.9 ± 303.7 ab	75.7 ± 4.5 f–i	7.95 ± 1.64 fgh	282.8 ± 209.0 c
	'Gloire'	27.95 ± 2.04 a–d	8.07 ± 0.89 d–h	721.2 ± 177.8 abc	1112.5 ± 243.2 ab	88.0 ± 4.1 a–g	9.53 ± 1.66 b–h	249.0 ± 85.8 c
	'du Lot'	-	-	-	-	-	-	-
	'Saltcreek'	18.42 ± 4.33 d	9.63 ± 1.16 a–g	810.0 ± 213.4 abc	822.0 ± 224.3 b	98.2 ± 2.7 a	9.79 ± 1.09 b–g	624.0 ± 424.0 abc
	'Shadi'	31.96 ± 13.92 a	8.33 ± 1.31 c–h	459.0 ± 243.0 c	852.9 ± 398.2 b	52.6 ± 15.9 k	8.00 ± 3.13 fgh	240.7 ± 150.2 c
	'Shanhe1'	29.90 ± 7.31 ab	7.82 ± 1.05 fgh	1242.8 ± 486.2 abc	1277.8 ± 493.1 ab	96.7 ± 4.3 a	11.01 ± 0.83 a–e	598.3 ± 275.0 abc
	'Shanhe2'	28.98 ± 4.77 abc	9.35 ± 1.47 a–h	724.5 ± 280.0 abc	927.1 ± 346.6 ab	77.9 ± 8.6 d–i	6.92 ± 1.08 h	509.8 ± 225.1 abc
	'Shanhe3'	29.52 ± 4.19 abc	7.84 ± 1.13 fgh	873.0 ± 402.5 abc	885.1 ± 405.9 b	98.0 ± 3.3 a	10.10 ± 0.91 a–g	447.5 ± 243.8 abc
	'Shanhe4'	28.07 ± 8.25 a–d	8.07 ± 0.76 d–h	747.9 ± 210.7 abc	998.3 ± 289.2 ab	75.0 ± 7.0 ghi	6.94 ± 1.01 h	355.6 ± 109.8 bc
	'SO4'	27.74 ± 2.38 a–d	10.02 ± 1.39 a–e	1361.0 ± 457.0 ab	1454.0 ± 440.1 a	93.0 ± 6.8 ab	10.09 ± 0.83 a–g	901.0 ± 340.7 ab
	'Wumao'	25.20 ± 3.09 a–d	7.65 ± 0.49 gh	1219.1 ± 334.3 abc	1383.5 ± 360.7 ab	86.3 ± 8.3 a–g	11.46 ± 1.62 abc	645.0 ± 213.7 abc

Note: Different lowercase letters within each column for each season represent significant differences at $p < 0.05$ by Tukey's test.

The trunk diameter increased by year, with the three-season average of these rootstocks ranging from 18 to 30.4 mm. The trunk diameter remained the smallest on 'Saltcreek', followed by 'Ganzin1' and 'du Lot', and the largest on 'Shadi'. Half of the rootstock varieties produced canes with a diameter of over 10 mm, while 'Shanhe1' and 'Shanhe3' developed canes with a diameter of less than 8 mm. Total cane length, i.e., the lignified shoot length per tree, ranged from 791.2 to 1861.1 cm with a mean of 1227.5 cm. Outstandingly, the total cane length for each vine of '1103P' was much longer than that of the other varieties, especially those of '1613C', 'Saltcreek', 'Shanhe4', 'Shadi', and 'Gloire'—less than half of it. Furthermore, the canes produced in 2016 were much longer (Table S3). The total shoot length shared a similar variation as the total cane length among rootstocks, of which the coefficient of variations (CVs) were 17% and 20.7%, respectively. The lignified proportion of these rootstocks was located at 64.6–96.1%, with a smaller variation than those of the above-mentioned length attributes. Its variation among seasons was much smaller. The highest lignified proportion of 96.1% was from 'Saltcreek', the shortest shoot producer; the lowest was from 'Shadi', followed by '1613C', another two of the varieties with the shortest canes. The internode length averaged 9.4 cm, ranging from 6.9 cm for 'Ganzin1' to 12 cm for '5BB', and it was smaller in 2017 than in the other two seasons. The pruning weight, as a final growth indicator, revealed the largest variance (CV = 25.3%) among tested rootstocks, with a range of 334.8–1066.7 g. Each plant produced an average of 706.1 g of canes, while 'Gloire' produced less than half of that. Annual pruning weight per plant for half of the varieties was over 735 g, among which '225Ru' produced the largest, followed by '5BB', 'SO4', etc. In addition, a much higher pruning weight was obtained in 2016—nearly 900 g per plant.

3.5. Correlations between Meteorological Data and Vine Growth

Considering the significant effect of the year on vine growth indicators, we summarized the meteorological data and analyzed their correlation with the growth indicators. The daily average temperature, as well as the relative humidity, changed similarly in the three years (Figure S1 and Table S1). The local annual precipitation in 2016, 831.9 mm, was 45% and 82% heavier than those in 2017 and 2018, respectively. Around 80% of the precipitation occurred from June to August. In total, sunshine duration reached 2006.5 h for the whole growing season in 2018, close to that in 2016, and nearly 180 h more than that in 2017.

The correlation between meteorological data and growth data was not significant (Figure 3). Despite that, the meteorological effects can be recognized on the figure. Rainfall contributes largely to nearly all growth indicators. The temperature positively correlates with cane diameter but negatively with other parameters, while the opposite is true for humidity and sunshine duration.

3.6. Principle Component Analysis (PCA) and Hierarchical Cluster Analysis (HCA) for the Rootstocks Based on the Growth Indicators

The PCA for growth parameters (trunk diameter, cane diameter, cane length, total shoot length, lignified proportion, internode length, and pruning weight) yielded seven principal components (PC) to explain 100% of the variance (Table S4). The first two PCs explained 73.6% of the total variance (Figure 4A). PC1 is highly influenced by longitudinal growth parameters (cane length, total shoot length, and internode length), pruning weight, and lignified proportion, while PC2 is highly influenced by lateral growth parameters, including cane diameter and trunk diameter. Variables within longitudinal or lateral parameters positively correlate with each other, and correlations between longitudinal variables are significant. Pruning weight also donates a smaller loading to PC2 and is highly correlated with cane diameter and longitudinal variables. No correlation exists between longitudinal and lateral parameters.

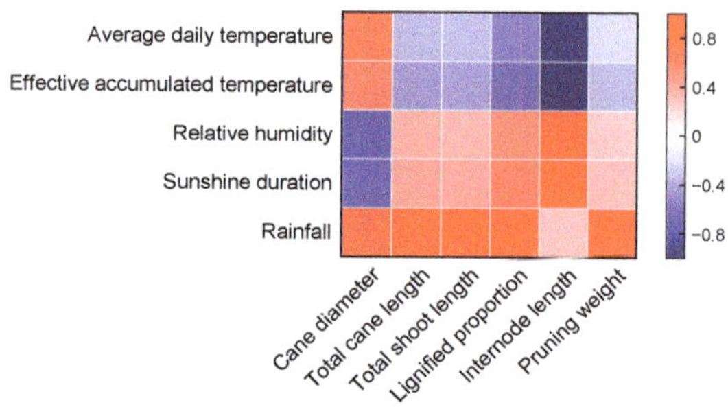

Figure 3. Heat maps of the correlations between growth-related traits of the adult vines and the meteorological factors.

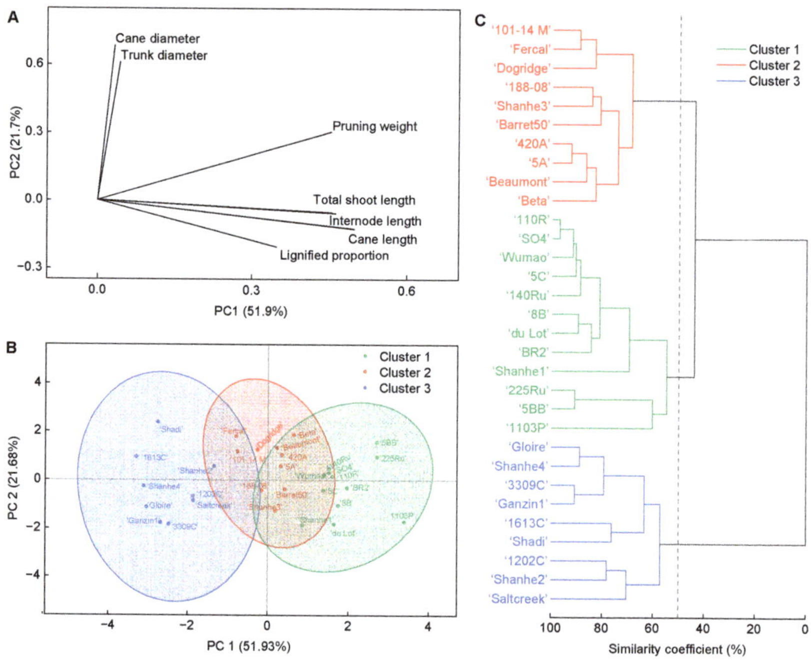

Figure 4. Principal component (PC) analysis of seven vigor-related variables for 31 grapevine rootstocks and hierarchical clustering (HC): (**A**) PC loading plot; (**B**) score plot; (**C**) dendrogram based on seven PCs. A confidence ellipse (95%) is drawn on each cluster on the score plot in a corresponding color.

Calculated scores obtained by multiplying factor scores by eigenvalues were subjected to hierarchical cluster analysis (HCA). The average linkage method and Euclidian distance were used to separate the rootstocks. The dendrogram was separated into three clusters at 50% similarity, and Clusters 1, 2, and 3 consisted of 12, 10, and 9 varieties, respectively (Figure 4C). The classification was then used to group the varieties on the score plot, which in turn revealed the relationship between these rootstocks (Figure 4B). Cluster 1, including '5BB', '225Ru', '1103P', etc., lies in the positive direction of PC1, the largest component donated by length- or weight-related parameters. Hence, varieties in cluster 1 are characterized by high vigor. Those rootstocks composing Cluster 3, such as 'Ganzin1', '3309C', 'Gloire', etc., distributed in the negative direction of PC1, are characterized by low vigor. Cluster 2 lies between Clusters 1 and 3 and should be classified as the medium-vigorous cluster, in which six rootstocks are scattered on the positive side of PC2, indicating their relatively high vigor in lateral growth. The confidence ellipse of Cluster 2 overlaps with that of Cluster 1 or Cluster 3, while barely overlapping between Clusters 1 and 3. Regardless, according to the figure, varieties in Cluster 1 are more vigorous than those in Cluster 3 (Figure 4B).

3.7. Genetic Background Effects on Tree Growth

Cultivars with the same parentage were grouped, and the groups containing one genotype were excluded. A total of eight groups were then compared on growth parameters (Figure 5). Raw data from each plant were used for the analysis. What stands out in the figure is the *V. berlandieri* × *V. riparia* group (in red), in which all the growth-related parameters reach the highest level. *V. berlandieri* × *V. rupestris* (in blue) is comparable to *V. berlandieri* × *V. riparia* on these traits. *V.* × *champinii* produced the largest cane diameter and lignified proportion and yielded a pruning weight as large as *V. berlandieri* × *V. riparia.* On the other hand, *V. rupestris* yielded the lowest pruning weight and length-related indicators. *V. vinifera* × *V. rupestris* had a smaller trunk and cane diameter. *V. amurensis* × *V. riparia* had the thickest trunk, but the thinnest cane.

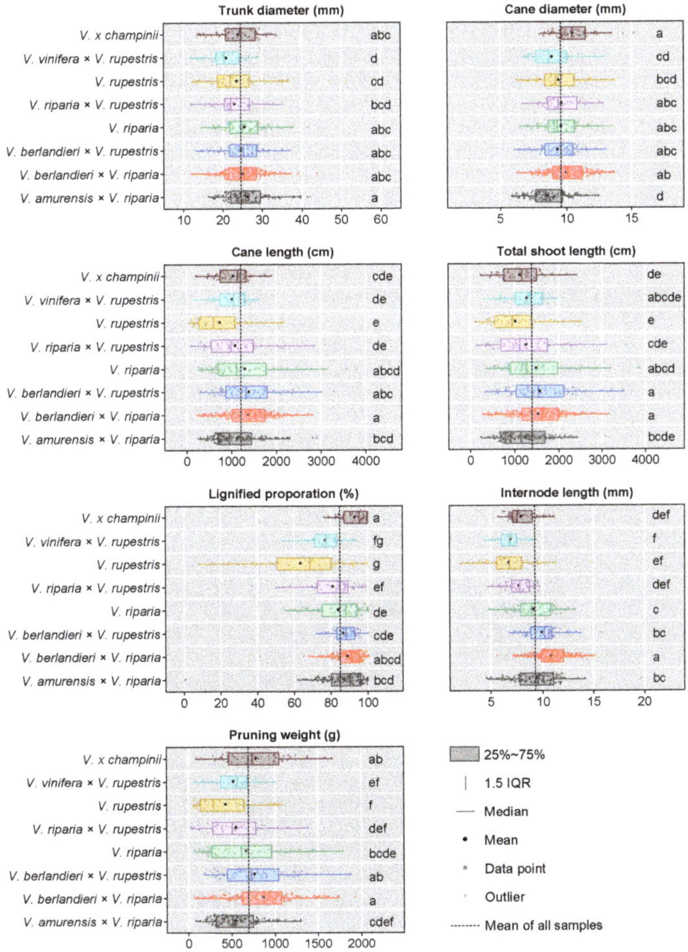

Figure 5. A comparison of the effects of eight parentages on the growth indicators of adult vines. Different lowercase letters on each figure indicate significant differences at $p < 0.05$ (adjusted by Bonferroni correction) by the nonparametric median test. Each data point corresponds to a single plant in one of the seasons. IQR, interquartile range.

4. Discussion

Well-formed callus tissue from the cambium shows as a ring on the bevel section of the cutting base where rooting can then occur, which ensures the nutrient uptake and vegetative growth of the cutting once it is planted in the field. Thus, cuttings with less callus may not generate adequate roots. However, the least-CFI gainer, '188-08', had more roots (1.6 roots per cutting) than some high-CFI genotypes such as '101-14M', '420A', or 'BR2' (Table 1). The irrelevance between CFI and rooting performance (Figure S2) might be attributed to the varying time requirements for callusing or rooting, i.e., some cultivars can form callus quickly or easily but may need more time for rooting than others, and vice versa. Waite et al. [27] reported excessive callus tissue impedes the xylem and phloem from forming across the graft union. The same goes for the perspective from Hartmann et al. [28] that callus tissue should not accumulate over 2–3 mm out from the graft union, and this might occur on those cuttings with a well-formed callus but sparse roots. On the other hand, cuttings rooted early tend to lose their roots when transplanted into the field, and re-rooting will cost the plant's stored reserves. In this case, more calluses but fewer roots may be the desired status for cuttings to grow in the field. 'Beta', '188-08', '1103P', and Shanhe series gained larger bud lengths and budbreak rates but smaller CFIs or root numbers, while the opposite occurred for rootstocks like '225Ru', 'Wumao', 'Saltcreek', or '8B' (Table 1), indicating a likely vegetal balance between the upper and lower parts of the cutting. It might be the reason for the slight negative correlation between CFI/root number and bud-related traits, despite the correlation being unnoticeable (Figure S2).

The nature of multiplication seems independent of growth because it did not promote vine growth in the field either in the current season or in adulthood (Figure S2). Even so, no clear relevance was observed between the growth indices of young and adult vines. The internode length could be a relatively stable trait because it showed a positive correlation between young vines and adult vines. Interestingly, the internode length for the current season is also correlated with the longitudinal growth and pruning weight of the adult vines (Figure S2), so it could be used as one of the potential indicators of growth vigor. A previous study by Köse et al. [29] on the current seasonal growth of grafted 'Merzifon Karasi' grapevines on nine rootstocks showed that the root numbers of '5BB', '140Ru', and 'SO4' were 19.5, 15.5, and 16.1, respectively, which was surprisingly close to the related results in the present study. Despite that, the roots of vines grafted on '8B' or '110R' are longer than those of the corresponding rootstocks we investigated. Apart from the soil type and regional climate, the effect of the scion on the rootstock could be one factor that caused the difference.

Our growth-based vigor evaluation for three consecutive seasons confirmed that '1103P' is with higher vigor, while 'Gloire' is with lower vigor—which has been acknowledged by many researchers [10,30,31]. This also enhanced the credibility of the vigor estimation and classification for the tested grape rootstocks. Despite that, our vigor definition for some rootstocks differed from the vigor rating summarized by Zhang et al. [9]. Therein, '3309C' was rated as medium-vigorous, and 'Dogridge' and 'Saltcreek' ('Ramsey') were rated as highly vigorous. These rootstocks were identified as less vigorous, even weak, in our observations. It seems that some rootstocks are more susceptible to growth conditions than others. Soil texture could affect vine vigor indirectly through cation exchange capacity, water holding capacity, root penetration, etc., and vine vigor is considered higher on gravelly soils than on silty loam soils [14,32]. The climatic condition is another major factor affecting shoot growth [33]. As shown in a multi-year and multi-site study by Dodson Peterson et al. [34], the conferred vigor (pruning weight) to Cabernet Sauvignon by '420A' in Mendocino La Ribera and Napa Rutherford was lower than that in the Sacramento Delta. The study also suggested that in most trial sites, '1103P' conferred high vigor while '3309C' conferred low vigor to the scion cultivars, which is consistent with our findings.

Based on the three seasonal growth performances, 31 rootstocks were finally classified into three groups (Figure 4B,C). Most rootstocks in Cluster 2 were also included in the confidence ellipse of Cluster 1 or Cluster 3, except for 'Dogridge'. Therefore, those rootstocks from Cluster 1 also covered by Cluster 2, for example, '140Ru', 'SO4', 'Wumao', '110R', '5C', and 'Shanhe1', could be less vigorous than those uniquely from Cluster 1, including '5BB', '225Ru', '1103P', 'BR2', '8B', and 'du Lot'. Similarly, 'Shanhe2' might be more vigorous than the others in Cluster 3. Interestingly, the Shanhe series, hybrids from *V. amurensis* and *V. riparia*, are distributed separately into three clusters. It might be a result of progeny segregation on the quantitative trait, as suggested in the transgressive segregation of 'Ramsey' × 'Gloire' progeny on canopy biomass [35].

The vigor comparison based on parentage suggested that genetic background did affect the vines' growth vigor. Vines with a *V. berlandieri* genetic background tend to be vigorous, whereas the opposite goes for those with a *V. riparia* ancestry (Figure 4). This corroborates the previous findings that rootstocks derived from the crosses of *V. berlandieri* with other species confer a higher vigor to scion than those from crosses of *V. riparia* [36,37]. *V. rupestris* is generally a species with vigorous growth [38], which is contrary to the result revealed in the present study. Apart from the environmental factors, a possible reason could be the smaller sample size—only two rootstocks were included in this species, of which du Lot showed a higher vigor and Shadi showed a lower vigor. Gautier et al. [39] pointed out that the younger age of the vines could be the reason for the difficulty in distinguishing the effects of the genetic background on conferring vigor to the scion. We infer that their limited data—acquired from a single season, with fewer vines and fewer genotypes included in the species could be another reason.

Relations between phenological periods and vine growth were not clear yet, or both altered in response to meteorological changes, as evidenced by the subtle changes in their relations across three seasons (Figure S3). Short-term weather changes, such as rising temperatures, might affect the phenological periods, which could explain the advanced flowering periods in 2017 (Figures 2 and S4). Meteorological factors even played a larger role than rootstock genotype in affecting vine growth, and rainfall seemed to be the major factor. The correlations between meteorological data and growth variables were high but not significant. Further collections of more seasonal data could be helpful to better understand the effect of each meteorological factor on vine growth.

5. Conclusions

The nature of multiplication or current seasonal growth varied largely among these rootstocks and was less connected with the growth indices of adult vines. Interestingly, the internode length of the new shoot developed on the cutting showed positive correlations with not only the internode length but also the cane length, shoot length, and pruning weight of the adult vines. Based on the growth measurements for three years, 31 rootstocks were separated into three clusters. Rootstocks such as '1103P', '5BB', and '225Ru' within Cluster 1 showed significantly higher vigor than those like 'Gloire', '3309C', and 'Ganzin1' within Cluster 3. This study also indicated that *V. berlandieri* could confer high vigor to the progeny.

Supplementary Materials: The following supporting information can be downloaded at: https://www.mdpi.com/article/10.3390/horticulturae9020241/s1. Figure S1: Summary of the monthly average meteorological data in the growing seasons of 2016–2018; Figure S2: Correlations among multiplication traits and growth parameters of grapevine rootstocks; Figure S3: Correlations between phenological periods and growth parameters of grapevine rootstocks for three individual seasons; Figure S4: Daily temperature changes before and after flowering in the growing seasons of 2016–2018; Table S1: Meteorological data during growing seasons in 2016–2018; Table S2: Basic composition of the soil; Table S3: Comparison of growth indicators among 31 grapevine rootstocks across three growing seasons (2016–2018); Table S4: Summary of PCA for vigor parameters.

Author Contributions: Conceptualization, Y.Y. and Z.G.; methodology, Y.Y. and Z.G.; software, Y.Y.; validation, N.J., M.L. and Y.S.; formal analysis, B.H.; investigation, B.H., C.L., Y.W., Q.G. and Y.Y.; resources, B.H. and Z.G.; data curation, N.J., Y.W. and Y.Y.; writing—original draft preparation, Y.Y.; writing—review and editing, Y.Y. and N.J.; visualization, Y.Y. and Q.G.; supervision, Z.G..; project administration, B.H.; funding acquisition, B.H. All authors have read and agreed to the published version of the manuscript.

Funding: This research was funded by the China Agriculture Research System (Grant No. CARS-29) and the Natural Science Foundation of Hebei Province (Grant No. C2021301036).

Data Availability Statement: Not applicable.

Acknowledgments: We would like to thank the National Grapevine Repository in Zhengzhou and Professor Chonghuai Liu for their help in offering partial multiplication materials.

Conflicts of Interest: The authors declare no conflict of interest.

References

1. Corso, M.; Bonghi, C. Grapevine rootstock effects on abiotic stress tolerance. *Plant Sci. Today* **2014**, *1*, 108–113. [CrossRef]
2. Walker, R.R.; Blackmore, D.H.; Clingeleffer, P.R. Impact of rootstock on yield and ion concentrations in petioles, juice and wine of Shiraz and Chardonnay in different viticultural environments with different irrigation water salinity. *Aust. J. Grape Wine Res.* **2010**, *16*, 243–257. [CrossRef]
3. Ferris, H.; Zheng, L.; Walker, M.A. Resistance of grape rootstocks to plant-parasitic nematodes. *J. Nematol.* **2012**, *44*, 377–386. [PubMed]
4. Pavloušek, P. Tolerance to lime-induced chlorosis and drought in grapevine rootstocks. In *Abiotic Stress—Plant Responses and Applications in Agriculture*; Vahdati, K., Leslie, C., Eds.; InTech: Vienna, Austria, 2013; pp. 277–305.
5. Fort, K.; Fraga, J.; Grossi, D.; Walker, M.A. Early measures of drought tolerance in four grape rootstocks. *J. Am. Soc. Hortic. Sci.* **2017**, *142*, 36–46. [CrossRef]
6. Cheng, J.; Li, H.; Wang, W.; Duan, C.; Wang, J.; He, F. The influence of rootstocks on the scions' aromatic profiles of *Vitis vinifera* L. cv. Chardonnay. *Sci. Hortic.* **2020**, *272*, 109517. [CrossRef]
7. Yin, Y.; Jia, N.; Li, M.; Liu, C.; Yuan, J.; Han, B.; Sun, Y.; Zhao, S.; Guo, Z. Rootstocks induce shifts in tree vigor, yield and berry quality of 'Summer Black' grapevines. *Eur. J. Hortic. Sci.* **2021**, *86*, 41–48. [CrossRef]
8. Jones, T.H.; Cullis, B.R.; Clingeleffer, P.R.; Rühl, E.H. Effects of novel hybrid and traditional rootstocks on vigour and yield components of Shiraz grapevines. *Aust. J. Grape Wine Res.* **2009**, *15*, 284–292. [CrossRef]
9. Zhang, L.; Marguerit, E.; Rossdeutsch, L.; Ollat, N.; Gambetta, G.A. The influence of grapevine rootstocks on scion growth and drought resistance. *Theor. Exp. Plant Physiol.* **2016**, *28*, 143–157. [CrossRef]
10. Cochetel, N.; Escudie, F.; Cookson, S.J.; Dai, Z.; Vivin, P.; Bert, P.F.; Munoz, M.S.; Delrot, S.; Klopp, C.; Ollat, N.; et al. Root transcriptomic responses of grafted grapevines to heterogeneous nitrogen availability depend on rootstock genotype. *J. Exp. Bot.* **2017**, *68*, 4339–4355. [CrossRef]
11. Lu, J.; Liu, C. Grapevine breeding in China. In *Grapevine Breeding Programs for the Wine Industry*; Reynolds, A., Ed.; Woodhead Publishing: Cambridge, UK, 2015; pp. 273–310.
12. Duan, C.Q.; Liu, C.H.; Liu, F.Z.; Wang, Z.Y.; Liu, Y.L.; Xu, L.M. Fruit scientific research in New China in the past 70 years: Grape. *J. Fruit Sci.* **2019**, *36*, 1292–1301. [CrossRef]
13. Liu, X.; Liu, C.; Fan, X.; Guo, D.; Zhang, G.; Sun, H. Construction of primary core collection of grape genetic resources. *J. Plant Genet. Resour.* **2012**, *13*, 72–76.
14. van Leeuwen, C. Terroir: The effect of the physical environment on vine growth, grape ripening and wine sensory attributes. In *Managing Wine Quality: Viticulture and Wine Quality*; Reynolds, A.G., Ed.; Woodhead Publishing: Cambridge, UK, 2010; pp. 273–315.
15. Khouni, S.; Laiadi, Z.; Bertazzon, N.; Angelini, E.; Migliaro, D. Preservation and sanitary status of Algerian grapevine germplasm: Management and improvement. *S. Afr. J. Bot.* **2023**, *153*, 346–356. [CrossRef]
16. Marín, D.; Miranda, C.; Abad, F.J.; Urrestarazu, J.; Mayor, B.; Villa-Llop, A.; Santesteban, L.G. Agronomic evaluation of eight 41 B × 110 Richter grapevine genotypes as rootstock candidates for Mediterranean viticulture. *Hortic. Plant J.* **2022**. [CrossRef]
17. Provost, C.; Campbell, A.; Dumont, F. Rootstocks impact yield, fruit composition, nutrient deficiencies, and winter survival of hybrid cultivars in eastern Canada. *Horticulturae* **2021**, *7*, 237. [CrossRef]
18. Bianchi, D.; Brancadoro, L. Water use efficiency and nutritional status of a new grapevine rootstock selection. *Horticulturae* **2021**, *7*, 503. [CrossRef]
19. Yuan, J.; Li, M.; Jia, N.; Liu, C.; Han, B.; Yin, Y.; Sun, Y.; Guo, Z.; Zhao, S. Evaluation of salt stress tolerance in twenty-one grape rootstocks. *Acta Agric. Boreali-Occident. Sin.* **2019**, *28*, 602–606. [CrossRef]
20. Xia, J.; Chen, C.; Liu, T.; Liu, C.; Liu, S.; Fang, J.; Shangguan, L. Germplasm resource evaluation and the underlying regulatory mechanisms of the differential copper stress tolerance among Vitis species. *Environ. Exp. Bot.* **2022**, *206*, 105198. [CrossRef]

21. Gao, Z.; Li, J.; Zhu, H.; Sun, L.; Du, Y.; Zhai, H. Using differential thermal analysis to analyze cold hardiness in the roots of grape varieties. *Sci. Hortic.* **2014**, *174*, 155–163. [CrossRef]
22. Li, Y.; Fu, Y.; Ma, Y.; Du, Y.; Zhai, H. Effects of waterlogging on characteristics of growth and photosynthesis in different grape rootstocks. *Sci. Agric. Sin.* **2013**, *46*, 995–1004. [CrossRef]
23. Zhang, J.L.; Cao, Z.Y.; Ma, J.F. Screening of cold-resistant seedlings of a Chinese wild grape (Vitis piasezkii Maxim var. pagnucii) native to loess plateau of eastern Gansu province, China, as rootstocks. *Sci. Hortic.* **2009**, *122*, 125–128. [CrossRef]
24. Zhang, P.; Leng, X.; Fan, X.; Liu, G.; Fang, J. Current status and research progress of grape rootstock germplasm resources. *SINO-Overseas Grapevine Wine* **2018**, *219*, 58–63. [CrossRef]
25. Li, X.; Dong, W.; Lin, J.; Wang, Z.; Yang, Q.; Chang, Y.; Zhang, Z. Structural, physiological and biochemical responses of Pyrus calleryana offspring to salt stress. *Eur. J. Hortic. Sci.* **2015**, *80*, 306–315. [CrossRef]
26. Coombe, B.G. Growth stages of the grapevine: Adoption of a system for identifying grapevine growth stages. *Aust. J. Grape Wine Res.* **1995**, *1*, 104–110. [CrossRef]
27. Waite, H.; Whitelaw-Weckert, M.; Torley, P. Grapevine propagation: Principles and methods for the production of high-quality grapevine planting material. *N. Z. J. Crop Hortic. Sci.* **2014**, *43*, 144–161. [CrossRef]
28. Hartmann, H.T.; Kester, D.E.; Davies, F.T.; Geneve, R.L. *Plant Propagation: Principles and Practices*, 5th ed.; Prentice-Hall Inc.: Englewood Cliffs, NJ, USA, 1990; p. 647.
29. Köse, B.; Karabulut, B.; Ceylan, K. Effect of rootstock on grafted grapevine quality. *Eur. J. Hortic. Sci.* **2014**, *79*, 197–202.
30. Rossdeutsch, L.; Schreiner, R.P.; Skinkis, P.A.; Deluc, L. Nitrate uptake and transport properties of two grapevine rootstocks with varying vigor. *Front. Plant Sci.* **2020**, *11*, 608813. [CrossRef]
31. Lecourt, J.; Lauvergeat, V.; Ollat, N.; Vivin, P.; Cookson, S.J. Shoot and root ionome responses to nitrate supply in grafted grapevines are rootstock genotype dependent. *Aust. J. Grape Wine Res.* **2015**, *21*, 311–318. [CrossRef]
32. Trought, M.C.T.; Dixon, R.; Mills, T.; Greven, M.; Agnew, R.; Mauk, J.L.; Praat, J.-P. The impact of differences in soil texture within a vineyard on vine vigour, vine earliness and juice composition. *OENO One* **2008**, *42*, 67–72. [CrossRef]
33. van Leeuwen, C.; Friant, P.; Chone, X.; Tregoat, O.; Koundouras, S.; Dubourdieu, D. Influence of climate, soil, and cultivar on terroir. *Am. J. Enol. Vitic.* **2004**, *55*, 207–217. [CrossRef]
34. Dodson Peterson, J.C.; Duncan, R.; Hirschfelt, D.; Ingels, C.; McGourty, G.; Smith, R.; Weber, E.; Wolpert, J.; Anderson, M.; Benz, J.; et al. Grape rootstock breeding and their performance based on the Wolpert trials in California. In *The Grape Genome*; Cantu, D., Walker, M.A., Eds.; Springer Nature: Cham, Switzerland, 2019; pp. 301–318.
35. Hugalde, I.P.; Aguero, C.B.; Barrios-Masias, F.H.; Romero, N.; Viet Nguyen, A.; Riaz, S.; Piccoli, P.; McElrone, A.J.; Walker, M.A.; Vila, H.F. Modeling vegetative vigour in grapevine: Unraveling underlying mechanisms. *Heliyon* **2020**, *6*, e05708. [CrossRef]
36. Galet, P.; Smith, J. *Grape Varieties and Rootstock Varieties*; Oenoplurimédia: Chaintré, France, 1998.
37. Bettiga, L.J. *Wine Grape Varieties in California*; University of California, Agriculture and Natural Resources: Oakland, CA, USA, 2003.
38. Keller, M. Taxonomy and anatomy. In *The Science of Grapevines*, 3rd ed.; Academic Press: New York, NY, USA, 2020; pp. 1–60.
39. Gautier, A.; Cookson, S.J.; Lagalle, L.; Ollat, N.; Marguerit, E. Influence of the three main genetic backgrounds of grapevine rootstocks on petiolar nutrient concentrations of the scion, with a focus on phosphorus. *OENO One* **2020**, *54*, 1–13. [CrossRef]

MDPI AG
Grosspeteranlage 5
4052 Basel
Switzerland
Tel.: +41 61 683 77 34

Horticulturae Editorial Office
E-mail: horticulturae@mdpi.com
www.mdpi.com/journal/horticulturae